Migration and Immobilization of Radionuclides in Nuclear Waste Disposal

Managing nuclear waste remains one of the most significant challenges for the nuclear industry in terms of both safety and environmental concerns. This book offers a thorough examination of the migration and immobilization of nuclear waste.

Focusing on the chemistry of radioactive elements in natural rocks, waters, and engineered barriers, the book details how key isotopes travel through or are retained by soils, clays, and groundwater. It also examines how chemical conditions, mineral surfaces, and colloidal particles control the movement of radionuclides and how glass, cement, and advanced ceramics can contain waste for long periods. The chapters in this book cover radioactivity essentials, waste generation from medical, energy, and industrial applications, and processing, transporting, and storage of waste. Accessible case studies—from Three Mile Island and Chernobyl to Fukushima—demonstrate how nuclear accidents unfold and how emergency responses can limit environmental harm.

This book will appeal to engineers, scientists, and researchers working in nuclear energy and waste management.

Binglin Guo is a professor and National Excellent Young Scholar at the School of Civil Engineering, Hefei University of Technology. He is the director of the Anhui Engineering Research Center for Low-Carbon Technology of Cementitious Materials. His research interests include solid waste utilization, environmental cement chemistry, and low-carbon cementitious materials.

Bin Ma is Marie Skłodowska-Curie Fellow and National Excellent Young Scholar. He is also a professor of environmental cement chemistry at North China Electric Power University in Beijing, China. His research interests include solid waste disposal chemistry and environmental cement chemistry, especially radioactive waste treatment.

Lei Wang is a Hundred-Talent Program Research Fellow, National Excellent Young Scholar, and Humboldt Fellow at the College of Energy Engineering, Zhejiang University. His research interests include sustainable waste-to-energy technologies, low-carbon hazardous waste treatment, and CO_2 sequestration and utilization in minerals.

Migration and Immobilization of Radionuclides in Nuclear Waste Disposal

Edited by Binglin Guo, Bin Ma, and Lei Wang

CRC Press
Taylor & Francis Group
Boca Raton London New York

CRC Press is an imprint of the
Taylor & Francis Group, an **informa** business

Designed cover image: © Shutterstock

First edition published 2027
by CRC Press
2385 NW Executive Center Drive, Suite 320, Boca Raton FL 33431

and by CRC Press
4 Park Square, Milton Park, Abingdon, Oxon, OX14 4RN

CRC Press is an imprint of Taylor & Francis Group, LLC

ISBN: 978-1-041-12625-6 (hbk)
ISBN: 978-1-041-17058-7 (pbk)
ISBN: 978-1-003-66579-3 (ebk)

DOI: 10.1201/9781003665793

Typeset in Minion
by Apex CoVantage, LLC

Contents

Contributors

Maria Boltoeva is a CNRS researcher at the Ions, Active Interfaces Laboratory of the Institute of Separation Chemistry in Marcoule, France. Her research focuses on the solution chemistry of actinides and lanthanides, as well as solvent extraction using alternative solvents.

Ping Chen is a post-doctoral researcher at the China Institute of Atomic Energy. His research focuses on the physical and chemical behavior of nuclides in geological media. To date, Ping Chen has published six journal articles, including one in the journal *Applied Clay Science* on reactive transport modeling of Tc, two in *Applied Geochemistry* on Fe sorption on illite and its impact on Tc transport, two others on uranium behavior on granite, and one on Sr transport in clay rock.

Paul Dutheil is a post-doctoral researcher in the Radiochemistry Group at the Hubert Curien Pluridisciplinary Institute (IPHC), CNRS and University of Strasbourg. His research interests encompass the fundamental chemistry of radioelements, novel methods for the separation and analysis of radionuclides, and the environmental behavior of radionuclides.

Yuxue Fan is a PhD student at the State Key Laboratory of Clean Energy Utilization, Zhejiang University.

Zhengye Feng is a PhD student at the State Key Laboratory of Chemistry for NBC Hazards Protection, Frontiers Science Center for Rare Isotopes, School of Nuclear Science and Technology, Lanzhou.

Binglin Guo is a professor and National Excellent Young Scholar at the School of Civil Engineering, Hefei University of Technology. He is the director of the Anhui Engineering Research Center for Low-Carbon Technology of Cementitious Materials. His research interests include solid waste utilization, environmental cement chemistry, and low-carbon cementitious materials.

Hong Li is a PhD student at the College of Environmental Science and Engineering, North China Electric Power University.

Qiu Li is a fellow of the Royal Society of Chemistry (FRSC) and professor of civil engineering materials at the State Key Laboratory of Silicate Materials for Architectures, Wuhan University of Technology. His research interests include cement chemistry, radioactive waste immobilization, cementitious materials durability, and ceramizable geopolymers.

Kui Liu is an assistant professor at the University of Strasbourg. She is conducting research in the Radiochemistry Group at IPHC. Her work focuses on the fundamental chemistry of radioelements, with projects/publications on radionuclide speciation, separation methods, and their applications in nuclear waste management.

Lu Liu is an assistant professor at the University of Strasbourg. She is conducting research in the Radiochemistry Group at IPHC. Her work focuses on the fundamental chemistry of radioelements, with projects/publications on radionuclide speciation, separation methods, and their applications in nuclear waste management.

Yang Liu is a PhD student at the School of Civil Engineering, Hefei University of Technology.

Bin Ma is Marie Skłodowska-Curie Fellow, National Excellent Young Scholar, and a professor of environmental cement chemistry at North China Electric Power University, Beijing, China. His research interests comprise solid waste disposal chemistry and environmental cement chemistry, with a particular focus on radioactive waste treatment.

Duoqiang Pan is a professor at Lanzhou University. His research focuses on interface interaction of radionuclides, transport and environmental risk of radionuclides, and radioactive waste management. He has more than 70 publications to his credit in *Nature Communications*, *Water Research*, and more.

Huyong Qin is a PhD student at the School of Civil Engineering, Hefei University of Technology.

Shi-Kuan Sun is a professor at Foshan University. His research focuses on the fundamental application of ceramic and glass–ceramic materials for immobilizing highly radioactive nuclear waste (such as uranium or plutonium).

Sheng-Heng Tan is at the Department of Radiochemistry, China Institute of Atomic Energy, Beijing, China.

Chen Wang is a PhD student at the State Key Laboratory of Clean Energy Utilization, Zhejiang University.

Cheng Wang is a PhD student at the School of Civil Engineering, Hefei University of Technology.

Lei Wang is Hundred-Talent Program Research Fellow, National Excellent Young Scholar, and Humboldt Fellow at the College of Energy Engineering, Zhejiang University. His research interests include sustainable waste-to-energy technologies, low-carbon hazardous waste treatment, CO_2 sequestration and utilization in minerals.

Xinpeng Wang, affiliated with Guangxi University, investigates nuclear fuel reprocessing. His work has been published in the *Journal of Hazardous Materials*, *Chemical Engineering Journal*, *Water Research*, and more.

Youna Wang is a PhD student at the State Key Laboratory of Clean Energy Utilization, Zhejiang University.

Hanqin Weng is at the Sun Yat-sen University. Hanqin's research focuses on radiation chemistry, especially on the fast reaction of active radicals produced by radiolysis. These reactions are important to elucidate what happens in a strong radiation field and how it generates long-term effects (*Radiation Physics and Chemistry*, 2025, 235, No. 112874; *Journal of Hazardous Materials*, 2023, No. 131852).

Hanyu Wu is a professor at Sun Yat-sen University. She focuses on environmental radiochemistry and the safe treatment and disposal of radioactive wastes and specializes in the migration, transformation of radionuclides, and radioactive pollution and control. Her research interests include radiation environmental protection and radiation environmental impact assessment. Her work focuses on radionuclide speciation and occurrence, surface/interface adsorption mechanisms, materials for radionuclide separation and enrichment, radionuclide migration and prediction under radioactive waste disposal conditions, remediation of uranium tailings and associated radioactive tailings, and risk analysis under accident conditions.

Qingyin Xia is a research scientist at the Beijing Research Institute of Chemical Engineering and Metallurgy. His research centers on uranium biogeochemistry, mineral–microbe interactions, and clay-based functional materials. He has published extensively in *Environmental Science & Technology* and *Geochimica et Cosmochimica Acta*.

Zhen Xu is an associate professor at Lanzhou University, specializing in environmental radiochemistry, nuclear fuel cycle, and materials.

Yuanlv Ye is a researcher and holds a PhD in radiochemistry. Yuanlv Ye specializes in nuclear waste management and software development, with extensive experience in nuclear fuel cycle and reactor radionuclide migration modeling.

Xiangbiao Yin is a professor and doctoral supervisor at the School of Nuclear Science and Technology, University of South China. His research interests include nuclear fuel cycle, environmental radiochemistry, and nuclear waste treatment. He has published over 90 SCI papers and applied for more than ten Chinese invention patents.

Shuyang Zhang is a PhD student at the State Key Laboratory of Clean Energy Utilization, Zhejiang University.

Jie Zheng, a master of science, senior engineer, and PhD candidate, specializes in organic chemistry synthesis and radiation-induced biological effects, with a strong focus on effective waste management in post-accident scenarios. Jie is affiliated with the 5th Electronics Research Institute of the Ministry of Industry and Information Technology in Guangzhou and the College of Chemistry and Chemical Engineering at Lanzhou University in Lanzhou.

Yi Zhou is a PhD student at the School of Nuclear Science and Technology, University of South China.

PART I

Overview of Nuclear Waste

CHAPTER 1

Natural Radioactivity and Anthropogenic Radioactivity

Shuyang Zhang, Yuxue Fan, Youna Wang, Chen Wang, Binglin Guo, Bin Ma, and Lei Wang

1.1 INTRODUCTION

The contemporary world confronts the dual challenge of climate change and energy security, which are deeply intertwined. Nuclear energy, as a key technology for the global energy transition, is ushering in an important period of revival. According to the "Nuclear Power Reactors in the World" report released by the International Atomic Energy Agency (IAEA) in August 2024, as of December 2023, there are 413 operating nuclear power units worldwide with a total installed capacity of approximately 371.5 million kW (Dong et al., 2025; Zhao & Zhao, 2023).

On the one hand, due to its advantages of high energy density and low carbon emissions, nuclear energy is regarded as one of the important means to achieve the goals of the Paris Agreement, playing a significant role in promoting the global carbon peak and carbon neutrality (Liu et al., 2023). Currently, nuclear energy accounts for approximately 10% of the global electricity supply. The International Atomic Energy Agency (IAEA) predicts that by 2030, nuclear power will have the potential to reduce global CO_2 emissions by 10%. At present, many countries have introduced a series of policies to leverage the advantages of nuclear energy in global carbon emission reduction. For instance, France and South Korea are increasing nuclear power generation by building new nuclear power projects and extending the service life of old ones.

On the other hand, with the acceleration of globalization and the escalation of regional conflicts in some areas, fossil fuel prices fluctuate frequently, and the energy supply chain crisis has further deteriorated, highlighting the importance of energy security. Nuclear energy, with its stable power generation capacity and highly localized fuel supply, has become a priority choice in the energy security strategies of various countries. Europe prioritizes nuclear energy deployment to mitigate natural gas dependency, which is evidenced by accelerated projects in France, the Czech Republic, and Poland.

DOI: 10.1201/9781003665793-2

For this reason, global policy support for nuclear energy has increased significantly in recent years. At the UAE COP29 Climate Conference held in 2024, more than 30 countries jointly committed to quadrupling global nuclear power installed capacity by 2050. In March 2024, the first Nuclear Energy Summit was held in Brussels, where leaders from more than 30 countries reached a number of consensuses, including promoting international cooperation, strengthening technological research and development, enhancing financial support, and optimizing nuclear waste management. The abovementioned series of measures demonstrates the determination of various countries to address global climate change and boost confidence in the global development of nuclear power.

However, we must also note that this powerful energy source, nuclear energy, is accompanied by radioactive by-products (Pyo et al., 2021). Unlike traditional fossil fuel wastes, improper handling of these radioactive by-products may lead to environmental risks, endanger human health, and thus trigger a series of social problems (Frigi et al., 2024). Therefore, it is crucial for modern society to understand the basic concepts related to radioactivity, quantify the hazards of nuclear waste, understand the technical means for nuclear waste disposal, and explore the conditions for the sustainable development of nuclear energy.

1.2 RADIOACTIVITY FUNDAMENTALS: DECAY, HALF-LIFE, AND UNITS

Before discussing radioactivity, let's briefly review the concepts of nuclides and isotopes. Nuclides, atoms characterized by specific proton (Z) and neutron (N) numbers, were defined to quantify atomic masses with precision. They are usually expressed as ${}^{A}_{Z}X$, where X is the element symbol, and A represents the mass number ($A - Z = N$). Nuclides can also be named in the X-A or ${}^{A}X$ format, which will be used in the subsequent sections of this paper. Several atoms with the same number of protons but different numbers of neutrons are called isotopes. They have basically the same chemical properties but exhibit different nuclear properties due to differences in atomic structure. For example, the hydrogen element has three atoms/nuclides: ${}^{1}_{1}H$, ${}^{2}_{1}H$ and ${}^{3}_{1}H$. They have 0, 1, and 2 neutrons, respectively, and are isotopes of each other. Isotopes have found extensive applications in various fields, including medicine, for diagnosing and treating health issues (Tan et al., 2024).

Elements that exist in only one nuclide in nature are called mononuclidic elements, such as beryllium, fluorine, aluminum, sodium, etc. There are a total of 20 elements that have no isotopes. For other polynuclidic elements, elements with an even number of protons can have more stable nuclides, generally no less than three. While elements with an odd number of protons usually have only one stable nuclide (generally no more than two), this is determined by the binding energy of nucleons. Scientific research shows that stable nuclides usually have a certain limit on the total number of nucleons (A $\leq$ 209), and the ratio of neutrons to protons should maintain a certain proportion (generally $N/Z \sim 1.5$). Nuclides with too many nucleons or an inappropriate N/Z ratio are often unstable and will transform into stable new nuclei by emitting particles or radiating rays. This lays the foundation for the subsequent discussion on radioactivity.

1.2.1 Radioactive Decay

The phenomenon that atomic nuclei spontaneously emit rays is called radioactivity. Isotopes that can spontaneously emit various rays are called radioactive isotopes. The process by which radioactive isotopes emit various rays and undergo nuclear transformation is called radioactive decay. The radioactive isotope before decay is called the parent nuclide, and the new isotope produced during the decay process is called the radiogenic isotope, or the daughter. The decay law conforms to the conservation of charge number, conservation of mass number, conservation of momentum, and conservation of energy.

The most common forms of spontaneous radioactive decay are the emission of alpha particles, beta particles, gamma rays, and neutrinos. In 1898, Rutherford discovered alpha and beta rays while studying radioactivity, and Becquerel verified their electrical properties in 1899. Later in 1900, Villard discovered gamma rays.

Alpha rays have the strongest ionization ability, relatively small radiation hazards, and the weakest penetration ability; the particles cannot even penetrate a piece of paper. Beta rays have moderate penetration ability and cannot pass through metal sheets. Gamma rays have the weakest ionization ability but the strongest penetration ability, and can only be blocked by thick lead plates or concrete walls.

According to the different emitted particles or radiation, radioactive decay mainly includes three basic types, such as alpha, beta, and gamma, whose essence is related to the adjustment of the proton–neutron ratio in the nucleus, accompanied by changes in the atomic number or mass number and energy release. Of course, in less common forms of radioactivity, fission fragments, neutrons, or protons may also be emitted.

1.2.1.1 Alpha Decay

This is the process by which an atomic nucleus spontaneously emits an alpha particle and transforms into another nucleus. For naturally radioactive isotopes, only heavy atomic nuclei with a mass number A greater than 140 can undergo alpha decay. This is particularly evident in radioactive isotopes with an atomic number Z greater than 82 and a mass number A greater than 209, which are dominated mainly by alpha decay. The nuclear equation for alpha decay is:

$$_{Z}^{A}\mathrm{X} \rightarrow {}_{Z-2}^{A-4}\mathrm{Y} + {}_{2}^{4}\mathrm{He} \tag{1}$$

A classic example involves early research on the radioactivity of uranium ores:

$$_{92}^{238}\mathrm{U} \rightarrow {}_{90}^{234}\mathrm{Th} + {}_{2}^{4}\mathrm{He} \tag{2}$$

1.2.1.2 Beta Decay

In the process of beta decay, a neutron in the nucleus will transform into a proton and an electron (i.e., a beta particle), and this process will also produce a neutrino (v_e) and an antineutrino ($\overline{v_e}$). Among them, the decay process that emits negative electrons is called beta minus decay, and the general formula is as follows:

$$_{Z}^{A}X \rightarrow {}_{Z+1}^{A}Y + {}_{-1}^{0}e + \overline{v_e} \tag{3}$$

Some classic examples are as follows:

$$ {}^{14}_{6}\mathrm{C} \rightarrow {}^{14}_{7}\mathrm{N} + {}^{0}_{-1}e + \overline{\nu_e} \tag{4} $$

$$ {}^{234}_{90}\mathrm{Th} \rightarrow {}^{234}_{91}\mathrm{Pa} + {}^{0}_{-1}e + \overline{\nu_e} \tag{5} $$

The decay process that emits positrons is called beta plus decay. The general formula is as follows:

$$ {}^{A}_{Z}\mathrm{X} \rightarrow {}^{A}_{Z-1}\mathrm{Y} + {}^{0}_{+1}e + \nu_e \tag{6} $$

1.2.1.3 γ Decay

When an atomic nucleus is in an excited state, it will return to the ground state by releasing gamma rays (high-energy electromagnetic radiation, such as radio waves, X-rays, etc.). Gamma decay occurs along with alpha or beta decay. During this process, both the mass number A and the atomic number Z of the atomic nucleus remain unchanged. The general equation is as follows:

$$ {}^{A}_{Z}\mathrm{X} \rightarrow {}^{A}_{Z}\mathrm{X} + \gamma \tag{7} $$

With research on artificial radionuclides, scientists have also successively discovered new decay modes such as electron capture, isomeric transition, neutron emission, proton decay, and spontaneous fission.

1.2.2 Hazard Time Scale: Half-Life

Half-life is a basic concept in physics that describes the decay rate of radionuclides. It refers to the time required for the number of atoms of a radioactive parent isotope to decay to half of the original number, denoted as $T_{1/2}$. This concept was first proposed by Ernest Rutherford in 1900 to characterize the decay law of radioactive substances. It is one of the important characteristic parameters of radionuclides that does not change with external conditions, element state, or mass. The half-lives of radioactive isotopes vary drastically, ranging from tens of millionths of a second to tens of billions of years. The half-lives of some nuclides are as follows:

TABLE 1.1 The Half-Lives of Some Nuclides (Bé et al., 2004–2013)

Nuclide	**Half life**
${}^{3}_{1}H$	12.312 ± 0.00025 a
${}^{11}_{6}C$	20.370 ± 0.00029 min
${}^{14}_{6}C$	5,700 ± 0.30 a
${}^{40}_{19}K$	(1.2504 ± 0.000030) × 10⁹ a

TABLE 1.1 (*Continued*) The Half-Lives of Some Nuclides (Bé et al., 2004–2013)

Nuclide	Half-life
${}^{90}_{38}Sr$	28.80 ± 0.007 a
${}^{99m}Tc$	6.0067 ± 0.000010 h
${}^{131}_{53}I$	8.0233 ± 0.000019 d
${}^{137}_{55}Cs$	30.05 ± 0.008 a
${}^{210}_{84}Po$	138.3763 ± 0.000017 d
${}^{211}_{84}Po$	0.516 ± 0.0003 s
${}^{212}_{83}Bi$	60.54 ± 0.006 min
${}^{215}_{84}Po$	$(1.718 \pm 0.0004) \times 10^{-3}$ a
${}^{216}_{84}Po$	0.150 ± 0.0005 s
${}^{220}_{86}Rn$	55.8 ± 0.03 s
${}^{226}_{88}Ra$	1600 ± 0.7 a
${}^{235}_{92}U$	$(704 \pm 0.1) \times 10^{6}$ a
${}^{238}_{92}U$	$(4.468 \pm 0.0005) \times 10^{9}$ a
${}^{239}_{94}Pu$	24,100 ± 0.11 a

The half-life determines the duration of the hazard of radioactive waste: the shorter the half-life of an isotope, the stronger its radioactivity. For example, in Table 1.1, ${}^{131}_{53}I$ is short-term medical waste, ${}^{137}_{55}Cs$ is intermediate-level radioactive waste, and ${}^{238}_{92}U$ is an object under supervision for 10,000 years. It is worth noting that while ${}^{131}_{53}I$ poses acute health risks due to its rapid absorption by the thyroid, ${}^{137}_{55}Cs$ presents long-term environmental concerns due to its persistence and bioaccumulation in organisms (Kamei-Ishikawa et al., 2013).

The half-life ($T_{1/2}$) is closely related to concepts such as the decay constant (λ) and average lifetime (τ). The decay constant is a parameter describing the decay rate, and the relationship between half-life and decay constant is given by the following equation:

$$T_{1/2} = \ln(2/\lambda) = 0.693/\lambda \tag{8}$$

The average lifetime τ refers to the average time a radioactive parent isotope exists before decaying, which is calculated as $\tau=1/\lambda$. Additionally, the radioactive activity $A(t)$ is the number of atomic nuclei decaying per unit time, which is given by the following equation:

$$A(t) = A(0)e^{-\lambda t} \tag{9}$$

For radionuclides with short half-lives, the half-life can be determined by measuring changes in radiation intensity (radioactive activity) over a period. For long-lived nuclides, the decay constant λ can be calculated by measuring their radioactive activity $A(t)$ over a short period, and the half-life can then be derived.

1.2.3 Radiation Units: Quantifying Radiation

The unit systems for radiation measurement are mainly divided into two categories: units describing the characteristics of the radioactive source itself (such as activity) and units describing the effect of radiation interaction with substances (such as exposure, absorbed dose, and dose equivalent). The following are the core units and their detailed introductions.

1.2.3.1 Radioactive Activity: Describing the Radioactive Source

Radioactive activity is defined as the number of atomic nuclei decaying per unit time in a radioactive source, denoted as $A(t)$, which characterizes the intensity of the radioactive source. Radioactive activity determines the strength of radioactivity and also follows an exponential law:

$$A(t) = \frac{-\mathrm{d}N}{\mathrm{d}t} = \lambda N(0)e^{-\lambda t} = A(0)e^{-\lambda t} \tag{10}$$

where $-\mathrm{d}N$ represents the number of nuclei decaying within a time interval $\mathrm{d}t$ from $N(t)$ nuclei of a certain type. Its international unit is the becquerel, denoted as Bq, with 1 Bq = 1 nuclear decay per second. The old special unit is the curie, denoted as Ci, where 1 Ci = 3.7×10^{10} decays per second (based on the activity of 1 gram of radium-226).

1.2.3.2 Absorbed Dose: Describing Energy Absorbed by Substances

Absorbed dose is defined as the average energy absorbed by a unit mass (M) of a substance, denoted as D. $D = E/M$, whose international unit is the gray (Gy): 1 Gy = 1 J/kg. The traditional unit is the rad, where 1 rad = 0.01 J/kg = 0.01 Gy.

1.2.3.3 Dose Equivalent: Describing the Risk of Radiation Biological Effects

Dose equivalent introduces the radiation weighting factor (w_r) on the basis of absorbed dose to quantify the relative hazard degree of different types of radiation to biological tissues, which is usually used for radiation protection and denoted as H. Its international unit is the sievert (Sv): 1 Sv = 1 J/kg. The calculation formula is as follows:

$$H = D \times w_{\mathrm{r}} \times N \tag{11}$$

where N is another correction factor, usually 1 for most cases. Its core significance is to reflect the difference in biological damage caused by different radiations at the same absorbed dose H.

1.2.4 Classification by Origin

Radioactivity can be classified into natural radioactivity and anthropogenic radioactivity based on its origin. Among them, anthropogenic nuclides are generally considered to have higher specific activity and bioavailability.

1.3 UBIQUITY OF NATURAL RADIOACTIVITY

Natural radioactivity is the phenomenon of spontaneous decay of radioactive isotopes in nature. Naturally radioactive nuclides are commonly found in nature. Trace amounts of radioactive substances are present in rocks, soil, air, water, and even living organisms (Manjulata et al., 2023; Islam et al., 2025; Adaikalam et al., 2024), causing the Earth to be filled with natural background radiation. In fact, all living organisms on Earth are exposed to low-dose natural radioactive environments (Saclier et al., 2020). Natural background radiation refers to the continuous ionizing radiation in nature, including internal and external radiation, which is spontaneously generated by the decay of natural radioactive nuclides and the action of cosmic rays. Although the global cosmic ray radiation varies with latitude, the magnitude of the change is not significant, with an average annual effective dose of 0.20 – 1.09 mSv (Hu & Zhang, 2003). The impact of naturally radioactive nuclides cannot be ignored. This section will comprehensively introduce two sources of naturally radioactive nuclides: primordial radioactive nuclides and cosmogenic radioactive nuclides.

1.3.1 Primordial Radionuclides Present since Earth's Formation

There are many naturally radioactive nuclides in the composition of the Earth, which appeared as early as the birth of the Earth and are also known as primordial radioactive nuclides. Naturally radioactive nuclides include uranium series, thorium series decay chain products, and crustal primordial nuclides such as ^{40}K. They are widely distributed in soil, water, rocks, and other environments on Earth, but mainly exist in the lithosphere. These nuclides undergo continuous decay, forming three naturally radioactive decay series, with nuclide $^{238}U(4n + 2)$, nuclide $^{235}U(4n + 3)$, and nuclide $^{232}Th(4n)$ as decay precursors (Huang, 2008). The detailed process is shown in Figures 1.1, 1.2, and 1.3. Typical intermediate products of the uranium series include radium-226 (^{226}Ra, first separated radioactive nuclide), radon 222 (^{222}Rn, radon gas, indoor carcinogen), polonium-210 (^{210}Po, a highly toxic alpha emitter). ^{235}U is a member of the actinide series and only accounts for a small proportion in natural uranium, accounting for 0.72% by weight. It is the only natural fissile nuclide. Radon-220, radium-228, polonium-212, and others in the thorium series are the main sources of environmental radiation. Radon is one of the nuclides in the uranium series radioactive decay chain, produced by the decay of radium isotopes. Its corresponding elemental substance is radon gas, which is a colorless, odorless, and tasteless inert gas that is soluble in water. Radon and its derivatives account for 55% of natural radiation exposure to the public (Cheng, 2008; Chang, 2002). Other independently existing natural radioactive nuclides are those in the Earth's crust with atomic numbers less than 81, such as ^{40}K, ^{87}Rb, etc. Their half-life is extremely long, and the decay products do not form a continuous series, so they do not belong to the three series mentioned earlier. ^{40}K only accounts for 0.0118% of potassium (K). Despite this small proportion, ^{40}K remains the main primordial single radioactive nuclide on Earth, as natural potassium (K) is abundant on Earth (Huang, 2008).

The radiation generated by natural radioactive nuclides varies greatly with different regions. In some regions of the world, known as high background areas, due to the high

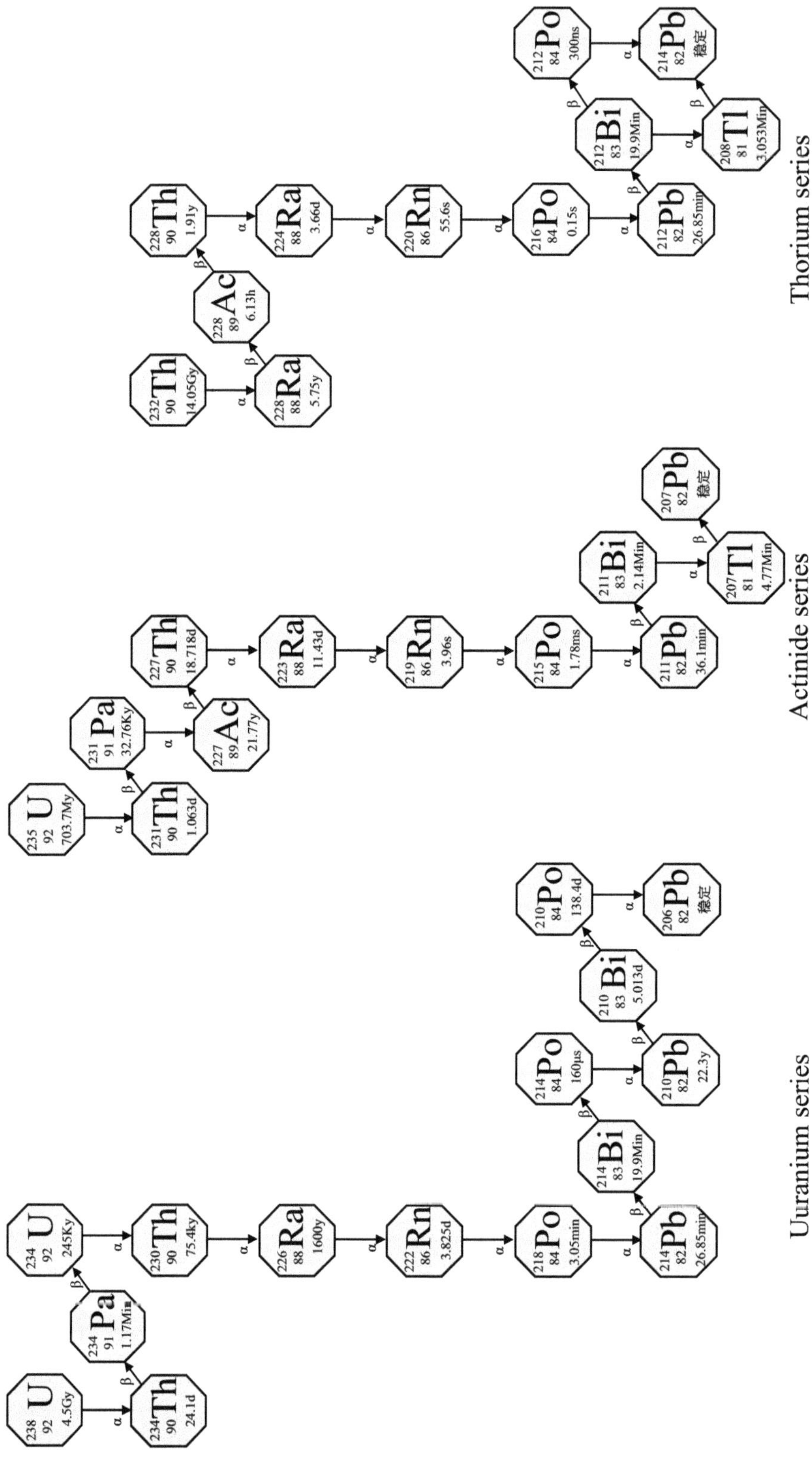

FIGURE 1.1 Decay Diagram of Uranium, Actinide, Thorium Series (Huang, 2008).

concentration of uranium and thorium in the surface layer, the surface gamma ray dose is higher than in general areas. The Yangjiang area in Guangdong Province is a famous high natural-radiation background area in China, with a per capita annual dose of up to 6.4 mSv/y, about three times the national average (approximately 2.4 mSv) (Sun et al., 2002; Wang & Liu, 2021).

The dose rate of background radiation may also increase due to the presence of certain mining areas, quarries, and springs (Xia, 2001). Granite is an equally important decorative material for modern people, mainly composed of minerals such as quartz and feldspar, containing long-lived radioactive nuclides such as ^{226}Ra, ^{232}Th, and natural potassium (Huang et al., 2004; Zhang, 2002), which is an igneous rock containing natural radioactivity of primordial radioactive nuclides. It is noteworthy that the radioactive nuclide ^{226}Ra contained in granite will decay into radon (Xie, 2015; Wu, 2004). The radioactive element content of different granite stones is shown in the following Table 1.2.

In the development and utilization of granite in the construction industry, the risks of radiation and radon release from granite should be considered (Kuzmanović et al., 2019; Wu, 2004). If a RAD7 radon meter is used to measure indoor radon concentration in a certain stone factory, the average radon concentration obtained is 34.75 Bq/m^3 (Xie, 2015). Phosphate ore, as an important chemical mineral raw material, has been applied in fields such as chemical engineering, light industry, and agriculture for over a hundred years. The uranium content in China's composite phosphate mines is high, especially in marine phosphate mines, and most of the world's phosphate mines also contain uranium. For example, the average content of thorium in the soil of the Yangshui mining area of Kaiyang Phosphate Mine is 11.34 μg/g, and the average content of uranium is 4.59 μg/g, both of which are higher than the average content of thorium (9 μg/g) and uranium (2 μg/g) in the world's soil and crustal abundance (uranium: 2.7 μg/g, thorium: 9.6 μg/g) (Cheng, 2012). This poses a risk of natural high-concentration irradiation for the mining and utilization of phosphate ore, especially for the phosphate fertilizer industry. Radon pollution in the living environment mainly comes from foundation soil, cracks, and pores. Radon can enter both work and living spaces through the space between the soil and the floor. After radon is produced in the soil, it can directly infiltrate into buildings through basements and floors, leading to the accumulation of radon in indoor air (Wang & Liu, 2021). In addition, radon in water is a potentially important source

TABLE 1.2 Content of Uranium and Thorium in Common Granites Used as Building Materials and Their Average Content in the Crust (Zhang, 2002)

Radioactive element		**Uranium**		**Thorium**	
Average crustal value (μg/g)		**3.0**		**12.0**	
		Average	range	Average	range
Light-colored series	Granite type (gray, white, light gray, light red)	4.5	1.0–12.0	18.0	5.0–20.0
	syenite type (light gray, rose color)	8.0	1.4–20.0	28.0	17.0–35.0
Light-colored series	Diorite andesite (dark gray, gray green, purple red brown)	2.3	0.5–7.0	9.0	2.0–20.0
Dark-colored series	Gabbro basalt type (gray black, black, gray green, dark purple)	1.0	0.2–3.0	6.0	0.5–10.0

of radon exposure. When the water source is groundwater, the radon concentration in the water supply system will also increase relatively. The radon concentration in the hot spring water of Pingshan Hot Springs in Hebei Province ranges from 93.6 to 121 Bq/L, which is 12.3 times higher than that of tap water (Zhao et al., 2012). The average radon content in typical urban groundwater in China is 11.8 Bq/L, ranging from 0.96 to 63.8 Bq/L. Approximately 37% of the water samples have radon levels exceeding the recommended drinking standard by the US EPA (11 Bq/L) (Wu, 2015). Radon concentrations in groundwater for domestic use in typical Chinese cities are summarized in Table 1.3.

1.3.2 Cosmogenic Radionuclides Produced by Cosmic Rays

Cosmic rays collide with atomic nuclei in the atmosphere, rock layers, and biosphere, undergo nuclear reactions, and generate a series of cosmogenic radioactive nuclides that enter the surface system through precipitation and biological cycles. Most of these nuclides are fragments generated in the form of scattering, while some are activated products produced by stable atoms interacting with neutrons or muons (Qu, 2006). For example, ^{14}C is produced by neutrons colliding with nitrogen atoms. Common cosmogenic radionuclides include ^{14}C, ^{7}Be, ^{3}H, and ^{22}Na, as shown in Table 1.4.

In terms of application, analyzing the content of cosmogenic radionuclides in samples can not only reconstruct natural processes such as past cosmic events, solar activity, and magnetic field changes, but also trace major events that occurred in the atmosphere and biosphere. The half-life of natural ^{32}Si is about 150 years, which has always been regarded as an ideal tracer for tracing time scales of 50–1,000 years, and has wide applications in disciplines such as oceanography, glaciology, hydrology, etc. (Zhou et al., 2015). The cosmogenic

TABLE 1.3 Radon Content in Groundwater for Domestic Use in Typical Chinese Cities

City	Number of samples	Mean value (Bq/L)	Range	City	Number of samples	Mean value	Range (Bq/L)
Beijing	14	12.4	1.6–25.0	Zhengzhou	4	4.0	3.4–4.5
Hohhot	10	7.01	1.0–11.4	Luoyang	4	27.4	22.7–34.2
Yinchuan	8	5.8	3.8–6.9	Shenyang	5	5.4	0.98–7.3
Shizuishan Mountain	2	11.8	8.0–15.7	Anshan	3	19.6	6.4–39.4
Xi'an	8	4.6	0.96–12.0	Jiamusi	5	13.0	11.6–16.9
Xianyang	4	8.7	1.2–19.0	Yichun	6	33.0	6.4–63.8

TABLE 1.4 Cosmogenic Radioactive Nuclide (UNSCEAR, 2000)

Nuclide name	Half-life	Nuclide name	Half-life
^{3}H	12.33 a	^{32}Si	172 a
^{7}Be	53.29 a	^{32}P	14.26 d
^{10}Be	1.51×10^{6} a	^{33}P	25.34 d
^{14}C	5730 a	^{35}S	87.51 d
^{22}Na	2.602 a	^{36}Cl	3.01×10^{5} a
^{26}Al	7.4×10^{5} a	^{37}Ar	35.04 d

nuclides ^{10}Be and ^{26}Al, due to their unique formation mechanism, stable geochemical behavior, and relatively long half-life, can effectively trace the evolution of the Earth's environment at different time scales (Li et al., 2004; Zhao & Zhao, 2023). The half-life of ^{14}C is approximately 5730 years. Radiocarbon dating is a technique that determines the age of a sample based on the degree of decay of carbon-14, making it possible to establish an absolute chronological sequence of archaeological cultures (Xia, 1977).

1.4 ANTHROPOGENIC RADIOACTIVITY: ORIGINS AND SCALE

Anthropogenic radioactivity refers to radionuclides and their radiation phenomena generated or released by human activities, such as the nuclear fuel cycle, medical applications, industrial irradiation, and nuclear weapons testing. Compared with natural radioactivity, it has two distinctive features: on the one hand, it may be abnormally concentrated in specific areas, such as the periphery of nuclear facilities, accident-contaminated areas, or waste disposal sites, forming local high hot spots; on the other hand, anthropogenically produced nuclides tend to break the long-term equilibrium of the natural radioactive decay chain, leading to a non-equilibrium decay chain between the parent and the daughter nuclides. That is, certain short-lived daughter nuclides may under-accumulate or rapidly decay to exhaustion. Given that the core of the management of anthropogenic radioactive substances lies in their final form—radioactive waste—it can be categorized as low-level waste (LLW), intermediate-level waste (ILW), or high-level waste (HLW) based on the intensity of radioactivity.

1.4.1 Power Applications

As the world's economy grows and the global population continues to increase, energy demand continues to rise, and nuclear energy has become an important part of the sustainable solution to meeting energy needs. It is thus evident that nuclear power generation occupies a crucial position in the global energy mix.

The production of nuclear energy is achieved primarily through controlled nuclear fission in reactors, in which fissionable isotopes of actinides, such as uranium-235, plutonium-239, and, in some cases, uranium-233, are used as fuel in the form of suitable compounds (or metal alloys). The entire life cycle of materials used for nuclear fuel—from ore mining to fuel fabrication, irradiation in reactors, and subsequent disposal of spent nuclear fuel—is known as the nuclear fuel cycle. The nuclear fuel cycle, in a general sense, begins with the mining of uranium and ends with the disposal of nuclear waste. The steps of mining and milling, conversion, enrichment, fuel fabrication, etc., constitute the "front end" of the nuclear fuel cycle, while the steps following the discharge of irradiated fuel from a reactor, such as interim storage of spent fuel, reprocessing, waste management, etc., are referred to as the "back end" of the nuclear fuel cycle. The nuclear fuel cycle flow is shown in Figure 1.2. Different stages of the nuclear fuel cycle produce different types of radioactive waste containing different nuclides. The main types and characteristics of radioactive waste at each stage of the nuclear fuel cycle are shown in Table 1.5 (IAEA, 2022).

In 2013, global inventories of high-level radioactive waste were estimated at approximately 250,000 metric tonnes. Estimates indicate Japan stored 17,000 metric tonnes of

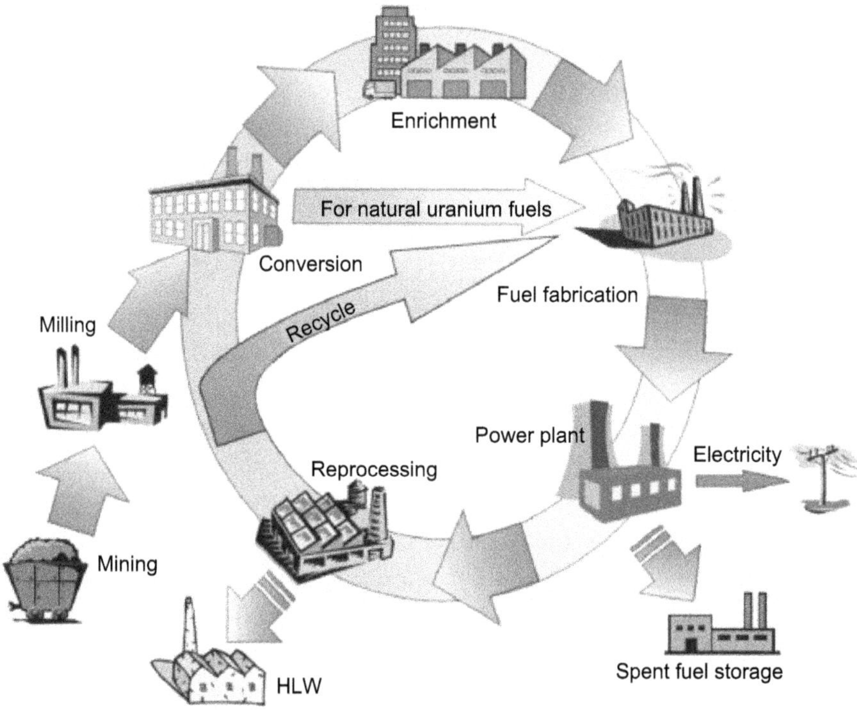

FIGURE 1.2 Entire Nuclear Fission Fuel Cycle (Zohuri, 2020).

TABLE 1.5 Main Types of Radioactive Waste at Various Stages of the Nuclear Fuel Cycle

Stage	Types of radioactive waste	Key nuclide	Half-life
Uranium ore purification	Tailings	^{238}U, ^{226}Ra	Hundreds of millions of years to 1,600 years
Uranium enrichment	Depleted uranium	^{238}U	4.5 billion years
Reactor operation	fission products	^{90}Sr, ^{137}Cs	28.8 to 30 years
	Actinide	^{238}Pu, ^{241}Am	24,000 years to 432 years
	activation product	^{60}Co, ^{63}Ni	5.3 to 100 years
Reprocess	Highly radioactive waste liquids	^{90}Sr, ^{137}Cs	28.8 to 30 years

HLW as of 2015, while the United States had accumulated over 90,000 metric tonnes by 2019. Russia's stockpile reached 23,000 metric tonnes of HLW by 2011, with annual additions of approximately 650 metric tonnes (Darda et al., 2021).

By the end of 2013, the European Union's projected radioactive waste inventory totaled 3,313,000 m^3, comprising approximately 74.04% LLW, 15.58% very-LLW, 10.20% ILW, and 0.18% HLW (Alwaeli & Mannheim, 2022).

1.4.2 Non-Power Applications

Radioisotopes are widely used in medical, industrial, and scientific research fields, and their use is inevitably accompanied by the generation of radioactive waste. The significant radiological, chemical, and biological hazards of these wastes, coupled with their large volume of generation, highly dispersed sources, and difficulties in large-scale disposal, pose a serious challenge to safety management.

Based on material estimates, Switzerland's inventory of low- and intermediate-level radioactive waste comprises the following: existing waste includes approximately 7,700 m^3 of operational waste from medical, industrial, and research facilities; approximately 27,000 m^3 of operational waste from nuclear power plants; and approximately 3,400 m^3 from reprocessing activities. Additionally, an estimated 29,600 m^3 of decommissioning waste is expected from currently operating nuclear power plants in the future, with a further 11,000 m^3 anticipated from the decommissioning of research facilities. In contrast, Hungary generates approximately 20–30 m^3 of low- and ILW annually, along with 1,000–3,000 disused sealed radioactive sources. These originate predominantly from small-scale sources outside the nuclear industry, chiefly medical, industrial, and research applications (OECD/NEA, 2006).

1.4.2.1 Medicine

Radioisotopes are widely used in medicine to diagnose and treat cancer patients by shrinking tumors and killing cancer cells. The main radioisotopes used in hospitals are technetium-99m (^{99m}Tc), iodine-131 (^{131}I), iodine-125 (^{125}I), iodine-123 (^{123}I), fluorine-18 (^{18}F), tritium-3 (^{3}H), and carbon-14 (^{14}C). Treatment for thyrotoxicosis and thyroid cancer is carried out orally by administration of radioactive iodine (^{131}I) in liquid or capsule form (Al Aamri et al., 2016), and contaminated gowns and biospecimen containers in isolation wards are characterized by moderate levels of radioactive waste. ^{99m}Tc is widely used for diagnostic imaging, including cardiac, bone, and other organ scans, and its use generates radioactive waste such as contaminated syringes, protective equipment, and absorbent materials contaminated with patient excreta, including cotton gauze and bed linen. The positron nuclide fluorine-18 generates discarded infusion lines and lead shielding containers after positron emission tomography (PET) scans, which are also subject to low-level radioactive waste standards.

Therapeutic diagnostics refers to an integrated approach to diagnosis and treatment using a combination of suitable molecularly targeted carriers and radionuclides. This can be achieved through the use of therapeutic radionuclides with simultaneous imaging capabilities that can be used for both therapeutic and diagnostic purposes. Therapeutic diagnostics is typically achieved through the use of diagnostic or therapeutic radionuclide-labelled molecular targeting vectors that act at the molecular level by targeting specific molecules. A classic example of therapeutic diagnostics is the use of ^{68}Ga-labelled tracers, where diagnostic use of ^{68}Ga can be accompanied by therapeutic use of therapeutic radionuclides labelled with the same tracer (e.g., ^{90}Y and ^{177}Lu) to personalize radionuclide therapy. The therapeutic radionuclides that are currently most commonly used or are expected to be widely used in the future are mainly radionuclides produced by nuclear reactors, including ^{47}Sc, ^{90}Y, ^{131}I, ^{166}Ho, ^{177}Lu, ^{188}Re and ^{213}Bi (Yeong et al., 2014).

Medical radioactive waste is primarily liquid, with only negligible amounts in solid or gaseous states. Medical radioactive liquid waste is a mixture of urine, feces, detergents, and radionuclides. Solid waste containing traces of radioactivity is in the form of syringes, needles, swabs, vials, contaminated gloves, and absorbent materials (Khan et al., 2010; Nasir et al., 2015). Clothing and utensils of patients who have taken high doses of radioisotopes (e.g., ^{131}I) constitute solid radioactive waste (Khan et al., 2010).

1.4.2.2 Industry

The main industrial applications of radioisotopes such as americium-241, antimony-124, barium-133, cobalt-57, cobalt-60, cesium-134, cesium-137, iridium-192, selenium-75, strontium-90, thulium-170, etc., are in manufacturing, chemical and mineral processing, coastal engineering, non-destructive testing, and borehole logging (Mushtaq, 2010).

In paper and steel sheet manufacturing, radioactive sources (e.g., strontium-90, cesium-137) can be used to measure the thickness of materials online, generating wastes such as contaminated work clothes, wiping materials, and discarded sealed sources. In Japanese paper companies, radioisotopes such as ^{85}Kr, ^{147}Pm, and ^{90}Cr are used as beta-ray emitters for paper weighing and moisture control. Radioisotope sensors are also used to measure thickness, color, differentiate composition, and detect the orientation of wood fibers (Yanagisawa et al., 2009). The petrochemical industry uses gamma rays (such as ^{60}Co) to monitor the liquid levels in storage tanks. However, during the decommissioning process, the equipment components that may become contaminated could cause pollution.

Industrial radiography is effective in detecting changes in material composition, measuring thickness, and locating hidden or defective components in assemblies. The process involves the use of a gamma or X-ray source on one side of the object to be inspected and a detector on the other side to produce a radiographic image. The energy of the radiation must be high enough to penetrate the object and expose the film. The radioactive waste generated by the process includes radioactively contaminated film, shielding materials, and discarded detection equipment.

1.4.2.3 Scientific Research

Experimental radioactive waste arises from the use of radioisotopes (e.g., ^{3}H, ^{14}C, ^{32}P, ^{125}I) or irradiators in nuclear physics, material science, and biomedical research. Typical products include nuclide-contaminated laboratory consumables (tubes, pipettes, protective gloves, etc.) as well as discarded radiolabeled reagents and biological samples (e.g., animal tissues, cell cultures). For example, radiocarbon-containing waste materials are generated in the biomedical field when ^{14}C isotopes are used to study metabolic pathways. Equipment maintenance and decommissioning waste comes from the maintenance and scrapping of scientific instruments (mass spectrometers, X-ray machines, etc.) or from laboratory modification and closure processes, and is typically represented by old instrumentation (e.g., light-emitting signage equipment) containing tritium-containing radioactive coatings, as well as radioactively contaminated building materials (broken concrete, piping systems). Nuclear technology research waste is generated in nuclear engineering simulation, radiochemical experiments (e.g., nuclear fuel cycle research, radiopharmaceutical development), and other scenarios, and is represented by special wastes such as low-concentration waste liquids, saturated ion-exchange resins, and nuclide-contaminated protective clothing.

Globally, approximately 118 nuclear research institutions and numerous academic facilities utilize various types of radioactive sources in their research and development activities, generating a significant quantity of radioactive waste. Taking Oak Ridge National Laboratory as an example, radioactive waste constitutes a major component of the facility's total waste output, originating from routine operations across multiple R&D facilities. The laboratory's ten primary radioactive waste streams include the Chemical Processing Plant, Fission Product Development Laboratory, Radioisotope Processing Area, Oak Ridge Research Reactor, Bulk Shielding Reactor, and similar installations, with an additional 12 to 15 high-temperature laboratories and numerous research laboratories generating smaller waste quantities. The laboratory categorizes its radioactive waste into four distinct classifications: (1) gas and airborne particulates; (2) process and low-level liquid wastes; (3) low-level and transuranic-contaminated solid wastes; and (4) radioactive hazardous wastes (co-contaminated).

1.4.3 Legacy Sources

Traditional anthropogenic sources of radiation constitute a significant source of radioactive material in the modern environment. One of the most notable contributors was the early development and frequent testing of nuclear weapons on a global scale, which produced huge quantities of radioactive fission products. Major nuclear accidents have resulted in more concentrated and intense radioactive releases at local and even regional scales. Inadequate management of radioactive waste, accumulated over long periods of operation of historical nuclear facilities, has also left a lasting legacy.

1.4.3.1 Early Nuclear Weapons Programs and Testing

Radioactive material from nuclear weapon explosions is the main source of man-made radionuclides in the environment. The first nuclear explosions carried out by mankind were at the end of the Second World War, when the United States tested a 19-kiloton nuclear device on 16 July 1945 and exploded nuclear weapons in Hiroshima and Nagasaki, Japan, on 6 and 9 August, respectively.

Between 1945 and the signing of the Comprehensive Nuclear-Test-Ban Treaty in 1996, more than 2,000 nuclear tests were conducted by the United States, the Soviet Union, the United Kingdom, France, and China, and one by India. More than 500 of these were above-ground tests; the largest was conducted by the Soviet Union, with a yield of more than 50 million tonnes (Holdstock & Waterston, 2000).

Among the radioisotopes released during nuclear tests, ^{137}Cs, ^{90}Sr, $^{239-240}Pu$, ^{241}Am, and ^{131}I stand out in terms of their significant impact on the environment and radiation exposure of the human body; these isotopes are found mainly at nuclear test sites around the globe. Since about two-thirds of the Earth's surface is covered by water, a large proportion of these radionuclides have been transferred to the marine environment, as is the case for isotopes such as ^{137}Cs and ^{90}Sr, and their negative impacts are mainly due to bioaccumulation through the food chain cycle. The indirect transfer of radionuclides to the Earth's layers through the food chain and their accumulation in living cells is another form of radioactive contamination of marine and terrestrial ecosystems. The most representative example of this is the ^{14}C isotope released into the atmosphere during nuclear tests, which is subsequently incorporated into carbon dioxide and then enters the marine environment through ocean–atmosphere gas exchange or through the process of photosynthesis (Prăvălie, 2014).

1.4.3.2 Nuclear Accidents

On April 26, 1986, the worst nuclear accident in history occurred in Chernobyl, in what is now Ukraine. One of the four reactors at the Soviet Chernobyl Nuclear Power Plant, located 129 km north of Kiev, exploded. As a result of the explosion, most of the radionuclides in the core, especially gaseous and volatile ones, were dispersed into the environment. As a result, 100% of the noble gases, such as ^{133}Xe, were released. About 60% of the ^{131}I nuclide (1,800,000 TBq) and about 30% of the ^{137}Cs nuclide (85,000 TBq) were released into the atmosphere. The large amounts of radioactivity released caused significant ground contamination in the vicinity of the plant. In addition, large amounts of radioactivity were lifted high in the atmosphere and transported throughout Europe, producing significant localized ground contamination depending on wind and precipitation factors. For example, about 4,000 TBq of ^{137}Cs was deposited on soil in Sweden.

The serious accident at the Fukushima Daiichi Nuclear Power Plant on March 11, 2011, caused immeasurable damage to the daily lives of local residents. As a result of the radioactive substances discharged from the nuclear power plant, the environment was severely and extensively contaminated, including houses, farmland, forests and the coastline, resulting in severe radiation exposure to the residents. Radioactive substances were discharged into the environment, the main radioactive substances being ^{131}I, ^{133}I, ^{134}Cs, ^{137}Cs, and rare gases (^{133}Xe, ^{85}Kr) (Tanaka, 2012). According to Japanese estimates, about 4,000 TBq of ^{137}Cs leaked into the sea in the first few months after the accident, although French experts estimated a much higher amount.

1.4.3.3 Waste Storage Leaks

Of the 113 sites in the United States that have been contaminated as a result of past nuclear weapons manufacturing, the most notable is the Hanford Site, where the B Reactor, the world's first plutonium-operated (plutonium-producing) nuclear reactor, was constructed. The process of producing nuclear weapons left the Hanford Site with a large amount of solid and liquid radioactive waste, leaving a potential environmental risk to the Columbia River Basin. Nuclear waste leaking from single underground storage tanks at the U.S. Department of Energy's (DOE) Hanford Site, or from piping entering/leaving these tanks, has been detected in the subsurface environment since the 1950s. Radionuclides in the Hanford tanks include uranium and transuranic elements such as neptunium, plutonium, and americium, as well as radioactive and stable isotopes of lighter elements resulting from fission and activation processes. Most on-site radioactive waste and nuclear materials remain in underground tanks or storage facilities. About half of the chemical waste remains in tanks, while the rest remains in soil, groundwater, and cemeteries (Gephart, 2010).

The Sellafield nuclear power station on the Cumbrian coast in the north-west of England is an important part of the UK's nuclear industrial heritage. Sellafield, including its smaller internal Winscot Lane nuclear power station, is a large integrated nuclear chemical facility. Sellafield and Winscotle have supported civil nuclear energy programmes since the late 1940s and were involved in the early production of plutonium for the UK's nuclear weapons program. Since 1952, the Sellafield nuclear site has been discharging authorized low-level radioactive aqueous waste into the East Irish Sea. Liquids from spent fuel reprocessing and wash water from fuel storage ponds are the main sources of this waste stream, which contains a range of actinides and fission products (e.g., cesium, uranium, neptunium, plutonium, and americium) (Ray et al., 2020). From 2019, after Sellafield reported storage unit leaks to the UK's National Office for Nuclear Research (ONR), the leaks worsened significantly, with 2.3–2.5 cubic meters of radioactive liquid leaking per day.

1.5 ENVIRONMENTAL IMPACTS

Anthropogenic releases can cause localized high-dose contamination through accidents or mismanagement. Acute accidents often bring about rapid, high-dose pollution, for instance, reactor explosions that cause immediate damage to ecosystems, whereas the improper handling of nuclear waste, liquid effluents, or contaminated agricultural products tends to result in long-term low-dose exposure. Although these two types of risks seem to represent opposite extremes, both can have significant and lasting impacts on the environment, from sudden ecosystem collapse to the gradual bioaccumulation of radionuclides in food chains.

1.5.1 Localized Contamination and Ecosystem Impacts

1.5.1.1 Localized Environmental Contamination Caused by Nuclear Accidents

After a nuclear accident, a large amount of radioactive nuclides causes high-dose local contamination in the surrounding area through various channels, including atmospheric diffusion and water seepage. Radioactive aerosols spread in the troposphere and settle on the ground with rainfall, forming a "radioactive fallout zone" centered on the nuclear power plant. ^{137}Cs, ^{131}I, and ^{90}Sr are the main polluting nuclides, which have long half-lives and strong mobility, and can remain in soil, water, and crops over extended periods, posing a continuing threat to the environment (UNSCEAR, 2000; JRC, 2016; IAEA, 2006).

The 1986 Chernobyl nuclear accident released more than 5×10^{18} Bq of radioactive nuclides, with ^{137}Cs settling at concentrations exceeding 1,000 kBq/m^2 in the core area of the accident. As shown in Figure 1.3, the contamination spread to Ukraine, Belarus, Austria (Bossew et al., 1995), and several other European countries, forming a highly contaminated area of more than 150,000 km^2(Steinhauser et al., 2014; Gofman, 2000). In comparison, the total amount of radioactivity released in the 2011 Fukushima Daiichi Nuclear Power Plant accident was relatively small. However, due to leakage from cooling water systems and heavy rainfall along the coast, a significant quantity of radioactive nuclides infiltrated the marine ecosystem, which was subsequently dispersed by ocean currents (Aoyama et al., 2020), as illustrated in Figure 1.4. Furthermore, ^{137}Cs concentrations, reaching levels in the tens of thousands of Bq/kg, were detected in surface soil, resulting in secondary impacts such as abandoned farmland and food export bans (Orita et al., 2016).

FIGURE 1.3 ^{137}CS Deposition Map in Europe Following the Chernobyl Accident (Steinhauser et al., 2014).

1.5.1.2 Destruction and Influence Mechanism to Ecosystem

Once radioactive nuclides enter the environment, they can exert radiological effects on multiple levels of ecosystems, primarily manifesting as cellular damage at the individual level, reproductive disorders at the population level, and structural disruption at the community level (UNSCEAR, 2020; IAEA, 2006). In particular, biologically active radioactive elements, such as ^{137}Cs and ^{90}Sr, are easily absorbed by plant roots or ingested by animals,

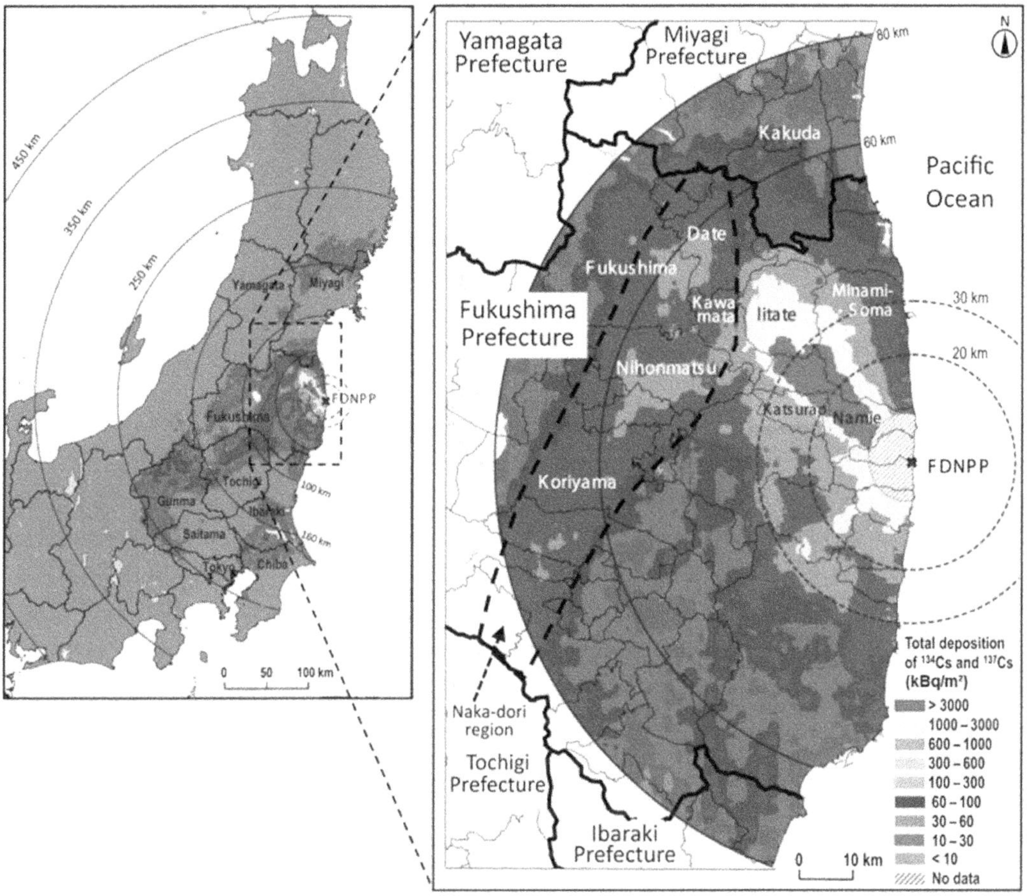

FIGURE 1.4 Dispersion of ^{137} CS in the North Pacific Ocean Following the Fukushima Accident (Steinhauser et al., 2014).

where they accumulate in the body, leading to physiological abnormalities, genetic mutations, and even population degradation (IAEA, 2015a).

In the Chernobyl accident, the high-intensity radiation released after the reactor explosion led to the rapid mortality of the pine forests encircling the core area within a few days. This area is referred to as the "Red Forest" and serves as a classic example of the devastating impact of radiation on ecosystems (IAEA, 2006). Within the accident-affected area, the population sizes of various bird species declined significantly, breeding rates decreased, and feather deformity rates increased. Additionally, insect and microbial diversity in some areas was significantly suppressed (Møller & Mousseau, 2007; Bonisoli-Alquati et al., 2015). The Fukushima nuclear accident also had a significant impact on marine ecosystems. Following the accident, the concentration of radioactive ^{137} Cs in fish along the Fukushima coast increased, with some deep-sea fish exhibiting concentrations hundreds of times higher than the background level in seawater. Radioactive accumulation was also observed in plankton, shellfish, and algae (Buesseler et al., 2012).

Once radioactive nuclides enter an ecosystem, they can undergo "biological magnification" through the food chain, gradually accumulating across different trophic levels. ^{137}Cs, which shares similar biological behavior with potassium ions, is easily absorbed by plants and enters higher trophic levels via herbivores; while ^{90}Sr mimics the behavior of calcium, easily depositing in bones and posing a long-term health threat to mammals (Møller & Mousseau, 2006; WHO, 2013). These nuclides not only possess long-range migration capabilities in the environment but can also form sustained cycles within ecosystems, resulting in "apparent decay" of contamination occurring far slower than their physical half-lives, which presents significant challenges for subsequent environmental remediation and risk control (Von Wehrden et al., 2013).

1.5.1.3 Long-Term Residual Issues and Repair Challenges

Long-lived radioactive nuclides, such as ^{137}Cs and ^{239}Pu, pose significant challenges to the environment due to their persistent radiotoxicity and complex migration behavior. These nuclides strongly adsorb onto soil particles and clay minerals, forming "fixed contamination" zones. However, under conditions such as heavy rainfall, snowmelt, or wildfires, they may re-migrate and re-enter surface water and the biosphere decades after deposition (UNSCEAR, 2020; Sakai et al., 2021). Studies of forest ecosystems around Chernobyl have shown that seasonal processes—such as litterfall and erosion—promote remobilization of ^{137}Cs, leading to detectably increased stream water levels even years post-accident (Goor & Thiry, 2004).

Efforts to remediate contaminated soil typically rely on methods such as topsoil stripping, chemical stabilization, and phytoremediation. Each technique has limitations: soil removal generates radioactive waste that requires further storage, while phytoremediation is often constrained by slow rates or poor efficacy in deeply contaminated scenarios (Yan et al., 2021). Additionally, decontamination efficiency varies significantly across different terrains. Additionally, the issue of radioactive nuclides transferring to fruits highlights the inefficiency of plant-based methods in ensuring food safety (Carini, 2009). Furthermore, decontamination efficiency varies significantly across different terrains. For example, forested or mountainous areas around Chernobyl and Fukushima remain untreated to this day due to logistical, economic, and ecological constraints (IAEA, 2006). Japan's decontamination program following the Fukushima nuclear accident involved removing approximately 16.5 million m^3 of topsoil. As of 2018, the cost of the program had exceeded 2.9 trillion yen (~27 billion USD) (Japanese Ministry of the Environment, 2018). The 30 km exclusion zone around Chernobyl remains closed and uninhabitable, while certain areas in Fukushima, despite extensive cleanup efforts, remain under restricted access (Steinhauser et al., 2014). The social consequences are profound: hundreds of thousands of people were forced to relocate, with many unable to return due to ongoing health concerns and the collapse of land values (WHO, 2013).

1.5.2 Global Effects and Long-Term Cumulative Consequences

1.5.2.1 Global Nuclear Tests and Radioactive Fallout

Between 1945 and 1980, approximately 520 atmospheric nuclear weapons tests released an estimated 950 PBq of ^{137}Cs and 1,000 PBq of ^{90}Sr into the environment, primarily through stratospheric injection (UNSCEAR, 2000; IAEA, 2006). Fallout from high-yield tests, such as the 1954 Castle Bravo explosion (yield: 15 Mt), was detectable worldwide and contributed significantly to global fallout inventories. Precipitation carried these radionuclides to the Earth's surface, creating a nearly uniform fallout layer that peaked around 1963, when average surface air concentrations of ^{90}Sr in the Northern Hemisphere exceeded 6 mBq/m^3, and deposition rates reached up to 500 Bq/m^2 annually in mid-latitude zones (UNSCEAR, 2000). Arctic ice cores and lake sediment records confirmed the timing and scale of this fallout (Almgren & Isaksson, 2006). Although the average cumulative effective dose to individuals from these events was relatively low (~2 mSv over several decades), ecosystems—particularly forest litter layers and peatlands—continue to retain ^{137}Cs inventories at levels exceeding 10–50 kBq/m^2, which can be re-mobilized during forest fires or erosion (IAEA, 2006; UNSCEAR, 2020).

1.5.2.2 Transnational Contamination Effects of the Nuclear Accident

Large-scale nuclear accidents, such as Chernobyl and Fukushima, have led to the transboundary spread of radioactive nuclides on an unprecedented scale, adding geographically uneven radioactive contamination to the global environmental inventory. The radioactive plume from the Chernobyl accident reached Scandinavia within 48 hours, with wet deposition levels in parts of Sweden measuring over 200 kBq/m^2 of ^{137}Cs (Åhman & Wright, 2001). This high deposition forced culling of contaminated reindeer and imposed long-term restrictions on the consumption of wild berries, mushrooms, and game meat—critical food sources for indigenous Sámi communities (Åhman & Wright, 2001; Rissanen & Rahola, 1990). Although the total release of radioactive nuclides from the Fukushima Daiichi Nuclear Power Plant Accident was relatively low, it still resulted in dual impacts of atmospheric deposition and direct liquid discharge into the ocean. Approximately 15 PBq of ^{137}Cs was released into the North Pacific (IAEA, 2015a), representing the largest single direct ocean discharge in history. Ocean currents and surface circulation facilitated rapid basin-wide transport; detectable concentrations of radioactive cesium (<10 Bq m^{-3}) reached the west coast of North America in less than two years (Smith et al., 2017). Oceanographic monitoring also showed measurable increases in ^{137}Cs concentrations in surface waters near Hawaii and in Arctic outflow waters.

Additionally, gamma monitoring networks in distant regions confirmed the airborne transport of radioactive nuclides from the Fukushima nuclear accident. Trace amounts of ^{131}I and ^{137}Cs were detected in meteorological and air samples in France, Canada, Brazil, and even the Southern Hemisphere (Masson et al., 2015). These findings indicate that the radioactive impact of nuclear accidents is not confined to national borders but reflects the interconnectedness of atmospheric and oceanic systems, thereby necessitating internationally coordinated monitoring and response strategies.

1.5.2.3 Long-Term Exposure, Ecological Persistence, and Societal Acceptance

Assessing long-term low-dose exposure to dispersed radioactive nuclides in the environment presents unique challenges. For example, Reiss (1961) famously revealed in their "baby teeth survey" that the levels of ^{90}Sr in the milk teeth of children born in 1963 were approximately 50 times higher than those in children born in 1950, a finding directly linked to atmospheric

TABLE 1.6 Long-Term Ecological Persistence of ^{137}Cs in Selected Environments

Environment type	Region	Typical ^{137}Cs level (kBq/m^2)	Notes/key parameter	References
Forest litter and soil	Chernobyl Exclusion Zone	20–50	Persistent topsoil inventory	(OECD–NEA, 2002; UNSCEAR, 2000)
Peatlands	Belarus	5–8	Radiocesium is concentrated in the organic layers	(Poręba & Bluszcz, 2007)
Agriculture fields	Fukushima (general region)	5–25	Depending on the distance from the release source	(Sakai et al., 2021)
Wild boar meat	Germany	-	Ecological half-life = 7.3 years	(Steinhauser et al., 2022)
Wild boar meat	Japan (Fukushima Prefecture)	-	Ecological half-life = 2.6 years	(Steinhauser et al., 2022)

nuclear testing. However, even in large-scale epidemiological retrospective studies, such as that by Hauptmann et al. (2020), evidence of increased risk of solid cancers at doses below 100 mGy remains inconsistent. Ecological studies are equally complex. Observations in Germany and Japan indicate that the ecological half-life of ^{137}Cs in wild boar meat can range from 2.6 to 7.3 years, highlighting its persistence in the food chain (Steinhauser et al., 2022). A summary of selected levels of ^{137}Cs in various ecosystems is provided in Table 1.6. Additionally, societal reactions to low-level contamination remain strong—a decade after the Fukushima nuclear accident, some markets still restrict the circulation of related foods (IAEA, 2015b). This underscores that the actual impact of radiation depends not only on exposure dose but also on its persistence in the environment and on societal trust.

1.6 CONCLUSION

Radioactivity, intrinsic to both Earth's geochemistry and anthropogenic nuclear technologies, presents multi-faceted environmental and societal challenges. Natural radionuclides—distributed through geological cycles—establish a baseline radiation exposure, while human activities have introduced non-equilibrium radioactive species (e.g., ^{137}Cs, ^{90}Sr, ^{239}Pu) with elevated bioavailability and persistence. The nuclear fuel cycle remains the dominant anthropogenic source, yielding HLWs requiring millennia-scale containment. Environmental impacts are starkly demonstrated by Chernobyl and Fukushima: ^{137}Cs deposition exceeding 1,000 kBq/m^2 caused long-term ecosystem disruption, including forest die-off ("Red Forest") and marine bioaccumulation. Remediation efforts face technical and economic constraints, exemplified by Japan's $27 billion soil decontamination program. Globally, legacy contamination from nuclear testing continues to re-mobilize through wildfires and erosion, sustaining ecological half-lives up to 7.3 years in key species.

Based on this information, the concept of nuclear waste is formally introduced: it refers to human-generated materials containing radionuclides without further utilization, typically characterized by "artificial introduction" and "activity concentrations significantly exceeding natural background levels." Subsequent sections will systematically elaborate on its generation mechanisms, potential risks, and management strategies.

In the writing process, generative artificial intelligence tools (DeepSeek-V3, ChatGPT-5, and Gemini-2) were used solely for language polishing, including grammar and spelling review, to enhance readability and linguistic quality. The authors have critically reviewed and validated all content and assume full responsibility for the academic integrity and originality of the work.

REFERENCES

Adaikalam, M.J., Shareef, N.Y., Khan, F.M., 2024. Bioaccumulation of natural radionuclides in edible molluscs and associated risk from the Ashtamudi Estuary, situated in a high background natural radiation zone (HBNRA), Kerala, southwest coast of India. *Reg. Stud. Mar. Sci.* 79, 103833.

OECD/NEA, 2006. Radioactive Waste Management Programmes in OECD/NEA Member Countries. *Radioactive Waste Management*. OECD Publishing, Paris.

Åhman, B., Wright, S.M., 2001. Effect of origin of radiocaesium on the transfer from fallout to reindeer meat. *Sci. Total Environ.* 278, 171–181.

Al Aamri, M., Ravichandran, R., Binukumar, J.P., Al Balushi, N., 2016. Therapeutic applications of radioactive (131)iodine: Procedures and incidents with capsules. *Indian J. Nucl. Med.* 31(3), 176–178.

Almgren, S., Isaksson, M., 2006. Vertical migration studies of 137Cs from nuclear weapons fallout and the Chernobyl accident. *J. Environ. Radioact.* 91, 90–102.

Alwaeli, M., Mannheim, V., 2022. Investigation into the current state of nuclear energy and nuclear waste management—a state-of-the-art review. *Energies* 15(12), 4275.

Aoyama, M., Tsumune, D., Uematsu, M., 2020. Distribution and evolution of Fukushima Dai-ichi derived 137Cs, 90Sr and 129I in surface seawater off the coast of Japan. *Environ. Sci. Technol.* 54, 1321–1330.

Bé, M.-M., Chisté, V., Dulieu, C., Browne, E., Chechev, V., Kuzmenko, N., Helmer, R., Nichols, A., Schönfeld, E., Dersch, R., 2004–2013. Table of Radionuclides, Monographie BIPM-5. *Bureau International des Poids et Mesures, Sèvres*, France.

Bonisoli-Alquati, A., et al., 2015. Abundance and genetic damage in barn swallows from Fukushima. *Sci. Rep.* 5, 16594.

Bossew, P., Falkner, T., Henrich, E., Kienzl, K., 1995. *Cäsiumbelastung der Böden Österreichs (Contamination of Austrian soils by Caesium-137).* Umweltbundesamt, Wien.

Buesseler, K.O., et al., 2012. Fukushima-derived radionuclides in the ocean and biota off Japan. *Proc. Natl. Acad. Sci. USA* 109, 5984–5988.

Carini, F., 2009. Radionuclide transfer to fruit in the IAEA TRS 364 revision. *J. Environ. Radioact.* 100, 752–756.

Chang, G.L., 2002. Radon and its hazards. *Uranium Geol.* 2, 122–128.

Cheng, X., 2012. *The Migration and Enrichment Patterns of Radioactive Nuclides in Phosphate Mining and Their Environmental Effects*. Chengdu University of Technology (Ph.D. thesis).

Cheng, Y.X., 2008. The hazards and control of radon and its derivatives in the environment. *Geoscience* 5, 857–868.

Darda, S.A., Gabbar, H.A., Damideh, V., Aboughaly, M., Hassen, I., 2021. A comprehensive review on radioactive waste cycle from generation to disposal. *J. Radioanal. Nucl. Chem.* 329(1), 15–31.

Dong, D., Wang, Z., Guan, J., Xiao, Y., 2025. Research on safe disposal technology and progress of radioactive nuclear waste. *Nucl. Eng. Des.* 435, 113934.

Frigi, E.L., Pazianotto, M.T., Federico, C.A., Andrade, E.R., 2024. Aircraft radiation exposure assessment in a radioactive environment: Ambient dose equivalents and effective doses. *Radiat. Meas.* 176, 107202.

Gephart, R.E., 2010. A short history of waste management at the Hanford Site. *Phys. Chem. Earth* 35(6–8), 298–306.

Gofman, J.W., 2000. Health effects on populations exposed to low-level radiation delivered gradually over time. In: *Proceedings of the 10th International Congress of the IRPA*. IRPA, Hiroshima, pp. 1–9.

Goor, F., Thiry, Y., 2004. Processes, dynamics and modelling of radiocaesium cycling in a chronosequence of Chernobyl-contaminated Scots pine plantations. *Sci. Total Environ.* 327, 181–197.

Hauptmann, M., Daniels, R.D., Cardis, E., et al., 2020. Epidemiological studies of low-dose ionizing radiation and cancer: Summary bias assessment and meta-analysis. *J. Natl. Cancer Inst. Monogr.* 2020, 188–200.

Holdstock, D., Waterston, L., 2000. Nuclear weapons, a continuing threat to health. *Lancet* 355(9214), 1544–1547.

Hu, T.H., Zhang, C.L., 2003. The issue of natural background radiation for the public. *J. Jinan Univ. (Nat. Sci. Med. Ed.)* 5, 30–33.

Huang, L.H., Xu, L.Y., Lin, M.R., Lai, S.K., Lin, C.P., 2004. Analysis of natural radioactive nuclide content in granite stone. *Occup. Health* 1, 99–101.

Huang, Z.J., 2008. Investigation and testing of radioactive pollution sources (part 1). *Radiat. Prot. Bull.* 2, 32–45.

IAEA, 2006. *Environmental Consequences of the Chernobyl Accident and Their Remediation.* International Atomic Energy Agency, Vienna.

IAEA, 2015a. *The Fukushima Daiichi Accident—Technical Volume 4.* International Atomic Energy Agency, Vienna.

IAEA, 2015b. *The Fukushima Daiichi Accident: Report by the Director General, Vol. 5—Post-Accident Recovery.* International Atomic Energy Agency, Vienna.

IAEA, 2022. *Status and Trends in Spent Fuel and Radioactive Waste Management.* IAEA Nuclear Energy Series No. NW-T-1.14 (Rev. 1).

Islam, S., Aranna, M.N., Ferdous, J., Bakshi, P.K., 2025. Natural and anthropogenic radionuclides in atmospheric air and assessment of inhalation dose over brick field areas, Dhaka, Bangladesh. *Radiochemistry* 67(1), 101–109.

Japanese Ministry of the Environment (MoE), 2018. *Environmental Remediation in Japan—Progress Reports.* Ministry of the Environment, Tokyo.

JRC, 2016. *The Fukushima Daiichi Accident.* Joint Research Centre, European Commission.

Kamei-Ishikawa, N., Ito, A., Tagami, K., Umita, T., 2013. Fate of radiocesium in sewage treatment process released by the nuclear accident at Fukushima. *Chemosphere* 93(4), 689–694.

Khan, S., Syed, A.T., Ahmad, R., Rather, T.A., Ajaz, M., Jan, F.A., 2010. Radioactive waste management in a hospital. *Int. J. Health Sci.* 4(1), 39.

Kuzmanović, P., Todorović, N., Nikolov, J., Hansman, J., Vraničar, A., Knežević, J., Miljević, B., 2019. Assessment of radiation risk and radon exhalation rate for granite used in the construction industry. *J. Radioanal. Nucl. Chem.* 321(2), 565–577.

Li, C.H., Ji, Y.L., Zhang, S.Q., Zhao, J.Q., 2004. Estimation of exposure time, erosion rate, and erosion thickness of erosion surfaces using cosmogenic nuclides Be-10 and Al-26. *J. China Univ. Pet. (Nat. Sci. Ed.)* 1, 1–4+137.

Liu, L.M., Guo, H., Dai, L.H., Liu, M.L., Xiao, Y., Cong, T.L., Gu, H.Y., 2023. The role of nuclear energy in the carbon neutrality goal. *Prog. Nucl. Energy* 162, 104772.

Manjulata, Y., Kumar, M.J., Ramola, R.C., 2023. Study of radionuclides in rocks samples from Ukhimath area and its correlation with soil and water data. *Chem. Afr.* 6(4), 2165–2173.

Masson, O., Ringer, W., Malá, H., Mikšová, J., Ashraf, M.A., Joseph, S., Steinkopff, T., 2015. Tracking of airborne radionuclides from the damaged Fukushima Dai-ichi nuclear reactors by European networks. *Environ. Sci. Technol.* 49, 11100–11108.

Møller, A.P., Mousseau, T.A., 2006. Biological consequences of Chernobyl: 20 years on. *Trends Ecol. Evol.* 21, 200–207.

Møller, A.P., Mousseau, T.A., 2007. Species richness and abundance of forest birds in relation to radiation at Chernobyl. *Biol. Lett.* 3, 483–486.

Mushtaq, A., 2010. Reactors are indispensable for radioisotope production. *Ann. Nucl. Med.* 24(10), 759–760.

Nasir, T., Ali, S., Farid, M., Shahbaz, M., Rizwan, M., Hannan, F., Ahmad, R., 2015. Infectious and radioactive waste management in a diagnostic and nuclear medicine centre. *Am. J. Environ. Prot.* 4(5), 245–250.

OECD-NEA, 2002. *Chernobyl: Assessment of Radiological and Health Impact—2002 Update of Chernobyl: Ten Years On*. Nuclear Energy Agency, Organisation for Economic Co-operation and Development, Paris.

Orita, M., Nakashima, K., Hayashida, N., Endo, Y., Yamashita, S., Takamura, N., 2016. Concentrations of radiocesium in local foods collected in Kawauchi Village after the accident at the Fukushima Dai-ichi Nuclear Power Station. *Sci. Rep.* 6, 28470.

Poręba, G., Bluszcz, A., 2007. Determination of the initial 137Cs fallout on the areas contaminated by Chernobyl fallout. *Geochronometria* 28, 57–62.

Prăvălie, R., 2014. Nuclear weapons tests and environmental consequences: A global perspective. *Ambio* 43(6), 729–744.

Pyo, J.Y., Um, W., Heo, J., 2021. Magnesium potassium phosphate cements to immobilize radioactive concrete wastes generated by decommissioning of nuclear power plants. *Nucl. Eng. Technol.* 53, 2261–2267.

Qu, L.M., 2006. *Research on the Evaluation of Radiation Environmental Quality in Qingdao Urban Area*. University of Chinese Academy of Sciences (Institute of Oceanology, Chinese Academy of Sciences) (Master's thesis).

Ray, D., Leary, P., Livens, F., Gray, N., Morris, K., Law, K.A., et al., 2020. Controls on anthropogenic radionuclide distribution in the Sellafield-impacted Eastern Irish Sea. *Sci. Total Environ.* 743, 140765.

Reiss, L.Z., 1961. Strontium-90 absorption by deciduous teeth: Analysis of teeth provides a practicable method of monitoring strontium-90 uptake by human populations. *Science* 134, 1669–1673.

Rissanen, K., Rahola, T., 1990. Radiocesium in lichens and reindeer after the Chernobyl accident. *Rangifer* 10, 55–61.

Saclier, N., Chardon, P., Malard, F., Konecny-Dupré, L., Eme, D., Bellec, A., Breton, V., Duret, L., Lefebure, T., Douady, C.J., 2020. Bedrock radioactivity influences the rate and spectrum of mutation. *eLife* 9, e56830.

Sakai, M., Tsuji, H., Ishii, Y., Ozaki, H., Takechi, S., Jo, J., Tamaoki, M., Hayashi, S., Gomi, T., 2021. Untangling radiocesium dynamics of forest-stream ecosystems: A review of Fukushima studies in the decade after the accident. *Environ. Pollut.* 288, 117744.

Smith, J.N., Aoyama, M., Hamilton, T.F., 2017. Arrival of the Fukushima radioactivity plume in North American continental waters. *Proc. Natl. Acad. Sci. USA* 114, 11092–11097.

Steinhauser, G., Berendes, O., Reddermann, J., 2022. Exemplifying the "wild boar paradox": Dynamics of cesium-137 contamination in wild boars in Germany and Japan. *J. Radioanal. Nucl. Chem.* 333, 1261–1272.

Steinhauser, G., Brandl, A., Johnson, T.E., 2014. Comparison of the Chernobyl and Fukushima nuclear accidents: A review of the environmental impacts. *Sci. Total Environ.* 470–471, 800–817.

Sun, Q.F., Qiu, Y.C.B., Tao, Z.F., Yuan, Y.L., Zou, J.M., Morishima, Y.Z., Kato, K., Cha, Y., Sugawara, N., Wei, L.X., 2002. Chronic exposure to natural high background radiation and excess risk of solid cancer in Yangjiang. *Chin. J. Radiol. Med. Prot.* 1, 55–58.

Tan, J.P., Clyde, C.W., Ng, C.C., Yeap, S.K., Yong, C.Y., 2024. Advancements in microbial-mediated radioactive waste bioremediation: A review. *J. Environ. Radioact.* 280, 107530.

Tanaka, S.I., 2012. Accident at the Fukushima Dai-ichi nuclear power stations of TEPCO—outline & lessons learned—. *Proc. Jpn. Acad. Ser. B* 88(9), 471–484.

UNSCEAR, 2000. *Sources and Effects of Ionizing Radiation*. United Nations Scientific Committee on the Effects of Atomic Radiation, New York.

UNSCEAR, 2020. *Sources, Effects and Risks of Ionizing Radiation*. United Nations Scientific Committee on the Effects of Atomic Radiation, New York.
Von Wehrden, H., Fischer, J., Brandt, P., Wagner, V., Kümmerer, K., Kuemmerle, T., Nagel, A., Olsson, O., Hostert, P., 2013. Consequences of nuclear accidents for biodiversity and ecosystem services. *Conserv. Lett.* 6, 277–285.
Wang, M.S., Liu, S.L., 2021. Discussion on the sources of radiation. *Chin. J. Radiol. Health* 30(2), 238–243.
WHO, 2013. *Health Risk Assessment from the Nuclear Accident after the 2011 Great East Japan Earthquake and Tsunami*. World Health Organization, Geneva.
Wu, R.N., 2004. The impact of granite building materials decoration on residential radiation. *Chin. J. Radiol. Med. Prot.* 3, 97–98.
Wu, Y.Y., 2015. *Study on the Impact of Domestic Groundwater on Indoor Radon and the Dose Contribution to Residents*. Chinese Center for Disease Control and Prevention (Master's thesis).
Xia, N., 1977. Carbon-14 dating and prehistoric archaeology in China. *Archaeology* 4, 217–232.
Xia, Y.H., 2001. Pay attention to the issue of increased natural exposure caused by human activities. *Radiat. Prot.* 1, 11–18.
Xie, R.N., 2015. *Study on Radon Release Rate of Granite and Indoor Radon Radiation Risk Assessment*. China University of Geosciences (Beijing) (Master's thesis).
Yan, L., Le, Q.V., Sonne, C., Yang, Y., Yang, H., Gu, H., Ma, N.L., Lam, S.S., Peng, W., 2021. Phytoremediation of radionuclides in soil, sediments and water. *J. Hazard. Mater.* 407, 124771.
Yanagisawa, K., Inoue, T., Hayakawa, K., Shiotari, H., Nakamura, Y., Matsuyama, K., Nagasawa, K., 2009. Economic scale of utilization of radiation in medicine in Japan. *Appl. Radiat. Isot.* 67(7–8), 1387–1391.
Yeong, C.H., Cheng, M.H., Ng, K.H., 2014. Therapeutic radionuclides in nuclear medicine: Current and future prospects. *J. Zhejiang Univ. Sci. B* 15(10), 845–863.
Zhang, S.B., 2008. Potential hazards of natural background radiation. *World Nucl. Geosci.* 3, 178–182+186.
Zhang, S.S., 2002. A trial talk on radioactivity of granite stone. *Geol. Fujian* 4, 221–225.
Zhao, M., Zhao, X., 2023. The comparative analysis of coastal site and inland site in nuclear power plants. *Adv. Civ. Eng. Tech.* 5(5), 1–3.
Zhao, M.Q., Cui, H.X., Shang, B., Liu, J.X., Liu, Z.L., Xue, W.H., 2012. Radon levels and dose contributions in Pingshan hot springs, Hebei Province. *Chin. J. Radiol. Health* 21(1), 30–32.
Zhou, P., Li, D.M., Liu, G.S., Cai, W.X., Huang, C.G., Fang, H.D., Ji, L.H., 2015. Study on the application of the radioactive isotope silico-32 in tracking ocean processes. *J. Isot.* 28(1), 7–19.
Zohuri, B., 2020. Nuclear fuel cycle and decommissioning. In: *Nuclear Reactor Technology Development and Utilization*. Woodhead Publishing, pp. 61–120.

CHAPTER 2

Nuclear Waste Generation and Potential Risks in the Nuclear Industry

Xinpeng Wang

2.1 OVERVIEW OF THE NUCLEAR INDUSTRY AND CONTEXT OF NUCLEAR WASTE GENERATION

Facing global challenges of energy shortage, environmental pollution, and climate change, the green and low-carbon transition has become the inevitable path to achieve "double carbon" goals. Nuclear energy, with its high energy density and significant carbon reduction potential, is seen as a key pillar of the future clean energy system. Studies show that nuclear power generation produces almost zero greenhouse gas emissions and no nitrogen oxides or particulate matter pollution, making it an optimal baseload power source (Xie et al., 2019; Liu et al., 2024). Since the Manhattan Project initiated human use of nuclear energy in the 1940s, the nuclear industry has evolved from primarily military applications to widespread civilian power generation. The International Energy Agency (IEA) forecasts that global nuclear power generation will increase by 82% by 2050 to meet energy demands from emerging economies and population growth, accelerating energy structure transformation. However, the rapid development of the nuclear industry is accompanied by significant environmental risks. During the nuclear fuel cycle and radioactive applications, radionuclides such as ^{235}U, ^{137}Cs, ^{90}Sr, and ^{129}I are inevitably released. Wastes with radioactive concentrations exceeding safety thresholds cause irreversible damage to ecosystems and pose long-term threats due to extremely long half-lives (some extending to tens of thousands of years) (Jun et al., 2021).

The nuclear industry chain encompasses multiple links, including uranium mining, fuel fabrication, nuclear power plant operation, and nuclear fuel reprocessing, each closely associated with nuclear waste generation. Radioactive wastes are systematically classified into three categories based on their radioactivity levels: low-level waste (LLW), intermediate-level waste (ILW), and high-level waste (HLW). LLW typically features relatively high

DOI: 10.1201/9781003665793-3

radioactivity clearance levels but contains limited quantities of long-lived radionuclides. In contrast, ILW, owing to the presence of long-lived radionuclides, necessitates more stringent containment and isolation strategies than near-surface disposal methods. HLW, on the other hand, is characterized by extremely high activity concentrations and generates substantial heat during radioactive decay.

2.2 MAIN CLASSIFICATION OF NUCLEAR WASTE

Proper disposal of nuclear waste is a key link in the safe use of nuclear energy, and scientific classification is the foundation of effective management. Nuclear waste is mainly divided into three types according to radioactivity, half-life, and potential hazard levels: LLW, ILW, and HLW (*Classification of Radioactive Waste*, 2009).

LLW typically exhibits radioactive activity below 4×10^6 Bq/kg and is characterized by weak radioactivity and a short decay period. Its sources are diverse, encompassing radioactive waste rocks from uranium mining, contaminated protective materials in the medical field (such as surgical gowns and gauze), and low-value radioactive consumables discarded from research laboratories (such as test tubes and pipettes). Since the radionuclides within LLW have half-lives ranging from months to years, conventional disposal methods, such as near-surface landfilling or cement solidification, can effectively isolate and facilitate the decay of radioactive substances, with relatively low technical complexity (*Scientific and Technical Basis for the Near Surface Disposal of Low and Intermediate Level Waste*, 2002).

ILW has radioactive activity ranging from 4×10^6 to 4×10^{11} Bq/kg. It mainly includes irradiated structural materials from reactor operations, nuclear fuel element claddings, and partially pre-treated medical or industrial radioactive liquid waste. Due to the longer half-lives of the radioactive substances in ILW, it requires medium-depth geological disposal and encapsulation in special containers for isolation (*Scientific and Technical Basis for the Near Surface Disposal of Low and Intermediate Level Waste*, 2002).

As the core challenge in nuclear waste management, HLW has radioactive activity exceeding 4×10^{11} Bq/kg, which is primarily derived from spent fuel discharged from reactors and high-radioactivity liquid waste from nuclear fuel reprocessing. It contains long-lived radionuclides such as ^{239}Pu and ^{99}Tc, with half-lives spanning tens of thousands of years or more, making it the focus and difficulty of safe disposal. Internationally, deep geological disposal is commonly adopted: HLW is solidified into glass or ceramic matrices and buried in stable geological layers hundreds of meters underground to isolate it from the biosphere and ensure long-term safety (*Policies and Strategies for Radioactive Waste Management*, 2009; Johnson et al., 2017).

Additionally, the International Atomic Energy Agency (IAEA) classifies nuclear waste into solid, liquid, and gaseous types according to physical properties, and establishes systematic management standards (Agency, 1984).

Solid waste refers to carriers of radioactive substances with fixed geometric forms, sourced from across the entire nuclear industry chain. These include radioactive waste rocks from uranium mining, metal components and concrete shielding structures from decommissioned nuclear facilities, radioactive drug-contaminated protective equipment in nuclear medicine (such as lead aprons and syringes), and waste resins/high-level glass solidified bodies from spent fuel reprocessing. IAEA standards stipulate that low-/intermediate-level solid waste must be compressed, encapsulated in metal containers, and landfilled in near-surface disposal sites equipped with engineering barriers; high-level solid waste must be placed in radiation-/corrosion-resistant special containers (e.g., titanium alloy or cast iron) and transported to stable geological layers more than 500 m underground for permanent sealing. Disposal sites are designed to meet million-year anti-seepage requirements.

Liquid waste comprises solutions, suspensions, or emulsions containing radionuclides and is classified into three activity levels by the IAEA. Its sources include reactor coolants, high-level liquid waste from fuel reprocessing, and low-level cleaning wastewater from medical/laboratory use. High-level liquid waste must undergo vitrification or solidification before disposal; low-level liquid waste is treated by ion exchange, evaporation concentration, etc., and can only be discharged after meeting the standards (total activity < 1 Bq/L).

Gaseous waste refers to gases or aerosols containing radionuclides generated during reactor gas emissions, fuel reprocessing volatilization, and medical radioactive drug use. The IAEA mandates strict purification processes: removing radioactive particles through high-efficiency particulate air (HEPA) filters, adsorbing iodine isotopes with activated carbon, and retaining tritium via catalytic oxidation. Discharge must comply with concentration limits (e.g., tritium < 100 Bq/m^3) and be diffused/diluted through tall chimneys.

The IAEA's classification by physical form, based on waste diffusion characteristics and treatment technology compatibility, aims to formulate precise strategies for different waste types. This approach ensures the long-term isolation of radioactive substances from the biosphere and helps establish a full-process regulatory system from waste generation to disposal.

2.3 MAIN SOURCES OF NUCLEAR WASTE

Nuclear waste is generated throughout the entire nuclear fuel cycle and multiple non-fuel cycle sectors, with its radioactive characteristics and output varying significantly by source. From an industrial perspective, the main generators of nuclear waste include the entire chain of uranium mining, nuclear energy utilization, nuclear technology applications, etc. The nuclear fuel cycle process, especially the reprocessing stage, is the main contributor to radioactive pollution.

2.3.1 Nuclear Fuel Cycle

The nuclear fuel cycle is a material conversion system integral to the nuclear energy industry chain, covering the entire life cycle from uranium resource extraction to nuclear fuel reutilization. As the main source of nuclear waste, this cycle generates materials that carry over 95% of the radioactive activity from global nuclear operations.

As illustrated in Figure 2.1, the conceptual diagram of an advanced nuclear fuel cycle encompasses uranium mining, hydrometallurgy, uranium enrichment, nuclear fuel assembly manufacturing, reactor operation, reprocessing, and ultimately the recycling stage. Uranium exploration and mining yields numerous radioactive ores of varying grades while also contaminating equipment and water. The uranium mining and fuel

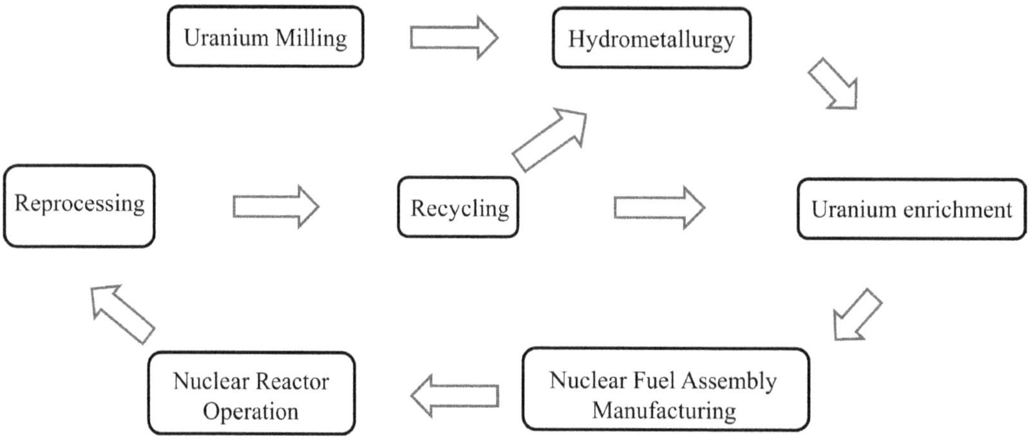

FIGURE 2.1 Advanced Nuclear Fuel Cycle Schematic.

element manufacturing processes generate solid, liquid, and gaseous nuclear waste that is predominantly composed of natural radionuclides. When a nuclear reactor operates, it produces liquid and solid nuclear waste that mainly consists of fission nuclides (^{60}Co, ^{63}Ni, etc.) (Zebroski & Levenson, 1974).

Irradiated nuclear fuel that can no longer be utilized in a nuclear reactor is termed spent fuel. Spent fuel contains newly formed ^{239}Pu and un-fissioned ^{235}U. To separate these two valuable substances, spent fuel must undergo reprocessing. Spent fuel reprocessing is the process of removing fission products from spent fuel and recovering fissile and convertible materials. If the spent fuel recovery process occurs in a thermal neutron reactor or fast neutron reactor, it is known as the "closed cycle" of nuclear fuel. Conversely, if the spent fuel is directly disposed of without undergoing reprocessing, it is referred to as the "once-through" cycle. The "once-through" cycle offers advantages such as a simple process, low cost, reduced risks of reprocessing leakage and transportation hazards, enhanced non-proliferation capabilities, and a low technical threshold, making it suitable for countries with abundant uranium resources. However, it suffers from a low uranium utilization rate, requires geological repositories for tens of thousands of years for disposal, is unable to separate actinides, is incompatible with fast reactor technology, and results in significant resource waste and environmental risks. In contrast, the "closed cycle" boosts the fuel utilization rate from 1% to 2% of the once-through cycle to over 60%, reduces the volume and toxicity of HLW, exhibits significant long-term economic benefits, is compatible with advanced reactor technology, and holds environmental protection and sustainability advantages, rendering it essential in the advanced nuclear fuel cycle (Rodríguez-Penalonga & Soria, 2017; Lamas, 2011; Park et al., 2011).

2.3.2 Sources of Non-Nuclear Fuel Waste

In addition to the nuclear fuel cycle, radioactive activities in many non-nuclear sectors also generate various types of nuclear waste. These wastes have low radioactive activity but come from dispersed sources, mainly distributed in medical treatment, scientific research, industrial applications, and decommissioned nuclear equipment scenarios.

The medical field is a major source of non-nuclear fuel waste. In nuclear medicine diagnosis and treatment, the use of radioactive drugs (such as reagents containing ^{99}Tc and ^{131}I) generates low-value radioactive-contaminated consumables, including syringes, infusion tubes, and medicine bottles. In tumor radiotherapy, components of radiotherapy equipment (such as ^{60}Co therapy machines) and radiation-contaminated protective gear must be managed as special waste due to residual radioactivity (Kurisingal et al., 2023).

Scientific research endeavors involving radioactive substances also contribute to non-nuclear fuel waste generation. Nuclear physics and radiochemistry laboratories in universities and research institutes produce experimental waste liquids containing radionuclides, such as solutions with ^{32}P and ^{14}C and discarded samples (irradiated metal, ceramic, and other materials) during radioactive tracer experiments and neutron activation analysis. In research related to accelerators and nuclear reactors, equipment components, once decommissioned, become waste containing neutron activation products (such as ^{60}Co and ^{63}Ni) as a result of long-term radiation activation. Additionally, discarded radioactive standard sources and detectors (such as scintillation detectors) in scientific research projects fall under the category of non-nuclear fuel radioactive waste, and their management must adhere strictly to nuclear safety specifications. The relative contributions of major radioactive waste sources are summarized in Table 2.1.

TABLE 2.1 Global Radioactive Waste Source Proportions (2024)

Source Category	Radioactivity Proportion	Waste Volume Proportion	Key Subsectors and Subproportions
Nuclear Fuel Cycle System	72%	0.2%	Uranium mining (15%): Low activity, 40% of volume. Nuclear power plant operation (40%): Medium-high activity, 15% of volume. Reprocessing (17%): HLW, 95% of radioactivity.
Medical Sector	18%	12%	Radioactive pharmaceuticals (12%): PET-CT isotopes account for 60%. Radiotherapy equipment (5%): Cobalt-60 waste sources. Diagnostic consumables (1%): Technetium-99m-containing waste.
Industrial and Research	8%	25%	Industrial non-destructive testing (4%): Iridium-192 flaw detection sources. Food irradiation (2%): Cobalt-60 device decommissioning. Research laboratories (2%): Isotope production waste.
Decommissioned Nuclear Facilities	5%	18%	Nuclear power plant decommissioning (3%): Pipes, concrete structures. Experimental reactor dismantling (1.5%): Activated materials. Nuclear facility decontamination (0.5%): Contaminated metals.
Others (Natural Radioactivity)	3%	45%	Oil and gas extraction (1.5%): Radium-containing drilling mud. Rare earth smelting (1%): Monazite tailings. Building materials (0.5%): Uranium-containing ore products.

The diversified applications of radioactive technologies in industrial production give rise to substantial non-nuclear fuel waste. In radioactive logging for oil and gas exploration, used logging source casings containing ^{137}Cs and ^{192}Ir, along with discarded instruments, require specialized disposal due to radioactive contamination.

The decommissioning of nuclear facilities and equipment represents a stage with intensive generation of non-nuclear fuel waste. During the decommissioning of nuclear research reactors, particle accelerators, and similar installations, concrete and metal components within building structures accumulate neutron activation products, such as ^{60}Co and ^{55}Fe, due to prolonged radiation activation. Once nuclear instruments and equipment (e.g., radioactive monitors and control rod drive mechanisms) are decommissioned, their internal circuit boards and mechanical components may retain radioactive contamination. Moreover, the soil and groundwater surrounding decommissioned nuclear facilities can become "contaminated waste" laden with radionuclides following long-term radiation penetration. These wastes typically exhibit high radioactivity and long half-lives, necessitating thorough decontamination and conditioning prior to disposal in specialized radioactive waste repositories.

Non-nuclear fuel waste stems from diverse sources and presents complex characteristics, with notable variations across medical, research, industrial, and decommissioning contexts. Effective management of such waste demands the establishment of a comprehensive, full-process regulatory system, closely integrating source classification and collection, intermediate treatment (including solidification and volume reduction), and final safe disposal (such as deep geological landfilling). Only through precise source identification and strict implementation of standards can we strike a balance between nuclear technology applications and radiation safety, protect the ecological environment and public health, and foster sustainable development within the nuclear field.

2.4 TREATMENT METHODS FOR NUCLEAR WASTE

In the peaceful utilization of nuclear energy, the generated radioactive waste is diverse. Therefore, based on the types of radioactive substances and the level of radioactivity contained in the waste, safe and reasonable methods must be adopted for proper disposal. In 1995, the International Atomic Energy Agency (IAEA) put forward nine basic principles for radioactive waste management, aiming to ensure the safety and controllability of the entire process of radioactive waste from generation to final disposal, while taking into account the needs of environmental protection, public health, and sustainable development (IAEA, 1995).

(1) Protection of human health: Radioactive waste management must ensure that the impact on human health reaches an acceptable level.

(2) Protection of the environment: Radioactive waste management must ensure that the impact on the environment reaches an acceptable level.

(3) Protection beyond national borders: Radioactive waste management should consider the possible impacts on the health of people and the environment beyond national borders.

(4) Protection of future generations: Radioactive waste management must ensure that the expected health impact on future generations is no greater than the acceptable level today.

(5) Burden on future generations: Radioactive waste management must ensure that it does not impose an undue burden on future generations.

(6) National legal framework: Radioactive waste management must be carried out within an appropriate national legal framework, with a clear division of responsibilities and provisions for independent regulatory functions.

(7) Control of radioactive waste generation: The generation of radioactive waste must be minimized as much as possible.

(8) Interdependence between radioactive waste generation and management: The interdependence between the various stages of radioactive waste generation and management must be given due consideration.

(9) Facility safety: The safety of radioactive waste management facilities during their service life must be ensured.

The IAEA's principles for radioactive waste management, with the core of "safety, responsibility, science, and collaboration," construct a complete framework from technical specifications to social governance, providing an operable guidance paradigm for countries around the world. With the development of nuclear energy technology, these principles will continue to evolve to balance the dual goals of energy demand and ecological security. This section will introduce various disposal methods according to the types of nuclear waste and describe their safety evaluations and current status.

2.4.1 Management and Disposal of Low-Level and Intermediate-Level Waste

The geological disposal of LLW and ILW commenced in the 1940s, and a relatively mature suite of disposal methods has since been developed. Currently, the commonly employed geological disposal techniques for solid LLW/ILW include land shallow burial, abandoned mine disposal, and deep geological disposal (Milnes, 1985).

2.4.1.1 Land Shallow Burial Method

Near-surface disposal of LLW/ILW involves the shallow burial of nuclear waste either at or below ground level, with or without the incorporation of engineering barriers and protective cover layers. The land shallow burial technology for LLW/ILW was first utilized in the United States in 1944, representing the earliest disposal approach globally. This technology has now reached a high level of maturity and is widely adopted by numerous countries (Ojovan & Steinmetz, 2022).

Based on the position of disposal projects relative to the ground surface, land shallow burial can be classified into near-surface disposal, mound-type shallow burial, and surface disposal. First, as illustrated in Figure 2.2, the near-surface disposal schematic sets

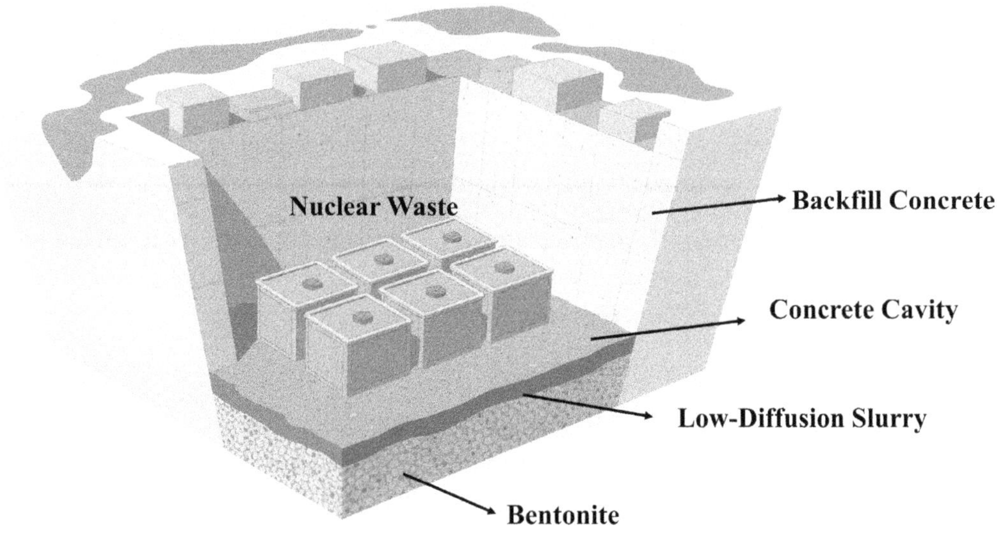

FIGURE 2.2 Schematic Cross-Section of Near-Surface Disposal.

the disposal depth at approximately 50–100 m below the surface. This depth is strategically chosen to minimize the migration of radioactive substances. In the disposal trenches, concrete structures are constructed to house the radioactive waste. The spaces between the nuclear fuel waste units are filled with materials like cement. Additionally, a mortar layer is established as a diffusion barrier outside the concrete structure, and a bentonite layer serves as a waterproof barrier beyond the mortar layer. Consequently, the nuclear fuel waste units, cement filling materials, concrete structures, diffusion barrier, and waterproof barrier collectively function as artificial barriers, while the low-level strata and rocks act as natural barriers, each fulfilling its distinct role. This type of disposal project is entirely situated underground and is well-suited for the disposal of LLW/ILW with relatively high specific radioactivity.

Mound-type shallow burial represents a semi-aboveground and semi-underground disposal approach. In this method, the majority of the disposal chambers are positioned above ground level, with their tops sealed using a combination of soil and reinforced concrete. Typically, disposal chambers above ground are designated for wastes with relatively low specific radioactivity, whereas those below ground can accommodate wastes with higher specific radioactivity. The French Manche disposal site, along with several sites in the United States, has employed this shallow burial technique for the disposal of low-level radioactive waste. This method is particularly suitable for handling low and intermediate-level radioactive waste characterized by relatively low specific radioactivity, offering a balance between cost-effectiveness and safety for less hazardous waste streams.

Surface disposal involves the containment of low and intermediate-level radioactive waste within closed reinforced concrete structures situated at ground level. However, due to its inherently lower safety profile compared to other disposal methods, this approach is less frequently utilized. The lack of sufficient shielding and isolation from the environment increases the risk of potential radioactive substance leakage, posing threats to human health and the ecosystem. As a result, its application is generally limited, and regulatory requirements for its implementation are often more stringent, emphasizing the need for enhanced monitoring and additional safety measures to mitigate risks.

2.4.1.2 Disposal in Abandoned Mines

Disposing of low- and intermediate-level radioactive waste in underground abandoned mines is considered a relatively safe disposal approach. Abandoned mine disposal entails utilizing natural, disused mines or artificially excavated mine cavities as disposal sites, and implementing engineering measures such as layered filling and sealed isolation to safely store low-level radioactive waste. By leveraging the existing geological spaces within mines, including vertical shafts and roadways, this method constructs a multi-barrier system in conjunction with backfill materials like bentonite and concrete. Through physical isolation and the natural decay of geological media, it effectively achieves long-term retardation of radionuclides.

This disposal method presents several notable advantages. It significantly reduces infrastructure costs due to the direct utilization of existing mine roadways and shafts, potentially saving more than 60% of construction expenses, while also facilitating resource recycling. Moreover, it benefits from the safety provided by natural geological barriers and can flexibly accommodate the characteristics of LLW. However, it also has inherent limitations. Old mines often face issues such as rock fractures, groundwater seepage, or corrosion of support structures. Given that most mines consist of heterogeneous rock masses, it is challenging for a single sealing material, such as concrete, to fully fill the fractures. To address this, a bentonite–polymer composite barrier is typically employed. Nevertheless, long-term groundwater erosion may lead to the formation of cracks, thereby increasing the risk of radioactive substance leakage. Additionally, since abandoned mines are generally located far from nuclear facilities, the waste requires long-distance transportation. This contrasts with the typical preference for on-site disposal of the large quantities of LLW and ILW, posing logistical challenges to the widespread application of this method.

2.4.1.3 Deep Geological Disposal

Deep geological disposal of LLW and ILW is a sophisticated technology that involves burying radioactive nuclear waste, including certain types of spent fuel, in stable geological layers situated 300–500 m beneath the Earth's surface. This method achieves permanent isolation of the waste through the implementation of a "multi-barrier system." The core principle underpinning this approach is to capitalize on the low permeability, high adsorbability, and long-term stability inherent in deep geological media, such as granite, salt rock, and claystone. These natural properties are complemented by engineered barriers, including robust metal containers and impermeable bentonite buffer layers, to effectively prevent the migration of radionuclides into the biosphere (Stewart, 1984).

The extremely low permeability of deep geological layers plays a crucial role in significantly extending the migration time of nuclides, often stretching it to hundreds of thousands of years, a duration far exceeding the radioactive decay period of most of the contained substances. Additionally, the montmorillonite minerals present in claystone exhibit remarkable adsorptive capabilities, capable of capturing more than 99% of radioactive cations. Moreover, the unique plastic flow characteristics of salt rock enable it to self-heal fractures, thereby further reducing the risk of radioactive substance leakage. By virtue of its deep subsurface location, this disposal method minimizes interference with the surface environment. However, the associated high costs, protracted construction cycles, and inherent geological uncertainties pose significant challenges that limit its widespread application.

2.4.2 Management and Disposal of HLW

HLW, encompassing spent fuel, vitrified HLW, residual cladding materials from reprocessing, and discarded silver adsorption materials, harbors a substantial quantity of long-lived radionuclides. As such, it demands highly specialized and long-term safe disposal strategies. Although deep geological disposal has advanced to the implementation stage as a promising technology for HLW disposal, the field continues to be characterized by ongoing research and exploration to refine and optimize these techniques.

The deep geological disposal of HLW entails placing solidified waste within artificial deep rock caverns, typically located more than 500 m underground. A comprehensive deep geological disposal repository generally comprises surface facilities and an underground disposal complex. The artificial barriers employed in HLW geological disposal are pivotal engineering elements designed to ensure the secure isolation of radionuclides. In concert with the natural geological barriers, they form an elaborate multi-barrier protection system. At the heart of this system are three key components: waste containers, buffer backfill materials, and sealing systems (Johnson et al., 2017; Birkholzer et al., 2012).

Nuclear waste containers serve as the primary safeguard against radioactive leakage. Constructed predominantly from corrosion-resistant specialty metals, they provide a durable containment solution. For example, the Swedish KBS-3V disposal system utilizes copper containers to encapsulate spent fuel. The exceptional corrosion resistance of copper allows these containers to maintain structural integrity and functionality for thousands of years, even in groundwater-rich environments. In the United States, alloy steel containers have been designed with an additional ceramic coating, enhancing their resistance to both radiation and chemical erosion. The buffer backfill materials enveloping the containers perform a multitude of essential functions. Nuclear-grade bentonite is the material of choice in most applications. Upon contact with water, bentonite swells to form a dense colloid, effectively blocking groundwater seepage and adsorbing radionuclides. The Finnish Onkalo disposal site exemplifies this approach, employing compacted bentonite blocks as the buffer layer. With a permeability coefficient as low as 10^{-12} m/s, this bentonite buffer significantly retards the migration of nuclides. Additionally, composite materials consisting of quartz sand, clay, and bentonite offer a balanced solution, optimizing both thermal conductivity and sealing performance. This helps mitigate the impact of the heat generated by the decay of HLW on the integrity of the barrier system.

Sealing systems are integral to the overall disposal strategy, tasked with closing off disposal roadways, shafts, and other open spaces to prevent groundwater intrusion and nuclide leakage. These systems are typically constructed from reinforced concrete, augmented with materials such as water-swellable rubber waterstops and epoxy sealants. The French Bure disposal site serves as a prime example, where multi-layer concrete sealing walls are installed at the entrance of roadways. Bentonite slurry is meticulously filled between the layers, creating a flexible-rigid composite sealing structure that ensures long-term sealing effectiveness. Through the synergistic effect of these artificial barriers and the natural properties of deep geological layers, the isolation time of radionuclides can be extended to tens of thousands of years. This provides a robust and reliable solution for the safe and long-term disposal of HLW, addressing one of the most significant challenges in the management of radioactive waste.

2.5 POTENTIAL RISKS IN THE NUCLEAR INDUSTRY

2.5.1 Safety Crisis Caused by Equipment Defects

As a pivotal sector at the intersection of energy and technology, the nuclear industry not only supplies efficient energy solutions but also propels advancements in medical research. However, this progress is accompanied by multi-faceted potential risks. Nuclear reactor operations inherently carry the risk of core meltdown. For instance, in the Chernobyl and Fukushima nuclear accidents, operational errors, equipment malfunctions, or external disasters disrupted reactor cooling systems, resulting in substantial releases of radioactive materials. The nuclear fuel cycle also presents significant hazards. During uranium mining, dust leakage can expose miners to radiation, increasing the risk of radiation-related diseases. Moreover, improper management and disposal of nuclear waste can lead to radionuclide leakage, posing severe threats to the surrounding environment and public health.

Nuclear industry equipment operates in a complex environment of high temperature, high pressure, and strong radiation for a long time. The defects and failures of the equipment itself are important factors causing nuclear safety risks, and their potential dangers are mainly reflected in equipment aging, design defects, key system failures, and other aspects.

Key equipment in nuclear facilities, including reactor pressure vessels and steam generators, endures long-term exposure to high-temperature, high-pressure, and intense radiation environments, resulting in a gradual degradation of material properties. For instance, the reactor pressure vessel head at the Davis-Besse Nuclear Power Plant in the United States once developed dents due to boric acid corrosion, posing a serious threat to equipment integrity. Similarly, the pipes of steam generators are susceptible to intergranular corrosion from prolonged exposure to these harsh conditions. In the event of pipe rupture, coolant leakage can occur, directly disrupting core cooling and posing significant safety hazards to nuclear facilities. This risk was a contributing factor to the failure of the emergency cooling system during the Chernobyl accident. Additionally, components within the cooling system, such as pipes and pumps, experience reduced reliability or even failure due to wear and fatigue after extended operation.

Design defects and manufacturing quality issues also pose significant concerns. Early nuclear reactor designs may have inherent flaws that could lead to a loss of reactivity control, potentially causing severe accidents under certain conditions. For example, if nuclear fuel transport containers are not designed to withstand extreme conditions like impacts and high temperatures, they pose a risk of radioactive leakage. During equipment manufacturing, lax quality control can result in metallurgical defects. As seen in the Flamanville Nuclear Power Plant in France, bubbles and inclusions were found in pressure vessel forgings, significantly weakening the equipment's pressure-bearing capacity and increasing the likelihood of rupture during operation.

Failures in key systems can trigger a series of catastrophic chain reactions. As the core mechanism ensuring the safe operation of nuclear reactors, malfunctions in the main pump or pipeline ruptures within the cooling system can prevent effective core cooling, leading to overheating. The Three Mile Island accident in the United States is a prime example, where a stuck circulation pump ultimately caused partial core meltdown. The containment vessel, serving as the last line of defense against radioactive material leakage, is vulnerable when its sealing devices age. Hardening of rubber waterstop strips, for instance, can compromise the vessel's integrity, as demonstrated by the forced shutdown of the San Onofre Nuclear Power Plant due to cracks in the containment vessel concrete. Failure of the spent fuel pool cooling system can cause rapid water temperature increases and boiling, resulting in radioactive steam leakage, as witnessed in the Fukushima nuclear accident.

Furthermore, specialized equipment, such as nuclear fuel handling machinery and radioactive waste treatment apparatus, harbors potential risks. The control systems of nuclear fuel handling manipulators are prone to failure in high-radiation environments, potentially causing errors in fuel rod handling and drop damage. In radioactive waste treatment equipment, failures of key components like thermocouples in HLW glass solidification furnaces can lead to melt solidification and pipeline blockages. False alarms or missed detections by monitoring instruments can mask real risks, preventing timely identification of potential hazards and delaying accident response. The Mihama Nuclear Power Plant in Japan experienced a situation where operators made misjudgments due to water level gauge failures, nearly resulting in serious consequences.

The risks associated with nuclear industry equipment are multi-faceted, encompassing equipment aging, design and manufacturing flaws, key system failures, and malfunctions in specialized equipment and monitoring instruments. Issues at any stage can potentially trigger severe nuclear safety accidents. Therefore, it is imperative to prioritize the safety management of nuclear industry equipment. Advanced detection technologies, stringent quality control measures, and a scientific operation and maintenance system are essential for mitigating equipment risks and ensuring the safe and stable operation of the nuclear industry.

2.5.2 Environmental Migration Risks

High-level radioactive waste is predominantly disposed of via deep geological repositories; however, the anti-seepage designs of these facilities harbor potential vulnerabilities. Groundwater acts as a pivotal medium for the migration of radioactive substances. In the event of aging sealing materials within the repository or alterations in the geological structure, radionuclides may infiltrate surrounding water bodies through groundwater, thereby contaminating vital groundwater sources. The Yucca Mountain HLW repository project in the United States serves as a prime example. Due to the intricate geological conditions and frequent groundwater activities in the project area, the initiative faced persistent controversies and numerous delays, ultimately failing to reach completion.

Emissions of gaseous waste can result in the atmospheric dispersion of radioactive substances. During nuclear reactor accidents or exhaust gas releases from reprocessing plants, radioactive gases such as ^{85}Kr and ^{133}Xe are often discharged into the atmosphere. These gases can then disperse over vast distances through atmospheric circulation. In the 1986 Chernobyl nuclear accident, a massive radioactive plume was carried northwestward by the wind, causing radioactive contamination across multiple European countries. Significantly elevated levels of radioactive substances were detected in Sweden, Finland, and other nations as a consequence.

Radioactive substances can bioaccumulate through the food chain. When water bodies and soil become radioactively contaminated, aquatic plants and crops absorb radionuclides. Subsequently, as fish, livestock, and other animals consume these plants, radioactive substances continue to accumulate within their bodies. Following the Fukushima nuclear accident in Japan, for instance, excessive levels of ^{137}Cs were detected in fish from the surrounding waters. If these contaminated fish enter the food market, they pose a direct threat to human health.

2.5.3 Long-Term Stability Risks

The site selection for HLW repositories necessitates consideration of geological stability over a timescale of tens of thousands of years. Nevertheless, the Earth's geological structure is in a state of perpetual transformation, and phenomena such as earthquakes, volcanic eruptions, and plate tectonics can potentially compromise the integrity of repositories. Even with rigorous geological exploration and evaluation during the site selection process, the potential impact of future geological changes on repository stability cannot be entirely mitigated. For example, seismic events may occur in certain regions millennia from now, potentially leading to the rupture of repository sealing layers and subsequent leakage of radionuclides.

Moreover, HLW generates substantial heat during its decay process. Inadequate heat dissipation can cause a significant increase in the internal temperature of the repository, which may trigger fracturing of the rock mass. Elevated temperatures can also accelerate the aging of packaging containers and the migration of radionuclides, thereby escalating the risk of nuclear waste leakage. Consequently, the design of HLW repositories must incorporate efficient heat dissipation mechanisms to maintain internal thermal stability. In addition to the aforementioned risks, human-induced factors, including transportation accidents, management oversights, and terrorist threats, also constitute significant components of the overall risks associated with the nuclear industry.

2.6 TECHNICAL RESPONSES AND MANAGEMENT OF NUCLEAR WASTE RISKS

2.6.1 Front-End Reduction Technologies

2.6.1.1 Advanced Reactor Design

New-generation advanced reactors, such as molten salt reactors (MSRs), can effectively reduce solid waste generation by employing novel nuclear fuels and coolants. MSRs utilize liquid fluoride salts that function both as coolants and nuclear fuel carriers. During operation, the online addition and removal of fuels are feasible, which not only decreases the amount of spent fuel but also enhances nuclear fuel utilization efficiency and mitigates the radioactive hazards of nuclear waste.

The core structural components of an MSR encompass the reactor primary system, auxiliary safety system, and thermo-hydraulic system. The reactor primary system is mainly composed of a molten salt fuel circuit, a graphite moderator, a main vessel, and a heat exchange system.

In the molten salt fuel circuit, molten salts like lithium fluoride and beryllium fluoride serve as carriers to dissolve uranium, thorium, and other fuels, forming liquid fuel salts that replace traditional solid fuel rods. These fuel salts circulate within the reactor core, fulfilling dual roles as "fuel" and "coolant." The core of the graphite moderator is filled with graphite blocks, which moderate neutrons and sustain the nuclear chain reaction. The high-temperature resistance of graphite aligns well with the chemical stability of molten salts, enabling the formation of a compact core structure. The main reactor vessel, fabricated from high-temperature-resistant alloys (e.g., Hastelloy), is equipped with a steam generator or an intermediate heat exchanger. Heat is transferred to the secondary loop via the flow of molten salt, facilitating steam generation.

As shown in Figure 2.3, the flow chart of an MSR illustrates its operational process. In the core reaction stage, the reactor acts as the energy source, and control rods are utilized to regulate the reaction rate. Purified salt, processed by the chemical treatment plant, enters the reactor and participates in the cycle along with the coolant salt. The coolant salt absorbs heat within the reactor. Regarding heat transfer and conversion, the heat-carrying

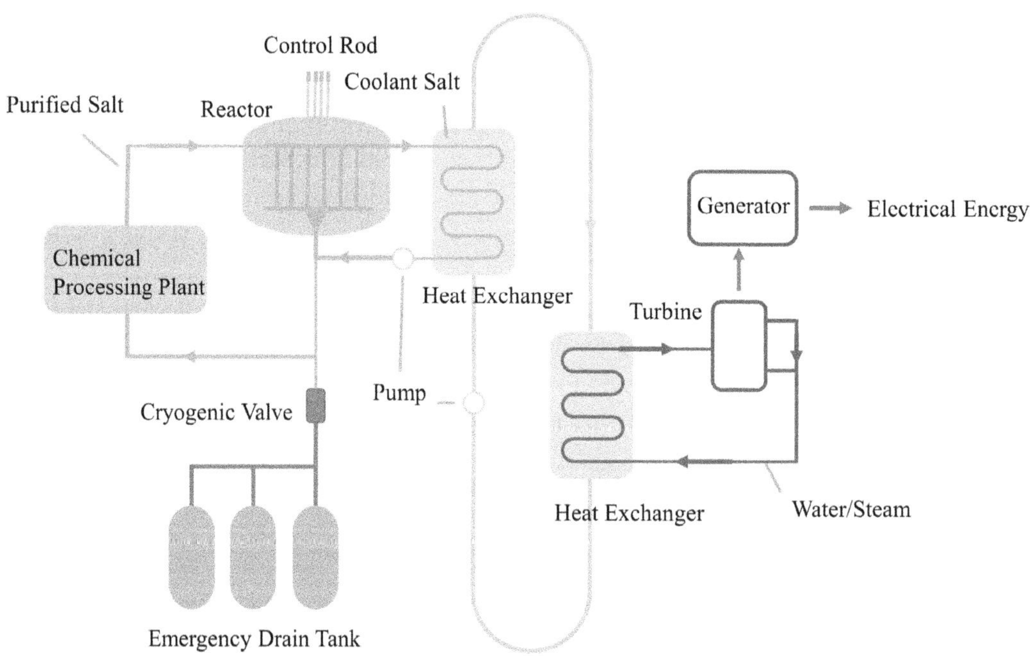

FIGURE 2.3 Flowchart of Molten Salt Reactor (MSR).

coolant salt flows into the heat exchanger, where it is pumped to complete the cycle and transfer heat to the water/steam working medium on the opposite side. The heated water/steam then enters another heat exchanger, driving the turbine, which in turn powers the generator to produce electricity. For safety protection, low-temperature valves and emergency drain tanks safeguard the system. In the event of an anomaly, the coolant salt can be discharged into the emergency drain tank through the low-temperature valve, averting potential dangers and ensuring the stable operation of the reactor. These components work in concert to construct the energy conversion and safety protection system of the MSR (Lamas, 2011).

2.6.1.2 Post-Treatment Technology Innovation and Waste Recycling Technology

Traditional U-Pu co-recovery post-treatment technology, while effective in separating uranium and plutonium, generates a substantial volume of high-level radioactive liquid waste. In contrast, emerging all-separation post-treatment technologies represent a significant advancement. These innovative methods are capable of separating a broader spectrum of radionuclides, enabling more precise and comprehensive treatment of nuclear waste. By doing so, they not only further reduce the volume of HLW but also decrease its radioactivity level, thereby facilitating subsequent disposal processes and minimizing long-term environmental risks.

One notable area of exploration is the high-temperature molten salt treatment of dry-process post-treatment waste. This approach aims to unlock recycling potential by leveraging the unique properties of molten salts. Through the development of novel materials and processes, radioactive waste can be transformed into a stable storage form. This not only mitigates the immediate radioactive hazards associated with the waste but also enables the recovery and reuse of valuable materials, effectively closing the loop on resource utilization within the nuclear fuel cycle.

The utilization of radionuclides such as plutonium-239 to produce mixed oxide (MOX) fuel stands as a pivotal strategy for nuclear waste recycling. MOX fuel, composed of a mixture of uranium and plutonium oxides, can be employed in pressurized water reactors, thereby facilitating the recycling of nuclear fuel. This practice significantly reduces reliance on natural uranium resources, a finite commodity, and simultaneously curbs the accumulation of plutonium in HLW streams. From an economic perspective, it optimizes resource utilization and potentially reduces long-term waste management costs. Environmentally, by decreasing the volume and radioactivity of waste requiring long-term disposal, it lessens the burden on geological repositories and associated environmental impacts, underscoring its dual benefits in promoting sustainable nuclear energy production.

2.6.2 Intermediate Treatment Technology

2.6.2.1 Solidification Technology

The solidification of radioactive waste is a crucial intermediate treatment process that transforms liquid waste into a stable solid form, known as a waste solidified body. Materials used to solidify liquid waste, referred to as solidification matrices, include glass, cement, asphalt, and other substances. Asphalt solidification, which uses asphalt as the matrix, and cement solidification are predominantly employed for treating medium to low radioactive liquid waste. In contrast, vitrification, or glass solidification, is the most common method for handling high-level radioactive liquid waste (Kerr, 1979).

Vitrification is currently the most widely applied solidification technology for HLW. This process involves mixing high-level radioactive liquid waste with glass raw materials and melting them at high temperatures to form a glass solidified body. In this body, radionuclides are immobilized within the glass network structure, thereby enhancing stability and safety. Ceramic solidification takes advantage of the high chemical stability and radiation resistance of ceramic materials to encapsulate radionuclides within a ceramic crystal structure. This method is suitable for treating HLW containing actinides. Cement solidification, commonly used for treating low- and intermediate-level radioactive waste (LLW and ILW), entails mixing radioactive substances with cementitious materials, such as cement, to form cement solidified bodies, which facilitates storage and disposal.

2.6.2.2 Radioactive Wastewater Treatment

Radioactive wastewater treatment technologies primarily encompass adsorption, precipitation, ion exchange, evaporation concentration, biotechnology, membrane separation, and photocatalysis, among others. The fundamental principle of radioactive wastewater treatment is to concentrate radionuclides into a smaller volume via a series of physical, chemical, and biological processes. This approach effectively reduces the nuclide concentration in the wastewater, and subsequent treatment can render the wastewater compliant with discharge or recycling standards.

Current research efforts regarding radioactive wastewater predominantly focus on two key aspects: enhancing existing treatment processes or optimizing related parameters and developing novel materials. In practical applications, the liquid waste generated from post-treatment plants typically exhibits complex compositions, high radioactivity levels, and large volumes. Consequently, when selecting treatment processes for comprehensive purification, future research should emphasize integrating multiple technologies (such as evaporation concentration combined with coagulation precipitation, or evaporation concentration coupled with ion exchange) with the pre-treatment of the waste liquid and the optimization of operating conditions. Such an integrated strategy aims to achieve optimal treatment outcomes while minimizing operational costs. However, the limitations and challenges associated with these combined processes still require further investigation and resolution (Wang, Zheng et al., 2025; Li et al., 2022; Chu et al., 2022; Liu et al., 2025; Ma et al., 2023).

2.6.2.3 Radioactive Waste Gas Treatment

The gaseous radioactive waste generated during nuclear reactor operation contains various radionuclides and chemical pollutants. Proper treatment of this waste is a critical step in ensuring nuclear safety and environmental sustainability. Currently, the core treatment technology system includes filtration, adsorption, absorption, distillation, and storage decay methods. In practical applications, combined processes are often employed to achieve efficient purification.

In terms of technological progress, the current standard process for reactor waste gas is as follows: first, short-lived radionuclides, such as certain isotopes of iodine, krypton, and xenon, are stored until their activity decays to acceptable levels. This is followed by deep treatment using room-temperature or low-temperature activated carbon adsorption technology. Finally, the treated gas is discharged into the atmosphere through a high chimney. In the field of iodine purification, mature technical routes have been established for retention decay, liquid scrubbing, and solid adsorption methods. For the treatment of ^{85}Kr, emerging technologies such as cryogenic distillation and liquefied fluorocarbon absorption are under research and application expansion. Additionally, the research and development of molecular recognition adsorption materials and membrane separation technologies provide new directions for the efficient separation and resource utilization of gaseous waste.

However, existing treatment technologies still have significant limitations. First, the emission method of ^{85}Kr based on high-altitude dilution does not fundamentally eliminate radioactive hazards. The uncertainty of environmental diffusion conditions may lead to risks of local radioactive contamination exceeding standards. Second, the complex process flow of post-treatment plants results in high construction and operation costs, restricting the large-scale application of these technologies. Furthermore, the problem of resource recovery efficiency is prominent, with a significant gap between the theoretical and actual recovery rates of some technologies, causing substantial waste of utilizable resources. Finally, the high-level radioactive liquid waste generated during the treatment process has a decay period of thousands of years. This not only increases the difficulty of intermediate storage and final disposal but also may further extend the radioactive pollution chain, forming new environmental risk sources (Wang, Liu et al., 2025; Zhang et al., 2025).

REFERENCES

Agency, International Atomic Energy, 1984. *Treatment of Low- and Intermediate-Level Radioactive Waste Concentrates: Report of a Panel.* International Atomic Energy Agency, Vienna.

Birkholzer, J., Houseworth, J., Tsang, C.F., Gadgil, A., Liverman, D.M., 2012. Geologic disposal of high-level radioactive waste: Status, key issues, and trends. *Soc. Sci. Electron. Publ.* 37(1), 79–+.

Chu, J., Huang, Q., Dong, Y., Yao, Z., Wang, J., Qin, Z., Ning, Z., Xie, J., Tian, W., Yao, H., Bai, J., 2022. Enrichment of uranium in seawater by glycine cross-linked graphene oxide membrane. *Chem. Eng. J.* 444, 136602.

Classification of Radioactive Waste, 2009. *General Safety Guide No. GSG-1.* International Atomic Energy Agency, Vienna.

IAEA, 1995. *The Principles of Radioactive Waste Management. Safety Series No. 111-F.* International Atomic Energy Agency, Vienna.

Johnson, B., Newman, A., King, J., 2017. Optimizing high-level nuclear waste disposal within a deep geologic repository. *Ann. Oper. Res.* 253(2), 733–755.

Jun, B.M., Lee, H.K., Park, S.B., Kim, T.J., 2021. Purification of uranium-contaminated radioactive water by adsorption: A review on adsorbent materials. *Sep. Purif. Technol.* 277, 119675.

Kerr, A.R., 1979. Nuclear waste disposal: Alternatives to solidification in glass proposed. *Science* 204(4390), 289–291.

Kurisingal, J.F., Yun, H., Hong, C.S., 2023. Porous organic materials for iodine adsorption. *J. Hazard. Mater.* 458, 131835.

Lamas, W.D.Q., 2011. An analysis of development and research on spent nuclear fuel reprocessing. *Energy Policy* 39(11), 6916–6924.

Li, W.-A., Peng, Y.-C., Ma, W., Huang, X.-Y., Feng, M.-L., 2022. Rapid and selective removal of Cs^+ and Sr^{2+} ions by two zeolite-type sulfides via ion exchange method. *Chem. Eng. J.* 442, 136377.

Liu, C., Xuan, K., Guo, Y., Jiang, H., Ma, W., Wang, X., Zhou, Z., Li, J., Guo, Y., 2025. Facilitated electron-hole separation and enhanced uranium(VI) capture via La-doped WO_3: Insights into oxygen vacancies and superior recyclability. *J. Hazard. Mater.* 491, 137878.

Liu, S., Tao, B., Zuo, B., Zheng, K., Abdelfattah, W., Bao, J., Yan, G., Wang, T., Wang, R., Li, Z., 2024. Function-oriented design principles for adsorbent materials of uranium extraction from seawater. *Chem. Eng. J.* 500, 151567.

Ma, H., Shen, M., Tong, Y., Wang, X., 2023. Radioactive wastewater treatment technologies: A review. *J. Radiat. Res. Appl. Sci.* 28(4), 1935–1952.

Milnes, A.G., 1985. *Geology and Radwaste.* Academic Press, London.

Ojovan, M.I., Steinmetz, H.J., 2022. Approaches to disposal of nuclear waste. *Energies* 15(20), 7804.

Park, B.H., Gao, F., Kwon, E.H., Ko, W.I., 2011. Comparative study of different nuclear fuel cycle options: Quantitative analysis on material flow. *Energy Policy* 39(11), 6916–6924.

Policies and Strategies for Radioactive Waste Management, 2009. *Specific Safety Guide No. SSG-1.* International Atomic Energy Agency, Vienna.

Rodríguez-Penalonga, L., Moratilla Soria, B.Y., 2017. A review of the nuclear fuel cycle strategies and the spent nuclear fuel management technologies. *Energies* 10(9), 1325.

Scientific and Technical Basis for the Near Surface Disposal of Low and Intermediate Level Waste, 2002. *Technical Reports Series No. 413.* International Atomic Energy Agency, Vienna.

Stewart, D.B., 1984. *Geochemical Aspects of Radioactive Waste Disposal.* Springer-Verlag, Berlin.

Wang, F., Zheng, Q., Tai, W.Y., Wu, Q., Li, X.L., Wu, Y.H.Z., Zheng, N.C., Ning, S.Y., Zeng, D.Q., Watabe, H., Wu, Y., Li, H., Wei, Y.Z., Yin, X.B., 2025. Facile synthesis of nanolayered manganese oxide for the efficient and selective removal of strontium(II) from nuclear wastewater. *Adv. Sci.* 12(3), 2408765.

Wang, Q., Liu, S., Sheng, Y., Chen, Z., Min, X., Zhou, Y., Zhao, B., Sun, T., 2025. An efficient Ni-based adsorbent for selective removal of ^{85}Kr and $^{14}CH_4$ in radioactive contaminants from nuclear process off-gas stream. *J. Hazard. Mater.* 482, 136596.

Xie, Y., Chen, C., Ren, X., Wang, X., Wang, H., Wang, X., 2019. Emerging natural and tailored materials for uranium-contaminated water treatment and environmental remediation. *Prog. Mater. Sci.* 103, 180–234.

Zebroski, E., Levenson, M., 1974. The nuclear fuel cycle. *High Ctry. News* 1(1), 101–130.

Zhang, W.X., Qu, M.N., Du, A.J., Sun, Q., 2025. Electric-field controlled switchable and efficient separation of radioactive Xe/Kr on borophene: A theoretical study. *ChemPhysChem* 26(2), e202400876.

PART II

Characteristics of Nuclear Waste

CHAPTER 3

Principles of Nuclear Waste Management

Huyong Qin, Cheng Wang, Yang Liu, and Binglin Guo

3.1 INTRODUCTION

Generally, nuclear waste refers to radioactive material generated from nuclear power plants and other applications of nuclear technology like nuclear medicine and research. According to the definition from the International Basic Safety Standards of Radiation Protection and Safety of Radiation Sources issued by the International Atomic Energy Agency (IAEA), radioactive waste is material that contains or has been contaminated with radionuclides. The activity concentrations of radionuclides in nuclear waste are greater than the established regulatory clearance levels and thus its further use is not considered. Due to the radioactivity being higher than clearance levels compared to non-nuclear waste, management of radioactive waste introduces a complex and highly regulated field within the nuclear industry. Due to its high radiotoxicity and radioactivity, nuclear waste is classified as hazardous waste that must be treated properly to protect the workers and the public from unacceptable risks (Bennett, 1996). Exposure to radiation emitted by nuclear waste can damage cells and tissues and cause cancer and other health issues. Nuclear waste remains radioactive and hazardous for very long periods, ranging from hundreds to millions of years depending on the radionuclide. This poses a challenge for safely isolating and containing nuclear waste. Proper management of this waste is crucial to minimize the risks to human health and the environment from radiation exposure (Yano et al., 2018). It is worth noting that the terms "radioactive waste" and "nuclear waste" are generally regarded as synonyms in the field of safety and waste management.

Compared with non-radioactive waste, nuclear waste requires more specific and strict waste management regulations. The key steps in nuclear waste management include waste classification, volume reduction, conditioning, storage, and disposal (Hare & Aikin, 1980; Drace et al., 2022; Kumar et al., 2022). The final step of nuclear waste management is disposal. This is the emplacement of nuclear waste into a facility or location with no intention of retrieval. Disposal is the most effective solution for long-term management of nuclear waste, and it is the global consensus among all countries that utilize

DOI: 10.1201/9781003665793-5

nuclear energy. Disposal involves the act of discarding waste, and ensuring the proper storage of waste in a stable and suitable facility is particularly crucial to prevent hazardous materials from entering the ecosystem. Therefore, the aim of nuclear waste disposal is to minimize the interference of radionuclides with the biosphere to a negligible degree. Considering radionuclides in nuclear waste exhibit high mobility and radiotoxicity with long half-lives in the ecosystem, a multiple-barrier approach that combines waste processing, engineered barriers, and appropriate siting is generally viewed as the most effective strategy (Allan & Nuttall, 1997). With proper management, the risks from nuclear waste can be minimized while also enabling the benefits of nuclear technologies.

In terms of nuclear waste disposal, it is necessary to use suitable containers or matrices to convert wastes into a durable solid form as the first step. This process facilitates the nuclear waste handling, transportation, storage, and finally disposal to engineered geological sites, aiming to minimize any risk of radionuclides being released into the ecosystem. According to the properties of nuclear waste, some countries also discharge limited amounts of gases and effluents into the environment for disposal (Xixi et al., 2022). Many issues determine the selection of disposal methods; both technical and administrative factors should be carefully considered. Nuclear waste management (NWM) comprises all administrative and operational activities involved in the handling, pre-treatment, treatment, conditioning, transport, storage, and disposal of radioactive waste (Drace et al., 2022).

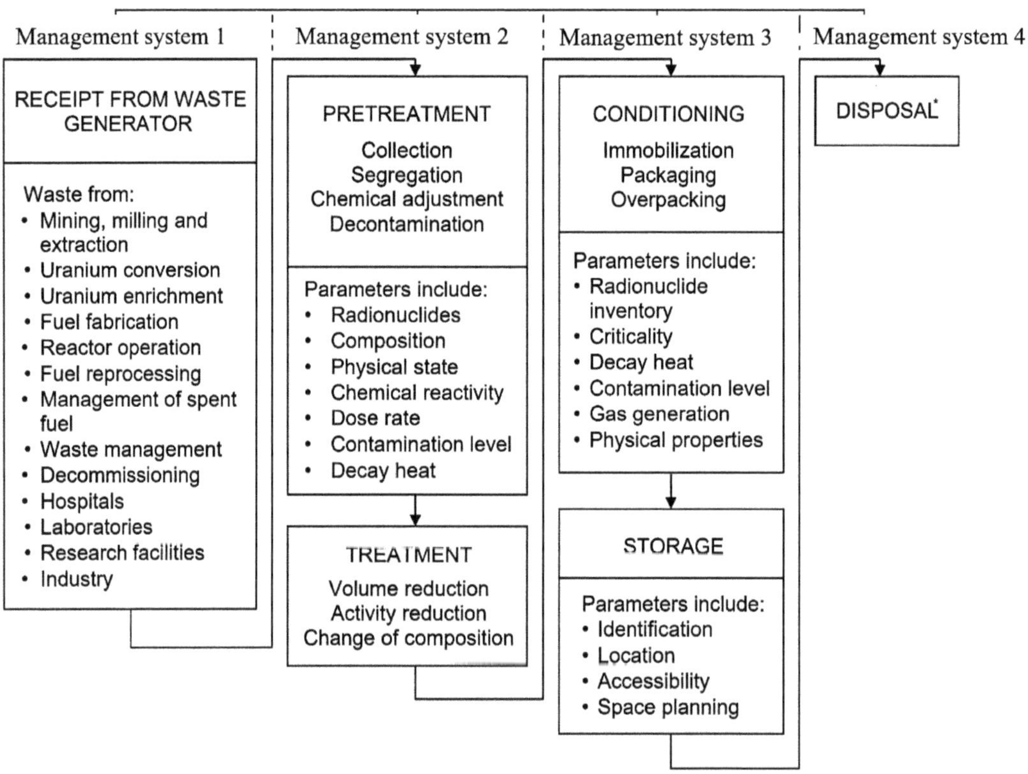

FIGURE 3.1 The Process of Radioactive Waste Management (Modified from Guo & Sasaki, 2022).

Decisions on disposal technology selection typically follow a graded approach. The following three main principles are used: (I) nuclear wastes are disposed of using the simplest disposal concept available, which is consistent with the hazards of waste and for which safety and environmental protection can be demonstrated (Drace et al., 2022). (II) The most hazardous wastes are disposed of using greater levels of engineering to provide for increased containment and/or are disposed of at greater possible depth to increase isolation from the surface environment (Hooper, 2010). (III) Where existing disposal facilities are available, consideration is given to first using them before developing new disposal facilities. Owing to the high risk associated with nuclear waste, it is necessary to properly manage the entire process of radioactive waste in order to reduce and minimize the hazards posed by its radioactivity to the environment and human beings. Therefore, the study of nuclear waste management is of great theoretical and practical significance.

3.2 SOURCES OF NUCLEAR WASTE

The sources of radioactive waste are crucial factors for nuclear waste management (Table 3.1). To date, nuclear power plants are the largest producers of radioactive waste (Laraia, 2015). They generate two main types of waste: low-level waste (substances with relatively low radiation levels) and spent nuclear fuel (SNF). Low-level waste consists of contaminated materials like protective clothing, tools, and equipment. SNF refers to the used nuclear fuel rods that are removed from reactors after several years of operation. Both types of waste are highly radioactive and require careful management. Research laboratories that utilize radioactive materials for medical, industrial, and scientific purposes also generate radioactive waste. This includes contaminated glassware, equipment, and materials used in experiments. However, the volumes of waste from research activities are much smaller compared to nuclear power plants (Rout et al., 2023). Medical facilities that utilize radioactive materials for diagnostic imaging and cancer treatment produce low-level radioactive waste (waste with moderate radiation levels). This includes contaminated materials like syringes, vials, protective clothing, and urine and fecal waste from patients. Most of the medical radioactive waste is short-lived. Military installations that operate nuclear reactors for powering submarines and aircraft carriers, as well as nuclear weapons programs, generate radioactive waste from spent fuel, decommissioned reactors, and

TABLE 3.1 Different Sources and Characteristics of Nuclear Waste

Source	Category	Feature	Material	Literature
Nuclear power plant	Low-level waste	Relatively low radiation levels	Protective clothing, tools, and equipment	(Laraia, 2015)
	SNF	Highly radioactive	Used nuclear fuel rods	
Research laboratory	Low-level waste	Small waste volume and moderate radiation levels	Contaminated glassware, equipment	(Rout et al., 2023)
Military installations	Radioactive waste	Nuclear waste is more complex and diverse	Spent fuel, decommissioned reactors and contaminated material	(Stabnikov et al., 2020)

contaminated materials. The waste from military nuclear applications tends to be more complex and varied (Stabnikov et al., 2020). The types and amounts of nuclear waste vary significantly depending on the source. However, all nuclear waste management programs must address the fundamental challenges of handling, storing, and isolating radioactive materials from the biosphere.

3.3 NUCLEAR WASTE CLASSIFICATION

The radioactivity level of the waste is the crucial factor to determine the management of radioactive waste. Thus, classification and labeling of nuclear waste is a crucial step of nuclear waste management. According to the new classification system (IAEA, 1994, 2009; Brookins, 2012), major waste categories can be classified according to elemental classification (actinium, thorium, protactinium, uranium, neptunium, plutonium, etc.), the type of material present (radioactive gaseous waste, radioactive liquid waste, and radioactive solid waste), the presence of alpha-emitting particles in the radioactive waste, whether the half-life is more than 20 years (transuranic waste and ordinary waste), and the level of radioactivity (very short-lived waste (VSLW), very low-level radioactive waste (VLLW), low-level radioactive waste (LLW), intermediate-level radioactive waste (ILW), spent fuel (SNF), and high-level radioactive waste (HLW)) (Figure 3.2).

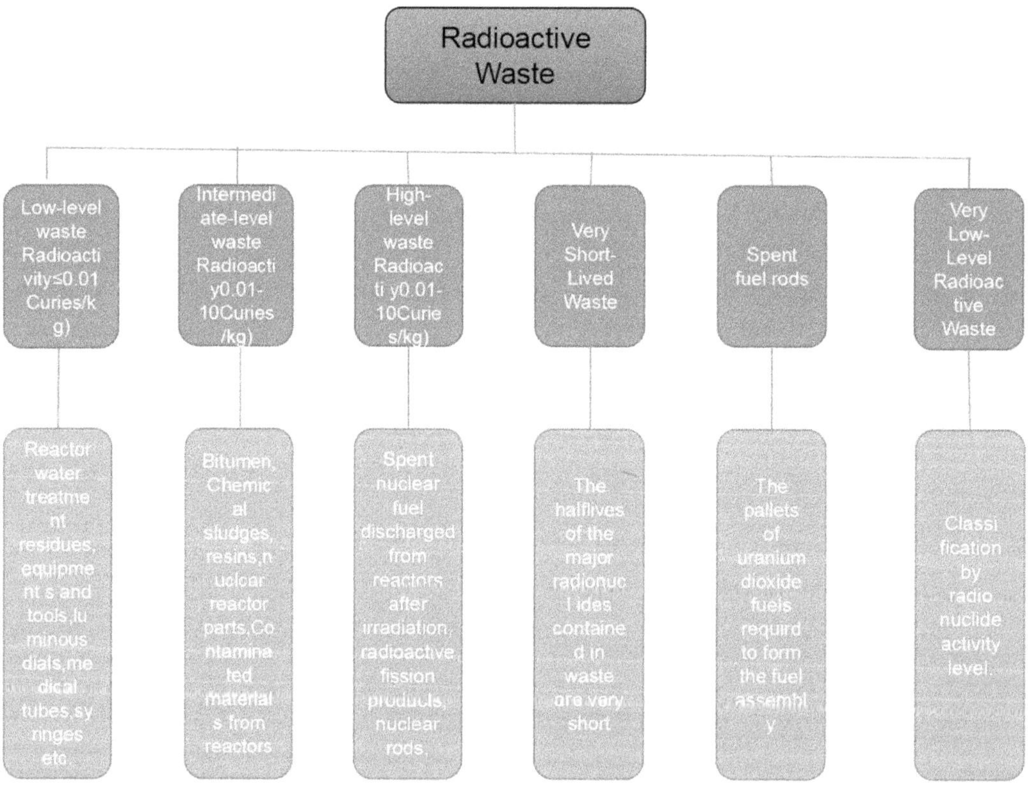

FIGURE 3.2 Classification of Nuclear Waste.

The following provides a more detailed explanation of the classification based on the level of radioactivity (Ackerman, 1991; Gera, 1974; Bennett, 1996):

- VSLW is radioactive waste that can be stored for decay over a limited period of no longer than a few years, with subsequent clearance from regulatory control, and typically includes radioactive wastes generated by application of nuclear energy in research and medicine.
- VLLW is radioactive waste that does not need a high level of containment and isolation and because of that is suitable for disposal in near-surface landfill-type facilities with limited regulatory control. Typical VLLW includes soil and rubble with low levels of activity concentration, which do not usually exceed 100 times the clearance levels for each of the radionuclides present. In some countries, VLLW is disposed of in specially designed disposal facilities, which can be in the form of earthen trenches with engineered covers, although it can be disposed of with LLW.
- LLW has limited amounts of long-lived radionuclides in it, and, because of that, it requires robust isolation and containment for periods of up to a few hundred years. LLW is suitable for disposal in engineered near-surface disposal facilities (NSDFs). LLW covers a very broad range of waste that contains long-lived radionuclides only at relatively low levels of activity concentration. LLW is generated in most facilities involved in nuclear power production, nuclear research, and nuclear medicine. It is common practice to dispose of LLW in NSDFs, although options for the disposal of LLW include simplified facilities such as engineered trenches or concrete vaults in which waste containers are placed. An engineered or earthen cap is then placed over the waste containers to minimize water infiltration. The NSDFs are subject to surveillance until the hazard associated with the nuclear waste has declined to acceptable (e.g., clearance) levels. Some countries prefer disposing of LLW in subsurface facilities or co-locating LLW with ILW or SNF in deeper facilities.
- ILW is radioactive waste that requires a greater degree of containment and isolation than that provided by NSDFs, although it needs no provision for radiogenic heat dissipation. ILW requires disposal at greater depths, ranging from tens of meters to one or even a few hundred meters. Moreover, disposal at depths of greater than several tens of meters is generally considered to be the most appropriate option for ILW. Co-disposal of ILW with SNF and HLW is an effective option considered in many countries. A precise boundary between LLW and ILW does not exist, although a limited content of 400 Bq/g on average and up to 4000 Bq/g for individual packages for long-lived alpha-emitting radionuclides has been adopted in many countries following the recommendation of the IAEA. The actual boundary limits are determined on a case-by-case basis from performance assessments and safety analysis reports and therefore cannot be applied generically to all facilities. For long-lived beta- and/or gamma-emitting radionuclides, such as ^{14}C, ^{36}Cl, ^{63}Ni, ^{93}Zr, ^{94}Nb, ^{99}Tc, and ^{129}I, the

allowable average activity concentrations may be considerably higher (up to tens of kBq/g), although they are always specific to the site and disposal facility.

- SNF: This is irradiated fuel from the nuclear reactor that can no longer be used. This waste is stored at the nuclear reactor sites (in spent fuel pools) for at least a year, if not decades, to allow the SNF to cool and serve as a radioactive shielding medium, after which the SNF can be transferred to the final disposal repository and finally transferred to the nuclear waste repository, such as a deep geologic repository.
- HLW is radioactive waste with levels of activity concentration high enough to require shielding in handling operations and that generates significant quantities of radiogenic heat due to the decay of nuclear waste radionuclides. Depending on the volume of waste, the heat generation levels of HLW are typically above a few W/m^3 for typical volumes of about 1 cubic meter. HLW can also refer to nuclear waste with large amounts of long-lived radionuclides that need to be considered when designing disposal facilities. Disposal in deep geological disposal facilities (GDFs) located in stable geological formations, usually several hundred meters or more below the surface, is the generally recognized HLW disposal option. While most countries with spent fuel and HLW are working toward national solutions, others, mainly for economic reasons, have indicated an interest in developing multi-national disposal facilities.

3.4 NUCLEAR WASTE HAZARD

The primary risks associated with human exposure to nuclear waste stem from radiation (Ayoub et al., 2023), encompassing both somatic and hereditary effects. Somatic effects encompass acute damage (resulting from the body's exposure to high doses within a short timeframe), chronic damage (arising from prolonged exposure to doses exceeding the permissible limit), and long-term effects (manifesting after 6 months of exposure to doses surpassing the permissible limit). Genetic effects are the effects of radiation on the germ cells of the individual recipient, causing them to undergo mutations that manifest in the offspring. Among the hereditary diseases caused by radiation are teratogenesis, premature births, intellectual disability, and minor metabolic disorders.

Radionuclides in nuclear waste radiate decay heat during the decay process deep underground (Allan & Nuttall, 1997), which is equivalent to adding a heat source to the underground media field. Primarily, the presence of the heat source alters the temperature distribution within the subsurface environment. This temperature variation, in turn, influences fluid viscosity and density, consequently impacting fluid movement. Concurrently, the chemical properties of certain substances deep underground change, directly affecting the migration of nuclides within the subsurface medium. Also, temperature fluctuations can induce the closure and opening of fractures, altering the subsurface stress field. The existence of heat sources results in significant environmental impacts, which are primarily localized in the near field. Simultaneously, leaching of nuclides from the solidified waste body may take place through the packaging container. The leached nuclides can enter the biosphere via groundwater migration and thereby influence the human ecosystem.

3.5 NUCLEAR WASTE MANAGEMENT AND PLANNING

3.5.1 Nuclear Waste Management

The life cycle of nuclear waste consists of several distinct phases, with the end of life being disposal (Figure 3.3), which indicates that before disposal, the waste typically undergoes a series of steps such as pre-treatment, treatment, conditioning, storage, transportation, and disposal, and that its properties are utilized throughout its life cycle. The life cycle of generating nuclear waste usually includes several different steps (Ojovan & Lee, 2005; Rahman et al., 2014; Ojovan, 2011): (1) Pre-treatment: This step consists of any operation before the treatment of the waste that permits the selection of technologies to be used further in the treatment of the waste, including treatment and conditioning. (2) Treatment:

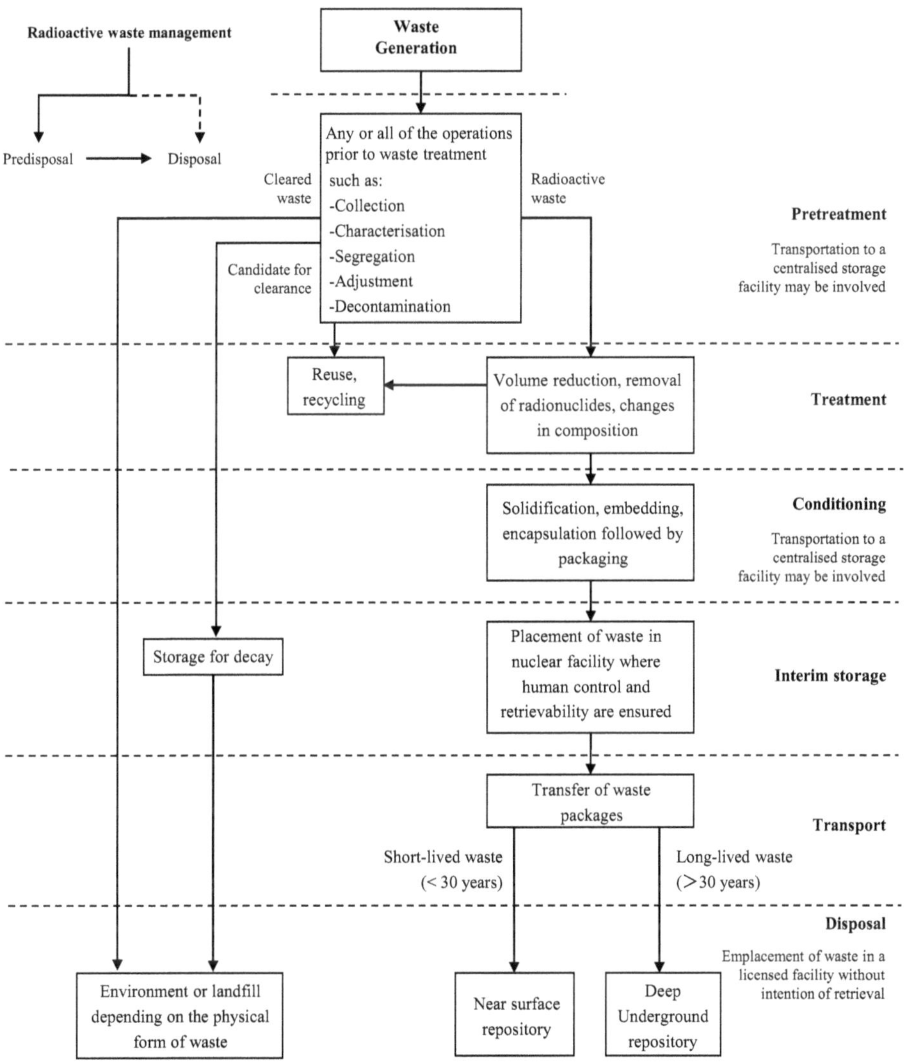

FIGURE 3.3 Nuclear Waste Life Cycle Diagram.

It consists of operations that improve the safety or economy of nuclear waste by altering its characteristics. While treated waste usually requires immobilization and/or backfilling, some treatments may result in appropriate waste forms. (3) Conditioning: This step produces a waste package suitable for handling, transportation, storage, and/or disposal. (4) Storage: It affords restrictions, isolation, environmental protection, and monitoring within a designated period (storage duration) to ensure the recyclability of used packages. (5) Transportation: It refers to the intentional physical movement of nuclear waste from one location to another, facilitated by specially designed packaging. (6) Disposal: This is the end of the life cycle of nuclear waste and envisages the placement of the waste in an appropriate facility with no intention of retrieval.

In addition, nuclear waste characterization is a necessary element of nuclear waste from the beginning to the end of its life cycle, including the determination of the physical, chemical, and radiological properties of the waste in order to determine whether it needs to be further adapted, handled, conditioned, or whether it is suitable for further treatment, processing, storage, and disposal. The six different stages of the nuclear waste life cycle are further described in the following section:

3.5.1.1 Nuclear Waste Pre-Treatment

Purification and concentration of radioactive waste (gases and liquids) and compression and incineration of solid waste materials: The purification, concentration, and disposal of liquid radioactive waste usually involve ion exchange, evaporation, coagulation and precipitation, reverse osmosis filtration, and other technologies to concentrate and reduce the volume of the waste liquid. The residual liquid is then concentrated and subjected to curing and disposal. The purified wastewater is then discharged into natural water bodies or reused. The concentration coefficient serves as a characterization parameter for the impact of radioactive waste liquid concentration treatment, representing the ratio of the waste liquid volume before and after the concentration process. Decontamination wash, on the other hand, is defined as the ratio of radioactive concentration before and after purification treatment of radioactive waste liquid (Hileman, 1982). Currently, it is common worldwide to reduce the volume of nuclear solid waste by compression and incineration, and the effect of the reduction is characterized by the volume reduction rate: the ratio of the volume of the waste before and after treatment. Of these, incineration is well-suited for combustible wastes and is classified into dry incineration and wet incineration. In the dry incineration process, waste is directly burned at high temperatures, while wet incineration involves the decomposition and treatment of waste in a moist environment. Each of these two technologies has its advantages, and the choice of the most suitable method depends on the nature of the waste and the treatment requirements. Before engaging in the compression, incineration, and dismantling of nuclear facilities, decontamination is commonly conducted. This process involves reducing or removing radioactive contaminants from an object, with the objective of lowering the radiation level of the contaminated material. Common methods for decontamination include mechanical cleaning, chemical decontamination, and electrochemical decontamination (Gonzalez, 1998).

3.5.1.2 Disposal of Radioactive Waste

The means of disposal of nuclear waste are multi-layered and multi-faceted, involving various aspects of physics, chemistry, and engineering. In general, nuclear waste comes in a variety of forms and is usually treated in different ways depending on the types of nuclides and the physical form of the waste, including treatments such as radioactive waste gas treatment, radioactive waste liquid treatment, radioactive solidification treatment, and methods for final disposal, whether in untreated or treated form. Of these, the solidification of radioactive waste is a primary method: it is the process of converting waste into solids, and commonly employed curing substrates include cement, artificial rock, asphalt, and other materials. Of these, conventional cement curing and asphalt curing are the primary techniques for treating low- and medium-radioactive waste streams, whereas glass curing is the most prevalent method for managing high-radioactive waste streams.

To achieve the solidification of radioactive waste liquids for the immobilization and long-term confinement of radionuclides, the curing product must possess the ability to resist damage. Furthermore, thermal, irradiation, mechanical, and chemical stability are frequently employed metrics to assess the ease of transportation, storage, and final disposal performance post-curing. These evaluations encompass processes such as waste concentration, denitrification, melting, drying, and annealing. The main curing treatments are as follows:

(1) Cement curing (Eskander et al., 2011): Cementitious curing is a method of curing wastes based on the hydration and water-hardening gelation of cement. Cement, serving as an inorganic binding material, typically undergoes hydration to produce a durable cementitious cured body, thereby fulfilling the objective of solidifying radioactive waste. Cement curing possesses several advantages. It is a straightforward process, it has low energy consumption, there is minimal equipment and operational investment, the operation is convenient, and it can be conducted at room temperature. However, cement curing also has significant drawbacks: first, it has a high radionuclide leaching rate, and second, there is a notable increase in the volume of the cured product. In recent years, a lot of improvements have been made to cement curing at home and abroad, which mainly includes the following three aspects: minimizing the volume of cured products, decreasing the leaching rate of nuclides, and exploring new applications for cement solidification.

(2) Asphalt curing (Subba Rao et al., 2022): Asphalt curing is a process whereby radioactive waste is uniformly encapsulated in asphalt, resulting in a cured product with robust resistance to leaching. Asphalt curing is suitable for solidifying low- and medium-release wastes, such as evaporation residues and chemical slurries. The distribution of radioactive materials in asphalt solidification is uniform, with a high curing capacity and a low radionuclide leaching rate. However, since asphalt is a combustible organic material that cannot withstand higher temperatures and intense radiation, asphalt curing is not suitable for solidifying wastes characterized by strong radiation, high heat release rates, and elevated oxidant content.

(3) Plastic curing (Ojovan & Lee, 2014): Plastic curing involves uniformly encapsulating radioactive waste in a thermoplastic or thermosetting plastic material, and the thermoplastic curing process is akin to asphalt curing, as both require heat for melting. The thermosetting curing process is like cement curing. There is a restriction on the water content of the waste liquid, and both processes necessitate pre-treatment and dewatering, or mixing and emulsification with an emulsifier. Plastic-cured nuclides have a low leaching rate and high package capacity, which is important for achieving long-term safe isolation, though their economic cost remains high.

(4) Glass curing (Ojovan et al., 2019): The inorganic material is mixed with radioactive waste according to a specific ratio of ingredients, then ground, melted, and poured at high temperatures. It then undergoes annealing and transforms into a stable glass-solidified body. There are two primary types of glass employed for solidifying high-level radioactive waste: borosilicate glass and phosphate glass, with borosilicate being the most utilized.

(5) Artificial rock curing (Han et al., 2023): Building on the concept of utilizing high-radionuclide natural minerals as nuclear fuel, which, even after millions or even hundreds of millions of years of crustal movement, retain their original structure, composition, and morphology without causing environmental or human harm, we can adopt the original rock structure to create artificial rock curing for geological disposal.

3.5.1.3 Conditioning of Radioactive Waste

Nuclear waste reconciliation is primarily concerned with controlling the generation of nuclear waste to minimize or optimize the amount and nature of the waste generated. Here are some ways to regulate nuclear waste:

(1) Fuel design: By changing the design of the nuclear fuel, the amount of nuclear waste produced can be reduced or regulated. For example, the use of uranium-enriched or reprocessing technologies can modify the combustion characteristics of nuclear fuel and reduce the amount of long-lived, highly radioactive nuclear waste produced.

(2) Fuel cycle: Advancing the nuclear fuel cycle through reprocessing and recycling of nuclear materials can reduce the volume and hazard of nuclear waste.

(3) Advanced reactor technologies: The use of advanced reactor technologies, such as fast neutron reactors or solid-bed reactors, can reduce the generation of nuclear waste and increase the efficiency of nuclear fuel utilization.

3.5.1.4 Radioactive Waste Storage

Interim storage of nuclear waste encompasses all types of retrievable short-term storage of nuclear waste under controlled conditions before final disposal, usually including pre-treatment interim storage, waiting storage, intermediate storage prior to disposal, and in situ containment. Among these, temporary storage methods for nuclear waste commonly include wet temporary storage and dry temporary storage. Spent fuel from nuclear power plants is often initially placed in wet temporary storage, where radioactive heat diminishes to a certain extent before being transferred out of the pool. Dry staging commonly refers to dry basement storage and dry container storage methods.

3.5.1.5 Transportation of Radioactive Waste

Moreover, due to the special nature of nuclear waste, transportation units often require the use of shielded mass containers with visible radiation markings on the outside of the container.

3.5.1.6 Nuclear Waste Disposal Scheme

(1) Ocean dumping (Templeton, 1997): This method is currently banned by the IAEA due to its serious, direct impact on humans and the environment, as radioactive substances in nuclear effluents can readily propagate through the food chain, elevating the risk of radiation-related diseases in humans.

(2) Near-surface disposal (Twining et al., 2011): Near-surface disposal primarily targets medium- or low-level radioactive waste, with a burial depth typically within 10 m from the ground. The safety monitoring period extends for 300 to 500 years. Near-surface cover and burial can be employed to effectively dispose of waste rock and tailings from uranium mine exploration, mining, and metallurgy that exhibit low radioactivity and short lifetimes, are present in large quantities, and are naturally radioactive. For nuclear power plant operations, nuclear fuel production, spent fuel reprocessing, and other activities generating medium- and low-radioactive short-lived solid waste, the typical approach involves long-term temporary storage or centralized near-surface cover and burial methods. For medium- and low-emission long-lived wastes, near-surface disposal utilizing dedicated repositories is necessary.

(3) Geological disposal (Turner et al., 2023): Geological disposal focuses on high-level and medium-level, long-lived wastes as the main targets for disposal. The shared characteristics of these wastes include long half-life, intense radioactivity, and typically high potential hazards. They need to be buried deep in the crustal rock layers hundreds or even thousands of meters below the ground. This helps in achieving permanent isolation of radioactive wastes from the biosphere and fulfills the goal of safe management and disposal, with certain suitable natural or man-made rock layers effectively stabilizing and containing these radioactive wastes. Currently, the geological disposal of high-level and medium-level long-lived wastes remains a challenging issue both domestically and internationally and is in a phase of ongoing exploration and research.

(4) Space disposal (Kim et al., 2016): Space disposal is the process of placing nuclear waste in a launch vehicle, sending it into space, and permanently storing it there. Because the launch accident rate of launch vehicles is usually about 2%, in order to solve the nuclear disaster caused by this, experts in various countries have designed

and constructed sealed containers using high-strength steel shells. These containers feature multi-layer insulation: an inner layer for shielding against radiation, an outer layer for prevention of corrosion, and a middle layer made up of a strong solid so as to ensure that the nuclear waste remains secure. Currently, this type of disposal is only conceptualized and has not been implemented.

(5) Spent fuel reprocessing (Paulillo et al., 2021): Currently, the most difficult of the nuclear wastes to manage, and at the same time the one with the longest and most hazardous half-life, is spent fuel. Spent fuel is thus a primary focus of nuclear waste management. Typically, spent fuel necessitates reprocessing for its ultimate disposal. Meanwhile, spent fuel reprocessing is also one of the important links to ensure the sustainable development of nuclear power. Through reprocessing, useful uranium and wye can be recovered from irradiated spent fuel, which can then be made into new fuel elements and returned to hot or fast reactors for use, thus greatly improving the utilization rate of uranium resources.

(6) Retention times and multibarrier approach of disposal: To achieve the safe long-term management of radioactive waste, repositories designed for storing or disposing of radioactive waste typically rely on multi-barrier systems to effectively isolate the waste from the biosphere. Multi-barrier systems typically consist of an engineered barrier system (EBS) and a natural geologic barrier provided by the host rock. The EBS consists of a waste matrix, buffers, containers, repacking or backfill, repository walls, and wall liners. The functions of the various barriers are generally consistent, starting with the containment of radionuclides and followed by restriction of their release into the accessible environment. The main functions of a geological repository are to isolate radioactive waste from near-surface processes and human activities, contain and retard the release of radionuclides as the waste form and engineered barriers gradually degrade, and protect the biosphere by dispersing and diluting any long-lived radionuclide flux that may eventually be released.

3.5.2 Nuclear Waste Scheme

According to the definition, nuclear waste management includes all administrative and operational activities involved in the pre-treatment, treatment, conditioning, storage, transportation, and disposal of radioactive waste. The planning of administration-related activities (Figure 3.4) plays a vital role in guaranteeing the security and effectiveness of operational activities.

The implementation of a nuclear waste management plan is based on the knowledge of the inventory of stockpiled nuclear waste streams, the projection of the nuclear waste streams that will be generated within the established integrated planning timetable, the assessment of the needs for different waste management facilities, the choice of technologies for the treatment of nuclear waste generated throughout its entire lifecycle (from cradle to grave) within the envisioned timetable, the estimation of the cost of the different waste management options during the planning period, and the development and funding modalities of the scenarios. An integrated approach to nuclear waste management is a

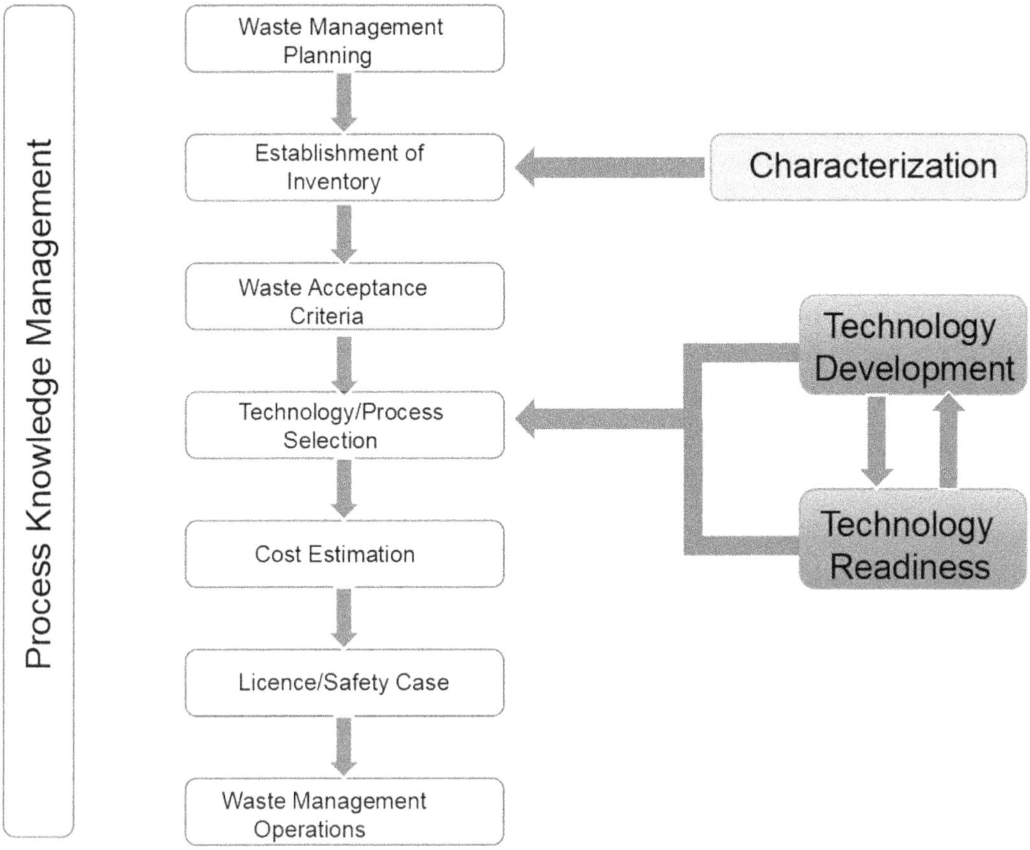

FIGURE 3.4 Schematic Diagram of Nuclear Waste Management.

logical and preferred optimization strategy used in its planning and implementation. An integrated nuclear waste management program should cover all stages from waste generation to final disposal, take into account the interdependence of these stages, and ensure that decisions made at one stage do not preclude options at subsequent stages. For example, waste generation is highly dependent on design, planning, and operations.

3.6 POLICIES FOR THE MANAGEMENT OF NUCLEAR WASTE

Policies on nuclear waste management have important national benefits; these policies vary considerably from one Member State to another. Some Member States have revisited their waste management programs; others see spent fuel as waste material; and still others see spent fuel as a resource from which they hope to extract fissile material. On July 19, 2011, the Council of the European Union adopted Council Directive 2011/70/Euratom, which establishes a Community framework for the responsible and safe management of spent fuel and radioactive waste. It establishes a public framework for proper nuclear waste management and provides internationally binding legal force for key recognized requirements and principles. The Radioactive Waste and Spent Fuel Management Directive requires the following conditions to be met: (1) Member States should have national policies for the proper management of SNF and NW. (2) Member States should develop and implement national programs for the proper management and safe decommissioning of nuclear waste, including within their territories. (3) Member States should establish a comprehensive and robust framework for the sound management and safe decommissioning of nuclear waste, which should be accompanied by a competent and autonomous regulatory body with adequate funding for the sound management and safe decommissioning of nuclear waste. (4) Information on nuclear waste should be readily available to ordinary citizens who are allowed to participate in the discussion. (5) Member States should conduct self-assessments and call for international peer reviews of their frameworks, authorities, and/or national programs at least once every 10 years.

According to the work of Ewing (Gonzalez et al., 1998), a typical national policy must include the following elements: clear safety and security objectives, the arrangements needed to manage nuclear waste resources, key methods for identifying nuclear waste management, export and import policies for nuclear waste, and provisions for public information as well as participation. In addition, the policy should clarify the role and responsibilities of the state in nuclear waste management.

3.7 MANAGEMENT PRINCIPLES FOR NUCLEAR WASTE

In order to ensure that the effects of nuclear waste are avoided as much as possible throughout the entire process of production and disposal, states follow certain principles and regimes in the management of nuclear waste, which are specified below (Bennett, 1996).

(1) Protection of human health: Nuclear waste management is primarily concerned with the protection of human health at a manageable level. Nuclear waste poses a radiological hazard, and exposure of workers and the public must be controlled within prescribed limits to achieve the lowest possible level.

(2) Protecting the ecological environment: Nuclear waste management must achieve environmental protection to a level deemed acceptable. Radionuclides are released into the environment and may affect living species other than humans. Nuclear waste disposal may have implications for the utilization of natural resources, while radioactive waste may be accompanied by non-radioactive hazardous substances. Therefore, the environmental protection requirements for nuclear waste management must be kept within the established limits for each category of radioactive waste.

(3) Consideration of transboundary movements: Nuclear waste management needs to take into account the adverse impacts on human health and the ecosystem in the event of transboundary movements.

(4) Protection of future generations and potential reduction of the burden on future generations: Nuclear waste management must guarantee that the anticipated health hazards for future generations are lower than what is deemed acceptable today. Simultaneously, it must ensure that what can be addressed today is indeed addressed, thereby minimizing any additional burden on future generations.

(5) Establishment of a national legal framework: Nuclear waste management must be underpinned by a comprehensive national legal framework, which includes the establishment of an independent auditing and control organization, as well as clear responsibilities and authorities. The state should enact laws and regulations concerning nuclear waste management, establish relevant institutions, define the allocation of responsibilities, implement the separation of auditing and management from operation, and ensure that nuclear waste management is subject to independent supervision and review.

(6) Process management of waste generation and quantities generated: Nuclear waste management must take into account the interdependence of the processes involved in generating nuclear waste. When considering any nuclear waste management activity, its impact on subsequent nuclear waste management activities, in particular on disposal, should be taken into account to implement whole-process management. In addition, the radioactivity and volume of nuclear waste are minimized as far as reasonably possible through optimal management, proper design and operation, and recycling and reuse.

3.8 CHALLENGES TO NUCLEAR WASTE MANAGEMENT

Nowadays, managing nuclear waste is regarded as one of the most complex and challenging environmental issues facing humanity. This is because nuclear waste poses serious health risks and remains radioactive for thousands of years, requiring extremely careful handling and isolation from the environment for centuries. There are significant technical, financial, political, and social challenges involved in safely and securely managing nuclear waste over such long timescales. One of the main technical challenges is finding suitable storage or disposal sites for nuclear waste. Nuclear waste needs to be isolated from the environment for hundreds to thousands of years, and it requires extremely stable and secure storage or disposal facilities. However, there are few suitable geological formations that can provide the necessary stability and isolation over such long periods. Finding appropriate sites and ensuring their long-term integrity is extremely difficult. Once sites are chosen, the waste needs to be transported there safely and securely, which also poses logistical challenges. Another major challenge is the longevity of nuclear waste. Radioactive elements in SNF remain hazardous for tens of thousands of years, far beyond the lifespan of any human institution or technology. This requires nuclear waste management strategies that can be maintained and monitored reliably for centuries. There are doubts about whether any human system can provide the necessary oversight, record-keeping, and institutional stability over such long timeframes. Even minor disruptions or failures could result in nuclear waste leaking into the environment. The costs of managing nuclear waste safely and securely over millennia are also extremely high. Nuclear waste management programs require massive upfront investments in storage facilities, transportation infrastructure, and monitoring systems. A meaningful nuclear waste management plan must be based on accurate estimates of costs and the resources required to meet waste management needs within a defined time frame. These plans call for the establishment of methodologies to conduct country-specific economic analyses of various nuclear waste options. It aims to assess the impact of variable waste generation, the total cost of addressing nuclear waste, how future costs will be estimated, the time required and the amount of funding needed, what technologies will ultimately be used in each step of the nuclear waste process, and whether technological solutions designed today will be socially acceptable in the future. Ongoing costs for maintenance, monitoring, and eventual disposal will also be significant for centuries. There are questions about whether governments and energy companies can commit the necessary financial resources reliably over such long periods.

Public acceptance and political willpower are also major challenges. Many communities are reluctant to host nuclear waste storage or disposal facilities due to fears about safety, environmental impacts, and stigma. Gaining public trust and acceptance for nuclear waste management programs requires extensive outreach, education, and transparency. There are also questions about whether political support for nuclear waste management programs can be sustained across changes in government and societal priorities. Any lack of political will could threaten funding and oversight of nuclear waste facilities. In summary, managing nuclear waste safely and securely for thousands of years poses complex technical, financial, political, and social challenges that humanity has never successfully overcome before. While some progress has been made in improving storage technologies and finding potential disposal sites, there are still major doubts about our ability to provide the necessary isolation, oversight, and resources reliably for centuries. Nuclear waste management will likely remain one of the most difficult environmental issues we face for the foreseeable future.

REFERENCES

Ackerman, D.J., 1991. *Transmissivity of Perched Aquifers at the Idaho National Engineering Laboratory, Idaho*. U.S. Department of the Interior, U.S. Geological Survey.

Allan, C.J., Nuttall, K., 1997. How to cope with the hazards of nuclear fuel waste. *Nucl. Eng. Des.* 176, 51–66.

Ayoub, A.A.T., Song, Y., Sagiroun, M.I.A., 2023. Radiological hazard assessment for humans from nuclear power plant effluent released under normal operation scenario (hypothetical case study). *Radiat. Phys. Chem.* 216, 111380.

Bennett, F.G., 1996. International basic safety standards for protection against ionizing radiation and for the safety of radiation sources. *J. Hazard. Mater.* 45(2), 282.

Brookins, D.G., 2012. *Geochemical Aspects of Radioactive Waste Disposal*. Springer, New York.

Drace, Z., Ojovan, M.I., Samanta, S.K., 2022. Challenges in planning of integrated nuclear waste management. *Sustainability* 14, 142114204.

Eskander, S.B., Abdel Aziz, S.M., El-Didamony, H., et al., 2011. Immobilization of low and intermediate level of organic radioactive wastes in cement matrices. *J. Hazard. Mater.* 190, 969–979.

Gera, F., 1974. The classification of radioactive wastes. *Health Phys.* 27(1), 113–121.

Gonzalez, A.J., 1998. A radiological legacy. Radioactive residues of the Cold War period. *IAEA Bull.* 40(4), 2–11.

Guo, B., Sasaki, K., 2022. Stabilization/solidification of radioactive waste in geochemical aspects. *Low Carbon Stabilization Solidif. Hazard. Wastes*, 469–482.

Han, W., Shu, X., Li, L., et al., 2023. Converting natural metamorphic granite and simulated actinide oxide to aluminosilicate glass for nuclear waste immobilization. *Ceram. Int.* 49(22), 36838–36849.

Hare, F.K., Aikin, A.M., 1980. Nuclear waste disposal: Technology and environmental hazards. In: *Nuclear Energy and the Environment*, pp. 168–199. Pergamon Press, Oxford.

Hileman, B., 1982. Nuclear waste disposal. *Environ. Sci. Technol.* 16(5), 271A–275A.

Hooper, A., 2010. *Crystalline Geological Repository Systems: Characterisation, Site Surveying and Construction Technologies and Techniques*. Woodhead Publishing, 121–152.

IAEA, 1994. Classification of Radioactive Waste. Safety Series No. 111-G-1.1. *International Atomic Energy Agency*, Vienna.

IAEA, 2009. Classification of Radioactive Waste. General Safety Guide No. GSG-1. *International Atomic Energy Agency*, Vienna.

Kim, H., Park, C., Kwon, O.J., 2016. Conceptual design of the space disposal system for the highly radioactive component of the nuclear waste. *Energy* 115, 155–168.

Kumar, P., Kumar, B., Singh, D., 2022. Radioactive waste management. In: *Hazardous Waste Management*, pp. 289–301. Elsevier.

Laraia, M., 2015. Radioactive contamination and other environmental impacts of waste from nuclear and conventional power plants, medical and other industrial sources. In: *Environmental Remediation and Restoration of Contaminated Nuclear and NORM Sites*, pp. 35–56. Woodhead Publishing, Cambridge.

Ojovan, M.I. (Ed.), 2011. *Handbook of Advanced Radioactive Waste Conditioning Technologies*. Woodhead Publishing, Cambridge.

Ojovan, M.I., Lee, W.E., 2005. *An Introduction to Nuclear Waste Immobilisation*. Elsevier, Oxford.

Ojovan, M.I., Lee, W.E., 2014. *An Introduction to Nuclear Waste Immobilisation*. 2nd ed. Frontiers.

Paulillo, A., Dodds, J.M., Palethorpe, S.J., et al., 2021. Reprocessing vs direct disposal of used nuclear fuels: The environmental impacts of future scenarios for the UK. *Sustain. Mater. Technol.* 28, e00278.

Rahman, R.O.A., Rakhimov, R.Z., Rakhimova, N.R., et al., 2014. *Cementitious Materials for Nuclear Waste Immobilization*. John Wiley & Sons, Hoboken.

Rout, A., Ramanathan, N., 2023. Treatment of laboratory generated radioactive organic liquid waste using a novel ionic liquid based resin: A feasibility study. *J. Mol. Liq.* 390, 123079.

Stabnikov, V., Ivanov, V., Vaseashta, A., 2020. Biotechnological immobilization of chemical, biological, and radioactive pollutants on land and infrastructure demolition waste after industrial accident, military action, or terrorist attack. In: *Bio-Based Materials and Biotechnologies for Eco-Efficient Construction*, pp. 377–394. Woodhead Publishing, Cambridge.

Subba Rao, T., Panigrahi, S., Velraj, P., 2022. Transport and disposal of radioactive wastes in the nuclear industry. In: *Microbial Biodegradation and Bioremediation*. 2nd ed., pp. 419–440.

Templeton, W.L., 1997. Research needed relative to radiological assessment of the dumping of radioactive wastes in the oceans. *Mar. Pollut. Bull.* 35(7–12), 374–380.

Turner, J.P., Berry, T.W., Bowman, M.J., et al., 2023. Role of the geosphere in deep nuclear waste disposal—An England and Wales perspective. *Earth-Sci. Rev.* 242, 104445.

Twining, J.R., Hughes, C.E., Harrison, J.J., et al., 2011. Biotic, temporal and spatial variability of tritium concentrations in transpirate samples collected in the vicinity of a near-surface low-level nuclear waste disposal site and nearby research reactor. *J. Environ. Radioact.* 102(6), 551–558.

Xixi, Z., Tongkun, Q., Yecheng, W., 2022. Optimal strategies for stakeholders of Fukushima nuclear waste water discharge in *Japan. Mar. Policy* 135, 104881.

Yano, K.H., Mao, K.S., Wharry, J.P., et al., 2018. Investing in a permanent and sustainable nuclear waste disposal solution. *Prog. Nucl. Energy* 108, 474–479.

CHAPTER 4

Nuclear Waste Processing, Transport, and Storage

Kui Liu

4.1 WASTE CONDITIONING AND IMMOBILIZATION TECHNOLOGIES

The initial phase of nuclear waste management involves the transformation of raw waste materials into chemically stable, mechanically strong forms suitable for long-term handling, transport, and disposal. This preparation stage is referred to as conditioning.

4.1.1 The Necessity of Conditioning: Enhancing Safety for Handling and Disposal

Conditioning is an essential step undertaken to change radioactive waste into a durable form that facilitates safe handling, transportation, storage, and final disposal. This process universally includes the immobilization of radionuclides within a stable matrix and enclosure in durable containers. Prior to conditioning, waste treatment techniques may be employed, such as compaction to reduce volume or filtration and ion exchange processes to remove specific radionuclides (US DOE, 2014; IAEA, 2022). Conditioning ensures the long-term chemical and physical integrity of the waste package, forming the initial element of the multi-barrier containment system.

4.1.2 Low- and Intermediate-Level Waste (LLW/ILW) Management

Low-level waste (LLW) typically consists of paper, rags, tools, and clothing containing small amounts of mostly short-lived radioactivity. While LLW accounts for approximately 90% of the total volume of radioactive waste, it comprises only about 1% of the overall radioactivity (IAEA, 2022). To maximize disposal efficiency (usually in near-surface facilities), LLW is frequently compacted or incinerated to reduce its bulk volume. Intermediate-level waste (ILW) contains higher levels of radioactivity and some long-lived radionuclides. Liquid LLW and ILW are commonly immobilized and solidified in matrices such as cement, which provides chemical stability and mechanical strength for containment (IAEA, 2022).

DOI: 10.1201/9781003665793-6

4.1.3 High-Level Waste (HLW) Conditioning: Vitrification

High-level waste (HLW), characterized by intense radioactivity and significant heat generation, requires the most stringent conditioning methods. The standard international approach involves vitrification, a process that yields an exceptionally stable waste form. Figure 4.1 shows the typical vitrified waste form development cycle, exemplified by the French Alternative Energies and Atomic Energy Commission (CEA), France (McCloy et al., 2024). This procedure first involves calcining or drying the liquid HLW before embedding it in a stable glass matrix. This glass form is engineered for extreme durability and leach resistance, serving as a robust primary containment barrier against radionuclide mobility once emplaced in a repository.

4.1.4 Reprocessing and Advanced Concepts

4.1.4.1 Standard Aqueous Reprocessing

The plutonium uranium reduction extraction (PUREX) process is the most widely adopted hydrometallurgy technique globally, forming the technical basis for nations employing a closed nuclear fuel cycle (e.g., France). It is designed to chemically separate usable uranium and plutonium (Lanham & Runion, 1949) from the bulk of highly radioactive fission products found in spent light water reactor (LWR) fuel.

The standard PUREX route involves the following key steps:

1. Preparation and Dissolution: Irradiated fuel, typically cooled for a period of at least five years, is first sheared into short segments (e.g., 5-cm length). The exposed fuel material is then dissolved in a nitric acid solution, yielding a nitrate solution containing uranium, plutonium, minor actinides, and fission products. Volatile fission products, such as iodine, krypton, and xenon, are released as off-gas and subsequently treated (Kumari et al., 2020; Hanson, 2015).

2. Solvent Extraction: The chemical separation relies on solvent extraction using a solution containing tributyl phosphate (TBP) in an organic phase. This process selectively extracts uranium and plutonium from the nitric acid solution, leaving most of the high-level fission products and minor actinides in the remaining aqueous waste stream (Kumari et al., 2020; Hanson, 2015).

3. Separation and Conversion: Uranium and plutonium are then chemically separated from each other. Plutonium is typically reduced (often using excess U^{4+}), causing it to transfer to the aqueous phase while uranium remains in the organic solvent. The purified uranium and plutonium nitrates are then concentrated, evaporated, and converted into stable oxide powders (UO_2 and PuO_2) for fabrication into new fuel, such as mixed oxide (MOX) fuel (Kumari et al., 2020; Hanson, 2015).

Modifications to PUREX have given rise to advanced concepts aimed at better managing or recycling minor actinides, such as the co-extraction of uranium and plutonium

FIGURE 4.1 Vitrified Waste Form Development Cycle, Exemplified by CEA France (McCloy et al., 2024).

(COEX) process, the selective separation of americium and curium (DIAMEX-SANEX) processes, and the grouped extraction of all actinides (GANEX) process (Ruiz-Lopez & Pescatore, 2015; Bell et al., 2012; Baron et al., 2001).

4.1.4.2 Advanced Processing Concepts

Current technological development is increasingly focused on advanced recycling and waste minimization strategies, collectively known as partitioning and transmutation (P&T). The strategic goal of P&T is to achieve a significant reduction in the mass, volume, total activity, and, critically, the effective half-life of the existing commercial spent nuclear fuel (SNF) stockpile slated for geological disposal (Kooyman, 2021; Gonzalez-Romero, 2011). This technology directly impacts the ultimate requirements for deep geological repositories (DGRs); by reducing the quantity of long-lived transuranic (TRU) elements, P&T seeks to mitigate the need for isolation over timescales approaching 1 million years, thereby improving the economic viability and simplifying the safety modeling of final disposal facilities.

A central methodology in this field is electrometallurgical pyroprocessing (Inoue & Koch, 2008; Fredrickson & Yoo, 2023). This technique operates through electrolytic processes within high-temperature molten salt baths, typically composed of chlorides or fluorides (e.g., LiCl-KCl).

4.1.4.3 Electrometallurgical Pyroprocessing

The process is designed to recover all long-lived actinides (including neptunium, americium, and curium, alongside uranium and plutonium) together. This collective recovery of TRU elements contrasts with older aqueous reprocessing methods that often isolated plutonium separately (Inoue & Koch, 2008; Fredrickson & Yoo, 2023). The general sequence involves dissolving the used fuel in the molten salt and then electrorefining to deposit the actinides onto a cathode, often a liquid cadmium electrode. Fission products, such as cesium and strontium, which are more chemically active, remain behind in the salt.

A key advantage of pyroprocessing is its inherent resilience to high radiation fields. It can be applied efficiently to high burn-up fuel that has undergone relatively little cooling, minimizing the duration required for interim storage and accelerating the recycling rate. The separated TRU elements are then prepared into ingots for subsequent fabrication into specialized fuel or transmutation targets (Simpson, 2018). These materials are intended for consumption in advanced burner reactors, which convert the long-lived actinides into shorter-lived fission products. The resulting fission products are subsequently recovered from the salt using processes like ion exchange in zeolite columns and are immobilized in a durable ceramic or vitreous waste form (Simpson, 2018). The decision to pursue pyrochemical conditioning methods is fundamentally linked to a closed fuel cycle strategy centered on fast reactor deployment, meaning waste management technology choice dictates the necessary supporting reactor infrastructure.

4.1.5 Divergent Global Fuel Cycle Strategies: Reprocessing or Direct Disposal

While all nuclear nations adhere to international safety standards, fundamental differences exist in how they approach the end of the nuclear fuel cycle, primarily revolving around the decision to recycle used fuel (a "closed" cycle) or treat it directly as waste (an "open" cycle).

1. Closed Fuel Cycle (Reprocessing Focus): Nations like France have historically adopted a policy of reprocessing all SNF, primarily via the PUREX process. This separation extracts usable uranium and plutonium, significantly reducing the volume and heat generation of the remaining waste, which is then vitrified as HLW for disposal (Paviet-Hartmann et al., 2009). China also adheres to a closed nuclear fuel cycle system, with reprocessing prioritized to secure its nuclear fuel supply. China has successfully completed hot tests at its reprocessing pilot plant and uses established HLW vitrification technology and LLW cementation methods (Cao et al., 2013). Advanced pyroprocessing techniques for P&T, developed in Russia and explored by the United States, China, Japan, and South Korea, represent the next evolution of the closed cycle, which aims to consume the long-lived actinides completely (Haynes & Chen, 2023; Yamana, 2005; Kulikova et al., 2021; Lin & Zhong, 2025).
2. Open Fuel Cycle (Direct Disposal Focus): Countries such as Sweden, Finland, Canada, and the United States have opted for the direct disposal of the entire SNF assembly, without chemically separating the fissile material (Sassani et al., 2010; De Saillan, 2010; Bruno et al., 2020). This strategy focuses on robust encapsulation (e.g., copper canisters in the KBS-3 method) to ensure isolation for 100,000 years or more. Japan, while historically supporting fuel recycling, is pursuing DGR siting through a phased, consent-based approach for the ultimate disposal of high-level radioactive waste and TRU wastes (Lawless et al., 2014).

TABLE 4.1 Waste Classification and Associated Conditioning Methods

Waste Category	Primary Hazard/ Half-Life	Volume Reduction Methods	Immobilization/ Conditioning Matrix	Source
LLW	Low radioactivity, mostly short-lived	Compaction, incineration	Cement, resin, bitumen	(US NRC, 1999)
ILW	Higher radioactivity, some long-lived nuclides	Compaction, filtration, ion exchange	Cement solidification	(US NRC, 1999)
HLW	High heat, high radioactivity, very long-lived actinides	Calcination/ drying (pre-treatment)	Vitrified glass matrix	(US NRC, 1999)
SNF	High heat, high radioactivity, TRU content	Partitioning/ separation (P&T)	Metallic ingots (for recycling) or copper canisters (for disposal)	(Simpson, 2018; McConnell et al., 2011)

4.2 SAFE AND SECURE TRANSPORTATION LOGISTICS

The safe transport of radioactive materials, particularly SNF and conditioned HLW, is globally recognized as one of the most rigorously regulated logistic operations.

4.2.1 International Regulatory Framework and Oversight

The safety of transporting radioactive materials rests on an internationally established regulatory framework, largely overseen and maintained by the International Atomic Energy Agency (IAEA). This framework is considered one of the most robust regulatory systems in existence, ensuring safety across all modes of transport worldwide. The regulatory requirements are exhaustive, specifying rules for package design and performance, marking, labeling, placarding of conveyances, necessary documentation, external radiation limits, and quality assurance programs (IAEA, 1996). Furthermore, certain shipments and package types require specific notification and approval from competent national authorities. National bodies, such as the US Nuclear Regulatory Commission (NRC), implement and enforce these standards, covering activities that include the licensing and physical movement of spent fuel (Maheras & Adkins, 2020).

4.2.2 Principles of Protective Packaging and Security

A graded approach is applied to packaging requirements, scaling the design stringency based on the inherent hazard of the contents. This ranges from standard commercial packaging for low-hazard materials to strict design criteria for contents with higher radioactivity levels (IAEA, 1996).

Physical protection is a critical element, acknowledging that safety (focused on unintended accidents or intrinsic hazards) must be closely linked with security (focused on intentional misuse or malicious acts by non-state elements). Effective nuclear security demands a continuous system of prevention, detection, and response to theft or sabotage. This requires sophisticated physical protection systems—integrating security hardware, procedural controls (e.g., guards), and facility design (layout)—all managed via a graded approach with increasing levels of defense-in-depth commensurate with the potential radiological consequences of failure. The IAEA actively assists its Member States in upgrading the physical protection of nuclear materials and associated facilities.

4.2.3 Engineering of Transport Casks (Type B Packages)

High-hazard materials like SNF must be transported in Type B packages, often referred to as transport casks. Figure 4.2 presents the Type B Transport Cask (Robatel Technologies, 2024). These casks are required to meet the stringent structural and containment standards of international transport regulations, such as the B(U) category. Many transport casks, such as the TN 1300, are designed to possess the necessary characteristics for dual use, serving reliably as interim storage casks as well as transport containers. This interoperability significantly reduces risks and costs by minimizing the number of times highly radioactive waste must be handled and transferred between systems.

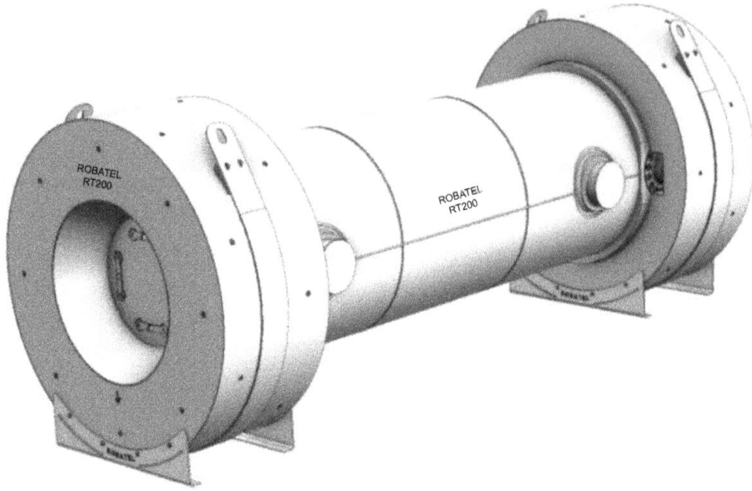

FIGURE 4.2 Type B Transport Cask (Robatel Technologies, 2024).

4.2.4 Containment Integrity under Accident Conditions

The engineering design of Type B casks is based on demonstrating containment under hypothetical accident conditions (HAC) of transport (IAEA, 1996; Robatel Technologies, 2024; IAEA, 2018). Regulatory compliance demands extreme conservatism in modeling. For instance, accident assessments assume a severe state of the fuel, postulating 100% rod failure and a release of 30% of the radioactive fission gases (primarily krypton-85, tritium, and iodine-129) into the cask cavity (IAEA, 1996; IAEA, 2018). The cask containment boundary must then prove leak-tight under this pressure.

The ultimate criterion for containment mandated by regulators, such as the US NRC (10 CFR 71.51), requires that the cask maintains a release rate of less than 1 A2/week following the HAC (IAEA, 1996; Robatel Technologies, 2024). To achieve this, the transportation cask must maintain a physical leak-tight boundary, defined by stringent standards like ANSI N14.5-1997. This standard sets the maximum allowable air standard leakage rate for the containment system at 1×10^{-7} ref cm^3/s (or 2×10^{-7} cm^3/s if tested with helium gas) (IAEA, 1996, 2018). This level of extreme engineering conservatism, which designs for failure modes far exceeding real-world probabilities, is the foundation of the exemplary safety record maintained over six decades of civil nuclear operation.

4.3 LONG-TERM STORAGE AND FINAL DISPOSAL

The final challenge in the nuclear fuel cycle is the long-term management of high-level, long-lived radioactive waste, requiring strategies that ensure containment over periods far exceeding the span of human civilization.

4.3.1 Interim Storage Solutions

Globally, about 400,000 tonnes of used fuel has been discharged from nuclear reactors worldwide (World Nuclear Association, 2025; Roma et al., 2024). The need for interim

TABLE 4.2 Key Regulatory and Safety Metrics for High-Level Waste Transportation

Metric/Criteria	Standard/Value	Regulatory Context	Significance
Containment Boundary Leak Rate	1×10^{-7} ref cm^3/sec	ANSI N14.5–1997, US NRC (10 CFR 71.51)	Defines the mandatory "leak-tight" condition for transport casks, ensuring integrity under all transport scenarios (IAEA, 1996, 2018).
Maximum Allowed Release Rate (Accident)	Less than 1 A2/week	US NRC (10 CFR 71.51)	Quantitative upper limit on the release of radioactivity following a hypothetical accident scenario.
Accident Gas Release Assumption	100% rod failure, 30% fission gas release	HAC modeling	Highly conservative assumption used to calculate pressure and potential release, validating the cask's robust design.
Regulatory Authority	IAEA, NRC	International and National Regulators	Provides the foundation for the highly secure and safe management of nuclear material logistics (IAEA, 1996, 2018; Maheras & Adkins, 2020).

storage, whether at reactor sites or centralized facilities (such as Sweden's CLAB, which was sited and built before centralized interim storage became controversial), is critical for providing a safe means of managing fuel while awaiting a permanent disposal solution. The US inventory alone totaled 86,000 metric tonnes of spent nuclear fuel stored at 119 sites as of 2019 and is expected to roughly double as the existing fleet of reactors completes its operational life (Roma et al., 2024; Messina, 2023).

The failure of governments to provide a designated disposal facility, such as the protracted stymieing of the Yucca Mountain project in the United States due to political opposition, has resulted in massive financial burdens, forcing tens of billions of dollars to be spent on research and exploration. This situation emphasizes that waste management is not just a technical problem, but an institutional and financial one (Crichlow, 2023).

4.3.2 Deep Geological Repositories (DGRs)

DGRs represent the definitive solution for highly toxic waste that cannot be further recycled, mandating isolation to prevent contamination of the air, ground, and underground water. The fundamental principle of a DGR is the creation of a facility that provides a high level of long-term isolation and containment without requiring ongoing human maintenance (Harithra et al., 2024). The chosen depth, typically several hundred meters (e.g., 500 m for Finland and France), ensures the waste is situated within a non-dynamic geological environment, shielded from potential surface perturbations, including future climate cycles, for timescales of 100,000 years or more (Näslund et al., 2013).

The safety case for DGRs relies on multiple layers of defense, the multi-barrier system (MBS), comprising both engineered barrier systems (EBS) and natural barrier systems (NBS).

Visualize a sector model showing the layers of protection designed to isolate the waste for millennia (Kohno & Kaneuji, 2024; Claret et al., 2018):

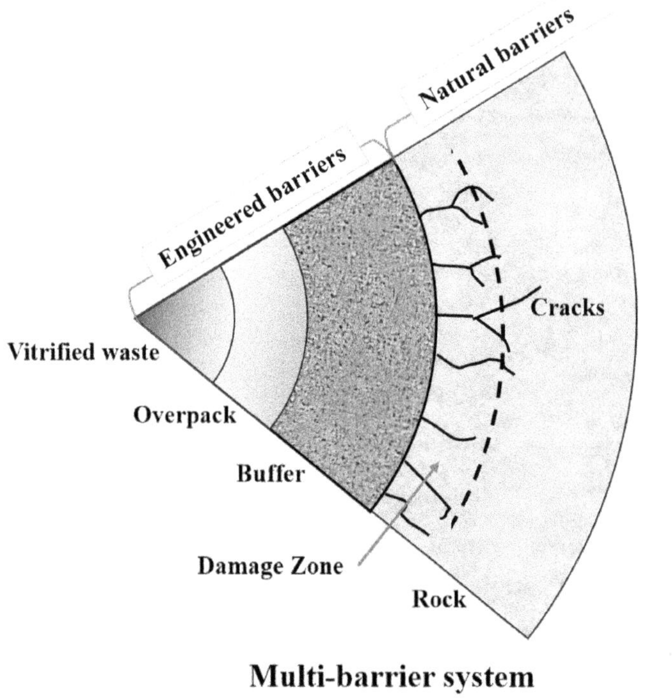

FIGURE 4.3 Multi-barrier system for geological disposal (concepted by Nuclear Waste Management Organization of Japan (NUMO)) (Kohno & Kaneuji, 2024).

1. Layer 1 (vitrified waste form): The primary internal barrier, either the solid vitrified glass (for HLW) or the spent fuel pellets themselves.

2. Layer 2 (overpack—disposal canister): The robust metal shell (e.g., copper or steel) that hermetically seals the waste. It is engineered to resist corrosion and mechanical stress for thousands of years.

3. Layer 3 (buffer material): A surrounding layer of highly dense, swelling clay (typically bentonite) that restricts the flow of groundwater and acts as a filter for any escaping radionuclides.

4. Layer 4 (host rock—natural barrier system): The surrounding deep, stable geological formation (e.g., granite, clay, or salt) characterized by extremely low permeability and chemical stability, providing the ultimate long-term isolation from the biosphere.

4.3.3 Geological and Hydrogeological Suitability

Selecting a DGR site requires strict adherence to demanding geological criteria. The host rock must be resilient to external forces over immense periods, particularly the mechanical and hydrogeological effects of one or more future glaciations, during which thick ice sheets would rest upon the rock (NEA, 1999).

Hydrogeologically, the objective is to locate the waste in environments characterized by deep groundwater with extremely low hydraulic conductivity (IAEA, 2003). In these settings, the pore water chemistry is often complex but stable, resulting from millennia of interaction between the water and rock-forming minerals. The design must account for the impact of heat generated by the waste, which increases saturation levels and consequently changes chemical equilibrium within the repository's near-field environment. Furthermore, interactions between the rock, groundwater (which may be saline or brine), and construction materials like concrete must be analyzed, as concrete leachate with a high pH value can initiate reactions that may either positively (iron passivation) or negatively (zinc dissolution) affect the durability of the EBS components (IAEA, 2003).

4.3.4 International DGR Case Studies and Implementation

Specific national strategies reflect different approaches to the fuel cycle and governance:

1. Finland and Sweden (KBS-3 Method): These countries are acknowledged global frontrunners, having achieved political and societal acceptance for final disposal sites (Thunberg, 2000). Figure 4.4 shows the KBS-3 Canister Emplacement. They employ the KBS-3 method, which involves direct disposal of SNF inside copper canisters embedded in bentonite clay buffer material within crystalline bedrock at a depth of 500 m (Sellin & Leupin, 2013). Finland's Onkalo facility, near the Olkiluoto Nuclear Power Plant, is the most advanced DGR, with construction underway and expected to receive its first spent fuel in the middle of the 2020s (Häikiö, 2022).

2. France (Cigéo Project): France, whose policy centers on reprocessing, is developing the *Centre Industriel de Stockage Géologique* (Cigéo) for the disposal of highly radioactive, long-lived waste produced by the reprocessing cycle. The facility, planned for

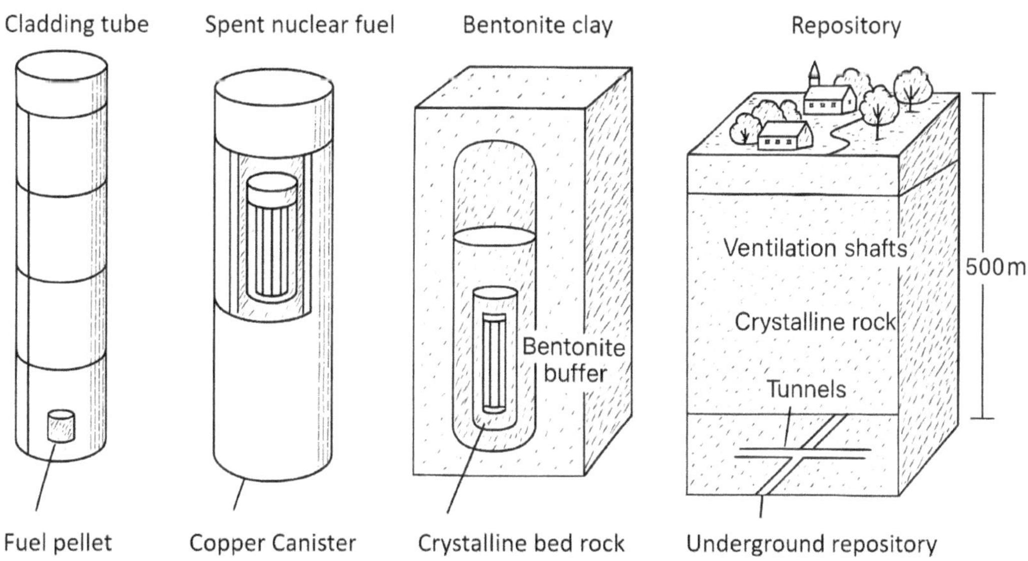

FIGURE 4.4 KBS-3 Canister Emplacement.

the Meuse/Haute-Marne region at a depth of 500 m, is designed for an operational life exceeding 100 years (Armand et al., 2015). A critical element of the French program, introduced following public debate, is an industrial pilot phase. This phase allows for the progressive testing of the facility's functions under real conditions, consolidating the demonstration of safety and reinforcing the features of reversibility and adaptability.

3. China (technically driven siting): China has adopted a technically driven national plan culminating in DGR construction around 2041–2050 for high-level radioactive waste and spent fuel not expected to be reprocessed. The current focus is on the Beishan area in Gansu Province, which has been identified as the preferred preselected area. An underground research laboratory (URL) (Figure 4.5) is currently under construction in Xinchang in the Beishan area at a depth of 560 m to verify geological suitability (Chen et al., 2023).
4. Japan (consent-based and phased siting): Japan's Nuclear Waste Management Organization (NUMO) is implementing a multi-stage, consent-based process for selecting a DGR site for high-level and TRU wastes (NUMO, 2023).
5. United States (consent-based siting): The United States has faced political difficulties in advancing its DGR plans, primarily the Yucca Mountain repository, which has been stalled by fierce state opposition despite being technically designated (Ross, 2000). In response, the US Department of Energy has pivoted to a consent-based

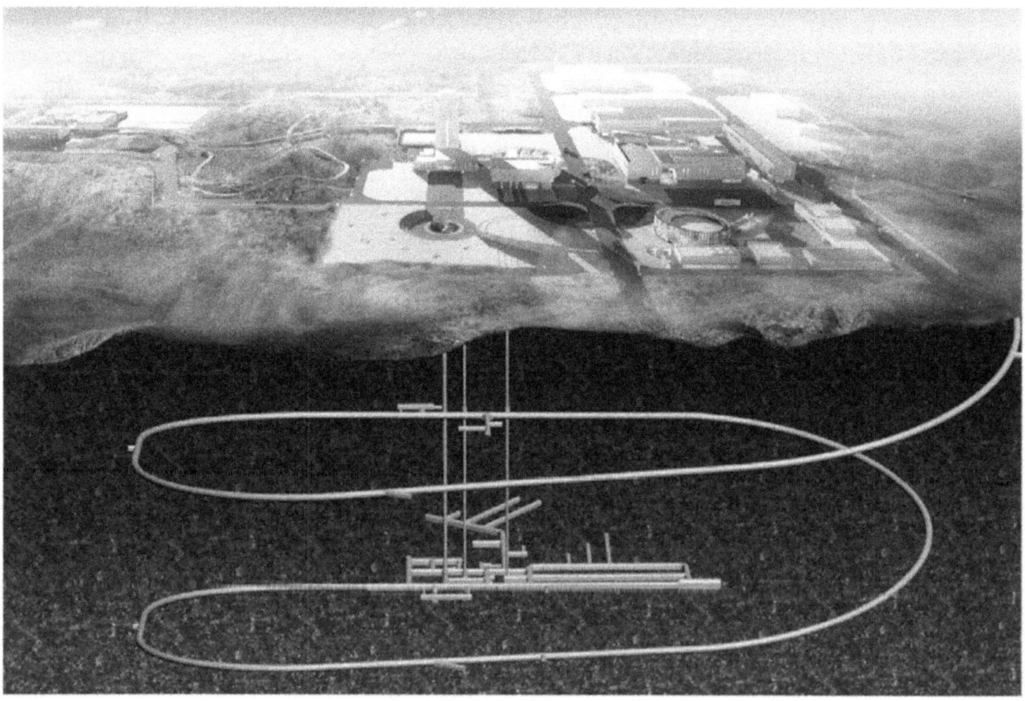

FIGURE 4.5 A 3D Perspective View of the Beishan URL.

TABLE 4.3 Comparative Status of DGR Programs

Country	Project/Model	DGR Focus	Host Rock/Depth	Key Policy/Siting Driver
Finland	Onkalo (KBS-3)	Direct Disposal (SNF)	Crystalline Bedrock/500 m	Direct final disposal; global frontrunner due to early societal consensus.
Sweden	KBS-3 Repository	Direct Disposal (SNF)	Crystalline Bedrock	Direct final disposal; political commitment and use of the KBS-3 method.
France	Cigéo	Reprocessing Waste (HLW)	Clay/Shale/500 m	Disposal of long-lived waste from mandatory reprocessing; features an industrial pilot phase for adaptability and confidence.
United States	Consent-Based Siting	Direct Disposal (SNF)	TBD	Shift from top-down siting (Yucca Mountain) to a consent-based approach due to political opposition.
China	Beishan DGR	Reprocessing Waste (HLW) and SNF	Granite (Beishan)/560 m (URL)	Closed fuel cycle; technically driven siting (URL construction underway).
Japan	NUMO Program	Reprocessing Waste (HLW) and TRU	TBD (R&D at Mizunami/ Horonobe)	Consent-based and phased siting; local veto power on progression.

Note: To be determined (TBD); research and development (R&D).

approach for selecting both consolidated interim storage facilities and two permanent geologic disposal facilities. This strategic shift acknowledges that successful implementation requires societal consent and demonstrates that political viability now supersedes technical suitability alone (Guo, 2025; White-Horton, 2020).

4.3.5 Long-Term Safety Assessment and Modeling

DGR safety assessments must account for future risks over immense timescales, necessitating projections of up to 1 million years for specific repositories (such as in Sweden) (Liakka et al., 2024). Predicting the evolution of the isolation system over these periods requires highly complex modeling (Calder et al., 2025).

The unpredictable nature of the climate system on such vast scales requires safety assessments to utilize a broad range of possible future climate scenarios. These must address extreme conditions, including high-end global warming trends over the next 100,000 years, periods of deep permafrost, and the formation of extensive ice sheets during glacial cycles (Näslund et al., 2013). These factors can significantly influence hydrological regimes, ground pressures, and the long-term chemical environment of the repository.

Further, studies must assess the impact of climate change on specific hydrological extremes. Projections often indicate a future that is likely to be wetter, requiring assessments of impacts on probable maximum precipitation (PMP) and intensity–duration–frequency (IDF) curves, especially for the 2050s and 2080s. By relying on multi-model ensembles and presenting results in terms of percentiles, regulators are able to select an acceptable level of risk for the design of surface facilities and hydrological controls, balancing investment against consequential environmental risks (Chapman & McCombie, 2003; Nývlt & Dobrovolný, 2015).

4.3.6 Societal Governance and Ethical Dimensions

The implementation of DGRs is inextricably linked to ethical considerations, primarily concerning the management of risk and transfer of burden to future generations. This has led to the emergence of policies centered on reversibility and retrievability (R&R) of the waste (Chapman & McCombie, 2023).

Reversibility refers to the institutional capacity to halt, delay, or modify decisions and actions related to disposal, while retrievability is the technical capability to physically remove the waste packages after they have been emplaced (Kermisch & Depaus, 2024). R&R is often included in national programs, not just for technical adaptation, but for ethical reasons and, critically, for public confidence. Public perception often views the technology as safer if an ultimate measure—retrieval—is known to be possible in the event of an unforeseen failure, however unlikely (Lehtonen, 2010).

The emphasis on retrievability often functions as a necessary component for achieving societal acceptance, acting as a "political price to pay" for gaining consensus, particularly in the immediate communities surrounding the proposed site (Kermisch & Depaus, 2024). This creates a nuanced tension in DGR design: while the system must ultimately rely on passive, permanent isolation without future human intervention, the political requirement for retrievability during the early operational phases necessitates features (like the industrial pilot phase in France) that maintain access, thus delaying the final, irreversible sealing of the containment system. This institutionalized delay acknowledges the ethical mandate to review disposal strategy before committing to geological permanence (Ruiz-Lopez & Pescatore, 2015).

4.4 CONCLUSION

The comprehensive management of nuclear waste is characterized by an interconnected system of technological innovation, regulatory conservatism, and socio-political evolution. Waste conditioning techniques, ranging from established cementation for ILW to high-tech vitrification for HLW, are continually being advanced by concepts like P&T, which aims to reduce the long-term radiotoxicity burden on final repositories through electrometallurgical recycling of actinides. The success of P&T could fundamentally alter the financial and scientific constraints placed on DGR siting and safety modeling. The choice between a closed fuel cycle (reprocessing), based on the PUREX process and adopted by nations like China and France, and an open fuel cycle (direct disposal), adopted by countries like Finland and the United States, fundamentally dictates a nation's conditioning requirements.

Transportation logistics are defined by a globally harmonized, exceptionally rigorous regulatory framework, supported by engineering conservatism that designs Type B casks to maintain leak-tight containment even under hypothetical, worst-case accident scenarios. This dedication to safety integrity is crucial, particularly given the growing need for flexible, multi-purpose storage and transport infrastructure globally.

Ultimately, the long-term solution rests on DGRs. Implementation requires mastering immense scientific complexity—modeling climate change impacts (including glaciation and extreme precipitation) over 1 million years—while simultaneously navigating complex societal governance issues. Global DGR projects vary widely in their approach, ranging from the technically driven, centralized program in China, which is currently constructing an URL, to the consent-based, locally focused siting strategies employed by the United States and Japan, where local government consent is essential for project progression. The failures encountered by projects based on top-down decision-making (like Yucca Mountain) have led to a critical policy shift toward consent-based siting and the incorporation of R&R provisions. The leadership demonstrated by Finland and Sweden, with their advanced DGR construction, affirms that political and social consensus on long-term isolation is as crucial a component of successful nuclear waste management as the integrity of the geological and engineered barriers themselves.

REFERENCES

Armand, G., Conil, N., Daoud, K., 2015. Development and new research program in the Meuse Haute-Marne Underground Research Laboratory (France). In: *ISRM Congress*. International Society for Rock Mechanics, pp. 1–8.

Baron, P., Rostaing, O., Dupont, M., 2001. Separation of the minor actinides: The DIAMEX-SANEX concept. In: *GLOBAL 2001*. American Nuclear Society, pp. 1–10.

Bell, K., Carpentier, C., Carrott, M., Geist, A., Gregson, C., Hérès, X., Magnusson, D., Malmbeck, R., McLachlan, F., Modolo, G., Müllich, U., Sypula, M., Taylor, R., Wilden, A., 2012. Progress towards the development of a new GANEX process. *Procedia Chem.* 7, 392–397.

Bruno, J., Duro, L., Diaz-Maurin, F., 2020. Spent nuclear fuel and disposal. In: *Advances in Nuclear Fuel Chemistry*. Elsevier, pp. 527–553.

Calder, N., Jensen, M., Smith, P., 2025. Supporting radioactive waste DGR safety assessments with detailed groundwater models. *Geosci. Can.* 52, 112–125.

Cao, B., Liu, Y., Wang, Z., 2013. Preliminary study on nuclear fuel cycle scenarios of China before 2050. *Energy Procedia* 39, 294–299.

Chapman, N., McCombie, C., 2003. Safety and security issues in deep geological disposal. *Waste Manag. Ser.* 3, 201–215.

Chapman, N., McCombie, C., 2023. Multinational repositories (MNRs) highlight the need for transnational information management in geological disposal. *Front. Nucl. Eng.* 2, 1208524.

Chen, L., Zhao, X., Liu, J., et al., 2023. Progress on rock mechanics research of Beishan granite for geological disposal of high-level radioactive waste in China. *Rock Mech. Bull.* 2, 100046.

Claret, F., Gaboreau, S., Guérin, R., 2018. Modeling the long-term stability of multi-barrier systems for nuclear waste disposal in geological clay formations. In: *Clay Confinement Systems for Geological Disposal*. OECD/NEA, pp. 395–451.

Crichlow, H., 2023. Looking very deep for disposal. *Nucl. Eng. Int.* 68, 40–43.

De Saillan, C., 2010. Disposal of spent nuclear fuel in the United States and Europe: A persistent environmental problem. *Harv. Envtl. L. Rev.* 34, 461–498.

Fredrickson, G.L., Yoo, T.-S., 2023. *Engineering Scale Pyroprocessing Activities in the United States*. Idaho National Laboratory Report INL/EXT-23-XXXX.

Gonzalez-Romero, E.-M., 2011. Impact of partitioning and transmutation on the high level waste management. *Nucl. Eng. Des.* 241, 3436–3444.

Guo, X., 2025. *Perspective Chapter: Safe Disposal and Storage of Nuclear Waste*. Colorado School of Mines Technical Note.

Häikiö, J., 2022. *Regulatory Oversight of Nuclear Safety in Finland: Annual Report 2021*. Radiation and Nuclear Safety Authority (STUK) Report.

Hanson, B., 2015. Process engineering and design for spent nuclear fuel reprocessing and recycling plants. In: *Reprocessing and Recycling of Spent Nuclear Fuel*. Elsevier, pp. 125–151.

Harithra, V., Sivakumar, P., Rajkumar, N., 2024. Long-term storage and safety of radioactive waste storage facilities. In: *Radioactive Pollutant: Sources, Issues and Remediation*. Springer, pp. 219–238.

Haynes, A.M., Chen, D., 2023. Pyroprocessing as a solution for green, sovereign energy in Southeast Asia. *J. Nucl. Mater. Manag.* 50, 5–15.

IAEA, 1996. *Regulations for the Safe Transport of Radioactive Material*, 1996 Edition. Safety Standards Series No. ST-1. International Atomic Energy Agency, Vienna.

IAEA, 2003. *Scientific and Technical Basis for Geological Disposal of Radioactive Wastes*. IAEA Technical Reports Series No. 413.

IAEA, 2018. *Regulations for the Safe Transport of Radioactive Material*, 2018 Edition. Safety Standards Series No. SSR-6 (Rev. 1). International Atomic Energy Agency, Vienna.

IAEA, 2022. *Status and Trends in Spent Fuel and Radioactive Waste Management*. International Atomic Energy Agency Report.

Inoue, T., Koch, L., 2008. Development of pyroprocessing and its future direction. *Nucl. Eng. Technol.* 40, 183–190.

Kermisch, C., Depaus, C., 2024. An ethical analysis of the impact of reversibility and retrievability provisions on well-being. *Prog. Nucl. Energy* 177, 105401.

Kohno, M., Kaneuji, Y., 2024. Experimental study on swelling and intrusive behavior of bentonite under various solutions and its simple model construction. *Acta Geotech.* 19, 5833–5853.

Kooyman, T.J.A., 2021. Current state of partitioning and transmutation studies for advanced nuclear fuel cycles. *Ann. Nucl. Energy* 157, 108239.

Kulikova, S.A., Kalmykov, S.N., Livshits, T.S., 2021. Perspective compounds for immobilization of spent electrolyte from pyrochemical processing of spent nuclear fuel. *Appl. Sci.* 11, 11180.

Kumari, I., Kumar, B.V.R., Khanna, A., 2020. A review on UREX processes for nuclear spent fuel reprocessing. *Nucl. Eng. Des.* 358, 110410.

Lanham, W., Runion, T., 1949. *PUREX Process for Plutonium and Uranium Recovery*. Oak Ridge National Laboratory Report ORNL-TM.

Lawless, W., Scherer, C.W., Flynn, J., 2014. Public consent for the geologic disposal of highly radioactive wastes and spent nuclear fuel. *Int. J. Environ. Stud.* 71, 41–62.

Lehtonen, M., 2010. Opening up or closing down radioactive waste management policy? Debates on reversibility and retrievability in Finland, France, and the United Kingdom. *Risk Hazards Crisis Public Policy* 1, 139–179.

Liakka, J., Lord, N.S., Kennedy-Asser, A., Lunt, D.J., Williams, C.J.R., Näslund, J.-O., 2024. Assessing future ice-sheet variability for long-term safety of deep geological repositories. *Adv. Geosci.* 65, 71–81.

Lin, R., Zhong, Z., 2025. Progress and development thinking of dry reprocessing technology for spent nuclear fuel from fast reactor. *At. Energy Sci. Technol.* 59, 41–53.

Maheras, S.J., Adkins, H.E., 2020. *Complete Documentation of Applicable Regulations and Regulatory Authority for Transportation*. Pacific Northwest National Laboratory, Richland, WA. PNNL-29718.

McCloy, J.S., Riley, B.J., Dixon Wilkins, M.C., et al., 2024. *International perspectives on glass waste form development for low-level and intermediate-level radioactive waste*. Mater. Today 80, 594–618.

McConnell, P., Hanson, B.D., Lee, M., Sorenson, K.B., 2011. Extended dry storage of used nuclear fuel: Technical issues: A USA perspective. *Nucl. Eng. Technol.* 43(5), 405–412.

Messina, B.C., 2023. *Environmental Injustice and Catastrophe: How Global Insecurities Threaten the Future of Humanity*. Walter de Gruyter, Berlin.

Näslund, J.-O., Brandefelt, J., Liljedahl, L.C., 2013. Climate considerations in long-term safety assessments for nuclear waste repositories. *Ambio* 42, 393–401.

NEA, 1999. *Radioactive Waste Management Geological Disposal of Radioactive Waste: Review of Developments in the Last Decade*. Nuclear Energy Agency Report.

NUMO, 2023. *The Nuclear Waste Management Organization of Japan's Pre-siting Safety Case Based on the Site Descriptive Model.* Nuclear Energy Agency Report NEA/NUMO-2023.
Nývlt, D., Dobrovolný, P., 2015. *Climate Stability of DGR Potential Sites.* Radioactive Waste Repository Administration Report, Prague.
Paviet-Hartmann, P., Benedict, B., Lineberry, M.J., 2009. Nuclear fuel reprocessing. In: *Nuclear Engineering Handbook.* CRC Press, pp. 333–384.
Robatel Technologies, 2024. *RT-200 Type B Cask Safety Analysis Report.* Robatel Internal Report.
Roma, A.C., Fishman, S., Hughes, C.T., 2024. *Nuclear Spent Fuel: Technology Is There for Disposal, But US Law Prevents Waste Options from Moving Forward.* Nuclear Waste Council Report.
Ross, D.P., 2000. Yucca mountain and reversing the irreversible: The need for monitored retrievable storage in a permanent repository. *Vt. L. Rev.* 25, 815–850.
Ruiz-Lopez, C., Pescatore, C., 2015. *Reversibility of Decisions and Retrievability of Radioactive Waste: An Overview of Regulatory Positions and Issues.* OECD/NEA Report.
Sassani, D.C., Freeze, G.A., Lee, K.H., 2010. *Shale Disposal of US High-Level Radioactive Waste.* Sandia National Laboratories Report SAND2010-XXXX.
Sellin, P., Leupin, O.X., 2013. The use of clay as an engineered barrier in radioactive-waste management: A review. *Clays Clay Miner.* 61(6), 477–498.
Simpson, M.F., 2018. Fundamentals of spent nuclear fuel pyroprocessing. *Nucl. Fuel Reprocess. Waste Manag.* 2, 27–54.
Thunberg, A.-M., 2000. Retrievability in an ethical perspective. In: *Retrievability of High Level Waste and Spent Nuclear Fuel. Proceedings of an International Seminar.* OECD/NEA, pp. 112–120.
US DOE, 2014. *Assessment of Disposal Options for DOE-Managed High-Level Radioactive Waste and Spent Nuclear Fuel.* U.S. Department of Energy Report.
US NRC, 1999. *Spent Fuel Storage and Transport Cask Decontamination and Modification.* NUREG/CR-XXXX.
White-Horton, J., 2020. *Safeguards for Deep Geological Repositories.* Oak Ridge National Laboratory Report ORNL/TM-2020/XXXX.
World Nuclear Association, 2025. *Radioactive Waste—Myths and Realities.* WNA Information Paper.
Yamana, H., 2005. Pyrochemical reprocessing developments in Japan: Overview and some topics. *Int. J. Nucl. Energy Sci. Technol.* 1, 139–147.

CHAPTER 5

Nuclear Waste Disposal and Migration in Geochemical Systems

Insights from Recent Advances in Mineral–Microbe Interactions

Qingyin Xia

5.1 INTRODUCTION

High-level radioactive waste (HLW) contains a wide spectrum of long-lived radionuclides, among which uranium (U) is of particular concern due to its radiotoxicity and chemical mobility under subsurface conditions. Geological disposal strategies, as endorsed by the International Atomic Energy Agency (IAEA) and many national waste management programs, rely on the concept of multiple barriers (Birkholzer et al., 2012). In practice, this approach combines engineered barriers, such as corrosion-resistant canisters and compacted bentonite buffers, with natural geological formations, including argillaceous rocks, crystalline host rocks, and salt deposits (Birkholzer et al., 2012). These barriers are designed to minimize radionuclide migration over geological timescales.

Clay-based materials, particularly bentonite, play a central role in this strategy because of their low permeability, high cation exchange capacity (CEC), and chemical stability (Sellin & Leupin, 2013). Yet, the long-term retention of uranium cannot be guaranteed solely by physical confinement. Instead, uranium's behavior is governed by complex geochemical and microbial interactions. Two processes are especially critical. One is sorption and surface complexation on mineral phases (Dong et al., 2012). The other is redox transformation, which can occur through structural Fe(II/III) in clays or through microbial metabolism (Chen et al., 2022; Lovley et al., 1991; Wang et al., 2024; Zhang et al., 2021; Zhang et al., 2022). Under repository conditions, uranium redox chemistry is dynamic. In oxidizing settings, it occurs mainly as the soluble uranyl ion, U(VI). Under reducing conditions, it is often present as

DOI: 10.1201/9781003665793-7

poorly soluble U(IV) minerals, such as uraninite. Several environmental factors control this speciation and mobility, including pH, redox potential, and the availability of electron donors and acceptors (Maher et al., 2013; Newsome et al., 2015; O'Loughlin et al., 2011).

Clay minerals such as montmorillonite, illite, nontronite, and kaolinite contribute significantly to uranium immobilization. Their surface reactivity and structural Fe content can influence U(VI) reduction and U(IV) phase stabilization (Chen et al., 2022; R. Li et al., 2024; Wang et al., 2024; Xia et al., 2022; Zhang et al., 2011; Zhang et al., 2009; Zhang et al., 2021; Zhang et al., 2022). Microbial activity adds further complexity. Dissimilatory metal-reducing bacteria, including *Geobacter*, *Shewanella*, and *Desulfovibrio*, can enzymatically reduce U(VI) to U(IV) (Boyanov et al., 2011; Lovley et al., 1991; Mohapatra et al., 2010; N'Guessan et al., 2008). Sulfate-reducing bacteria (SRB) generate sulfide, which in turn influences uranium speciation (Boyanov et al., 2011). Microbes also change local geochemistry. They can shift pH through respiration, affect ion exchange and clay swelling, and release extracellular polymeric substances (EPSs) that coat mineral surfaces (Tourney & Ngwenya, 2014). These changes modify sorption and surface complexation pathways. Such mineralogical, chemical, and microbial processes are tightly coupled. Their interactions are dynamic under repository conditions and must be assessed as a whole rather than in isolation.

Although repository designs aim to maintain anaerobic and reducing environments that stabilize uranium as immobile U(IV), several processes threaten this stability (Moon et al., 2007; Sharp et al., 2011; Shi et al., 2025; Stewart et al., 2013). Groundwater recharge (Moon et al., 2007) or canister corrosion (Stewart et al., 2013) may introduce oxidants that re-oxidize U(IV) to U(VI). Colloid-facilitated transport can also carry nano-particulate U(IV) over considerable distances (Shi et al., 2025). Uranium geochemistry in repositories should therefore be viewed as a dynamic equilibrium shaped by both engineered barriers and natural processes. Migration can occur through several pathways. In permeable zones, uranium may move by advection with groundwater (Zachara et al., 2016). In low-permeability clays, diffusion driven by concentration gradients dominates (Neuzil, 2019). Colloids such as clays, Fe oxides, and organic matter can transport associated uranium (Shi et al., 2025). In addition, redox cycling may convert reduced U(IV) back into mobile U(VI). A clear understanding of these coupled mechanisms is essential for evaluating long-term uranium retention and for ensuring the safety of geological disposal strategies.

5.2 INTERACTION OF URANIUM WITH CLAY MINERALS

Clay minerals play a central role in both natural geological formations and engineered barriers designed for nuclear waste disposal (Sellin & Leupin, 2013). Their distinctive physicochemical properties, including large specific surface area, high cation exchange capacity, and chemical reactivity, make them highly effective in retarding radionuclide migration (Sellin & Leupin, 2013). In the case of uranium, clays can provide multiple immobilization pathways such as surface adsorption and ion exchange, inner- and outer-sphere complexation, precipitation of secondary uranium phases, and reductive transformation of U(VI) to U(IV) (Chen et al., 2022; Wang et al., 2024; Zhang et al., 2021; Zhang et al., 2022). The incorporation of bentonite, dominated by montmorillonite, into engineered barrier systems is motivated by these retention capabilities, although effectiveness depends strongly on uranium speciation, pH, ionic strength, and the mineralogical and redox states of the clay matrix.

Different classes of clay minerals vary in structure and reactivity. Smectites such as montmorillonite (Bishop et al., 2011) and nontronite (Bishop et al., 2011) are expandable 2:1 layer silicates with high cation exchange capacity, while illites (Zhang et al., 2012) are non-expandable 2:1 clays with lower exchange capacity. Kaolinites (Biswas et al., 2023), with their 1:1 structure, exhibit relatively low surface reactivity, and chlorites (Bishop et al., 2011) often contain abundant Fe(II)/Mg. Among these groups, nontronite and montmorillonite have received the most attention in uranium studies because their redox-active structural iron and high sorption capacity make them especially effective for immobilization.

U(VI) interacts with clay surfaces primarily through surface complexation. At the edges of clay layers, hydroxyl groups such as ≡Al–OH, ≡Si–OH, and ≡Fe–OH provide binding sites, while interlayer regions support ion exchange (Li et al., 2023; K. J. Li et al., 2024). Two sorption modes are typically observed. Outer-sphere complexes are electrostatic, weak, and reversible (Nekipelov & Zamaraev, 1985). Inner-sphere complexes involve direct coordination with surface oxygens and are therefore stronger and more stable (Warren et al., 2012). Sorption behavior is highly pH-dependent, with maximum uptake typically occurring between pH 5 and 8 (Crawford et al., 2017; Katsoyiannis, 2007). At higher pH, carbonate complexation dominates and suppresses surface binding (Katsoyiannis, 2007). Isotherm behavior is often described by Langmuir or Freundlich models. However, competing ions can reduce sorption capacity. Cations such as Na^+, Ca^{2+}, and Mg^{2+} occupy exchange sites (Stanley & Wilkin, 2019), while anions such as sulfate, silicate, or phosphate stabilize uranium in solution (Bachmaf et al., 2008).

Beyond surface complexation, structural Fe(II) within the octahedral sheets of smectites like nontronite can serve as an abiotic reductant of U(VI), yielding nano-crystalline uraninite (UO_2) (Zhang et al., 2011; Zhang et al., 2009). The efficiency of this process is controlled by factors such as the Fe(II)/Fe(III) ratio, clay crystallinity, solution pH, the presence of electron shuttles, and the availability of labile U(VI) species (Gorski et al., 2012a; Gorski et al., 2012b; Gorski et al., 2013; Xie et al., 2018). This reduction pathway represents a powerful mechanism for long-term uranium immobilization, provided that the U(IV) solids remain stable under repository redox conditions.

Over the geological timescales relevant to repositories (10^3–10^5 years), clay minerals themselves are subject to transformation. Processes such as illitization reduce swelling capacity and cation exchange potential, while Fe redox cycling modifies the Fe(II)–Fe(III) balance, and dissolution–reprecipitation reactions alter sorption site availability (Kim et al., 2004; Kim et al., 2019). These changes, driven by thermal and geochemical conditions, affect both sorption and reduction capacities of clays.

Altogether, clay minerals provide robust sorption and reduction pathways that significantly enhance uranium retention under repository-relevant conditions. Their performance, however, must be evaluated in the context of long-term mineralogical evolution and coupled biogeochemical processes, which ultimately determine the persistence of uranium immobilization.

5.3 MICROBIAL INFLUENCE ON URANIUM SPECIATION AND MOBILITY

Microorganisms exert a profound influence on uranium cycling in subsurface environments, particularly within the context of geological repositories for nuclear waste. Their activities shape uranium speciation and mobility through a wide spectrum of direct and indirect processes (Belli & Taillefert, 2018; Boyanov et al., 2011; Lovley et al., 1991; Mohapatra et al., 2010; Suzuki & Banfield, 1999; Ulrich et al., 2011; Wilkins et al., 2007). Central among these is the enzymatic reduction of soluble U(VI) to insoluble U(IV), a reaction that dissimilatory metal-reducing bacteria (DMRB) such as *Geobacter*, *Shewanella*, *Desulfovibrio*, and *Anaeromyxobacter* carry out during anaerobic respiration (Alessi et al., 2014; Bernier-Latmani et al., 2010; Boyanov et al., 2011; Newsome et al., 2014). In this pathway, electrons derived from donors such as acetate, lactate, or hydrogen are transferred via outer-membrane cytochromes, conductive pili, or redox shuttles to uranium, producing nano-crystalline UO_2 that is sparingly soluble and geochemically stable. This process, first described by Lovley and colleagues in 1991, forms the basis for bioreduction strategies in uranium-contaminated systems (Lovley et al., 1991).

Even microbes not directly engaged in uranium reduction can alter its speciation through indirect mechanisms. Sulfate-reducing bacteria generate sulfide, precipitating uranium as sulfide phases or promoting FeS formation, which facilitates U(VI) reduction (Lee et al., 2014). Other microorganisms reduce Fe(III) to Fe(II), which subsequently drives abiotic uranium reduction, thereby further enhancing overall immobilization (Zhang et al., 2011; Zhang et al., 2009). Redox-active metabolites such as flavins and quinones further extend microbial influence beyond the organisms that act as direct reducers (Wolf et al., 2009).

Microbial interactions with uranium also occur through biosorption and bioaccumulation. Cell wall functional groups, S-layer proteins (Fagan & Fairweather, 2014), and EPS (Tourney & Ngwenya, 2014) provide abundant binding sites. Carboxyl, phosphate, and hydroxyl groups are especially important for complexing U(VI) (Guo et al., 2009). This passive process is usually rapid but strongly affected by pH, ionic strength, competing ions, and biomass density (Guo et al., 2009). In some cases, microbes accumulate uranium inside the cell, which is less common and often reflects a response to toxicity (Suzuki & Banfield, 2004). A related pathway is biomineralization, in which microbial activity induces the precipitation of secondary uranium phases. Depending on local geochemical conditions, the products may include uraninite, autunite, chervetite, or amorphous UO_2 (Zhang et al., 2020). Microbial release of phosphate often helps to stabilize these phases (Zhang et al., 2020). Although biogenic minerals are typically nano- or poorly crystalline, they can persist for long periods under reducing conditions.

Biofilms represent another important microbial strategy that modifies uranium's behavior. Communities embedded within EPS not only concentrate sorption sites but also alter local redox and pH conditions, creating confined microenvironments conducive to uranium immobilization (Tourney & Ngwenya, 2014; Zhao et al., 2022). EPS itself can bind uranium and occasionally act as a redox mediator (Tourney & Ngwenya, 2014; Xu et al., 2024). Studies have shown that biofilm-coated sediments can significantly retard uranium mobility, even under flowing groundwater conditions (Xu et al., 2024). Such microenvironments are especially relevant in clay-rich repositories, where microbial colonization of pore spaces may persist over long timescales (Bradford et al., 2013).

The role of microbial communities must also be understood dynamically, as their composition and activity evolve over thousands of years (Konopka, 2009). Repository-relevant conditions of limited substrates, redox gradients, temperature and pressure regimes, and the restricted porosity of bentonite clays shape microbial ecology (Stone et al., 2016). Metagenomic and transcriptomic surveys of uranium-contaminated aquifers and clay systems have identified key groups, including *Geobacter*, *Anaeromyxobacter*, and *Pseudomonas*, as central players in uranium cycling, capable of nitrate and sulfate reduction, methanogenesis, and metal reduction (Yan et al., 2016).

Beyond fundamental understanding, microbial processes are increasingly considered in engineered applications. Field-scale bioremediation has demonstrated that stimulating native microbial communities with acetate or lactate can accelerate U(VI) reduction (Wu et al., 2006; Wu et al., 2007). Emerging strategies propose bioaugmentation with selected U-reducing strains, construction of synthetic consortia, or the development of biofilms on clay and iron oxide surfaces as targeted uranium immobilization technologies.

Taken together, microbial activity profoundly affects uranium redox speciation, sorption, and long-term stability in geological repositories. Direct enzymatic reduction (Alessi et al., 2014; Bernier-Latmani et al., 2010; Boyanov et al., 2011; Newsome et al., 2014) provides a robust immobilization pathway, while biosorption (Guo et al., 2009), biomineralization (Zhang et al., 2020), and biofilm formation (Tourney & Ngwenya, 2014; Xu et al., 2024) offer additional retention mechanisms. Indirect influences mediated through Fe, S, and organic cycling further underscore the necessity of treating microbial interactions as integral components of repository geochemistry (Lee et al., 2014). To ensure repository performance, these biological processes must be incorporated into reactive transport models and long-term safety assessments, recognizing microbes as active agents that both stabilize and, under certain conditions, remobilize uranium.

5.4 COUPLED CLAY–MICROBE SYSTEMS: INTEGRATED BIOGEOCHEMISTRY OF URANIUM MOBILITY

In natural and engineered repository environments, uranium's behavior is rarely controlled by a single mechanism but instead emerges from the coupled interactions between clay minerals and microbial communities (Zhang et al., 2011; Zhang et al., 2009). These interactions generate non-linear geochemical feedback loops that strongly influence uranium's speciation, mobility, and long-term fate (Zhang et al., 2011; Zhang et al., 2009). Clay minerals provide reactive surfaces, redox-active sites, and significant diffusion control, while microorganisms mediate redox reactions, alter mineral surfaces, and modify local geochemical conditions (Cuadros, 2017). Together, these processes create complex multiphase systems that evolve dynamically over repository timescales of 10^3 to 10^5 years.

The interplay between clays and microbes can be either synergistic or antagonistic, depending on environmental conditions (Cuadros, 2017). For example, reduction of microbial Fe(III) produces Fe(II), which in turn enhances abiotic reduction of U(VI) on clay surfaces, thereby strengthening immobilization (Chen et al., 2022; Wang et al., 2024; Zhang et al., 2011; Zhang et al., 2009; Zhang et al., 2021; Zhang et al., 2022). Similarly, the production of EPS promotes uranium retention through biosorption and biofilm formation

(Tourney & Ngwenya, 2014; Xu et al., 2024; Zhao et al., 2022). These pathways stabilize uranium by driving U(VI) reduction to U(IV) or enhancing surface sorption. However, destabilizing processes may also arise. Microbial metabolism can shift pH and undermine clay buffering capacity, leading to desorption and increased uranium mobility (Cui et al., 2023). The balance between these stabilizing and destabilizing influences is sensitive to changes in geochemical gradients, microbial succession, and repository evolution over time.

Conceptual models of these coupled processes demonstrate how microbial respiration and clay redox buffering interact to control the fate of uranium (Claret et al., 2022; Leupin et al., 2017; Sellin & Leupin, 2013). Laboratory studies have shown that microbes such as *Shewanella* and *Geobacter*, in the presence of Fe-bearing clays like nontronite, can accelerate U(VI) reduction while simultaneously influencing the distribution of Fe(II) and Fe(III) (Chen et al., 2022; Wang et al., 2024; Zhang et al., 2011; Zhang et al., 2009; Zhang et al., 2021; Zhang et al., 2022). These interactions underscore the importance of treating uranium mobility as a system-level property, not as the outcome of isolated reactions. Over repository timescales, the dominant processes change (Claret et al., 2022; Leupin et al., 2017; Sellin & Leupin, 2013). In the early years, sorption and redox buffering play the main role. Later, microbial colonization and Fe cycling become more important. At longer times, mineral transformations control the system, leading either to stabilization or to remobilization of biogenic uranium phases.

Feedback loops arising from these interactions may reinforce stability or promote remobilization. On one hand, microbial Fe(III) reduction supplies Fe(II), driving U(VI) reduction and subsequent precipitation of U(IV) phases that are sparingly soluble under anoxic conditions (Chen et al., 2022; Wang et al., 2024; Zhang et al., 2011; Zhang et al., 2009; Zhang et al., 2021; Zhang et al., 2022). Biofilm development further enhances retention by increasing sorption capacity and creating protective microenvironments (Tourney & Ngwenya, 2014; Xu et al., 2024; Zhao et al., 2022). On the other hand, geochemical perturbations such as pH changes or mineralogical alterations may destabilize previously immobilized uranium, especially over extended time periods (Sharp et al., 2011; Zhong et al., 2005). Sequential Fe redox cycling, for instance, can repeatedly reset uranium mobility states, resulting in alternating phases of stabilization and release (Noël et al., 2019; Stewart et al., 2009).

The recognition of these coupled clay–microbe processes has significant implications for repository science. Long-term performance assessments must account for dynamic biogeochemical trajectories that span decades to millennia rather than assuming short-term equilibrium. Reactive transport models need to integrate microbial kinetics, mineral evolution, and redox cycling to capture the feedback-driven nature of uranium mobility. Experimental studies in underground laboratories have highlighted the potential for biofilm formation, Fe cycling, and clay mineral transformations to significantly alter radionuclide behavior over long durations (Brookshaw et al., 2012; Brown et al., 1994).

Despite advances, major knowledge gaps remain. The stability and crystallinity of biogenic U(IV) phases under repository-relevant conditions are uncertain, as is the extent to which microbial succession influences long-term mineralogical evolution. Kinetic parameters describing multi-component sorption and desorption under dynamic conditions remain poorly constrained, limiting the predictive capacity of current models. These uncertainties highlight the need for integrated experimental and modelling approaches that explicitly couple microbial, mineralogical, and geochemical processes.

Overall, uranium mobility in geological repositories is governed by non-linear, coupled interactions between clay minerals and microorganisms. These systems create feedback loops that may either immobilize or remobilize uranium, depending on evolving chemical and biological conditions. To ensure the safety of geological disposal, repository design must not only rely on the inherent retention capacity of clays but also anticipate and manage the dynamic microbial influences that shape uranium's long-term fate.

5.5 ENVIRONMENTAL AND EXPERIMENTAL CASE STUDIES OF URANIUM MIGRATION

Field observations and long-term experiments provide essential validation for mechanistic insights into uranium geochemistry. While laboratory studies offer precise control over variables, in situ experiments and natural analogue sites capture the real-world complexity of repository environments. These systems highlight the combined effects of groundwater chemistry, microbial activity, mineralogy, and scale-dependent transport phenomena, operating over distances from centimeters to hundreds of meters and timescales extending to millennia. Together, they provide critical lessons for predicting uranium mobility in repository-relevant settings.

Investigations in European clay formations demonstrate the importance of diffusion-limited transport and strong mineral sorption. In the Opalinus Clay of Switzerland, deep sedimentary claystones retain uranium effectively, with diffusion coefficients on the order of 10^{-13} m^2/s, and immobility is largely attributed to sorption on illite edge sites under reducing conditions (Wersin et al., 2008). Similarly, in the French Callovo-Oxfordian argillite, uranium is mainly mineral-associated, with calcite precipitation and redox buffering further limiting its mobility (Jacquier et al., 2013; Tournassat et al., 2007). In contrast, Belgium's Boom Clay presents higher concentrations of dissolved organic carbon, which increase the apparent mobility of uranium compared with other clay formations, though the underlying sorption matrix remains effective in attenuating transport (Maes et al., 2003; Maes et al., 2002).

Complementary insights come from crystalline host rock environments where bentonite buffers are used to compensate for high fracture permeability. At Sweden's Forsmark and Äspö Hard Rock Laboratory, uranium retention has been observed primarily in bentonite, while microbial activity within granitic fractures and compacted buffers has been documented (Pedersen, 1997; Pedersen et al., 2000). Long-term monitoring has revealed that zones of Fe(II) production often coincide with U(VI) reduction, underscoring the role of coupled microbial–mineral redox processes in repository performance. Comparable findings emerge from China, where Gaomiaozi (GMZ) bentonite, rich in montmorillonite, is being tested as an engineered barrier material for the Beishan granite repository site (Sun et al., 2018; Tan et al., 2024). Laboratory studies indicate that Fe(II)-bearing nontronite promotes U(VI) reduction and that indigenous microbial populations contribute to uranium immobilization under anoxic conditions (Zhang et al., 2021; Zhang et al., 2022).

Natural analogues outside repository programs further demonstrate uranium mobility mechanisms. At the Nevada Test Site in the United States, uranium from underground nuclear tests migrated over distances exceeding 1 km, carried as nano-phase particles associated with colloids such as clay, silica, and carbon (Um & Papelis, 2002; Utsunomiya et al., 2009). This finding provided critical evidence that U(IV), long considered immobile, can, under certain conditions, be transported in the subsurface. Large-scale simulation experiments also provide validation of repository concepts. In Spain's FEBEX project, compacted bentonite effectively slowed U(VI) migration under saturated conditions, while other engineered mock-up tests have examined the role of gas intrusion and swelling clays in shaping uranium transport pathways (Martin et al., 2006; Villa-Alfageme et al., 2015).

Taken together, these case studies reveal that uranium migration is highly site-specific and strongly influenced by mineralogy, redox conditions, hydrology, and biological activity. Diffusion and sorption dominate in dense clay formations such as Opalinus and Callovo-Oxfordian, while microbial processes and redox cycling emerge as important controls in bentonite-buffered crystalline systems. Colloid-facilitated transport provides an additional pathway for uranium mobility, even for reduced U(IV) phases, challenging assumptions about its long-term stability. The overarching lesson is that predictive models of repository performance must integrate mineralogical reactivity, redox buffering, microbial contributions, and colloidal pathways to capture the coupled processes that ultimately govern uranium's fate over geological timescales.

5.6 SPECTROSCOPIC AND MODELING APPROACHES TO URANIUM SPECIATION AND MIGRATION

Understanding uranium's behavior in subsurface environments relevant to nuclear waste repositories requires analytical approaches that can capture processes across multiple scales. Traditional bulk chemical analyses provide concentrations but cannot resolve uranium's oxidation states, coordination environments, or specific binding mechanisms. To overcome this limitation, researchers increasingly combine advanced spectroscopic techniques with computational and reactive transport modelling, generating a more mechanistic and predictive framework for uranium geochemistry. Spectroscopy reveals molecular-level processes such as uranium's redox state and local bonding environment, while modelling extrapolates these mechanisms to system behavior over decades to millennia.

Among spectroscopic methods, X-ray absorption techniques have been central to advances in uranium studies. Extended X-ray absorption fine structure (EXAFS) provides information on uranium's coordination shells and distances to neighboring atoms, allowing differentiation of sorption geometries and characterization of U(IV) solids (Allen et al., 1996). X-ray absorption near-edge structure (XANES) complements this by identifying uranium oxidation states and tracking redox transformations in situ (Kalkowski et al., 1987). Together, these techniques have become benchmarks for determining uranium speciation in complex mineral–microbial systems. Additional spectroscopic methods also provide indirect insights. Mössbauer spectroscopy does not probe uranium directly, but it quantifies Fe(II)/Fe(III) ratios in clays such as nontronite (Charlet et al., 2023), which strongly affects uranium reduction (Luan & Burgos, 2012; Luan et al., 2014; Tsarev et al.,

2016). Infrared and Raman spectroscopy add further information by resolving functional group interactions and characteristic uranyl vibrations (Alcorn et al., 2019).

Spectroscopic insights are most powerful when paired with quantitative models. Surface complexation models (SCMs) describe uranium sorption onto clay minerals by parameterizing site densities, binding constants, and electrostatic effects (Gao et al., 2024). These models, validated against EXAFS and XANES data, allow uranium sorption processes to be incorporated into larger geochemical simulations. At the atomic scale, density functional theory (DFT) and molecular dynamics (MD) simulations provide mechanistic insights into sorption geometries, hydration shells, and diffusion in hydrated clay interlayers (Chen et al., 2019). These computational results not only illuminate reaction pathways but also inform parameterization of larger-scale models.

For repository-scale predictions, reactive transport models (RTMs), such as PHREEQC (a pH–redox–equilibrium geochemical code written in C/C++), TOUGHREACT (a reactive transport simulator based on Transport of Unsaturated Groundwater and Heat), PFLOTRAN (Parallel FLOw and TRANsport), and CrunchFlow reactive transport software, integrate chemical reactions with physical transport processes (Arora et al., 2015; Steefel et al., 2015). These codes require inputs, including diffusion coefficients, sorption parameters, and thermodynamic constants, and are increasingly adapted to include microbial redox kinetics. When calibrated with spectroscopic and molecular-scale data, RTMs provide powerful tools for evaluating uranium migration across heterogeneous geological systems and over geological timescales.

Recent work emphasizes the integration of spectroscopy and modeling as a means of ensuring mechanistic fidelity. Spectroscopy constrains the local binding environments and oxidation states that SCMs and RTMs must capture, while molecular modelling provides parameters for reactions that cannot be measured directly. This combined approach reduces uncertainties and strengthens the predictive capacity of long-term safety assessments. Nonetheless, limitations remain. Spectroscopic methods often rely on synchrotron facilities and provide bulk-averaged signals, while models face challenges in representing non-equilibrium kinetics and microbial contributions. Computational approaches are constrained by system size and resource demands. Bridging these gaps remains a priority.

Future directions include developing spectroscopy-informed surface complexation models, improving microbial integration into reactive transport simulations, and using machine learning to reduce uncertainty in long-term predictions. Advances in portable field-based spectroscopic tools and incorporation of microbial community data also promise to improve in situ monitoring. Ultimately, combining experimental spectroscopy, molecular-scale computation, and system-level transport modelling provides the most robust pathway toward predicting uranium's behavior in repository environments. This integration enables mechanistic understanding at the atomic scale to inform dynamic simulations at repository scales, ensuring a more reliable basis for nuclear waste isolation strategies.

5.7 KNOWLEDGE GAPS, OUTLOOK, AND POLICY IMPLICATIONS

Despite decades of research, significant uncertainties remain in predicting uranium's behavior over the long timescales relevant to nuclear waste repositories. These uncertainties arise from the intricate coupling of geochemical, microbiological, mineralogical, and hydrological processes, all of which interact dynamically across spatial and temporal scales. A major challenge lies in the long-term stability of uranium redox states, particularly biogenic U(IV), as well as the fate of non-uraninite phases and mixed-valence species that may persist under repository conditions. Uncertainties also surround the persistence and activity of microbial communities in compacted clay environments, including their ability to drive coupled Fe, S, and U cycling and to establish biofilms over extended timescales. Additional complexity is introduced by colloid-facilitated transport, where U(IV)-bearing nano-particles and clay or carbon colloids may provide unexpected pathways for mobility even under otherwise favorable geochemical conditions. These factors collectively highlight the difficulty of scaling laboratory observations to repository systems, where heterogeneity in mineralogy, porosity, and redox fronts, combined with long-term mineral transformations such as illitization or dissolution–reprecipitation, may lead to outcomes not captured in controlled experiments.

Policy frameworks require robust performance assessments that demonstrate containment of radionuclides for tens to hundreds of thousands of years. International regulatory agencies emphasize conservative treatment of uncertainties and call for transparent models validated by multiple lines of evidence, including underground laboratory studies and natural analogues. Increasingly, safety cases incorporate microbial influences, recognizing that repository environments are not inert but host dynamic, evolving ecosystems. Field experiments in underground laboratories such as Grimsel, Mont Terri, and Bure provide critical validation, while analogues such as uranium ore bodies and the Oklo natural reactor offer rare insights into uranium's behavior over geological timescales.

To reduce uncertainties and improve repository design, future research should prioritize long-term characterization of microbial–mineral interactions, including metagenomic and transcriptomic studies to predict microbial pathways and in situ investigations of biofilm formation in clay matrices. Real-time monitoring technologies, including biosensors, isotopic tracing, and portable spectroscopy, could provide direct feedback on uranium speciation under repository conditions. Advances in modeling are equally crucial: multiscale frameworks that couple molecular simulations, surface complexation models, and reactive transport codes must be developed, with uncertainty quantified through data assimilation and machine learning approaches. Another promising direction lies in the design of engineered barrier materials with optimized mineralogical properties, such as clays enriched in structural Fe(II) to promote uranium reduction, or materials engineered to limit undesired microbial colonization.

Policy recommendations emerging from these insights emphasize the integration of biogeochemistry into safety cases, requiring explicit treatment of microbial processes and their impact on uranium redox transformations and mobility. Regulatory frameworks must remain flexible, allowing for updates as new scientific evidence emerges, while international collaboration and data-sharing platforms will accelerate the convergence of standards and best practices.

The emerging picture is that uranium migration in repositories cannot be described solely as a geochemical process; it is inherently biogeochemical. The long-term success of geological disposal depends on acknowledging repositories as dynamic, evolving systems shaped by coupled chemical and biological processes. Future strategies must therefore move beyond the notion of static containment toward an adaptive perspective of co-evolution with complexity. Only through integrated research bridging chemistry, microbiology, mineralogy, hydrology, and policy can the enduring safety of nuclear waste repositories be ensured.

REFERENCES

Alcorn, C.D., Cox, J.S., Applegarth, L.M., Tremaine, P.R., 2019. Investigation of uranyl sulfate complexation under hydrothermal conditions by quantitative Raman spectroscopy and density functional theory. *J. Phys. Chem. B* 123, 7385–7409.

Alessi, D.S., Lezama-Pacheco, J.S., Stubbs, J.E., Janousch, M., Bargar, J.R., Persson, P., Bernier-Latmani, R., 2014. The product of microbial uranium reduction includes multiple species with U (IV)-phosphate coordination. *Geochim. Cosmochim. Acta* 131, 115–127.

Allen, P.G., Shuh, D.K., Bucher, J.J., Edelstein, N.M., Palmer, C.E.A., Silva, R.J., Nguyen, S.N., Marquez, L.N., Hudson, E.A., 1996. Determinations of uranium structures by EXAFS: Schoepite and other U(VI) oxide precipitates. *Radiochim. Acta* 75, 47–53.

Arora, B., Şengör, S.S., Spycher, N.F., Steefel, C.I., 2015. A reactive transport benchmark on heavy metal cycling in lake sediments. *Comput. Geosci.* 19, 613–633.

Bachmaf, S., Planer-Friedrich, B., Merkel, B.J., 2008. Effect of sulfate, carbonate, and phosphate on the uranium (VI) sorption behavior onto bentonite. *Radiochim. Acta* 96, 359–366.

Belli, K.M., Taillefert, M., 2018. Geochemical controls of the microbially mediated redox cycling of uranium and iron. *Geochim. Cosmochim. Acta* 235, 431–449.

Bernier-Latmani, R., Veeramani, H., Vecchia, E.D., Junier, P., Lezama-Pacheco, J.S., Suvorova, E.I., Sharp, J.O., Wigginton, N.S., Bargar, J.R., 2010. Non-uraninite products of microbial U (VI) reduction. *Environ. Sci. Technol.* 44, 9456–9462.

Birkholzer, J., Houseworth, J., Tsang, C.-F., 2012. Geologic disposal of high-level radioactive waste: Status, key issues, and trends. *Annu. Rev. Environ. Resour.* 37, 79–106.

Bishop, M.E., Dong, H., Kukkadapu, R.K., Liu, C., Edelmann, R.E., 2011. Bioreduction of Fe-bearing clay minerals and their reactivity toward pertechnetate (Tc-99). *Geochim. Cosmochim. Acta* 75, 5229–5246.

Biswas, B., Islam, M.R., Deb, A.K., Greenaway, A., Warr, L.N., Naidu, R., 2023. Understanding iron impurities in Australian kaolin and their effect on acid and heat activation processes of clay. *ACS Omega* 8, 5533–5544.

Boyanov, M.I., Fletcher, K.E., Kwon, M.J., Rui, X., O'Loughlin, E.J., Löffler, F.E., Kemner, K.M., 2011. Solution and microbial controls on the formation of reduced U (IV) species. *Environ. Sci. Technol.* 45, 8336–8344.

Bradford, S.A., Morales, V.L., Zhang, W., Harvey, R.W., Packman, A.I., Mohanram, A., Welty, C., 2013. Transport and fate of microbial pathogens in agricultural settings. *Crit. Rev. Environ. Sci. Technol.* 43, 775–893.

Brookshaw, D., Pattrick, R., Lloyd, J., Vaughan, D., 2012. Microbial effects on mineral–radionuclide interactions and radionuclide solid-phase capture processes. *Mineral. Mag.* 76, 777–806.

Brown, D.A., Kamineni, D.C., Sawicki, J.A., Beveridge, T.J., 1994. Minerals associated with biofilms occurring on exposed rock in a granitic underground research laboratory. *Appl. Environ. Microbiol.* 60, 3182–3191.

Charlet, L., Tournassat, C., Grenèche, J.-M., Wersin, P., Géhin, A., Hadi, J., 2023. Mössbauer spectrometry insights into the redox reactivity of Fe-bearing phases in the environment. *J. Mater. Res.* 38, 958–973.

Chen, J., Min, F.-F., Liu, L.-Y., Liu, C.-F., 2019. Mechanism research on surface hydration of kaolinite, insights from DFT and MD simulations. *Appl. Surf. Sci.* 476, 6–15.

Chen, Y., Zhang, L., Wang, S., Zeng, Q., Xia, Q., Li, R., Guo, D., Pan, Z., Dong, H., 2022. Combined effects of humic substances and clay minerals on U (VI) bioreduction. *Geochim. Cosmochim. Acta* 338, 181–198.

Claret, F., Dauzeres, A., Jacques, D., Sellin, P., Cochepin, B., De Windt, L., Garibay-Rodriguez, J., Govaerts, J., Leupin, O., Monlopez, A., 2022. Modelling of the long-term evolution and performance of engineered barrier system. *EPJ Nucl. Sci. Technol.* 8, 15.

Crawford, S.E., Lofts, S., Liber, K., 2017. The role of sediment properties and solution pH in the adsorption of uranium (VI) to freshwater sediments. *Environ. Pollut.* 220, 873–881.

Cuadros, J., 2017. Clay minerals interaction with microorganisms: A review. *Clay Miner.* 52, 235–261.

Cui, Q., Zhang, Z., Beiyuan, J., Cui, Y., Chen, L., Chen, H., Fang, L., 2023. A critical review of uranium in the soil-plant system: Distribution, bioavailability, toxicity, and bioremediation strategies. *Crit. Rev. Environ. Sci. Technol.* 53, 340–365.

Dong, W., Tokunaga, T.K., Davis, J.A., Wan, J., 2012. Uranium (VI) adsorption and surface complexation modeling onto background sediments from the F-Area Savannah River Site. *Environ. Sci. Technol.* 46, 1565–1571.

Fagan, R.P., Fairweather, N.F., 2014. Biogenesis and functions of bacterial S-layers. *Nat. Rev. Microbiol.* 12, 211–222.

Gao, P., Liu, X., Guo, Z., Tournassat, C., 2024. Bridging the gap between molecular simulations and surface complexation modeling for heterogeneous surfaces: A case study with uranium and arsenic adsorption on clay minerals. *Environ. Sci. Technol.* 58, 22733–22743.

Gorski, C.A., Aeschbacher, M., Soltermann, D., Voegelin, A., Baeyens, B., Marques Fernandes, M., Hofstetter, T.B., Sander, M., 2012a. Redox properties of structural Fe in clay minerals. 1. Electrochemical quantification of electron-donating and -accepting capacities of smectites. *Environ. Sci. Technol.* 46, 9360–9368.

Gorski, C.A., Klüpfel, L., Voegelin, A., Sander, M., Hofstetter, T.B., 2012b. Redox properties of structural Fe in clay minerals. 2. Electrochemical and spectroscopic characterization of electron transfer irreversibility in ferruginous smectite, SWa-1. *Environ. Sci. Technol.* 46, 9369–9377.

Gorski, C.A., Klüpfel, L., Voegelin, A., Sander, M., Hofstetter, T.B., 2012. Redox properties of structural Fe in clay minerals. 2. Electrochemical and spectroscopic characterization of electron transfer irreversibility in ferruginous smectite, SWa-1. *Environ. Sci. Technol.* 46, 9369–9377.

Gorski, C.A., Klüpfel, L.E., Voegelin, A., Sander, M., Hofstetter, T.B., 2013. Redox properties of structural Fe in clay minerals: 3. Relationships between smectite redox and structural properties. *Environ. Sci. Technol.* 47, 13477–13485.

Guo, Z., Li, Y., Wu, W., 2009. Sorption of U (VI) on goethite: Effects of pH, ionic strength, phosphate, carbonate and fulvic acid. *Appl. Radiat. Isot.* 67, 996–1000.

Jacquier, P., Hainos, D., Robinet, J., Herbette, M., Grenut, B., Bouchet, A., Ferry, C., 2013. The influence of mineral variability of Callovo-Oxfordian clay rocks on radionuclide transfer properties. *Appl. Clay Sci.* 83, 129–136.

Kalkowski, G., Kaindl, G., Brewer, W.D., Krone, W., 1987. Near-edge x-ray-absorption fine structure in uranium compounds. *Phys. Rev. B* 35(6), 2667–2677.

Katsoyiannis, I.A., 2007. Carbonate effects and pH-dependence of uranium sorption onto bacteriogenic iron oxides: Kinetic and equilibrium studies. *J. Hazard. Mater.* 139, 31–37.

Kim, J., Dong, H., Seabaugh, J., Newell, S.W., Eberl, D.D., 2004. Role of microbes in the smectite-to-illite reaction. *Science* 303, 830–832.

Kim, J., Dong, H., Yang, K., Park, H., Elliott, W.C., Spivack, A., Koo, T.-H., Kim, G., Morono, Y., Henkel, S., 2019. Naturally occurring, microbially induced smectite-to-illite reaction. *Geology* 47, 535–539.

Konopka, A., 2009. What is microbial community ecology? *ISME J.* 3, 1223–1230.

Lee, S.Y., Cha, W.S., Kim, J.-G., Baik, M.H., Jung, E.C., Jeong, J.T., Kim, K., Chung, S.Y., Lee, Y.J., 2014. Uranium (IV) remobilization under sulfate reducing conditions. *Chem. Geol.* 370, 40–48.

Leupin, O.X., Bernier-Latmani, R., Bagnoud, A., Moors, H., Leys, N., Wouters, K., Stroes-Gascoyne, S., 2017. Fifteen years of microbiological investigation in Opalinus Clay at the Mont Terri rock laboratory (Switzerland). In: *Mont Terri Rock Laboratory, 20 Years: Two Decades of Research and Experimentation on Claystones for Geological Disposal of Radioactive Waste*. Springer, pp. 345–356.

Li, K.J., Kurniawan, A., Christidis, G.E., He, J.Y., Zhou, C.H., 2024. Interfacial interactions controlling adsorption of metal cations on montmorillonite. *Am. Mineral.* 109, 633–655.

Li, K.-W., Lu, H.-L., Nkoh, J.N., Xu, R.-K., 2023. The important role of surface hydroxyl groups in aluminum activation during phyllosilicate mineral acidification. *Chemosphere* 313, 137570.

Li, R., Zhang, L., Chen, Y., Xia, Q., Liu, D., Huang, Y., Dong, H., 2024. Oxidation of biogenic U (IV) in the presence of bioreduced clay minerals and organic ligands. *Environ. Sci. Technol.* 58, 1541–1550.

Lovley, D.R., Phillips, E.J.P., Gorby, Y.A., Landa, E.R., 1991. Microbial reduction of uranium. *Nature* 350, 413–416.

Luan, F., Burgos, W.D., 2012. Sequential extraction method for determination of Fe (II/III) and U (IV/VI) in suspensions of iron-bearing phyllosilicates and uranium. *Environ. Sci. Technol.* 46, 11995–12002.

Luan, F., Gorski, C.A., Burgos, W.D., 2014. Thermodynamic controls on the microbial reduction of iron-bearing nontronite and uranium. *Environ. Sci. Technol.* 48, 2750–2758.

Maes, A., Bruggeman, C., Geraedts, K., Vancluysen, J., 2003. Quantification of the interaction of Tc with dissolved boom clay humic substances. *Environ. Sci. Technol.* 37, 747–753.

Maes, N., Moors, H., Wang, L., Delècaut, G., De Cannière, P., Put, M., 2002. The use of electromigration as a qualitative technique to study the migration behaviour and speciation of uranium in the Boom Clay. *Radiochim. Acta* 90, 741–746.

Maher, K., Bargar, J.R., Brown Jr., G.E., 2013. Environmental speciation of actinides. *Inorg. Chem.* 52, 3510–3532.

Martin, P., Barcala, J., Huertas, F., 2006. Large-scale and long-term coupled thermo-hydro-mechanic experiments with bentonite: The FEBEX mock-up test. *J. Iber. Geol.* 32, 259–282.

Mohapatra, B.R., Dinardo, O., Gould, W.D., Koren, D.W., 2010. Biochemical and genomic facets on the dissimilatory reduction of radionuclides by microorganisms–A review. *Miner. Eng.* 23, 591–599.

Moon, H.S., Komlos, J., Jaffé, P.R., 2007. Uranium reoxidation in previously bioreduced sediment by dissolved oxygen and nitrate. *Environ. Sci. Technol.* 41, 4587–4592.

Nekipelov, V., Zamaraev, K., 1985. Outer-sphere coordination of organic molecules to electrically neutral metal complexes. *Coord. Chem. Rev.* 61, 185–240.

Neuzil, C., 2019. Permeability of clays and shales. *Annu. Rev. Earth Planet. Sci.* 47, 247–273.

Newsome, L., Morris, K., Lloyd, J.R., 2014. The biogeochemistry and bioremediation of uranium and other priority radionuclides. *Chem. Geol.* 363, 164–184.

Newsome, L., Morris, K., Shaw, S., Trivedi, D., Lloyd, J.R., 2015. The stability of microbially reduced U (IV); impact of residual electron donor and sediment ageing. *Chem. Geol.* 409, 125–135.

N'Guessan, A.L., Vrionis, H.A., Resch, C.T., Long, P.E., Lovley, D.R., 2008. Sustained removal of uranium from contaminated groundwater following stimulation of dissimilatory metal reduction. *Environ. Sci. Technol.* 42, 2999–3004.

Noël, V., Boye, K., Kukkadapu, R.K., Li, Q., Bargar, J.R., 2019. Uranium storage mechanisms in wet-dry redox cycled sediments. *Water Res.* 152, 251–263.

O'Loughlin, E.J., Boyanov, M.I., Antonopoulos, D.A., Kemner, K.M., 2011. Redox processes affecting the speciation of technetium, uranium, neptunium, and plutonium in aquatic and terrestrial environments. In: *Aquatic Redox Chemistry*. ACS Publications, pp. 477–517.

Pedersen, K., 1997. Microbial life in deep granitic rock. *FEMS Microbiol. Rev.* 20, 399–414.

Pedersen, K., Motamedi, M., Karnland, O., Sandén, T., 2000. Cultivability of microorganisms introduced into a compacted bentonite clay buffer under high-level radioactive waste repository conditions. *Eng. Geol.* 58, 149–161.

Sellin, P., Leupin, O.X., 2013. The use of clay as an engineered barrier in radioactive-waste management: A review. *Clays Clay Miner.* 61, 477–498.

Sharp, J.O., Lezama-Pacheco, J.S., Schofield, E.J., Junier, P., Ulrich, K.-U., Chinni, S., Veeramani, H., Margot-Roquier, C., Webb, S.M., Tebo, B.M., 2011. Uranium speciation and stability after reductive immobilization in aquifer sediments. *Geochim. Cosmochim. Acta* 75, 6497–6510.

Shi, X., Wei, X., Yang, M., Qi, T., Li, J., Zhang, C., Wang, J., Xu, Z., Pan, D., Wu, W., 2025. Colloid mediated transport of uranium(IV) in sulfide-rich aquifers: An overlooked pathway for uranium remobilization under reductive conditions. *J. Hazard. Mater.* 496, 139522.

Stanley, D.M., Wilkin, R.T., 2019. Solution equilibria of uranyl minerals: Role of the common groundwater ions calcium and carbonate. *J. Hazard. Mater.* 377, 315–320.

Steefel, C.I., Appelo, C.A.J., Arora, B., Jacques, D., Kalbacher, T., Kolditz, O., Lagneau, V., Lichtner, P., Mayer, K.U., Meeussen, J., 2015. Reactive transport codes for subsurface environmental simulation. *Comput. Geosci.* 19, 445–478.

Stewart, B.D., Girardot, C., Spycher, N., Sani, R.K., Peyton, B.M., 2013. Influence of chelating agents on biogenic uraninite reoxidation by Fe(III) (hydr)oxides. *Environ. Sci. Technol.* 47(1), 364–371.

Stewart, B.D., Nico, P.S., Fendorf, S., 2009. Stability of uranium incorporated into Fe (hydr) oxides under fluctuating redox conditions. *Environ. Sci. Technol.* 43, 4922–4927.

Stone, W., Kroukamp, O., McKelvie, J., Korber, D., Wolfaardt, G., 2016. Microbial metabolism in bentonite clay: Saturation, desiccation and relative humidity. *Appl. Clay Sci.* 129, 54–64.

Sun, Z., Chen, Y.-G., Cui, Y.-J., Xu, H.-D., Ye, W.-M., Wu, D.-B., 2018. Effect of synthetic water and cement solutions on the swelling pressure of compacted Gaomiaozi (GMZ) bentonite: The Beishan site case, Gansu, China. *Eng. Geol.* 244, 66–74.

Suzuki, Y., Banfield, J.F., 1999. Geomicrobiology of uranium. *Rev. Mineral. Geochem.* 38, 393–432.

Suzuki, Y., Banfield, J.F., 2004. Resistance to, and accumulation of, uranium by bacteria from a uranium-contaminated site. *Geomicrobiol. J.* 21, 113–121.

Tan, Z., Fu, C., Henderson, M.J., Dai, X., Cheng, J., Xie, J., Zhu, S., Yan, M., 2024. Stability of Gaomiaozi bentonite: Interlayer hydration structure and aggregation morphology. *Clays Clay Miner.* 72, e5.

Tournassat, C., Gaucher, E.C., Fattahi, M., Grambow, B., 2007. On the mobility and potential retention of iodine in the Callovian–Oxfordian formation. *Phys. Chem. Earth* 32, 539–551.

Tourney, J., Ngwenya, B.T., 2014. The role of bacterial extracellular polymeric substances in geomicrobiology. *Chem. Geol.* 386, 115–132.

Tsarev, S., Waite, T.D., Collins, R.N., 2016. Uranium reduction by Fe (II) in the presence of montmorillonite and nontronite. *Environ. Sci. Technol.* 50, 8223–8230.

Ulrich, K.-U., Veeramani, H., Bernier-Latmani, R., Giammar, D.E., 2011. Speciation-dependent kinetics of uranium(VI) bioreduction. *Geomicrobiol. J.* 28, 396–409.

Um, W., Papelis, C., 2002. Geochemical effects on colloid-facilitated metal transport through zeolitized tuffs from the Nevada Test Site. *Environ. Geol.* 43, 209–218.

Utsunomiya, S., Kersting, A.B., Ewing, R.C., 2009. Groundwater nanoparticles in the far-field at the Nevada test site: Mechanism for radionuclide transport. *Environ. Sci. Technol.* 43, 1293–1298.

Villa-Alfageme, M., Hurtado, S., El Mrabet, S., Pazos, M.C., Castro, M.A., Alba, M.D., 2015. Uranium immobilization by FEBEX bentonite and steel barriers in hydrothermal conditions. *Chem. Eng. J.* 269, 279–287.

Wang, S., Chen, Y., Pan, Z., Liu, J., Ding, Y., Wang, Y., Liu, D., Wu, S., Hu, D., Li, R., 2024. Effects of EDTA and bicarbonate on U (VI) reduction by reduced nontronite. *Environ. Sci. Technol.* 58, 23031–23041.

Warren, J.J., Lancaster, K.M., Richards, J.H., Gray, H.B., 2012. Inner-and outer-sphere metal coordination in blue copper proteins. *J. Inorg. Biochem.* 115, 119–126.

Wersin, P., Soler, J., Van Loon, L., Eikenberg, J., Baeyens, B., Grolimund, D., Gimmi, T., Dewonck, S., 2008. Diffusion of HTO, Br–, I–, Cs+, 85Sr2+ and 60Co2+ in a clay formation: Results and modelling from an in situ experiment in Opalinus Clay. *Appl. Geochem.* 23, 678–691.

Wilkins, M., Livens, F., Vaughan, D., Beadle, I., Lloyd, J., 2007. The influence of microbial redox cycling on radionuclide mobility in the subsurface at a low-level radioactive waste storage site. *Geobiology* 5, 293–301.

Wolf, M., Kappler, A., Jiang, J., Meckenstock, R.U., 2009. Effects of humic substances and quinones at low concentrations on ferrihydrite reduction by Geobacter metallireducens. *Environ. Sci. Technol.* 43, 5679–5685.

Wu, W., Carley, J., Fienen, M., Mehlhorn, T., Lowe, K., Nyman, J., Luo, J., Gentile, M.E., Rajan, R., Wagner, D., 2006. Pilot-scale in situ bioremediation of uranium in a highly contaminated aquifer. 1. Conditioning of a treatment zone. *Environ. Sci. Technol.* 40, 3978–3985.

Wu, W., Carley, J., Luo, J., Ginder-Vogel, M.A., Cardenas, E., Leigh, M.B., Hwang, C., Kelly, S.D., Ruan, C., Wu, L., 2007. In situ bioreduction of uranium (VI) to submicromolar levels and reoxidation by dissolved oxygen. *Environ. Sci. Technol.* 41, 5716–5723.

Xia, Q., Jin, Q., Chen, Y., Zhang, L., Li, X., He, S., Guo, D., Liu, J., Dong, H., 2022. Combined effects of Fe(III)-bearing nontronite and organic ligands on biogenic U(IV) oxidation. *Environ. Sci. Technol.* 56, 1983–1993.

Xie, J., Wang, J., Lin, J., Zhou, X., 2018. The dynamic role of pH in microbial reduction of uranium (VI) in the presence of bicarbonate. *Environ. Pollut.* 242, 659–666.

Xu, G., Yang, H., Han, J., Liu, X., Shao, K., Li, X., Wang, G., Yue, W., Dou, J., 2024. Regulatory roles of extracellular polymeric substances in uranium reduction via extracellular electron transfer by Desulfovibrio vulgaris UR1. *Environ. Res.* 262, 119862.

Yan, X., Luo, X., Zhao, M., 2016. Metagenomic analysis of microbial community in uranium-contaminated soil. *Appl. Microbiol. Biotechnol.* 100, 299–310.

Zachara, J.M., Chen, X., Murray, C., Hammond, G., 2016. River stage influences on uranium transport in a hydrologically dynamic groundwater-surface water transition zone. *Water Resour. Res.* 52, 1568–1590.

Zhang, D., Chen, X., Larson, S.L., Ballard, J.H., Knotek-Smith, H.M., Ding, D., Hu, N., Han, F.X., 2020. Uranium biomineralization with phosphate—biogeochemical process and its application. *ACS Earth Space Chem.* 4, 2205–2214.

Zhang, G., Burgos, W.D., Senko, J.M., Bishop, M.E., Dong, H., Boyanov, M.I., Kemner, K.M., 2011. Microbial reduction of chlorite and uranium followed by air oxidation. *Chem. Geol.* 283, 242–250.

Zhang, G., Senko, J.M., Kelly, S.D., Tan, H., Kemner, K.M., Burgos, W.D., 2009. Microbial reduction of iron(III)-rich nontronite and uranium(VI). Geochim. *Cosmochim. Acta* 73, 3523–3538.

Zhang, J., Dong, H., Liu, D., Fischer, T.B., Wang, S., Huang, L., 2012. Microbial reduction of Fe(III) in illite–smectite minerals by methanogen Methanosarcina mazei. *Chem. Geol.* 292–293, 35–44.

Zhang, L., Chen, Y., Xia, Q., Kemner, K.M., Shen, Y., O'Loughlin, E.J., Pan, Z., Wang, Q., Huang, Y., Dong, H., 2021. Combined effects of Fe (III)-bearing clay minerals and organic ligands on U (VI) bioreduction and U (IV) speciation. *Environ. Sci. Technol.* 55, 5929–5938.

Zhang, L., Dong, H., Li, R., Liu, D., Bian, L., Chen, Y., Pan, Z., Boyanov, M.I., Kemner, K.M., Wen, J., 2022. Effect of siderophore DFOB on U (VI) adsorption to clay mineral and its subsequent reduction by an iron-reducing bacterium. *Environ. Sci. Technol.* 56, 12702–12712.

Zhao, C., Wang, Y., Cheng, M., Zhang, H., Yang, Y., Liu, N., Liao, J., 2022. Performance and mechanism of anaerobic granular sludge enhancing uranium immobilization via extracellular polymeric substances in column reactors and batch experiments. *J. Clean. Prod.* 363, 132517.

Zhong, L., Liu, C., Zachara, J.M., Kennedy, D.W., Szecsody, J.E., Wood, B., 2005. Oxidative remobilization of biogenic uranium (IV) precipitates: Effects of iron (II) and pH. *J. Environ. Qual.* 34, 1763–1771.

CHAPTER 6

Nuclear Disasters and Practical Accident Management

Yuanlv Ye and Jie Zheng

6.1 INTRODUCTION

The pursuit of nuclear technology, whether for energy or defense, has introduced a unique category of technological risk, necessitating a precise and stratified terminology to accurately communicate the nature and severity of any unintended occurrence. This is particularly important given the dual-edged nature of nuclear technology, which can both benefit and pose significant threats to humanity. At the core of this glossary is the term "nuclear event," which refers to any unforeseen occurrence within a nuclear facility or during the transit of radioactive materials. This term encompasses a range of incidents, from minor deviations to severe accidents, as defined by the International Nuclear Event Scale (INES). Within this spectrum, a critical distinction is made between a nuclear incident and a nuclear accident, as defined by the International Nuclear and Radiological Event Scale (INES). An incident refers to an operational deviation or failure in safety provisions that, while indicating a potential vulnerability, results in no actual off-site radiological release or measurable health or environmental consequences (IAEA, 2013). In contrast, a nuclear accident, such as the catastrophic Chernobyl Disaster in 1986, involves the actual release of radioactive materials into the environment or a significant overexposure of individuals to radiation, leading to definite, albeit often localized, health or environmental effects. The INES is typically used to classify such events, with accidents rated at Level 4 or above. When the scale of an accident is so vast that it overwhelms local and national response capabilities, resulting in widespread, long-term contamination, significant socio-economic disruption, and profound human and ecological devastation, it is rightfully classified as a nuclear accident (Ohtsuru et al., 2015); The 1986 Chernobyl accident and the 2011 Fukushima Daiichi accident are quintessential examples of major nuclear accidents, both of which were rated at the maximum Level 7 on the INES scale because of their catastrophic and long-lasting impacts

DOI: 10.1201/9781003665793-8

on the nuclear industry, public health, and the environment (Magwood IV et al., 2021; Bricker, 2014; Saenko et al., 2011).

To establish a uniform international standard for evaluating the safety implications of nuclear and radiological incidents, the International Atomic Energy Agency (IAEA) collaborated with the Nuclear Energy Agency (NEA) of the OECD to create the INES. This scale classifies events into seven levels, ranging from 0 (deviation) to 7 (major accident), with each level indicating increasing severity and potential impact on safety. The INES is a logarithmic classification system ranging from Level 0 to Level 7 and categorizes events according to three areas of impact: effects on people and the environment (off-site effects), effects on radiological barriers and controls within the facility (on-site effects), and degradation of defense-in-depth safety provisions (IAEA, 2013). Levels 1–3 are termed "incidents" (anomaly, incident, and serious incident), which are characterized by varying degrees of equipment failure, worker overexposure, or minimal off-site release. Levels 4–7 are classified as "accidents" with increasing severity: Level 4 (accident with local consequences), Level 5 (accident with wider consequences), Level 6 (serious accident), and Level 7 (major accident), reserved for events involving a major release of radioactive material with widespread health and environmental effects, requiring long-term countermeasures (IAEA, 2013).

The meticulous examination of past nuclear accidents, such as the catastrophic Chernobyl and Fukushima Daiichi nuclear accidents, is essential for the evolution of nuclear safety protocols. Forensic analysis of past failures provides invaluable, real-world data that is otherwise unattainable. This research is paramount for several reasons. First, it directly informs the improvement of future nuclear waste management systems. Understanding how containment failed, materials melted, and contaminants migrated under extreme accident conditions allows engineers to design more robust geological repositories and storage casks (IAEA, 2015b; Kajan, 2016). Second, accidents serve as extreme, full-scale experiments that enhance the predictive modeling of radionuclide behavior in the environment. The

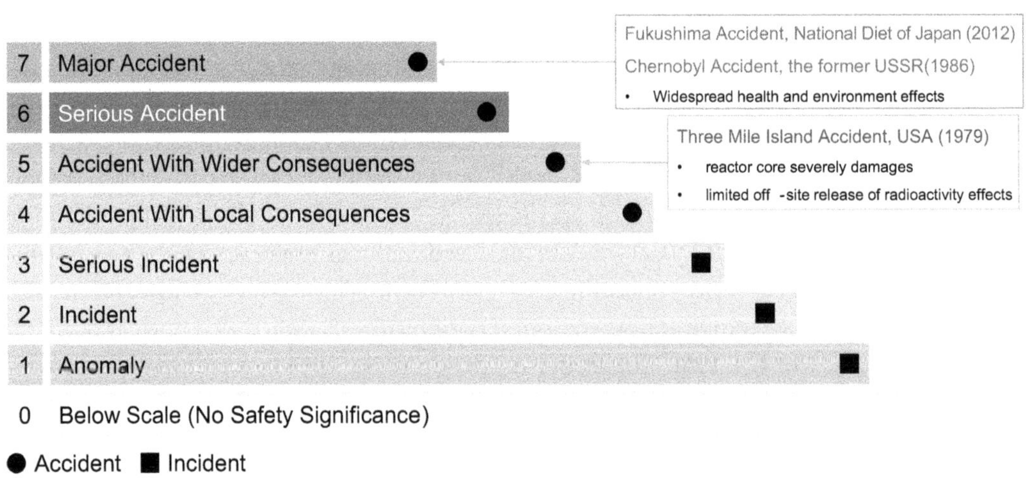

FIGURE 6.1 International Nuclear and Radiological Event Scale (INES) (IAEA, 2013).

dispersion patterns from Chernobyl and Fukushima have profoundly refined our understanding of atmospheric and oceanic transport, soil-to-plant transfer, and the long-term bioavailability of key isotopes like ^{137}Cs and ^{131}I (IAEA, 2015a; Saenko et al., 2011). Finally, and most crucially, these analyses are indispensable for perfecting Emergency Preparedness and Response frameworks. Lessons learned from evacuation challenges, communication breakdowns, and public health interventions during past accidents are directly applied to update protocols, train responders, and protect populations (Bennett et al., 2006), thereby ensuring that the profound cost of past mistakes translates into a safer nuclear future.

6.2 OVERVIEW OF MAJOR HISTORICAL NUCLEAR DISASTERS

Historical evidence consistently shows that severe nuclear accidents rarely result from isolated technical failures or individual human mistakes, but rather from a combination of technical issues, human errors, and inadequate management. Failures often arise from a complex interplay of vulnerabilities across various domains, including technical design flaws, human decision-making under stress, organizational culture deficiencies, and systemic regulatory and oversight failures. A comprehensive and multidisciplinary understanding of these deeply intertwined root causes is absolutely essential for developing more robust safety protocols, fostering a resilient safety culture, and ultimately preventing future catastrophes (IAEA, 1986).

Technical failures frequently initiate sequences of events that lead to accidents, with other factors subsequently amplifying the impact. The Chernobyl disaster was exacerbated by inherent design flaws in the Reaktor Bolshoy Moshchnosti Kanalnyy (RBMK; high-power channel-type reactor) design, including the positive void coefficient, which led to significant instability at low power levels, as well as progressive equipment aging, exemplified by reactor pressure vessel embrittlement and widespread corrosion of essential piping, and critically insufficient system redundancy. The Fukushima Daiichi Disaster starkly illustrated this when its emergency diesel generators and electrical switchgear—all co-located in a tsunami-inundation zone—were simultaneously disabled, resulting in a complete station blackout and subsequent core meltdowns (IAEA, 1986; National Diet of Japan, 2012). At the same time, human and organizational factors significantly exacerbate these technical vulnerabilities. Operator errors, which often occur under high-stress conditions and stem from insufficient training, unrealistic procedures, or misleading instrumentation, interact with safety cultures to suppress questioning, discourage the reporting of anomalies, and normalize deviations from safety protocols. The Chernobyl and Fukushima nuclear accidents highlighted the critical role of safety culture and organizational factors in preventing catastrophic failures, showing how their neglect, coupled with technical issues, can lead to accidents. Inadequate cross-shift communication, limited information sharing with management, and insufficient training for scenarios beyond design basis further diminish the capacity to respond effectively during escalating crises (IAEA, 1986; National Diet of Japan, 2012).

External hazards beyond original design parameters represent another major threat. The 2011 tsunami that overwhelmed Fukushima Daiichi's sea walls—exceeding height assumptions by nearly 10 m—demonstrates how extreme natural events can trigger accidents (National Diet of Japan, 2012). Moreover, malevolent acts such as military attacks or cyber intrusions, highlighted by the occupation of the Zaporizhzhia and Chernobyl sites, introduce evolving risks poorly addressed in conventional safety plans (Przybylak, 2024). Underpinning these direct causes are systemic failures: regulatory capture, lack of independence, and economic or political pressures that prioritize production or national prestige over safety, undermining the entire safety architecture and allowing preventable vulnerabilities to persist (IAEA, 1986; National Diet of Japan, 2012).

In conclusion, the doctrine of "defense-in-depth" relies on multiple independent layers of protection. The catastrophic breaches at Chernobyl and Fukushima exemplify how a combination of technical flaws, operator mistakes, natural calamities, and regulatory shortcomings can culminate in disastrous outcomes. Truly resilient nuclear safety requires a holistic culture of technical excellence, operational transparency, robust regulation, and

FIGURE 6.2 The Chernobyl-4 Reactor Several Months after the Accident (IAEA, 1986).

FIGURE 6.3 Fukushima Daiichi Nuclear Accident (IAEA, 2015a).

humble preparedness for both natural and human-induced extremes (IAEA, 1986; National Diet of Japan, 2012; Przybylak, 2024).

6.2.1 Behavior of Radionuclides in Nuclear Accidents

In nuclear reactors, the most commonly employed fuel is composed of cylindrical pellets of uranium dioxide (UO_2) or mixed plutonium and uranium oxides (MOX) (Choppin et al., 2013). These pellets are encapsulated within fuel rods, which are assembled into fuel assemblies to form the reactor core. During operation, nuclear energy is generated through the process of nuclear fission, where a fissile nucleus, such as uranium-235 or plutonium-239, absorbs a neutron and subsequently splits into two smaller atoms, known as fission products (FPs). This process releases a significant amount of energy, primarily in the form of kinetic energy and gamma radiation, as well as additional neutrons that can sustain a chain reaction if not controlled. Each fission event of a uranium-235 nucleus releases approximately 200 MeV of energy, primarily in the form of kinetic energy carried by the high-velocity fission fragments. The distribution of these fragments is highly asymmetric in mass, a phenomenon illustrated in Figure 6.4, with over 30 possible fission pathways yielding around 60 different FPs (Alameri & Alkaabi, 2020). Many of these products are radioactive, leading to the accumulation of a complex mixture of elements within irradiated fuel and contributing to both decay heat and radiological hazards.

As clearly illustrated in Figure 6.4, the atomic mass distribution of FPs spans a remarkably wide range, reflecting the physical complexity of nuclear fission. The quantity of the accumulated products is directly proportional to the number of fission events that have occurred per unit mass of fuel, a parameter known as burn-up. In modern light water reactors (LWRs), this burn-up can reach levels as high as 6% of the initial heavy metal (uranium/plutonium) content, leading to a significant buildup of diverse FPs within the fuel matrix (Konings et al., 2015). Their relative abundances vary systematically with burn-up; on average, the inventory consists of approximately 8% volatile noble gases (xenon and krypton isotopes), about 10% noble metals from the platinum group (such as Ru, Pd, Rh), and roughly 25% comprise elements from the lanthanide and actinide series, which form solid oxides (Konings et al., 2015). The potential release of these FPs, particularly the radioactive isotopes such as ^{131}I, ^{137}Cs, and ^{90}Sr, alongside ^{131}I, is of primary environmental and health concern during nuclear accidents due to their radiotoxicity, environmental mobility, and potential for long-term contamination. For instance, ^{137}Cs, a radioactive isotope of cesium, is highly reactive and can explode upon contact with water, posing significant risks to the environment and human health. Strontium-90, another radioactive isotope, can affect calcium levels in human bones and teeth, leading to bone diseases and cancers. These isotopes can remain in the environment for extended periods, posing long-term contamination risks.

All modern nuclear power plants (NPPs) are designed around the fundamental safety principle of *defense-in-depth*, which employs multiple, independent layers of protection to prevent the release of radioactive materials. Central to this strategy are four key physical

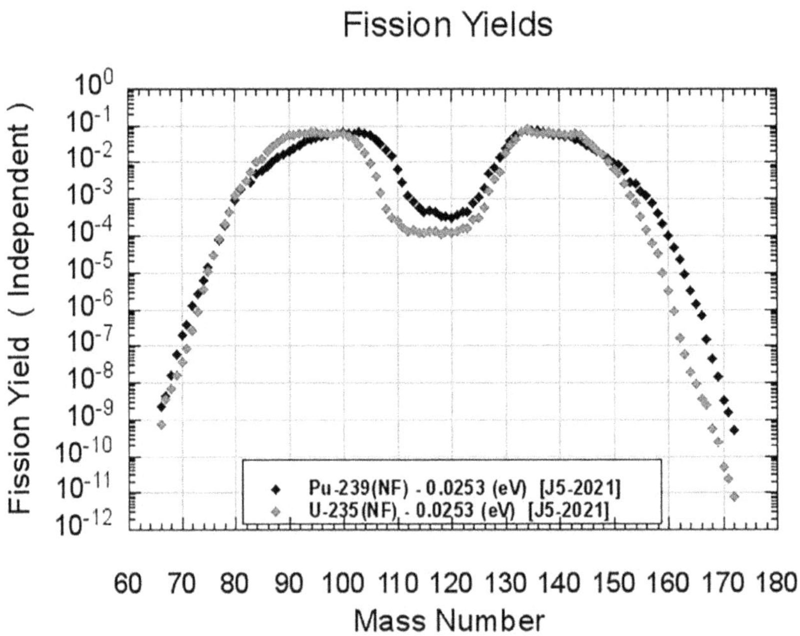

FIGURE 6.4 Fission Product Yield by Fission of ^{235}U (red) and ^{239}Pu (black) by Thermal Neutrons (JAEA, 2021).

barriers that sequentially impede the escape of FPs. The initial barrier is the fuel matrix, commonly composed of uranium dioxide (UO_2) or mixed oxide (MOX), which retains the majority of solid FPs. The second barrier is the *fuel cladding*, a robust zirconium alloy tube that encases the fuel pellets, sealing in gaseous and volatile FPs. The third barrier is the *reactor pressure vessel (RPV)*, a massive steel structure designed to withstand high temperatures and pressures, containing the entire fuel core and primary coolant. The fourth and final barrier is the *containment building*, a heavily reinforced, leak-tight concrete and steel structure that encloses the reactor vessel and primary systems, intended to isolate any radioactive release from the environment (Sehgal, 2012b). Each barrier must be breached before FPs can escape, and the failure modes—such as cladding rupture due to overheating or containment overpressure—involve distinct physical and chemical phenomena at each stage. Although this multi-barrier system is highly effective, historical accidents have demonstrated that the simultaneous failure of all barriers, though extremely rare, can lead to significant environmental contamination. The release phases of FPs from the fuel are categorized into five distinct stages, as detailed in Table 6.1, with experimental evidence supporting the differentiation of these phases under various conditions.

The release of FPs during a severe nuclear accident is systematically categorized into distinct phases based on the underlying physical and chemical phenomena, each marked by specific failure events and release mechanisms (Soffer et al., 1995). The initial coolant activity phase commences with a loss-of-coolant accident (LOCA), such as a pipe rupture, which can lead to a rapid cool-down and depressurization of the primary circuit. This phase extends until the first fuel rod fails, a critical point where the integrity of the reactor core is compromised. During this period, FPs are primarily transported by the circulating coolant, but the fuel and cladding remain largely intact. Upon cladding failure, the gap activity phase begins, leading to the swift discharge of volatile FPs like noble gases, iodine, and cesium. These elements had been building up in the space between the fuel pellet and the cladding prior to the breach. While a significant fraction of the inventory is still retained within the fuel matrix at this stage, the more volatile species enter the reactor coolant system and subsequently the containment. This phase ends as temperatures rise sufficiently to initiate the release of less-volatile species directly from the fuel grains.

As core degradation progresses, the *early in-vessel phase* begins, characterized by the loss of core geometry, melting of fuel and structural materials, and the formation of corium. The significant release of volatile and semi-volatile FPs is observed under conditions of high temperatures and oxidative environments, as evidenced by studies on irradiated UO_2. If the

TABLE 6.1 Release Phases and Their Duration for PWRs and BWRs (Soffer et al., 1995)

Release phase	PWR duration	BWR duration
Coolant activity	10–30 s	30 s
Gap activity	0.5 h	0.5 h
Early in-vessel	1.3 h	1.5 h
Ex-vessel	2 h	3 h
Late in-vessel	10 h	10 h

Note: PWR = pressurized water reactor; BWR = boiling water reactor; s = seconds; h = hours.

reactor vessel fails, the accident enters the *ex-vessel phase*, wherein molten corium interacts with concrete in the reactor cavity, generating combustible gases and aerosolizing additional FPs through vigorous corium–concrete interactions. This phase releases substantial amounts of volatiles and some non-volatiles, and terminates only when the debris cools sufficiently to halt further release (Nourbakhsh, 1993; Soffer et al., 1995). Concurrently, the *late in-vessel phase* may occur, involving the re-vaporization of previously deposited FPs from hot surfaces in the reactor cooling system (RCS), which can contribute to delayed source terms (Nourbakhsh, 1993). A simplified schematic of this multi-phase release and transport process is provided in Figure 6.5 (Sehgal, 2012a).

The *source term* refers to the quantity, timing, and characteristics of hazardous radioactive materials released into the environment from a nuclear facility during an accident. A related and more specific concept is the *containment source term*, which denotes the total inventory of FPs present within the containment structure at any given time following the initiation of an accident (Soffer et al., 1995). It is critical to recognize that the radioactive inventory contributing to the source term arises from two primary sources: FPs generated within the nuclear fuel and activation products (APs) resulting from neutron activation of materials in the reactor core (such as structural components, coolant, and impurities). Both categories of radionuclides can be mobilized and released under severe accident conditions. Quantifying the source term—including the total activity, isotopic composition, release rates, and, crucially, the chemical and physical form (e.g., aerosol, vapor, or particulate) of the released nuclides—is a foundational element of accident risk evaluation

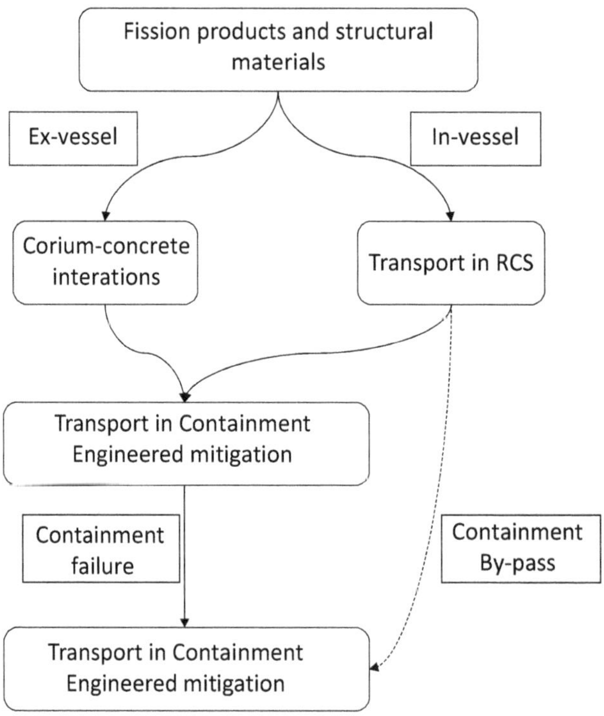

FIGURE 6.5 Paths of Fission Product Release and Transport (Sehgal, 2012a).

and subsequent emergency response planning. Unlike conventional industrial accidents, a nuclear reactor accident is uniquely characterized by the release of radioactive substances such as radioactive iodine, cesium, and inert gases, which can lead to complex environmental transport and significant radiological consequences, including thyroid disorders, multi-organ damage, and long-term cancer risks. Therefore, precise source-term evaluation, particularly an understanding of release mechanisms from degraded fuel matrices, is crucial and remains one of the central challenges in severe accident research (Sehgal, 2012a).

6.2.2 Fission Product Classification and Release

The volatility of radionuclides, including fission and APs, is a key physicochemical property that influences their release and dispersion in the event of a nuclear accident. For instance, radioactive iodine and cesium isotopes, with their distinct volatility characteristics, can enter the human body through various pathways, posing significant health risks. Highly volatile FPs, such as iodine-131 and cesium-137, which have half-lives of approximately 8 days and 30 years, respectively, exhibit early release from the fuel matrix as temperatures increase. These elements, along with noble gases like xenon-133, are of particular concern due to their potential to cause significant health effects and environmental contamination. These elements readily vaporize or condense into aerosols, contributing dominantly to the early volatile source term and posing immediate radiological risks due to their mobility and biological relevance. In contrast, semi-volatile species (e.g., antimony, silver) and non-volatile radionuclides (e.g., strontium-90, cerium, plutonium isotopes) require significantly higher temperatures or extended periods of heating for release, often occurring in parallel with fuel degradation phenomena such as cladding failure, fuel liquefaction, or molten pool formation. The release kinetics of these less volatile species are thus closely tied to the severe accident progression timeline.

Beyond release timing, volatility profoundly influences the physical and chemical behavior of radionuclides: it affects their interaction with structural materials, chemical speciation under different atmospheric conditions, and eventual aerosol formation and deposition processes within the primary system and containment. These factors collectively determine the environmental dispersal potential, inhalation hazards, and long-term contamination patterns. Consequently, a systematic classification of radionuclides by volatility—such as that detailed in Table 6.2—constitutes an essential tool for refining severe accident models, improving source term estimations, supporting containment

TABLE 6.2 Fission and Activation Product Volatility According to Sehgal (2012a) and (Soffer et al., 1995)

Group	Representative fission or APs
Noble gasses	Kr, Xe,
Volatile FPs	I, Br, Cs, Te, Rb
Semi-volatile FPs	Mo, Rh, Tc, Pd
Low-volatile FPs	Ru, Eu, Sr, Nb, La, Ce
Non-volatile FPs	Actinides, Zr, Nd

safety analyses, and ultimately guiding effective emergency preparedness and mitigation strategies.

The quantity of fission and APs released from nuclear fuel during an accident is governed by a complex interplay of several key parameters, with *temperature*, *atmospheric conditions*, and *chemical form* being among the most critical.

Temperature serves as the primary driving force for the release of noble gases and volatile FPs (e.g., I, Cs, Te) (Sehgal, 2012a; Hastings et al., 1985). These elements are released readily during the initial *gap release phase* following cladding failure and continue to be emitted throughout subsequent in-vessel and ex-vessel phases (as outlined in Table 6.2). The release mechanism for gaseous FPs is generally described as a two-step process: first, atomic or bubble *diffusion* through the fuel grain interior toward the grain boundaries (Kleykamp, 1985); followed by the accumulation, coalescence, and eventual *migration* of these bubbles to the free volumes of the fuel rod. Once the rod's integrity is compromised, these gases are rapidly vented into the reactor coolant system.

Oxidizing or reducing conditions within the reactor core profoundly influence the release of semi-volatile and low-volatility elements. For instance, the release of ruthenium (Ru) and molybdenum (Mo) is significantly enhanced under oxidizing atmospheres, whereas elements such as barium (Ba), strontium (Sr), lanthanum (La), europium (Eu), and rhodium (Rh) exhibit higher release rates under reducing conditions (Sehgal, 2012a; Pontillon & Ducros, 2010; Ducros et al., 2013). These behaviors have been quantitatively demonstrated in integral experimental programs such as PHÉBUS, and in separate-effect tests like the VERCORS series (Pontillon & Ducros, 2010; Ducros et al., 2013).

The *chemical form* and *microstructure* of the elements within the fuel matrix further dictate their release behavior. Noble metal FPs (e.g., Ru, Mo, Tc, Pd, Rh) tend to form metallic *white inclusions*—precipitates that influence their retention and release kinetics (Konings et al., 2015; Kleykamp, 1985). In contrast, elements like barium and zirconium are often found in oxidized states (e.g., Ba^{2+}, Zr^{4+}) (Geiger et al., 2016), affecting their volatility and interaction with other materials. *Rare earth elements (REs)*, such as cerium and lanthanum, typically dissolve in the fuel matrix as oxides (e.g., forming solid solutions like UO_2–REO_2 or heterogeneous regions with UO_2–RE_2O_3) (Kleykamp, 1985; Mariani et al., 2011), which impacts their release temperatures and modes during severe accident conditions.

The burn-up level and physical state of nuclear fuel are critical factors that significantly impact the kinetics and the total quantity of FPs released during an accident, as evidenced by theoretical studies on FP behavior. As nuclear fuel undergoes higher burn-up, it experiences microstructural alterations, including increased grain boundary porosity and a higher density of radiation-induced defects. These changes can facilitate the diffusion and release of specific radionuclides. Furthermore, under oxidizing conditions at high temperatures, UO_2 fuel can undergo oxidation to U_3O_8, accompanied by a substantial volumetric expansion of approximately 30%(Hunt et al., 1994). This transformation often leads to fuel fragmentation and pulverization, drastically increasing the surface area exposed to the oxidizing environment. As a result, the oxidation of uranium and the mobilization of FPs are accelerated, leading to notably higher release of both volatile and non-volatile species (Cox et al., 1990). This process is particularly critical for the release of otherwise low-volatility elements, such as ruthenium, since the fracturing of the fuel and the penetrating oxidative atmosphere promote the formation of volatile oxides (Hunt et al., 1994; Barrand et al., 1999).

Chemical interactions with cladding and structural materials also play a key role in retarding or altering the release of certain FPs. For example, elements like barium (Ba), tellurium (Te), and antimony (Sb) can be temporarily retained within the zirconium-based cladding through chemical reaction or dissolution. However, these retained species may be re-released during further oxidation of the cladding materials (Miwa et al., 2015). Additionally, the *fuel type* itself dictates release behavior: MOX fuel, with its heterogeneous microstructure and localized plutonium-rich agglomerates, often demonstrates higher release rates for certain nuclides compared to standard UO_2 fuel. This is attributed to the locally elevated burn-up and differing thermochemical properties within the MOX matrix (Sehgal, 2012a).

Given the multitude of coupled phenomena—including thermochemistry, oxidation kinetics, material interactions, and fuel microstructure evolution—the system governing FP and AP release is inherently complex. A comprehensive understanding requires not only the study of individual element behavior under varied atmospheric and thermal conditions but also the integrated analysis of their interdependent release pathways.

6.3 NUCLEAR ACCIDENT CASE STUDIES: RADIONUCLIDE RELEASE, MIGRATION, AND WASTE MANAGEMENT

The 1986 Chernobyl accident involved a catastrophic steam explosion and fire, releasing massive quantities of radionuclides (e.g., ^{131}I, ^{137}Cs) directly into the atmosphere, resulting in widespread atmospheric dispersion across Europe. In contrast, the 2011 Fukushima Daiichi Disaster, triggered by an earthquake and tsunami, led to core meltdowns and hydrogen explosions, characterized by significant initial atmospheric releases followed by persistent marine contamination from water ingress and deliberate discharges. Both accidents necessitated immense off-site waste management efforts, with waste from Chernobyl entombed on-site while Fukushima faces ongoing water treatment and storage challenges.

6.3.1 The Chernobyl Disaster

On April 26, 1986, Reactor 4 of the Chernobyl Nuclear Power Plant exploded during a safety test, resulting in the worst nuclear accident in history. The explosion led to 31 immediate deaths and an estimated 93,000 indirect fatalities due to radiation exposure, with 270,000 people developing cancer as a result. Operators, rushing to complete the test, violated procedures by disabling safety systems and over-withdrawing control rods while running the RBMK-1000 reactor at low power. The Chernobyl nuclear accident was caused by a critical design flaw in the RBMK-1000 reactor type, specifically a positive void coefficient, where the formation of steam led to increased reactivity and a subsequent power surge. As the coolant pumps slowed, steam bubbles formed in the reactor core, increasing reactivity because of the RBMK reactor's positive void coefficient and triggering a rapid power surge. The attempted emergency shutdown failed because the control rods were inserted too slowly, initially increasing reactivity. Within seconds, steam and hydrogen explosions ravaged the reactor, releasing a massive amount of radioactive material. The accident caused 31 immediate deaths, 28 later from acute radiation sickness, and spread contamination across Europe. Analysis showed that avoiding any one of the operational errors might have prevented the accident, highlighting the vital importance of safety culture, human factors, and robust design in nuclear engineering.

6.3.1.1 The Chernobyl Source Term

The catastrophic explosion of the core at Chernobyl Unit 4 on April 26, 1986, led to the initial, massive release of radioactive materials into the atmosphere. The release persisted over an extended period through multiple stages, each characterized by distinct compositions and quantities of radionuclides (IAEA, 1986). From April 26 to May 1, the release rate declined, likely due to the extinguishing of graphite fires and initial mitigation efforts. The release in this phase consisted mainly of finely dispersed nuclear fuel particles and white combustion aerosols, which exhibited a radionuclide composition similar to that of nuclear fuel. From May 1 to May 5, the release rate surged. Initially, volatile radionuclides, especially iodine isotopes, were predominant, followed by particles with a composition similar to that of fuel composition. This surge was attributed to the decay heat, which elevated fuel temperatures beyond 2000 °C. On May 6, the release abruptly decreased, likely owing to chemical interactions with materials deposited in the core. Measurable releases continued through late May, with notable peaks around days 20–21 and 25–30 after the accident (Buzulukov & Dobrynin, 1993).

Table 6.3 summarizes the quantitative estimates of the magnitude of the releases from the Chernobyl-4 Reactor. Based on various data, including measurements of ^{137}Cs in fuel masses inside the reactor sarcophagus and in the global fallout, the total core cesium released to the environment is estimated to be 33% ± 10%, or 2.3 ± 0.7 MCi (Belyaev et al., 1991). ^{131}I release was estimated at 50–60% of its initial inventory (40–50 MCi), corresponding to 30–35 MCi when accounting for decay. The ratio of gaseous to aerosol ^{131}I was approximately 3:7 in the first two weeks but shifted to about 10:1 by the third week (Buzulukov & Dobrynin, 1993).

Compared with the revised source terms of LWRs, the release fractions of volatile species (inert gases, iodine, and cesium) at Chernobyl were similar in magnitude. However, the non-volatile releases, estimated at 3.5% ± 0.5% of the core fuel inventory, far exceeded the typical estimates for pressurized water reactors due to the explosion and fire. Importantly, unlike engineered containment in Western reactors, Chernobyl's releases were directly atmospheric (IAEA, 1986; Belyaev et al., 1991).

TABLE 6.3 Estimated Releases from the Chernobyl-4 Disaster (Belyaev et al., 1991)

Noble Gases	**100%**	**190 ± 20**	**MCi**
^{131}I	55 ± 5%	45 ± 5	MCi
^{137}Cs	33% ± 10%	2.3 ± 0.7	MCi
^{90}Sr, ^{90}Y	4%	2.8 ± 0.8	MCi
Fuel	3.5 ± 0.5%		

6.3.1.2 Dispersion of the Chernobyl Disaster

As shown in Figure 6.6, the Chernobyl Nuclear Power Plant is located in the Kyiv region of Ukraine between 51° and 52° north latitude, just 15 km from the southern border of the Gomel region in Belarus (Izrael et al., 1990). The western part of Russia's Bryansk region lies about 150 km from the plant. The area is relatively flat, with elevations not exceeding 200 m. The climate is moderate with warm summers and mild winters, and the region experiences an average annual rainfall that has shown a statistically significant increasing trend, ranging from 20 to 24 inches (500 to 650 mm).

Following the catastrophic explosion at the Chernobyl Nuclear Power Plant on April 26, 1986, a thermal plume dispersed radioactive isotopes, including volatile forms of iodine and cesium, to altitudes ranging from 6 to 9 km. The accident, which released a significant amount of radioactive material into the atmosphere, had severe consequences for the surrounding environment and populations, with iodine-131 and cesium-137 isotopes contributing to widespread contamination across Europe. On April 27, dose rates measured 5 to 10 km from the reactor at a height of 200 m were about 1 rad/h, decreasing to 0.5 rad/h by April 28 (Izrael et al., 1990). On April 26, the radioactive plume was lifted and carried northwest by wind, passing over Latvia, Scandinavia, Lithuania, and northern Poland. The next day, wind direction shifted to the west and southwest, carrying the plume over southern Poland, Switzerland, Italy, Austria, southern Germany, and France. By the fifth day, winds again turned northwest, moving the cloud over central Germany, the Netherlands, and the UK. Eventually, the wind shifted northeast, dispersing fallout over central Russia. Some radioactive material was carried by jet streams to the United States and other countries. Trace amounts of ^{131}I were detected in Japan and the US by May 5 (Jaworowski & Kownacka, 1988).

The Chernobyl Disaster affected over 20 countries, with an estimated 20 billion people exposed to radiation, as reported by the Chinese Academy of Social Sciences. Deposition patterns were complex and varied, depending on particle density as well as weather conditions. The largest particles, mainly nuclear fuel particles, were deposited within 100 km of the reactor. Across this near zone, the former Soviet Union, and Europe, contamination levels depended heavily on rainfall during the plume's passage. Beyond the former USSR, following the Chernobyl Disaster, the Sami people in northern Sweden experienced severe consequences, with their reindeer herds becoming highly contaminated and deemed unfit for human consumption. Ukraine, Belarus, and Russia were most severely contaminated (see Figure 6.6) (Izrael et al., 1990; Marples, 1996). The three main contaminated zones, or "hot spots," are the Central, Bryansk-Belarus, and Kaluga-Tula-Orel areas (see Table 6.4).

The Central hot spot formed during the initial active release phase and was concentrated to the west and northwest. ^{137}Cs surface soil activity exceeded 1.0 $\mu Ci/m^2$ over large areas in Ukraine's Kyiv, Zhitomir, Chernigov, Rovno, and Lutsk regions, as well as in Belarus' Gomel and Brest regions. The most heavily contaminated area was within 30 km of the reactor, where ^{137}Cs levels often exceeded 40 $\mu Ci/m^2$. ^{137}Cs is a radioactive isotope with a half-life of approximately 30 years, and it has been known to cause significant environmental contamination as seen in various nuclear incidents. Beyond 30 km, significant contamination also occurred west and northwest of the reactor in the Gomel, Kyiv, and Zhitomir

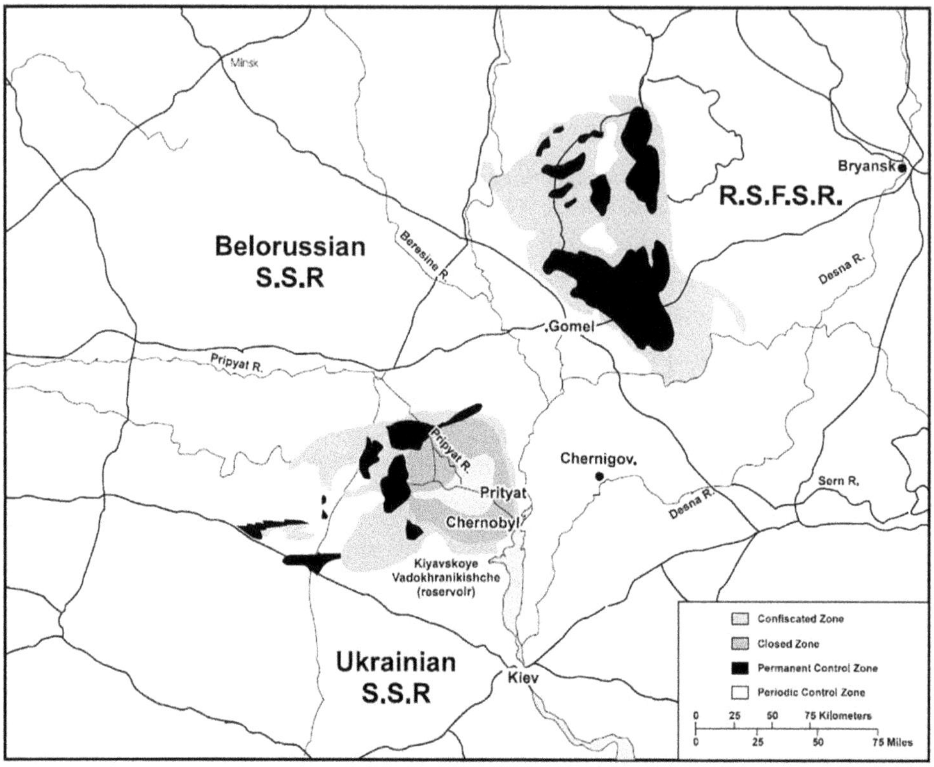

FIGURE 6.6 Radiation Hot Spots Resulting from Chernobyl Nuclear Power Plant Accident (Izrael et al., 1990).

TABLE 6.4 Characteristics of Hot Spots Resulting from the Chernobyl Disaster (Izrael et al., 1990)

Hot Spot	Direction	^{137}Cs Soil Surface Activity (μCi/m^2)	Initial Dose Rate (mrad/h)	1991 (5 yr) Dose Rate (mrad/h)
Central	W-NW	up to >40	1 to 200	0.005 to 1
Bryansk-Belarus	Centered approximately 200 km N–NE of the reactor	>130	0.3 to 30	0.0005 to 0.5
Kaluga-Tula-Orel	Centered approximately 500 km NE of the reactor	<0.16	0.3 to 3	0.0005 to 0.05

regions. Initial gamma dose rates (1 m above ground) ranged from 1 to 200 mrad/h. By 1991, these had decreased to 0.005–1 mrad/h.

The Bryansk-Belarus hot spot, located about 200 km north-northeast of Chernobyl, resulted from rainfall on April 28–29, 1986, in Russia's Bryansk region and the adjacent Gomel and Mogilev regions of Belarus. In the most contaminated areas of the Chernobyl Disaster zone, the surface activity of ^{137}Cs was comparable to that in the Central zone, with some villages in Mogilev exceeding 130 μCi/m^2 and Zaborye village in Bryansk reaching 110 μCi/m^2. Initial airborne dose rates ranged from 0.3 to 30 mrad/h. By 1991, they had fallen to 0.005–0.5 mrad/h. Due to the dominance of long-lived ^{137}Cs, this region is referred to as a "cesium hot spot."

The Kaluga-Tula-Orel hot spot, about 500 km northeast of the reactor in Russia, is also a "cesium hot spot." Like the Bryansk-Belarus area, it was formed by rainfall on April 28–29, but with lower ^{137}Cs contamination levels, below 0.16 μCi/m^2. Initial dose rates were 0.3–3.0 mrad/h, declining to 0.005–0.05 mrad/h by 1991.

Beyond these three major hot spots, many other areas in the European part of the former USSR had ^{137}Cs contamination levels mainly in the range of 1–5 μCi/m^2. In total, within the former Soviet Union, an area of approximately 3,100 km^2 had ^{137}Cs levels >40 μCi/m^2; 7,200 km^2 had 16–40 μCi/m^2; 17,600 km^2 had 5–16 μCi/m^2; and 103,000 km^2 had 1–5 μCi/m^2(Izrael et al., 1990).

In areas where ^{137}Cs contamination exceeded 140 μCi/m^2, the total ^{137}Cs activity was about 1.0 MCi, including 0.3 MCi within 40 km of the reactor (Izrael et al., 1990). Including larger areas with ^{137}Cs levels below 1.0 μCi/m^2, the total ^{137}Cs deposition over the former USSR was approximately 1.3 MCi, with 95% deposited in Europe and 5% beyond Europe (east of the Ural Mountains).

Most of the released ^{90}Sr was deposited near the accident site. In fact, areas with ^{90}Sr surface activity exceeding 3.0 μCi/m^2 were almost entirely within 30 km of the plant. Regions with ^{90}Sr levels above 1.0 μCi/m^2 were almost entirely within 100 km. Only a few locations in the Bryansk-Belarus hot spot had ^{90}Sr concentrations between 1.0 and 3.0 μCi/m^2.

Information on plutonium isotope contamination is limited due to measurement difficulties. The only "plutonium hot spot" with ^{239}Pu and ^{240}Pu surface activity exceeding 0.1 μCi/m^2 was entirely within 30 km of the site. In the Bryansk-Belarus and Kaluga-Tula-Orel hot spots, plutonium activity ranged from 0.002 to 0.02 μCi/m^2 and 0.002 to 0.008 μCi/m^2, respectively. While ^{137}Cs and ^{90}Sr concentrations correlated well in these regions, no clear correlation was found between Pu and either ^{137}Cs or ^{90}Sr levels.

6.3.1.3 The Chernobyl Disaster Waste

Cleanup activities over the past three decades have resulted in the generation of a vast quantity of radioactive waste, which is presently managed within 17 storage or disposal facilities inside the Exclusion Zone (see Table 6.5) (IAEA, 2022). These facilities collectively contain over 13 million m^3 of material. The future demolition of Reactor 4 and the original Shelter Structure is expected to produce additional low-level waste (LLW) for on-site disposal, while intermediate- and high-level waste (ILW and HLW) will be placed in long-term storage within the New Safe Confinement (NSC). The final disposition of fuel-containing wastes, particularly spent nuclear fuel, remains undecided, with countries like China considering both reprocessing and direct disposal strategies. and the precise location of all fuel debris in and around the site is still not fully agreed upon.

As indicated in Table 6.5, the largest component of the waste consists of contaminated soils. A significant complicating factor is the presence of fine fuel particles within some of these soils—a plume of such debris was deposited unevenly across the zone, primarily to the west and north of the NPP, stretching up to 100 km away in a narrow band (IAEA,

TABLE 6.5 Estimated Total Amount of Radioactive Waste within the Chernobyl Exclusion Zone (Oskolkov et al., 2011)

Type of storage facility	Number of facilities	Waste volume (m^3)	Activity (TBq)
Disposal sites	3	631,519	5,420
Temporary storage facilities	9	1,296,588	1,840
Shelter object (including industrial site)	1	662,500	740,000
NPP storage facilities	4	19,794	385
Exclusion Zone soils	–	11,000,000	8,130
Total	17	13,610,401	755,775

TABLE 6.6 Detailed Breakdown of Wastes Resulting From The Chernobyl Disaster (IAEA, 2022)

Radioactive Material	RW Type	Quantity	Mass (t)/Volume (m^3)
Shelter Object—Fuel-Containing Materials			
Fuel assemblies (fresh and spent)	HLW	177 pieces	20.3
Lava-fuel containing masses	HLW	—	1,250
Graphite	HLW/ILW	—	≈700
On-site and Exclusion Zone Storage			
Buried nuclear fuel	HLW	—	≈0.6
Bulk soil (gravel, sand, and crushed stone)	LLW	—	137,000 m^3
Concrete and blocks	LLW	—	131,000 m^3

2022). The most prevalent radioactive contaminants are FP radionuclides, such as ^{3}H, ^{14}C, ^{129}I, ^{90}Sr, and ^{60}Co, as seen in the case of the Fukushima nuclear wastewater.

A more detailed classification of waste materials is provided in Table 6.6. In the aftermath of the Chernobyl Disaster, the initial waste collection and removal of contaminated materials began within the first 18 months post-accident, occurring simultaneously with the erection of the sarcophagus (Shelter) over the damaged reactor and the decontamination of the surrounding industrial area. During the most hazardous phase of the Chernobyl Disaster, workers faced the daunting task of retrieving core and reactor debris, which was emitting extremely high levels of radiation. Military personnel using shielded armored vehicles gathered scattered debris into metal containers. The contaminated topsoil was bulldozed to the boundaries of the damaged reactor building and covered with concrete slabs. The shift durations were brief, and the equipment underwent decontamination after each use. Some HLW containers were initially buried, but later a system of "pioneer walls" was developed around the reactor building. The waste containers were strategically placed behind the Shelter Structure and integrated by layering gravel, sand, and concrete, contributing to the robustness of the Chernobyl Shelter.

Temporary storage was used for topsoil from areas that were not sealed with concrete. This HLW, which also contained fuel and reactor debris, was collected in 1-m^3 containers, some of which were placed in an unfinished LLW disposal facility on-site (Samojlenko & Golubev, 1989). In total, over 500,000 m^2 of topsoil was removed. A particularly dangerous task was the manual clearance of reactor core material, which included the removal of 16.5

tonnes of fuel and 182 tonnes of graphite, a substance that plays critical roles in nuclear reactors such as moderation and reflection, from the roofs of adjacent buildings, including Unit 3. Materials were either pushed into the reactor void or collected and removed using a crane or helicopter (Samojlenko & Golubev, 1989).

Major internal decontamination was also required in Unit 3, and Units 1–2 were contaminated through the ventilation systems. Between 1987 and 1988, an extensive cleanup operation was conducted at the Chernobyl nuclear site, involving over 24 million m^2 of facilities and 6 million m^2 of surrounding areas. This monumental task resulted in the removal of 38,000 tons of contaminated equipment and the safe containment of over 16,000 tons of radioactive waste. Following the accident, cleanup operations concentrated on decontaminating areas within a 30-km radius, particularly those within a 5-km perimeter of the NPP, encompassing Pripyat, due to the high levels of surface contamination. Many contaminated buildings were demolished. Large-scale decontamination of the town was unsuccessful, with cleaned areas quickly becoming re-contaminated (IAEA, 2022).

6.3.2 The Fukushima Daiichi Disaster

The Fukushima Daiichi Disaster was triggered by the catastrophic Great Tohoku Earthquake and subsequent tsunami on March 11, 2011, leading to a severe nuclear accident. The tsunami, surpassing the plant's protective measures, resulted in a prolonged loss of off-site power and failure of emergency diesel generators, leading to core meltdowns in Units 1–3 and subsequent hydrogen explosions. The Fukushima Daiichi reactors (Units 1–5 utilized MARK-I type containments, Figure 6.7) are boiling water reactors (BWRs) (Theriault, 2016). A significant design aspect of the MARK-I containment is its relatively small volume and the location of its suppression pool (torus). This design influenced the accident's progression and the strategy for managing radioactive releases.

The Fukushima Daiichi Disaster was initiated on March 11, 2011, by a magnitude 9.0 earthquake and a subsequent tsunami reaching heights of approximately 15 m, which overpowered the plant's protective measures. The tsunami resulted in a prolonged station blackout (SBO), which disabled all power supplies for Units 1–4. Following the magnitude 9.0 earthquake and subsequent tsunami on March 11, 2011, the Fukushima Daiichi Nuclear Power Plant experienced a catastrophic failure of its core cooling systems. The isolation condensers in Unit 1 and the reactor core isolation cooling systems in Units 2 and 3 were compromised, leading to the meltdown of the reactor cores in Units 1, 2, and 3 within hours. The meltdown in Unit 1 occurred approximately 15 hours post-earthquake, while Unit 2 experienced a meltdown around 101 hours later, on March 15. Unit 3's meltdown took place approximately 60 hours after the earthquake, on March 14. Hydrogen explosions subsequently destroyed the reactor buildings of Units 1, 3, and 4. Containment venting and failures at the Fukushima Daiichi Nuclear Power Plant resulted in significant radioactive releases, with estimates reaching up to 900,000 terabecquerels into the atmosphere, as reported by Tokyo Electric Power Company (TEPCO). The accident highlighted the vulnerability of the MARK-I containment design and the severe consequences of extended SBO.

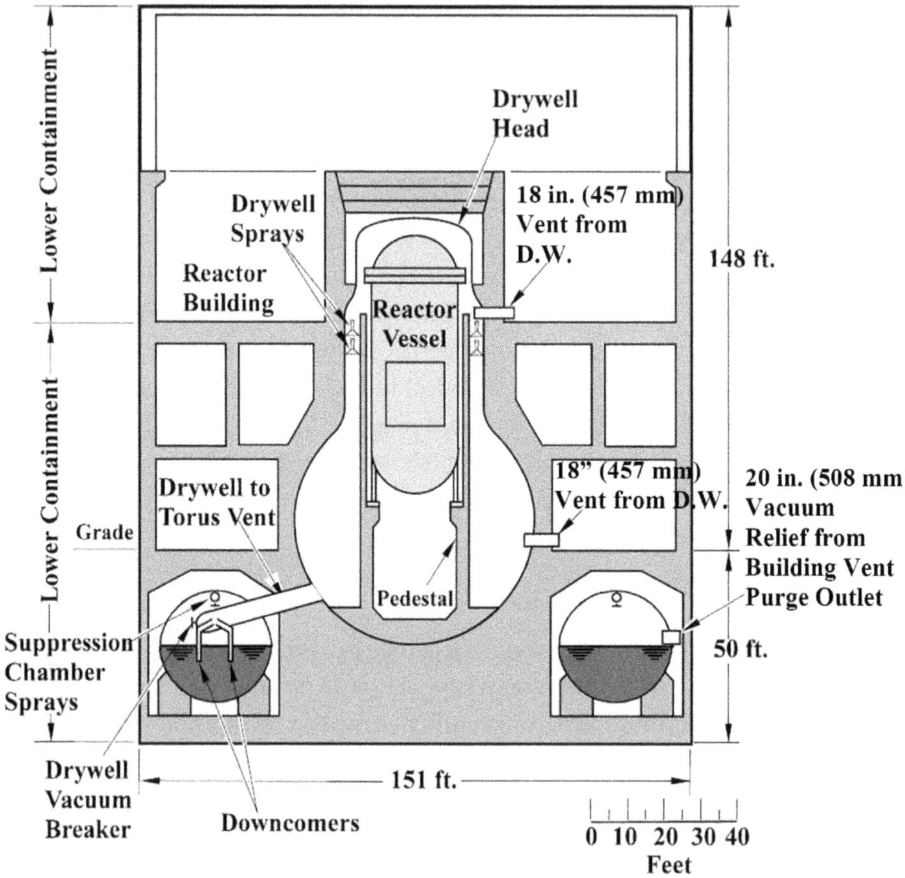

FIGURE 6.7 BWR Mark I Containment (IAEA, 2015).

Note: D.W. = drywell.

6.3.2.1 The Fukushima Daiichi Source Term

During the Fukushima Daiichi Disaster, Unit 1 of the Fukushima Daiichi Nuclear Power Plant, which used standard UO_2 pellets (400 assemblies), experienced an almost complete core meltdown. Unit 2, utilizing standard UO_2 pellets (587 assemblies), and Unit 3, employing both standard UO_2 pellets (512 assemblies) and uranium-plutonium mixed MOX pellets (32 assemblies), experienced severe damage to over 80% of their cores (see Figure 6.8) (IAEA, 2015). The Fukushima Daiichi Disaster resulted in a significant dispersion of radioactive materials, with the release being influenced by meteorological conditions and containment failure modes. The primary sources of the radioactive wastewater include the original primary coolant from the reactors, water continuously injected after the accident to cool the molten core debris, and large amounts of infiltrated groundwater. This wastewater carries a significant load of molten debris fragments, along with FPs and transuranic nuclides released from them, resulting in a highly complex composition.

The Fukushima Daiichi Disaster source term (see Table 6.7 for core inventories of Units 2 and 3) has resulted in wastewater containing large quantities of FPs and highly toxic

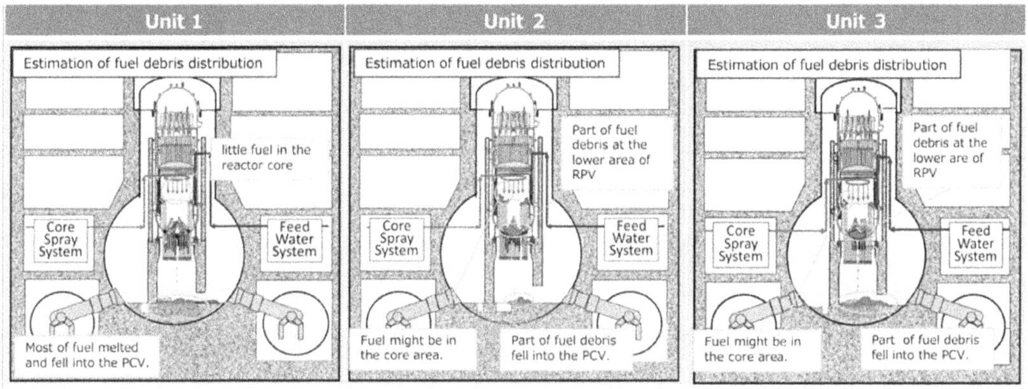

FIGURE 6.8 TEPCO and International Research Institute for Nuclear Decommissioning (TEPCO and IRID, 2017) Locating Fuel Debris inside the Unit 3 Reactor using Muon Measurement Technology at Fukushima Daiichi Nuclear Power Station (IAEA, 2015).

transuranic nuclides such as plutonium and curium being released after core damage. The wastewater contains a diverse array of radionuclides with varying half-lives, ranging from short durations to those exceeding millions of years, indicating a complex mixture of radioactive substances that pose significant environmental and health risks. Among these, radionuclides that undergo pure beta decay (e.g., ^{3}H, ^{90}Sr) or pure alpha decay (e.g., ^{238}Pu, ^{242}Cm) can form radioactive colloids with high mobility, which are difficult to identify using conventional measurement methods. Furthermore, Unit 3 used MOX fuel, resulting in a 20% higher plutonium inventory in its core compared to Unit 2, which used standard UO_2 pellet fuel. Plutonium-containing wastewater readily forms electrically neutral radioactive colloids with strong migration capacity.

6.3.2.2 Dispersion of the Fukushima Daiichi Disaster

The Fukushima Daiichi Disaster released radionuclides into the environment, as assessed by multiple models (Eslinger et al., 2014; Stohl et al., 2012; Buesseler et al., 2011). Following the Fukushima Daiichi Disaster, atmospheric releases primarily dispersed eastward over the Pacific Ocean, with minor northwest deposition on land, as evidenced by the detection of radioactive cesium in the northern Pacific and the Bering Sea. Direct oceanic discharges, such as those from the Fukushima Daiichi Disaster, have led to severe contamination. For instance, radioactive cesium levels in sea fish from the Fukushima Daiichi Nuclear Power Plant's port were found to be 180 times the legal limit, as reported by Nippon Hoso Kyokai (NHK), Japan's public broadcaster.

6.3.2.2.1 Releases Multiple assessments using mathematical models and computer codes have estimated radionuclide releases from the Fukushima Daiichi Disaster (Eslinger et al., 2014; Stohl et al., 2012; Buesseler et al., 2011). Early atmospheric releases from nuclear accidents, such as the one at Fukushima, included noble gases like ^{85}Kr (half-life 10.76 years) and ^{133}Xe (5.25 days), which contributed to external

TABLE 6.7 Estimated Core Inventories of Fukushima Daiichi Units 2 and 3 (IAEA, 2015)

Nuclides	Units 3	Units 2	Ratio
^{134}Cs	2.52E+17	2.76E+17	−9%
^{136}Cs	8.18E+16	8.17E+16	0%
^{137}Cs	2.41E+17	2.55E+17	−5%
^{131}I	2.33E+18	2.34E+18	0%
^{133}I	4.90E+18	4.90E+18	0%
^{132}Te	3.37E+18	3.36E+18	0%
^{127}Te	1.78E+17	1.74E+17	2%
^{127m}Te	1.33E+16	1.21E+16	10%
^{140}Ba	4.35E+18	4.35E+18	0%
^{125}Sb	1.57E+16	1.65E+16	−5%
^{95}Nb	3.26E+18	2.88E+18	13%
^{89}Sr	2.35E+18	2.21E+18	6%
^{90}Sr	1.81E+17	1.91E+17	−5%
^{140}La	4.42E+18	4.43E+18	0%
^{95}Zr	3.68E+18	3.40E+18	8%
^{238}Pu	5.53E+15	4.57E+15	21%
^{239}Pu	1.04E+15	8.83E+14	18%
^{240}Pu	1.36E+15	1.04E+15	31%
^{241}Pu	3.15E+17	2.81E+17	12%
^{110m}Ag	5.99E+15	6.41E+15	−7%
^{85}Kr	2.95E+16	3.11E+16	−5%
^{131m}Xe	2.60E+16	2.57E+16	1%
^{133m}Xe	1.40E+17	4.67E+18	−97%
^{133}Xe	4.67E+18	1.37E+18	241%
^{86}Rb	4.35E+15	4.50E+15	−3%
^{125m}Te	5.00E+15	5.38E+15	−7%
^{99m}Tc	3.86E+18	3.88E+18	−1%
^{99}Mo	4.42E+18	4.43E+18	0%
^{95m}Nb	3.29E+16	3.03E+16	9%
^{124}Sb	7.99E+14	7.91E+14	1%
^{123}Sn	1.39E+15	1.33E+15	5%
^{125}Sn	1.06E+16	1.03E+16	3%
^{111}Ag	1.12E+17	1.07E+17	5%
^{132}I	3.43E+18	3.43E+18	0%
^{151}Sm	9.08E+14	8.87E+14	2%
^{239}Np	4.30E+19	4.32E+19	0%
^{244}Cm	2.71E+15	3.21E+15	−16%
^{237}U	1.28E+18	1.32E+18	−3%
^{156}Eu	3.12E+17	3.21E+17	−3%
^{90}Y	1.85E+17	1.95E+17	−5%
^{147}Pm	4.07E+17	4.19E+17	−3%
^{242}Cm	1.04E+17	8.94E+16	16%
^{106}Ru	8.48E+17	8.71E+17	−3%
^{103}Ru	3.25E+18	3.00E+18	8%
^{144}Ce	2.27E+18	2.31E+18	−2%

exposure and environmental contamination. The short-lived iodine-131 (with a half-life of 8.02 days) presented a risk of thyroid exposure when inhaled or ingested, as seen in the aftermath of the Fukushima Daiichi Disaster. In contrast, the long-lived isotopes cesium-134 (half-life of 2.06 years) and cesium-137 (half-life of 30.17 years) contributed to both external and internal exposure risks. Strontium, ruthenium, and actinides (e.g., plutonium) were also released in smaller quantities. Neutrons detected near the plant on March 13 suggested spontaneous fission from damaged fuel (IAEA, 1991; Kaul et al., 1996; Pflugbeil, 2006).

The Fukushima Daiichi Disaster released an estimated 6.4–32.6 PBq of ^{85}Kr, 400–11,000 PBq of ^{133}Xe, 140–200 PBq of ^{131}I, and 12–16 PBq of ^{137}Cs into the atmosphere, which is roughly one-tenth of the radioactive releases from the Chernobyl Disaster. Most radionuclides dispersed over the North Pacific, limiting ground verification (Buesseler et al., 2011). Direct oceanic releases peaked in April 2011, with ^{131}I at 10–20 PBq and ^{137}Cs at 1–6 PBq (up to 15 PBq per some estimates (IAEA, 2015)).

6.3.2.2.2 Dispersion The dispersion of radionuclides from the Fukushima Daiichi Disaster was assessed using a combination of theoretical models, including Fick's second law and Gaussian plume models, and extensive environmental measurements of ^{131}I, ^{134}Cs, and ^{137}Cs in air, soil, seawater, sediments, and biota. These models considered factors such as atmospheric turbulence, cloud cover, solar radiation, wind speed, and terrain to predict the spread of radioactive materials. Guided by prevailing winds, atmospheric transport has primarily carried radioactive releases eastward over the Pacific Ocean, with potential global implications. Figure 6.9 exemplifies atmospheric modeling for ^{137}Cs, showing a sharp decrease in activity concentration with increasing distance from the plant (IAEA, 2015). Sensitive global monitors detected trace-level radioactivity as far as Europe and North America, though the impact on global background radiation was negligible.

Oceanic dispersion was dominated by the Kuroshio Current and the North Pacific gyre, which transported and greatly diluted radionuclides eastward (Wu et al., 2013). Activity was detected in distant marine life, such as bluefin tuna (Madigan et al., 2012), but at minimal levels. Figure 6.10 presents various oceanic models simulating the distribution of ^{137}Cs in seawater, all confirming very low activity concentrations that decreased markedly with distance from the release point, as evidenced by the study on the Fukushima Daiichi Disaster's impact on the northwest Pacific Ocean.

6.3.2.2.3 Deposition Deposition density, measured in Bq/m^2, quantifies radionuclide activity deposited on Earth's surface. For oceanic deposition of ^{137}Cs, models were used for estimation (Figure 6.11), though accurately determining the amount of atmospherically released ^{137}Cs deposited on the ocean remains challenging (Rypina et al., 2013). Pre-accident global ^{137}Cs background from nuclear testing was substantial, estimated at 290 ± 30 PBq (1970) and ~69 PBq in the North Pacific (Aoyama et al., 2006).

The primary terrestrial deposition from radioactive releases on March 12, 14, and 15, 2011, at the Fukushima Daiichi Nuclear Power Plant occurred northwest of the facility, involving significant amounts of ^{131}I, ^{134}Cs, and ^{137}Cs, as detailed in the global survey

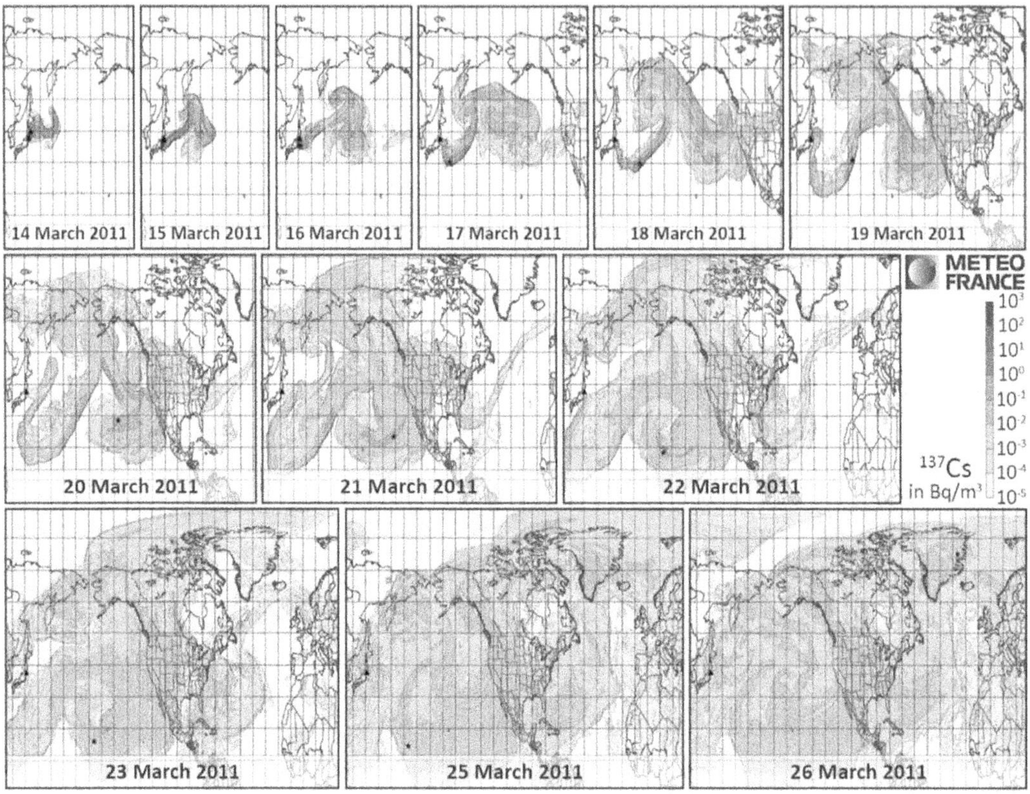

FIGURE 6.9 Results from One of the Global Models of the Atmospheric Dispersion of ^{137}Cs (IAEA, 2015).

findings. Patterns were highly heterogeneous, influenced by precipitation, topography, and the distinct physical-chemical properties of iodine and cesium. Total ^{137}Cs deposition in the worst-hit northwestern zone was estimated at 2–3 PBq (IAEA, 2015). Deposition density decreases over time through physical decay, environmental processes (e.g., weathering, solubilization), and remediation. Figure 6.12 shows the aerial measurement of the ambient dose equivalent rate in the northwest region (IAEA, 2015).

Following the Fukushima Daiichi Disaster, the levels of ^{137}Cs deposition northwest of the plant were extraordinarily high, with localized hotspots reaching up to 10^7 Bq/m^2, an average of around 10^6 Bq/m^2, and a prefectural average of approximately 10^5 Bq/m^2. Levels decreased significantly outside the core area to ~10^4 Bq/m^2 and were generally below 10^3 Bq/m^2 in most of Japan (IAEA, 2015). Following the Fukushima Daiichi Disaster in 2011, initial ^{131}I deposition reached levels of approximately 3×10^6 Bq/m^2, but its concentration rapidly diminished due to the short half-life of iodine-131. The presence of ^{137}Cs leads to protracted human exposure, adding to global background levels from past nuclear tests, which were historically as high as 4000 Bq/m^2 in the mid-1960s in the Northern Hemisphere and had decayed to ~2000 Bq/m^2 by 2000 (UNSCEAR, 2000).

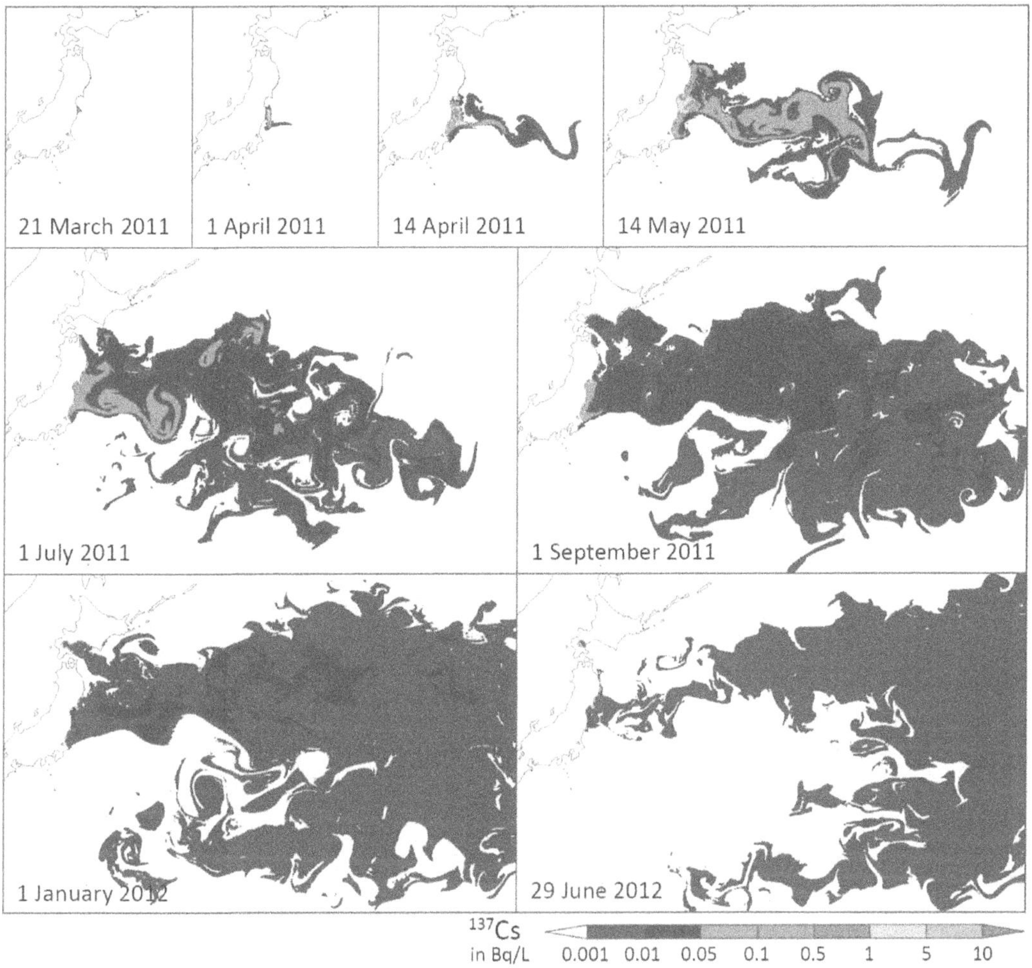

FIGURE 6.10 Oceanic Models Have Been Used to Estimate the Activity Concentration of ^{137}Cs in Seawater. (a) An Example of Modeling Contaminated Waters from March 21, 2011, to June 29, 2012 (Masumoto et al., 2012). (b) The Horizontal Distribution of ^{137}Cs in Surface Waters Was Simulated for the Period between April 14 and 26, 2011, as Part of a Broader Study on the Dispersion of Radioactive Materials Following the Fukushima Disaster (Honda et al., 2012).

6.3.2.3 The Fukushima Daiichi Disaster Waste

By the end of 2011, the emergency phase of the Fukushima Daiichi NPP accident had been managed, resulting in a cold shutdown of the reactors. This milestone marked the initiation of the primary on-site stabilization and off-site remediation efforts. Historically, the absence of specific legislation to govern the disposal of accident waste and radioactively contaminated materials has been a significant issue, as evidenced by the aftermath of the Fukushima Daiichi Disaster, and the existing near-surface waste disposal facility for radioactive waste from normal operation at the nuclear power station is not available for the disposal of radioactively contaminated materials. The wastes generated in the accident,

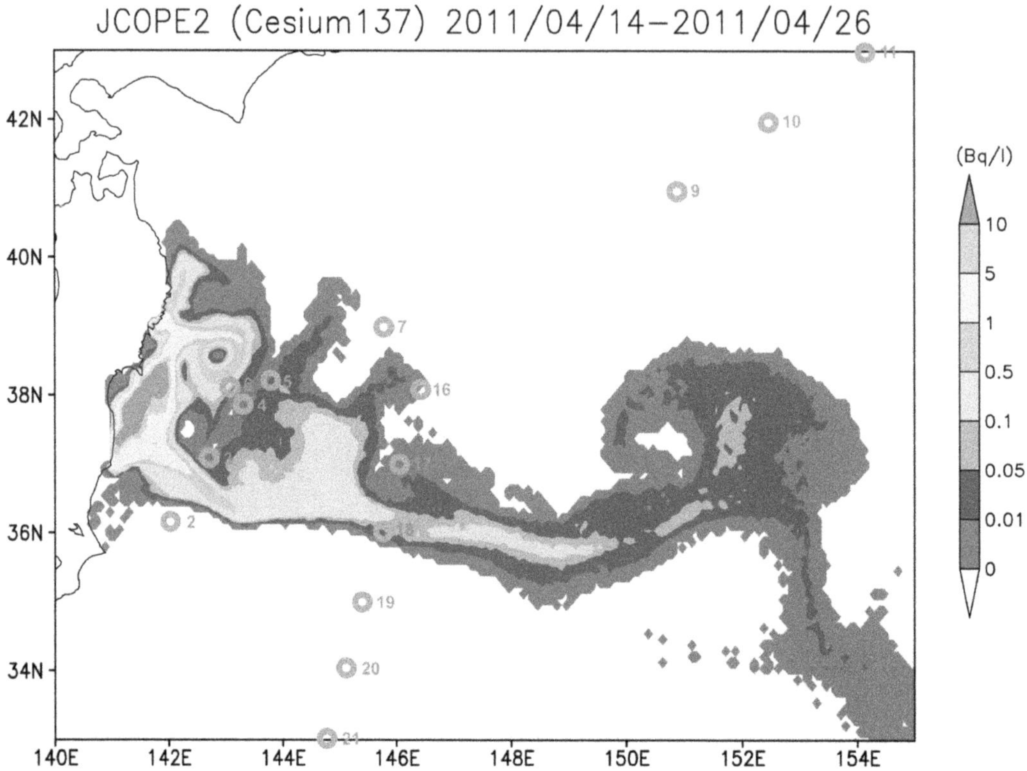

FIGURE 6.10 (Continued)

including the nuclear debris analyzed and the radioactive waste quantified, are summarized in Table 6.8.

6.3.2.3.1 Off-Site Remediation Wastes Remediation activities in Fukushima have centered on decontaminating residential and agricultural areas to reduce radiation exposure, generating significant radioactive waste. A structured management strategy includes: temporary storage near collection points; volume reduction through incineration and soil washing; and final disposal based on radioactivity levels. Approximately 800,000 tonnes of contaminated waste and an additional 160,000 tonnes of designated waste (>8,000 Bq/kg) have been handled. Waste is segregated by cesium activity, with lower-activity materials managed as municipal waste, while higher-activity wastes require specialized disposal. Near Fukushima Daiichi, an interim storage facility (ISF) is planned to hold up to 22 million m^3 of treated soil and waste, with a transitional storage facility in the area having a maximum capacity of approximately 22 million m^3. Final disposal methods for high-activity wastes remain under development.

6.3.2.3.2 On-Site Decommissioning Wastes Significant quantities of solid radioactive waste have been produced as a result of the stabilization and decommissioning efforts at the

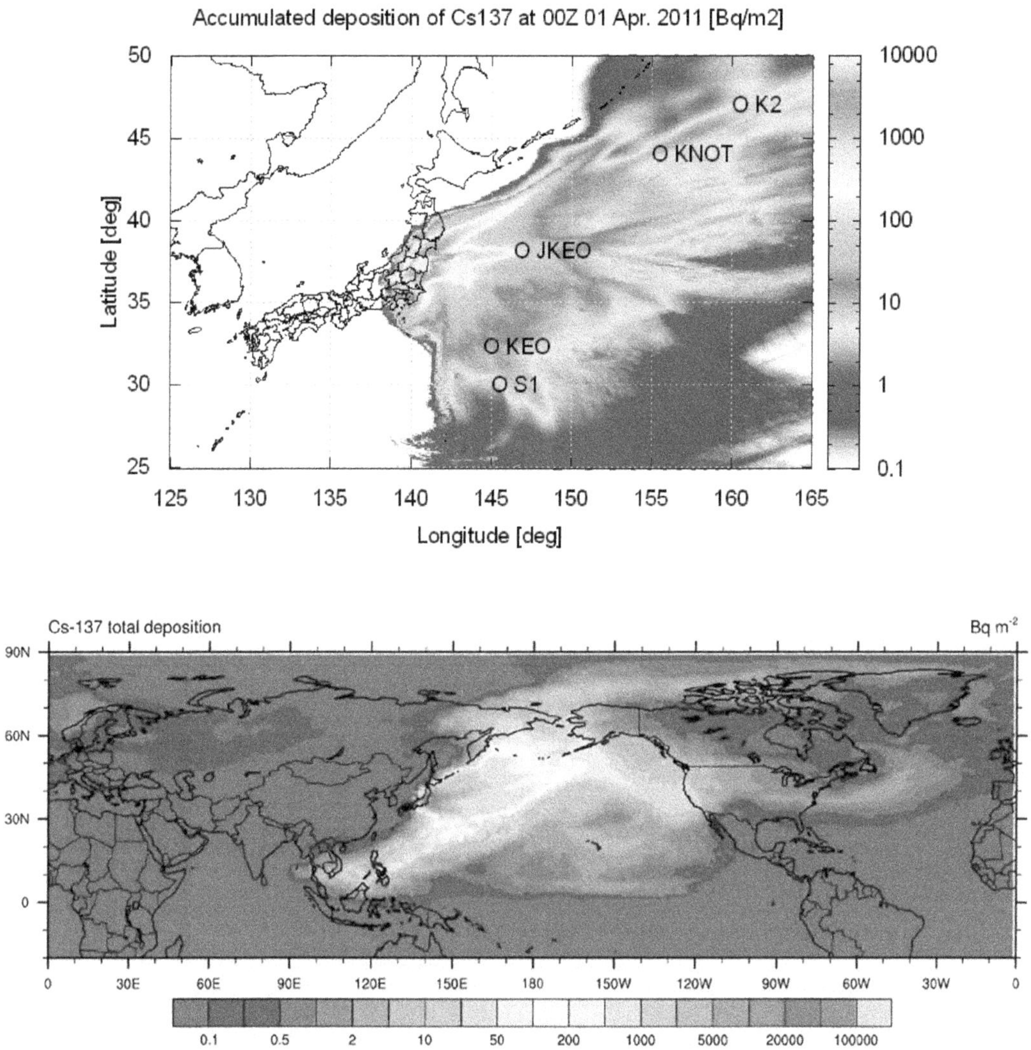

FIGURE 6.11 Various Models Have been Utilized to Estimate the Oceanic Deposition Density of ^{137}Cs, with Units Expressed in Bq m^{-2}. (a) Modelling the Cumulative Aeolian Input through April 1, 2011 (Honda et al., 2012). (b) An Example of the Ensemble-Averaged ^{137}Cs Deposition (March 11–31, 2011) (Bailly du Bois et al., 2014).

compromised Fukushima Daiichi reactors. By mid-2015, approximately 160,000 m^3 of debris and 83,000 m^3 of felled trees were stored on-site in various structures (see Figure 6.13), segregated by dose rate and material type. The decommissioning of the Fukushima Daiichi Nuclear Power Station is a complex, multi-layered project expected to span 30 to 40 years, involving multiple stages, as detailed in the TEPCO timeline. Conservative estimates, in line with studies on the Fukushima Daiichi Nuclear Power Plant incident, project that approximately 560,000 m^3 of contaminated material will ultimately be generated. A centralized storage facility with a capacity of 160,000 m^3 is planned, with the volume difference

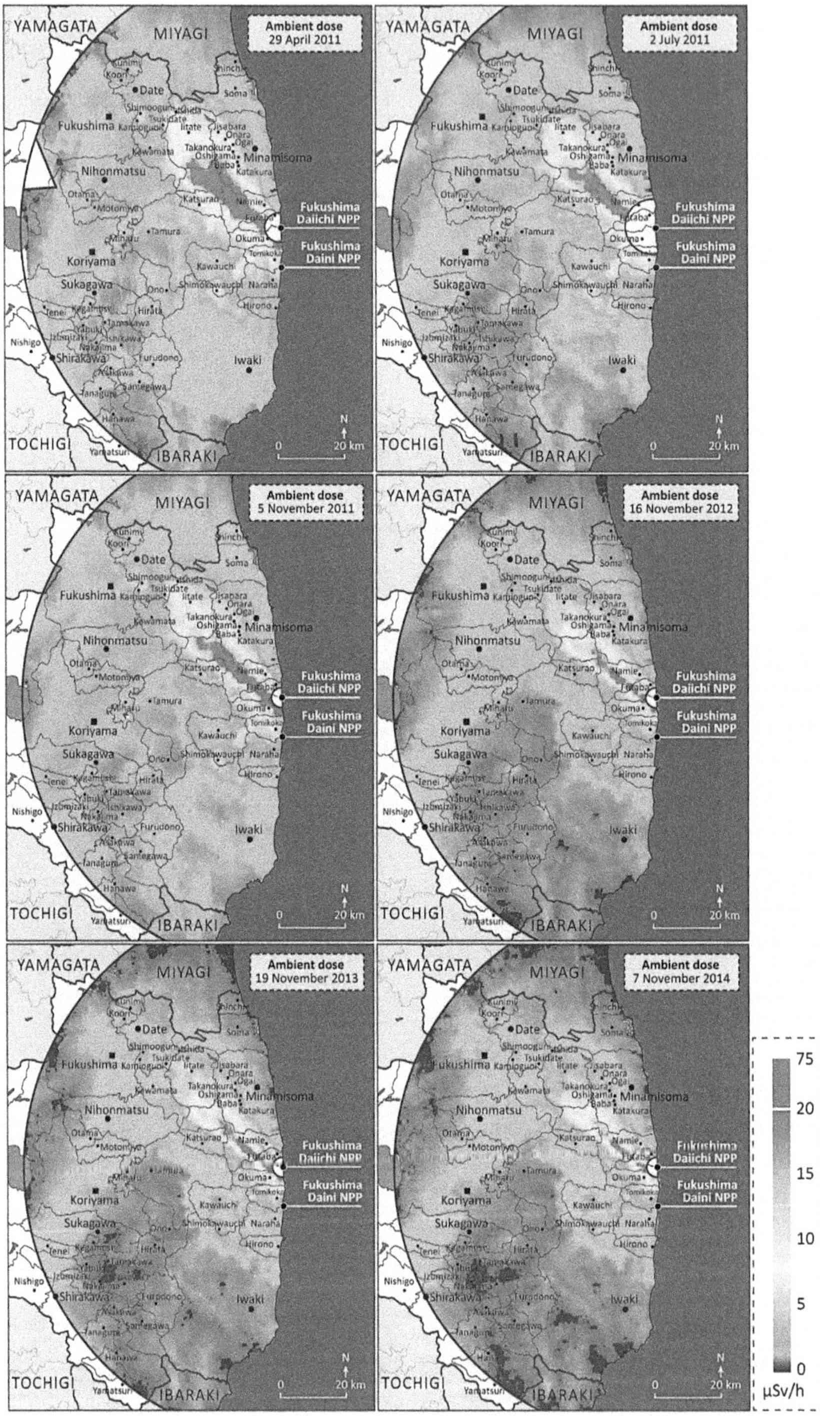

FIGURE 6.12 Measured Aerial Ambient Dose Equivalent Rate (in μSv/h) Resulting from Deposits from the Releases That Spread in Areas to the Northwest of the Plant (IAEA, 2015).

TABLE 6.8 Solid and Liquid Wastes Resulting From the Fukushima Daiichi Disaster

Waste Type	Volume/Mass	Activity/Notes
	On-Site Solid Waste	
Rubble	268,200 m^3	<0.1 mSv/h
Felled trees	134,200 m^3	
	Off-Site Waste	
Contaminated soil	~14 million m^3	
Incineration ash	20,000 m^3	>100,000 Bq/kg
Designated wastes	260,612 m^3	>8,000 Bq/kg
	Liquid Waste	
Treated water	1,201,434 m^3	Stored on-site

to be addressed through waste segregation, volume reduction, and recycling. An on-site incinerator began operation in 2016 to support these efforts.

6.3.2.3.3 Liquid Waste Management The management of large volumes of contaminated water remains a significant challenge, with daily treatment capacities ranging from 150,000 m^3 to 115 million m^3 across various facilities. While the advanced liquid processing system (ALPS) removes most radionuclides, it is unable to remove tritium, thus preventing its discharge. Over 1 million tons of treated water are stored on-site in tanks, with capacity expected to be exceeded by 2022 (Magwood IV et al., 2021). Secondary wastes from treatment are stored in concrete containers, with vitrification under evaluation for volume reduction.

International collaboration with organizations such as the International Ocean Garbage Partnership (IOGP), the European Union, and the United Nations Environment Programme (UNEP) continues to address unprecedented waste management challenges through initiatives like the "Marine Plastic Waste Management Plan," "Marine Plastic Waste Management Projects," and "Marine Plastic Waste Cleanup Actions." Cleanup and decommissioning will persist for decades.

6.4 SUMMARY

The Chernobyl (1986) and Fukushima Daiichi (2011) accidents represent the most severe nuclear crises, each rated Level 7 on the INES scale. Chernobyl's steam explosion and fire released massive radionuclides directly into the atmosphere, causing widespread contamination in Europe. Fukushima, triggered by an earthquake and tsunami, led to core meltdowns and hydrogen explosions, resulting in significant atmospheric releases and prolonged marine contamination via water ingress and deliberate discharges.

Both accidents released volatile FPs (e.g., noble gases, ^{131}I, ^{137}Cs) and less-volatile nuclides (e.g., ^{90}Sr, Pu isotopes), with dispersion influenced by meteorology and ocean currents.

FIGURE 6.13 Decontamination and Collection of Soil in an Agricultural Area and a Temporary Soil Storage Facility. Courtesy of Okuma Town (IAEA, 2022).

Chernobyl's deposition was the highest northwest of the site, forming "hot spots" with ^{137}Cs levels exceeding 10^7 Bq/m^2. Fukushima's terrestrial contamination was also concentrated northwest, while oceanic dispersion via the Kuroshio Current diluted radionuclides in the Pacific.

Waste management remains a critical challenge. Chernobyl generated ~13.6 million m^3 of waste, including contaminated soil, debris, and fuel-containing materials, which was managed within the Chernobyl Exclusion Zone. Fukushima produced vast amounts of contaminated water (over 1.2 million m^3 stored on-site) and solid waste (e.g., ~14 million m^3 of soil, incineration ash). Ongoing efforts include water treatment, waste segregation, and development of interim storage facilities, with decommissioning expected to take decades. Lessons from both accidents underscore the need for robust containment, improved emergency response, and international cooperation in waste management and environmental remediation.

REFERENCES

IAEA, 2013. *INES: The International Nuclear and Radiological Event Scale User's Manual*. 2008 Edition. International Atomic Energy Agency, Vienna.

Alameri, S.A., Alkaabi, A.K., 2020. Fundamentals of nuclear reactors. In: *Nuclear Reactor Technology Development and Utilization*. Elsevier.

Aoyama, M., Hirose, K., Igarashi, Y., 2006. Re-construction and updating our understanding on the global weapons tests 137 Cs fallout. *J. Environ. Monit.* 8, 431–438.

Bailly du Bois, P., et al., 2014. *A Review of the Model Comparison of Transportation and Deposition of Radioactive Materials Released to the Environment as a Result of the Tokyo Electric Power Company's Fukushima Daiichi Nuclear Power Plant Accident*. Sectional Committee on Nuclear Accident, Committee on Comprehensive Synthetic Engineering, Science Council of Japan, Tokyo.

Barrand, R., Dickson, R., Liu, Z., Semeniuk, D., 1999. *Release of Fission Products from CANDU Fuel in Air, Steam and Argon Atmospheres at 1500–1900 Deg C: The HCE3 Experiment.*

Belyaev, S.T., Borovoi, A.A., Demin, V.F., Rimsky-Korsakov, A.A., Kheruvimov, A.N., 1991. The Chernobyl source term. In: *Proceedings of the Seminar on Comparative Assessment of the Environmental Impact of Radionuclides Released During Three Major Nuclear Accidents: Kyshtym, Windscale, Chernobyl*, Vol. 1. Commission of the European Communities, Luxembourg, pp. 71–91.

Bennett, B., Repacholi, M., Carr, Z., 2006. *Health Effects of the Chernobyl Accident and Special Health Care Programmes*. World Health Organization.

Bricker, M.K. (Ed.), 2014. *The Fukushima Daiichi Nuclear Power Station Disaster: Investigating the Myth and Reality*. Routledge, London.

Buesseler, K., Aoyama, M., Fukasawa, M., 2011. Impacts of the Fukushima nuclear power plants on marine radioactivity. *Environ. Sci. Technol.* 45, 9931–9935.

Buzulukov, Y.P., Dobrynin, Y.L., 1993. Release of radionuclides during the Chernobyl accident. In: Merwin, S.E., Balonov, M.I. (Eds.), The Chernobyl Papers. *Volume 1: Doses to the Soviet Population and Early Health Effects Studies*. Research Enterprises, Richland, WA, pp. 3–21.

Choppin, G., Liljenzin, J.-O., Rydberg, J., Ekberg, C., 2013. *Radiochemistry and Nuclear Chemistry*. 4th ed. Academic Press, Oxford.

Cox, D., Hunt, C., O'Connor, R., Barrand, R., Iglesias, F., 1990. High-temperature oxidation behaviour of UO2 in air and steam. In: *High-Temperature Oxidation and Sulphidation Processes*. Elsevier.

Ducros, G., Pontillon, Y., Malgouyres, P., 2013. Synthesis of the VERCORS experimental programme: Separate-effect experiments on fission product release, in support of the PHEBUS-FP programme. *Ann. Nucl. Energy* 61, 75–87.

Eslinger, P.W., et al., 2014. Source term estimation of radioxenon released from the Fukushima Daiichi nuclear reactors using measured air concentrations and atmospheric transport modeling. *J. Environ. Radioact.* 127, 127–132.

Geiger, E., Bes, R., Martin, P., Pontillon, Y., Ducros, G., Solari, P., 2016. Insights on fission products behaviour in nuclear severe accident conditions by X-ray absorption spectroscopy. *J. Nucl. Mater.* 471, 25–33.

Hastings, I., Hunt, C., Lipsett, J., 1985. Release of short-lived fission products from UO2 fuel: Effects of operating conditions. *J. Nucl. Mater.* 130, 407–417.

Honda, M.C., et al., 2012. Dispersion of artificial caesium-134 and-137 in the western North Pacific one month after the Fukushima accident. *Geochem. J.* 46, e1–e9.

Hunt, C., Cox, D., Iglesias, F., 1994. Fission-product release during accidents—an accident management perspective. *Nucl. Eng. Des.* 148, 205–216.

IAEA, 1986. *Summary Report on the Post-Accident Review Meeting on the Chernobyl Accident, Safety Series No. 75-INSAG-1*. IAEA Vienna.

IAEA, 1991. *The International Chernobyl Project: Technical Report: Assessment of Radiological Consequences and Evaluation of Protective Measures*. International Atomic Energy Agency, Vienna.

IAEA, 2015a. *The Fukushima Daiichi Accident—Technical Volume 4: Radiological Consequences*. International Atomic Energy Agency, Vienna.

IAEA, 2015b. *The Fukushima Daiichi Accident—Technical Volume 5: Post-accident Recovery*. International Atomic Energy Agency, Vienna.

IAEA, 2022. *Experience in the Management of Radioactive Waste After Nuclear Accidents: A Basis for Preplanning*. IAEA Vienna.

Izrael, Y.A., et al., 1990. *Chernobyl: Radioactive Contamination of the Environment*. Gidrometeoizdat, Leningrad.

JAEA, 2021. *Graph of Fission Product Yields*. Japan Atomic Energy Agency.

Jaworowski, Z., Kownacka, L., 1988. Tropospheric and stratospheric distributions of radioactive iodine and cesium after the Chernobyl accident. *J. Environ. Radioact.* 6, 145–150.

Kajan, I., 2016. *Transport and Containment Chemistry of Ruthenium under Severe Accident Conditions in a Nuclear Power Plant*. PhD thesis, Chalmers University of Technology, Gothenburg.

Kaul, A., Landfermann, H., Thieme, M., 1996. One decade after Chernobyl: Summing up the consequences. *Health Phys.* 71, 634–640.

Kleykamp, H., 1985. The chemical state of the fission products in oxide fuels. *J. Nucl. Mater.* 131, 221–246.

Konings, R.J.M., Wiss, T., Beneš, O., 2015. Predicting material release during a nuclear reactor accident. *Nat. Mater.* 14, 247–252.

Madigan, D.J., Baumann, Z., Fisher, N.S., 2012. Pacific bluefin tuna transport Fukushima-derived radionuclides from Japan to California. *Proc. Natl. Acad. Sci. USA* 109, 9483–9486.

Magwood IV, W.D., et al., 2021. *Fukushima Daiichi Nuclear Power Plant Accident, Ten Years on: Progress, Lessons and Challenges*. Organisation for Economic Co-operation and Development.

Mariani, R.D., et al., 2011. Lanthanides in metallic nuclear fuels: Their behavior and methods for their control. *J. Nucl. Mater.* 419, 263–271.

Marples, D.R., 1996. *Belarus: From Soviet Rule to Nuclear Catastrophe*. Palgrave Macmillan, London.

Masumoto, Y., et al., 2012. Oceanic dispersion simulations of 137Cs released from the Fukushima Daiichi nuclear power plant. *Elements* 8, 207–212.

Miwa, S., et al., 2015. Research program for the evaluation of fission product and actinide release behaviour, focusing on their chemical forms. *Energy Procedia* 71, 168–181.

National Diet of Japan, 2012. *The Official Report of the Fukushima Nuclear Accident Independent Investigation Commission*. The National Diet of Japan, Tokyo.

Nourbakhsh, H., 1993. *Estimate of Radionuclide Release Characteristics into Containment Under Severe Accident Conditions. Final Report*. Nuclear Regulatory Commission, Washington, DC (United States).

Ohtsuru, A., et al., 2015. Nuclear disasters and health: Lessons learned, challenges, and proposals. *Lancet* 386, 489–497.

Oskolkov, B.Y., et al., 2011. Radioactive waste management in the Chernobyl Exclusion Zone: 25 years since the Chernobyl Nuclear Power Plant accident. *Health Phys.* 101, 431–441.
Pflugbeil, S., 2006. Chernobyl–looking back to go forwards: The September 2005 IAEA conference. *Med. Confl. Surviv.* 22, 299–309.
Pontillon, Y., Ducros, G., 2010. Behaviour of fission products under severe PWR accident conditions: The VERCORS experimental programme—Part 2: Release and transport of fission gases and volatile fission products. *Nucl. Eng. Des.* 240, 1853–1866.
Przybylak, J., 2024. Nuclear power plants in war zones: Lessons learned from the war in Ukraine. *Secur. Def. Q.* 46, 84–103.
Rypina, I.I., et al., 2013. Short-term dispersal of Fukushima-derived radionuclides off Japan: Modeling efforts and model-data intercomparison. *Biogeosciences* 10, 4973–4990.
Saenko, V., et al., 2011. The Chernobyl accident and its consequences. *Clin. Oncol.* 23, 234–243.
Samojlenko, Y.N., Golubev, V., 1989. *Special Zone Territory Decontamination.* Dezaktivatsiya Territorii Osoboj Zony.
Sehgal, B.R., 2012a. Fission product release and transport. In: *Nuclear Safety in Light Water Reactors.* Academic Press, Oxford, pp. 425–517.
Sehgal, B.R., 2012b. Light water reactor safety: A historical review. In: *Nuclear Safety in Light Water Reactors.* Academic Press, Oxford, pp. 1–88.
Soffer, L., Burson, S.B., Ferrell, C.M., Lee, R.Y., Ridgely, J.N., 1995. *Accident Source Terms for Light-Water Nuclear Power Plants.* NUREG-1465. U.S. Nuclear Regulatory Commission, Washington, DC.
Stohl, A., et al., 2012. Xenon-133 and caesium-137 releases into the atmosphere from the Fukushima Dai-ichi nuclear power plant: Determination of the source term, atmospheric dispersion, and deposition. *Atmos. Chem. Phys.* 12, 2313–2343.
TEPCO, IRID, 2017. Locating Fuel Debris inside the Unit 3 Reactor Using a Muon Measurement Technology at Fukushima Daiichi Nuclear Power Station. *Tokyo Electric Power Company Holdings, Inc. and International Research Institute for Nuclear Decommissioning*, Tokyo.
Theriault, K., 2016. Boiling water reactors. In: *Nuclear Engineering Handbook.* CRC Press.
UNSCEAR, 2000. *Sources and Effects of Ionizing Radiation. Volume I: Sources.* UNSCEAR 2000 Report. United Nations.
Wu, J., Zhou, K., Dai, M., 2013. Impacts of the Fukushima nuclear accident on the China Seas: Evaluation based on anthropogenic radionuclide 137Cs. *Chin. Sci. Bull.* 58, 552–558.

CHAPTER 7

Radiochemistry of the Key Radionuclides in High-Level Waste

Lu Liu, Paul Dutheil, and Maria Boltoeva

7.1 INTRODUCTION

High-level radioactive waste (HLW) represents one of the most challenging categories of radioactive materials generated by the nuclear fuel cycle. It arises primarily as a byproduct of nuclear fission reactions within reactors. According to the International Atomic Energy Agency (IAEA) classification scheme, HLW is defined as waste containing activity concentrations high enough to generate substantial heat from radioactive decay or containing significant quantities of long-lived radionuclides that must be considered in the design of disposal facilities (International Atomic Energy Agency [IAEA], 2009).

National nuclear fuel cycle strategies vary across countries, leading to differences in the types and forms of HLW produced. Figure 7.1 illustrates two main approaches. In some countries, such as Canada, Finland, and Sweden, spent fuel assemblies are classified directly as HLW and prepared for disposal without further chemical treatment (the so-called open fuel cycle). In contrast, countries such as France, China, and Japan adopt a closed fuel cycle, in which spent fuel is reprocessed to recover fissile materials, particularly uranium and plutonium, for potential reuse in fuel fabrication (International Atomic Energy Agency [IAEA], 2005). While reprocessing reduces the overall volume of HLW, it generates highly radioactive liquid residues that must be immobilized in durable waste forms, most commonly borosilicate glass, prior to disposal. Consequently, HLW exists in two principal forms: (1) spent fuel assemblies designated for direct disposal and (2) solidified residues resulting from the reprocessing of spent fuel.

Irrespective of origin, HLW contains a complex inventory of radionuclides. These include actinides, which are primarily residual components of nuclear fuel and products of neutron capture reactions, as well as a broad range of fission products and activation products. As these radionuclides decay, additional daughter radionuclides appear, further complicating

DOI: 10.1201/9781003665793-9

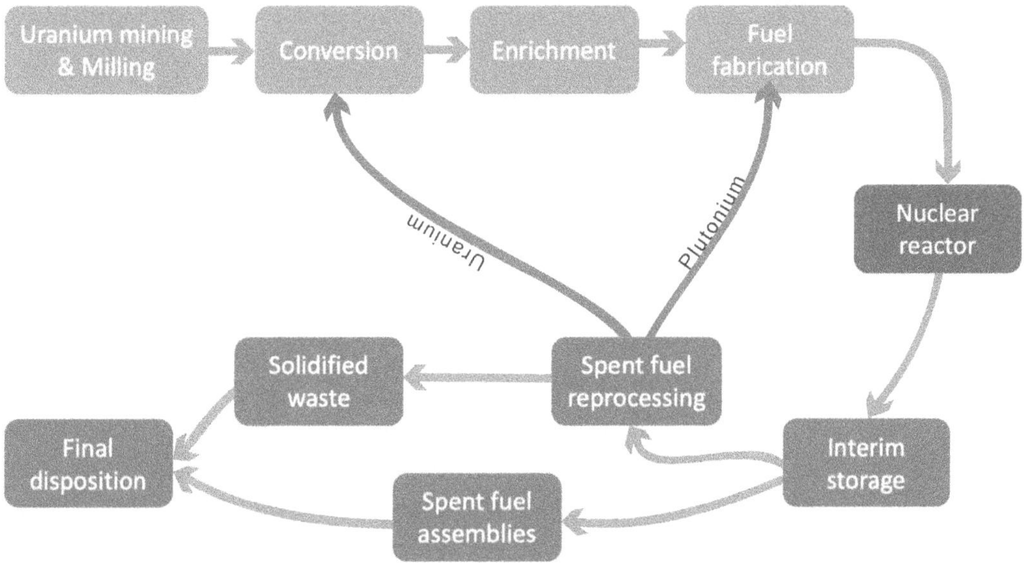

FIGURE 7.1 Schematic Representation of the Nuclear Fuel Cycle, Highlighting the Two Principal Strategies and the Corresponding Forms of High-Level Waste.

the inventory. Together, these radionuclides determine the radiological impact of HLW across the long timescales relevant to disposal. Because of their long-lived radiotoxicity, they must be isolated from the biosphere until they decay to levels considered radiologically harmless.

It is broadly accepted at the technical level that, at this time, deep geological disposal represents the safest and most sustainable option as the end point of the management of HLW (The Council of the European Union, 2011). This concept involves the emplacement of HLW in engineered facilities located several hundred meters to more than 1 km below the Earth's surface, within geologically stable host formations such as clay and granite. Repository safety relies on a multi-barrier system composed of engineered and natural barriers designed to delay and limit radionuclide migration, including the waste form, canisters, buffer materials, and the surrounding host rock. The performance of such systems must be demonstrated over timescales ranging up to 1 million years. It is generally acknowledged that the principal pathway for radionuclide release from a repository to the accessible environment will be transport by groundwater, predominantly as dissolved species, but also potentially as colloids or particulates. Because institutional control cannot be assured over such extended durations, confidence in repository performance ultimately depends on a robust scientific understanding of the chemical and physical processes that govern radionuclide behavior.

Among the processes controlling radionuclide transport, the chemistry of HLW radionuclides plays a central role. Chemical speciation, governed by redox transformations, complexation with inorganic ligands (e.g., carbonate, hydroxide, sulfate), and interactions with natural organic matter, dictates key processes such as solubility, sorption to mineral surfaces, mobility in groundwater, and incorporation into secondary solid phases. A comprehensive understanding of these processes is therefore essential for predicting radionuclide migration from a repository and for evaluating the long-term safety of disposal strategies.

The objective of this chapter is to provide an overview of the radiochemistry of key radionuclides in HLW, with particular emphasis on their behavior in aqueous systems. Section 7.2 examines the radionuclides present in HLW and identifies those selected as "key" radionuclides for detailed discussion. Section 7.3 reviews their fundamental chemical properties, including redox characteristics, speciation, and interaction with inorganic and organic ligands. Finally, Section 7.4 presents conclusions and a short outlook on remaining challenges.

7.2 KEY RADIONUCLIDES IN HLW

When discussing key radionuclides in HLW, it is important to recognize that no universal or definitive list can be established. The relative significance of radionuclides depends on multiple factors, including waste type, burn-up, cooling time, disposal concept, geologic and geochemical conditions, and radionuclide transport mechanisms. Consequently, different studies (Barney & Wood, 1980; Jarzemba & Pickett, 1995) have proposed varying sets of "key" radionuclides, each based on criteria tailored to specific repository designs and management strategies.

In general, three aspects are most commonly considered to identify radionuclides of concern:

(1) Inventory: the abundance of the radionuclide in HLW.

(2) Biological toxicity: the potential for harm due to chemical properties or radiation dose.

(3) Leachability and mobility: the potential of release from the waste matrix and subsequent migration in groundwater systems.

In the following section, the relative importance of radionuclides in HLW is reviewed with reference to these three criteria.

7.2.1 Inventory of Radionuclides in HLW

The inventory of specific radionuclides in HLW depends on several factors, including reactor type, fuel burn-up, cooling time after discharge, and whether the waste is managed as direct spent fuel or as reprocessed waste.

For light water reactor (LWR) spent fuel with a typical burn-up of 50 gigawatt days per tonne of initial heavy metal (GWd/tHM), the average composition is approximately 93.4% uranium (including ~0.8% ^{235}U), 5.2% fission products, 1.2% plutonium, and 0.2% minor transuranic elements such as neptunium, americium, and curium (Feiveson et al., 2011). From a radiological perspective, however, the relative importance of these radionuclides evolves over time due to radioactive decay. Xu et al. (Xu et al., 2005) evaluated the time-dependent contributions of individual radionuclides to spent fuel radioactivity. Figure 7.2 illustrates these changing contributions for LWR fuel at 50 GWd/tHM. In

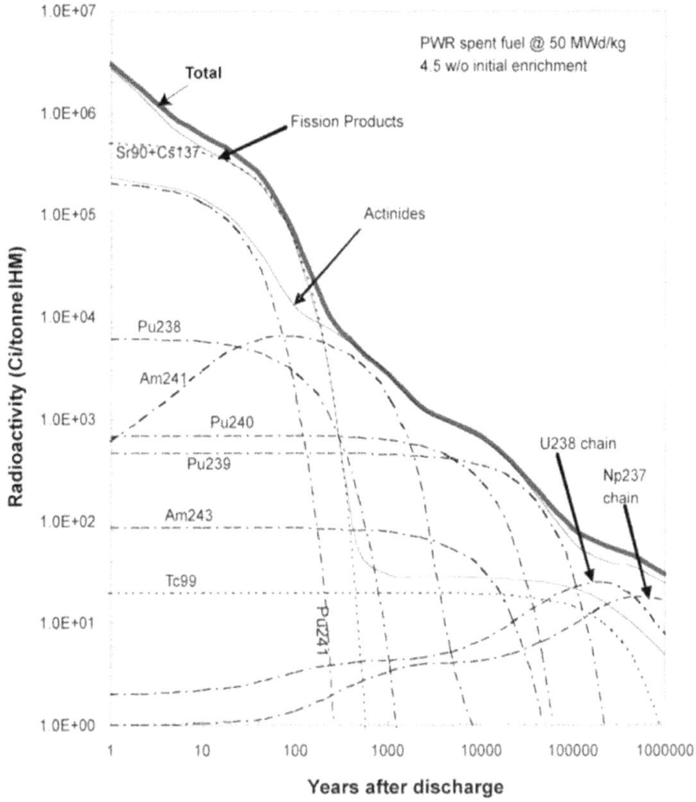

FIGURE 7.2 Evolution of Radioactivity for PWR Spent Fuel with 50 MWd/kg Burn-Up. (Reprinted from Xu et al. (2005)), Copyright 2005, with Permission from Taylor & Francis Group.

the short term (up to ~200 years), relatively short-lived fission products, primarily ^{137}Cs and ^{90}Sr, dominate the activity inventory, but their influence declines rapidly with decay. Between a few hundred and a few thousand years, ^{241}Am becomes the principal contributor. By ~100,000 years, long-lived actinides emerge as the dominant contributors. At even longer timescales, activity is shared among actinides, residual fission products, and decay products of ^{238}U and ^{237}Np, which are added to the overall radiological burden.

Uranium and plutonium account for the majority of spent fuel mass and include fissile radionuclides such as ^{235}U and ^{239}Pu. Both uranium and plutonium can be separated and recycled through chemical reprocessing for new fuel fabrication or other applications. The most widely used method is the plutonium uranium reduction extraction (PUREX) process (Herbst et al., 2011).

Meleshyn and Noseck (Meleshyn & Noseck, 2012) reported on radionuclide and stable element inventories in vitrified HLW produced at La Hague, Orano, between June 1989 and January 2007 and delivered to Germany. These inventories, summarized in Table 7.1, indicate that uranium isotopes remain the predominant actinides after reprocessing, followed by neptunium, americium, curium, and plutonium. Among long-lived fission products (half-lives >30 years), cesium isotopes contribute most significantly to radioactivity, followed by ^{90}Sr, ^{99}Tc, ^{93}Zr, ^{126}Sn, and ^{79}Se.

TABLE 7.1 Total Inventories of Actinides and Fission Products (Half-Lives >30 years) in Vitrified Waste Delivered by Orano to Germany in standard vitrified waste canister (CSD-V) Canisters, as Reported by Orano.

Radionuclides	Mass at the time of vitrification, in kg	Activity at the time of vitrification, in TBq
^{238}U	2536	/
^{241}Am	2095	/
^{237}Np	1708	/
^{243}Am	620	/
^{244}Cm	148	/
^{235}U	20.3	/
^{239}Pu	19.6	/
^{236}U	9.16	/
^{240}Pu	8.75	/
^{245}Cm	7.76	/
^{241}Pu	3.37	/
^{242}Pu	2.16	/
^{238}Pu	0.665	/
^{234}U	0.385	/
^{137}Cs	/	16,870,000
^{90}Sr	/	10,730,000
^{99}Tc	/	2,624
^{93}Zr	/	304
^{135}Cs	/	101
^{126}Sn	/	110
^{79}Se	/	58.1
^{107}Pd	/	19.6

Data from (Meleshyn & Noseck, 2012).

7.2.2 Biological Toxicity

The quantity of each radionuclide in waste provides a first measure of its potential importance, but inventory alone is insufficient to rank radionuclides by relative hazard. Toxicity to humans varies significantly among isotopes and depends on retention in the body, radiosensitivity of the affected tissues, the target organ, physical half-life, and the type and energy of emitted radiation. One used indication of relative toxicity is the concentration guideline values for radionuclides in drinking water, as recommended by the World Health Organization (WHO) (WHO, 2011). These limits reflect safe intake levels and provide an indirect indicator of radiological hazard.

WHO guidance levels for most HLW-related radionuclides fall within the range 0.1–100 Bq/L (Table 7.2). Actinides and certain decay products of ^{238}U (^{210}Pb and ^{226}Ra) have the lowest limits (0.1–1 Bq/L), reflecting their high biological toxicity. By comparison, most long-lived fission products have higher guideline values, indicating lower relative toxicity. An important exception is ^{129}I, which is more radiologically significant due to its long half-life and selective fixation in the thyroid.

TABLE 7.2 World Health Organization (WHO) Guidance Levels for Selected Radionuclides* in Drinking Water (WHO, 2011).

Guidance levels	Radionuclides
0.1 Bq/L	^{210}Pb, ^{229}Th, ^{231}Pa, ^{248}Cm
1 Bq/L	^{129}I, ^{226}Ra, $^{230/232}$Th, $^{232/233/234/235/236}$U, ^{237}Np, $^{238/239/240/242/244}$Pu, $^{241/242m/243}$Am, $^{244/245/246/247}$Cm, $^{249/251}$Cf
10 Bq/L	^{238}U, ^{241}Pu, ^{137}Cs, ^{90}Sr
100 Bq/L	^{99}Tc, ^{93}Zr, ^{135}Cs,

*Note: Only HLW-related radionuclides with half-lives >30 years are listed. ^{126}Sn, ^{79}Se, and ^{107}Pd are not listed in the WHO guidance tables.

7.2.3 Leachability and Mobility of Radionuclides

The leachability and mobility of radionuclides are critical factors in assessing the long-term safety of geological repositories, as they control both the release of radionuclides from waste forms and their transport in the surrounding geosphere.

For spent UO_2 fuel, radionuclide release depends strongly on the chemical properties and partitioning behavior of each element. An instantaneous release occurs upon waste package failure, involving fission gases (e.g., Xe and Kr) and volatile elements (e.g., I, Cs, and Cl) that have migrated during reactor operation to grain boundaries and the pellet–cladding gap. The long-term release of most fission products and actinides is controlled by the alteration and dissolution of the fuel matrix, primarily UO_2 (Bruno & Ewing, 2006; Ewing, 2015; Shoesmith, 2000). For vitrified HLW (borosilicate glass), radionuclide release is highly element-dependent. Results from laboratory experiments, large-scale mockups, and in situ studies show that, under the anoxic and reducing conditions expected in most repositories, the solubility of long-lived radionuclides is generally very low, particularly for actinides (Gin et al., 2017).

Once released from the waste matrix, radionuclide transport is governed by interactions with groundwater and host rock, characterized by a balance of mobilizing and immobilizing processes, including solubility control, sorption onto mineral surfaces, incorporation into secondary phases, and colloid association. In general, actinides such as Pu, Am, Np, and Cm are strongly retained under reducing conditions. By contrast, long-term radiological risks are dominated by a small set of mobile fission and activation products—^{129}I, ^{99}Tc, ^{79}Se, ^{36}Cl, and ^{14}C—which occur mainly as anionic species. Because these anions interact only weakly with mineral surfaces, they display relatively high mobility and thus represent the critical contributors to long-term safety assessments (Grambow, 2008).

7.2.4 Selected Radionuclides

Broadly speaking, the selection of key radionuclides in this chapter is guided by two complementary perspectives:

(1) Recycling and reuse of fissile materials.

(2) Medium- and long-term radiological risk, associated with the release of radionuclides into the biosphere.

Uranium and plutonium isotopes are of high interest due to their potential for recycling of fissile isotopes (^{235}U and ^{239}Pu) and their contribution to long-term activity. Americium isotopes, notably ^{241}Am and ^{243}Am, are present in significant quantities in reprocessed waste and dominate the radiological hazard over timescales of thousands to hundreds of thousands of years. Long-lived and mobile fission products such as ^{79}Se, ^{99}Tc, and ^{129}I are also of particular concern. ^{137}Cs and ^{90}Sr strongly influence HLW radioactivity in the first few hundred years after disposal, but their relatively short half-lives mean that their long-term significance is limited. They are therefore not a primary focus of this chapter. Table 7.3 summarizes the selected radionuclides and their characteristics.

7.3 CHEMISTRY OF KEY RADIONUCLIDES

For each element, multiple isotopes may exist, differing in their nuclear properties such as half-life, radiation type, and decay energy, while retaining identical chemical properties. Thus, the chemistry of a radionuclide is essentially the same as that of its stable or other radioactive isotopes. In certain cases, radioactive decay can locally modify environmental

TABLE 7.3 Selected Key Radionuclides in HLW

Isotope	Half-life (years)	Decay type (Energy)	Main detection methods*	Main origin in HLW	Remarks
^{79}Se	2.95×10^5	β^- (E_{max} = 151 keV)	LSC, ICP-MS	Fission product	Moderate biological toxicity, high mobility
^{99}Tc	2.111×10^5	β^- (E_{max}=294 keV)	LSC, ICP-MS	Fission product	Moderate biological toxicity, long-term dose contribution, and high mobility
^{129}I	1.57×10^7	β^- (E_{max} = 154 keV)	LSC, ICP-MS	Fission product	High biological toxicity, high mobility
^{235}U	7.038×10^8	α (E = 4.397 MeV)	α-Spectrometry, ICP-MS	Residual fissile uranium	High biological toxicity, long-term dose contribution, and fissile material
^{238}U	4.468×10^9	α (E = 4.198 MeV)	α-Spectrometry, ICP-MS	Main residual uranium isotope	High biological toxicity, dominant inventory in HLW, and long-term dose contribution
^{239}Pu	2.411×10^4	α (E = 5.157 MeV)	α-Spectrometry, ICP-MS	Neutron capture in ^{238}U	High biological toxicity, high inventory, long-term dose contribution, and fissile material
^{240}Pu	6.561×10^3	α (E = 5.168 MeV)	α-Spectrometry, ICP-MS	Neutron capture in ^{239}Pu	High biological toxicity, high inventory, and long-term dose contributor
^{241}Am	432.6	α (E = 5.486 MeV)	α-Spectrometry, ICP-MS	Decay of ^{241}Pu	High biological toxicity, high inventory, and long-term dose contribution
^{243}Am	7370	α (E = 5.275 MeV)	α-Spectrometry, ICP-MS	Neutron capture in ^{242}Pu	High biological toxicity, high inventory, and long-term dose contribution

*LSC = liquid scintillation counting. ICP-MS = inductively coupled plasma mass spectrometry.
Nuclear data retrieved from the ENDF/B-VIII.0 database (Brown et al., 2018).

conditions, for example, through the radiolysis of water or surrounding media, which may result in modified apparent behavior. Unless otherwise specified, the following sections focus on the general chemistry of the elements rather than isotope-specific features.

7.3.1 General Chemistry of Actinides

The actinide series corresponds to elements of atomic number $89 \leq Z \leq 103$ and is located in the f-block of the periodic table. The chemistry of actinide (An) elements is mainly defined by the characteristic filling of the 5f orbitals. In late actinides (after americium), the filled subshells $6s^2$ and $6p^6$ shield the partially filled 5f orbitals, in analogy with their lighter 4f homologs, lanthanides. This results in a chemical behavior similar to the lanthanides, showing mainly +3 oxidation state (except for Nobelium) and a tendency to engage in ionic bonding rather than covalent or dative bonding. The 5f subshells are increasingly shielded with increasing nuclear charge, which results in a lower availability of f-electrons for chemical bonding as Z increases. However, in early actinides, the filled $6s^2$ and $6p^6$ subshells are not as contracted as those the heavy actinides. As a consequence, 5f orbitals are delocalized and extend radially close to the 6d, 7s, and 7p orbitals with small energy differences, which facilitates their hybridization. This enables them to engage more readily in metal–ligand or covalent bonding and results in a higher diversity of oxidation states. Consequently, light actinides display a transition metal-like behavior when compared to elements of the series above Am. The resulting diversity of oxidation states observed for early actinides (from +2 to +7) is illustrated in the upper portion of Figure 7.3. The increase of nuclear charge throughout the series is leading to a contraction of the 5f orbitals with their successive filling. This effect is termed the actinide contraction, in analogy with the similar effect observed for the lanthanide series. It leads to the progressive decrease of ionic size, which can be observed when comparing the ionic radii of actinides in a given oxidation state (see Figure 7.3).

In aqueous solution, stable actinide ions appear in the form of An^{3+} and An^{4+} for trivalent and tetravalent actinides, whereas they take the form of *trans*-dioxo cations AnO_2^{+} and AnO_2^{2+} in their pentavalent and hexavalent states, respectively. The latter are called actinyl cations and are formed by immediate hydrolysis with water in oxidation states +5 and +6. Actinyl cations are known for actinides from uranium to Am and display a symmetrical and linear geometry of the type O=An=O. This O=An=O motif often persists in complexes or the solid state. The An–O bonds in actinyl cations are covalent, with shorter and stronger bonds in AnO_2^{2+} than in AnO_2^{+} (Edelstein et al., 2010). An–O bonds also show a decrease of bond strength in the order U > Np > Pu > Am for hexavalent ions AnO_2^{2+}. In actinyl cations, some electron density is transferred from the An atom to O atoms. It follows that charges are distributed between the central An atom, which bears a partial positive charge, and the O atoms, which bear partial negative charges. This yields effective charges (z_{eff}) larger than the expected formal charge of the cation. The resulting z_{eff} are decreasing in the following order $An^{4+} > AnO_2^{2+} \geq An^{3+} > AnO_2^{+}$, with the values of +3.0 to 3.3 and +2.2 for AnO_2^{2+} and AnO_2^{+}, respectively (G. R. Choppin & Jensen, 2010; G. R. Choppin & Rao, 1984).

In aqueous solution, actinides form complexes where the metal–ligand bonds are predominantly of an ionic nature (electrostatic) due to the shielding of 5f orbitals. Actinides

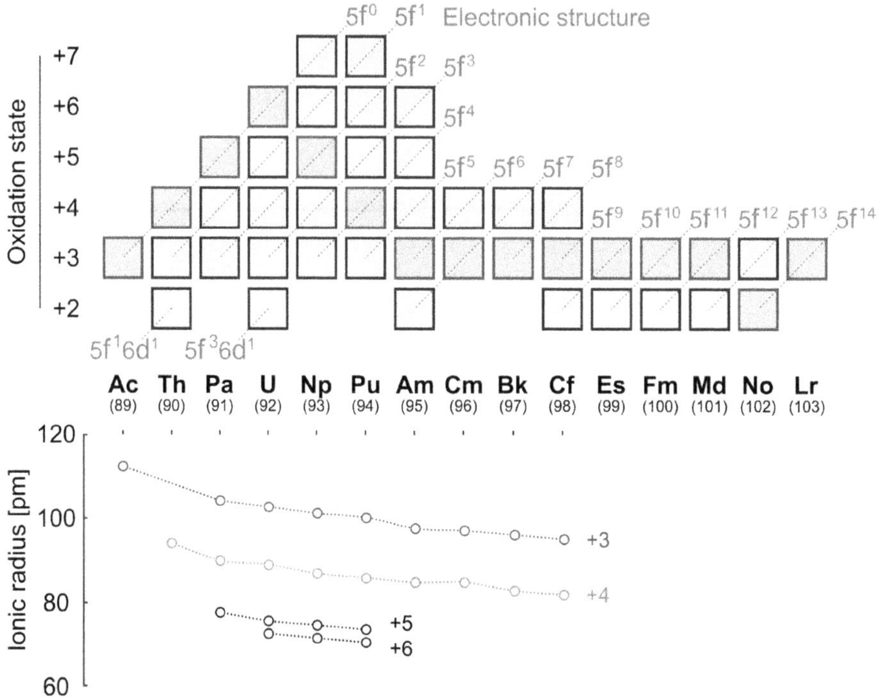

FIGURE 7.3 Scheme Showing the Oxidation States of Actinide Elements (According to (Baker, 2022)), with Their Associated Electron Configuration in the Gas Phase (Based on [Rn]$7s^2$ Filled Subshell Structure (Edelstein et al., 2010)). The Most Stable Oxidation States are Highlighted in Light Gray in Print and in Light Blue in the Online Color Version. The Effective Ionic Radii as Functions of the Elements and Their Oxidation State are Shown in the Lower Part of the Scheme (from X-Ray Crystal Structure in Their Coordination Number 6, According to (Shannon, 1976)).

are hard Lewis acids, and this results in strong interactions with ligands containing hard Lewis base donor atoms such as fluoride or oxygen, in agreement with Pearson hard and soft acids and bases (HSAB) theory. Small and weakly polarizable ligands of high negative charge typically form strong complexes with actinides. In the absence of steric constraints, the electrostatic interactions that drive the formation of actinide complexes imply that the formation of bonds by ligands depends highly on the charge of the ions, and therefore, on their oxidation state. As a result, actinides are forming increasingly stable complexes with increased z_{eff} in the following sequence: $AnO_2^+ < An^{3+} \leq AnO_2^{2+} < An^{4+}$. The variation of ionic radius (and charge) is also substantially changing the strength of the interaction of actinides with ligands. Therefore, increasingly strong complexes are generally formed with a given ligand when going down the actinide series in a given oxidation state. The qualitative strengths of complexes with simple inorganic ions follow the given sequence:

$$PO_4^{3-} > CO_3^{2-} > OH^- > SiO(OH)_3^- > HPO_4^{2-} > F^- > SO_4^{2-} > H_2PO_4^- > H_3PO_4 > NO_3^- > Cl^-.$$

This is based on the 1st complexation constant with U(IV), in good agreement with results obtained for Np(V), Am(III), or Th(IV) as presented in Table 7.4 (Altmaier et al., 2013). This general trend can be inferred for other actinides/oxidation states based on the previously discussed charge-controlled complex formation mechanisms.

Actinide ions also have a high tendency to hydrolyze due to their high acidity. The strength of hydrolysis formation depends on z_{eff}. Therefore, it follows the same trend as the strength of complex formation for the different ions: $AnO_2^+ < An^{3+} \leq AnO_2^{2+} < An^{4+}$. Some An^{4+} ions start to form hydrolyzed species at pH ≤ 2, whereas AnO_2^+ only start to hydrolyze at pH ≥ 8. The hydroxides are often observed in the form of polynuclear complexes, which tend to form very stable polymers. Actinide hydroxides are highly insoluble species and precipitate even at low concentrations, especially in the +4 oxidation state. Hydrolysis often overcomes the complex formation with other ligands in neutral to basic solutions, causing precipitation of actinides from the solution, except in carbonate media, which typically increase their solubility under these pH conditions (G. R. Choppin & Jensen, 2010; Edelstein et al., 2010). The thermodynamic data of early actinides, which are relevant for waste disposal, have been thoroughly compiled and critically reviewed in Grenthe et al. (2021).

7.3.1.1 Uranium

Uranium (U) is the fourth element of the actinide series ($Z = 92$). Uranium can be found in oxidation state +2 to +6. In aqueous solution, uranium is typically found in its most stable form, the hexavalent UO_2^{2+} ion (Grenthe et al., 2010). However, it can also be present as U^{3+}, U^{4+}, or UO^{2+} ions, U^{4+} being the most stable of these three species. U^{3+} and U^{4+} are only present in reducing conditions, U^{3+} is rather unstable as it is quickly oxidized by water and air to U^{4+}, which is stable in the presence of water, but is slowly oxidized to UO_2^{2+} by air. UO_2^+ is also unstable, as it easily disproportionates to U^{4+} and UO_2^{2+}. The standard

TABLE 7.4 Stability Constant for the Formation of (1,1) Complexes of Np(V), Am(III), U(VI), and Th(IV)

	NpO_2^+	Am^{3+}	UO_2^{2+}	Th^{4+}
$PO_4^{3}-$			13.23±0.15	
CO_3^{2-}	4.96 ± 0.06	8.0 ± 0.4	9.94 ± 0.03	
OH^-	2.7 ± 0.7	6.8 ± 0.5	8.75 ± 0.24	11.5 ± 0.5
$SiO(OH)_3^-$		8.13 ± 0.18	7.97 ± 0.10	
$HPO_4^{2}-$	2.95 ± 0.10		7.24 ± 0.26	
F^-	1.2 ± 0.3	3.4 ± 0.4	5.16 ± 0.06	8.87 ± 0.15
SO_4^{2-}	0.44 ± 0.27	3.30 ± 0.15	3.15 ± 0.02	6.17 ± 0.32
$H_2PO_4^-$		3.0 ± 0.5	3.26 ± 0.06	5.59 ± 0.32
H_3PO_4			0.76 ± 0.15	1.89 ± 0.31
NO_3^-		1.33 ± 0.20	0.30 ± 0.15	1.3 ± 0.2
Cl^-		0.24 ± 0.03	0.17 ± 0.02	1.7 ± 0.1

(Adapted with permission from Altmaier et al. (2013), Copyright 2013 American Chemical Society.)

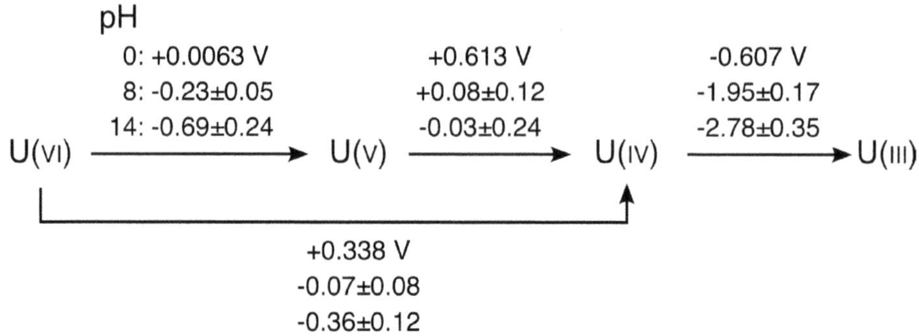

FIGURE 7.4 Latimer Diagram of Main Oxidation States of Uranium in Aqueous Solutions at Different pH. Electrode Potentials Taken from (G. Choppin et al., 2013).

reduction potentials associated with reactions of U^{3+} to UO_2^{2+} are shown in the Latimer diagram presented in Figure 7.4.

UO_2^{2+} readily hydrolyzes in solution above pH 3, and U^{4+} starts to hydrolyze at pH < 1. The resulting hydroxide species can be polynuclear and have low solubility. $UO_2^{2+}$and U^{4+} form strong complexes with inorganic oxoanions and fluorides, with increasing strength in the following sequence: $SO_4^{2-} < F^- < OH^- < CO_3^{2-} < PO_4^{3-}$ (see Table 7.4). The solubility of uranium in aqueous solution is mainly governed by hydrolysis, which forms insoluble phases or by the presence of inorganic ligands, such as carbonate, which overrides hydrolysis by forming especially stable soluble complexes even at neutral or alkaline pH. Strong complexes of low solubility are also formed with phosphates. Organic carboxylate ligands like acetate or oxalate also form strong complexes with $UO_2^{2+}$and U^{4+}. Tributylphosphate (TBP) is another noteworthy ligand, which forms complexes with UO_2^{2+} in nitrate media, in the form $UO_2(NO_3)_2(TBP)_2$. Uranium complexes with TBP are soluble in organic solvents and can be readily extracted during solvent extraction processes in a very selective manner since most fission products and transuranics do not forming extractable complexes with TBP. This made TBP a cornerstone in the nuclear fuel cycle, both for uranium ore processing and for nuclear fuel reprocessing (e.g., in the PUREX process) (Herbst et al., 2011).

The most common uranium oxide is UO_2, where U is in the +4 oxidation state, and is the main form used for the preparation of nuclear fuel. Other oxides also exist upto UO_3, where U is in the +VI oxidation state, with intermediate $U0_{2+x}$ structures such as U_4O_9, U_30_7, U_2O_5, or U_30_8. Uranium oxides are soluble at low pH in mineral acids like HCl, HNO_3, $HClO_4$, or H_2SO_4 (Grenthe et al., 2010). However, UO_2 is less soluble at more moderate pH, and under such pH conditions, the dissolution of UO_2 is mostly governed by corrosion processes. The oxidation of UO_2 to the more soluble UO_3 increases considerably the solubility of U in aqueous solution (see Figure 7.5). Therefore, the presence of oxidizing agents such as O_2 or water radiolysis products like H_2O_2 and radical species increases the corrosion rate and the dissolution of UO_2 (Ewing, 2015; Shoesmith, 2000). Species that form strong complexes with U, especially carbonates, also increase the dissolution rate of UO_2.

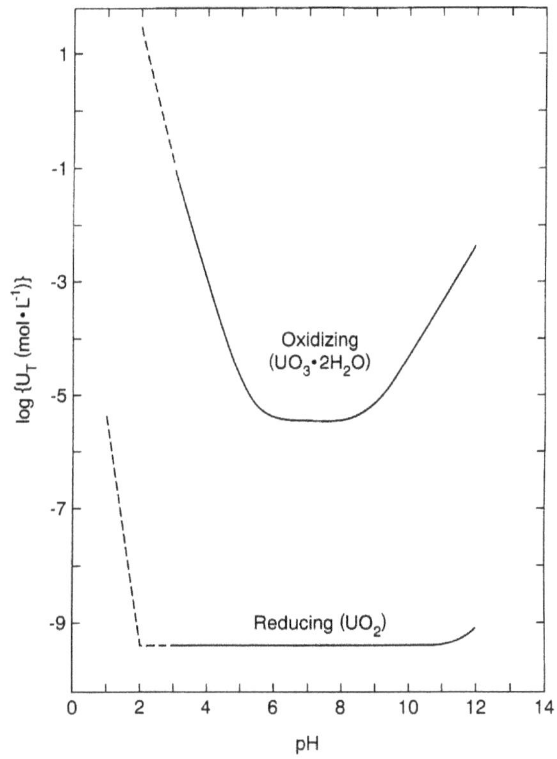

FIGURE 7.5 Solubilities of Uranium Dioxide (UO_2) and Schoepite ($UO_3{\cdot}2H_2O$) as a Function of pH at 25°C (Reprinted from Shoesmith (2000), Copyright 2000, with Permission from Elsevier).

7.3.1.2 Plutonium

Plutonium (Pu) is the sixth element of the actinide series ($Z = 94$). Pu exists in oxidation states +3 to +7. In acidic and neutral aqueous solution, Pu is found in the form of Pu^{3+} and Pu^{4+} ions for oxidation states +3 and +4, respectively, and as PuO_2^{+} and PuO_2^{2+} plutonyl ions for oxidation states +5 and +6, respectively. Pu^{3+} is stable in the presence of air and water, whereas Pu^{4+} is stable only in concentrated acid solution (pH << 1) and slowly disproportionates to Pu^{3+} and PuO_2^{2+} in dilute acid solution. PuO_2^{+} is most stable in mildly acidic conditions ($3 < pH < 6$) and disproportionates to Pu^{4+} and PuO_2^{2+} in other pH conditions. PuO_2^{2+} is stable in acidic to near-neutral conditions (Altmaier et al., 2019; Clark et al., 2010; Edelstein et al., 2010). The reduction and oxidation reactions associated with Pu(III)/Pu(IV) and Pu(V)/Pu(VI) redox couples proceed easily and reversibly, but the reduction and oxidation reactions involved in the breaking or formation of Pu–O bonds are thermodynamically irreversible and, thus, more difficult. The standard reduction potentials for some of these couples are indicated in Figure 7.6. All standard potentials are very close to one another in acidic solution, explaining that oxidation states +3 to +6 can co-exist simultaneously in the same solution. This makes Pu chemistry challenging to study, as it depends largely on its oxidation state. Although less soluble than in acidic solution, Pu(V) and Pu(VI) are also stable in alkaline solution under the form of hydroxo species such as

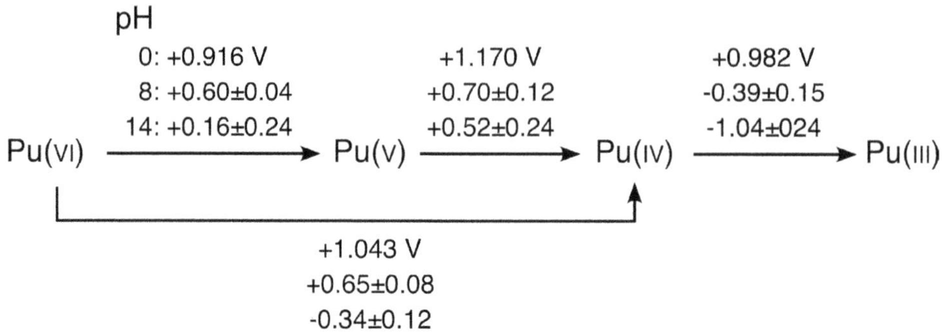

FIGURE 7.6 Latimer Diagram of the Main Oxidation States of Pu in Aqueous Solutions at Different pH. Electrode Potentials Taken from G. Choppin et al. (2013).

$PuO_2(OH)_2^-$ and $PuO_2(OH)_4^{2-}$ (for the +5 and +6 oxidation states, respectively). Pu(VII) exists only in very alkaline solutions, where it can be stabilized as $PuO_4(OH)_2^{3-}$ ion, under highly oxidizing conditions.

The general increased tendency for hydrolysis is consistent with the increase of z_{eff} according to: $PuO_2^+ < Pu^{3+} < PuO_2^{2+} < Pu^{4+}$, with Pu^{4+} readily undergoing hydrolysis at pH < 1. Pu^{3+}. Pu^{4+} ions display an increased tendency to hydrolyze when compared to lighter actinides (i.e., Np, U, Pa, or Th) due to their comparatively smaller ionic radius and higher nuclear charge. When Pu undergoes hydrolysis, it forms various hydroxide/oxyhydroxide complexes, with Pu(IV) and Pu(VI) also able to form polynuclear hydroxide complexes through hydroxo- or oxo-bridges. The main stable hydrolysis complexes are shown in Figure 7.7.

The high tendency of Pu for hydrolysis, combined with the low solubility of the resulting hydrolysis products, has a significant impact on the general solubility of Pu in aqueous solutions (see Figure 7.8). Furthermore, the Pu(IV) polymers are able to further condense to form large cluster units (Altmaier et al., 2019; Haire, 2019). The resulting nm- to µm-sized particles are readily forming colloidal solutions (*sol*). These colloidal suspensions can be relatively stable and have a significant impact on the Pu behavior in solutions since they can induce the formation of gel, transform to macro crystallites of Pu oxides (Virot et al., 2022), or provide colloid transport pathway for Pu in the environment (Walther & Denecke, 2013).

Pu follows the general trend observed for actinides when it comes to complex formation. It forms strong complexes with hard Lewis bases such as oxygen donor ligands (e.g., PO_4^{3}–, CO_3^{2-}, OH^-, and carboxylate acids) or fluoride. Increasingly, strong complexes are formed when z_{eff} increases following the previously described trend for the hydrolysis of Pu as a function of its oxidation states. Pu^{3+} and Pu^{4+} form stronger complexes than the lighter actinides in the same oxidation states for a given ligand. Besides hydroxide, Pu forms especially stable complexes with carbonate in its oxidation states +3, +5, and +6. Carbonate complexes result in a substantial increase in solubility for Pu in mildly alkaline aqueous solutions by preventing the formation of hydroxides, with carbonates being displaced by hydroxides only in highly alkaline conditions. Pu forms stable and insoluble

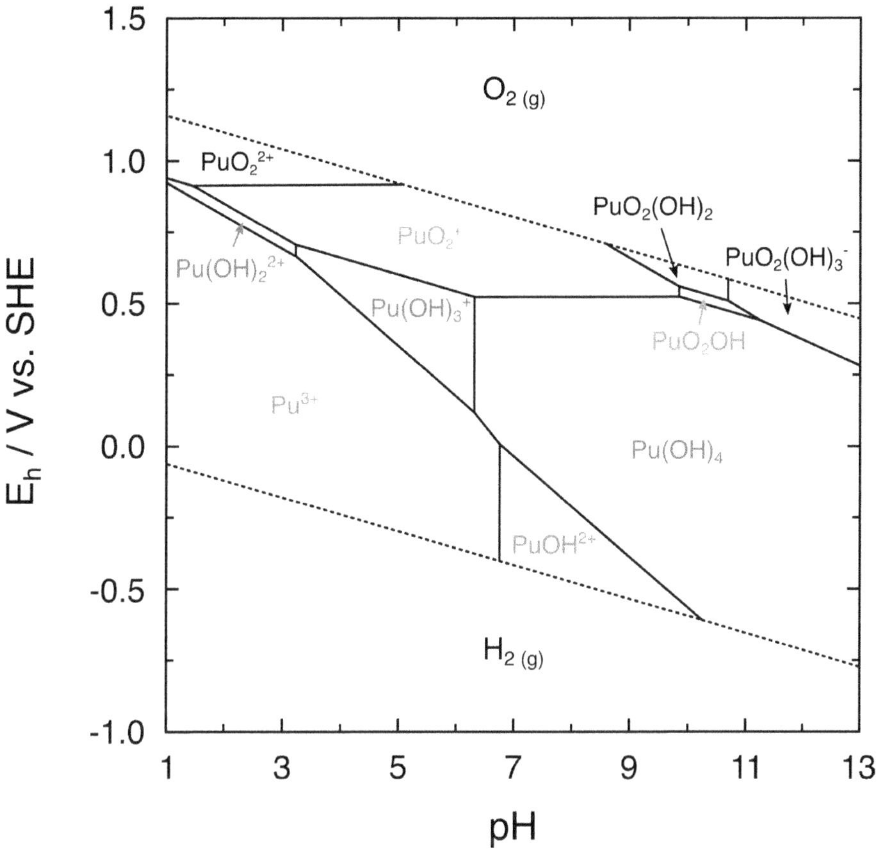

FIGURE 7.7 Eh–pH Diagram of Aqueous Pu Species Calculated for the $H^+/Na^+/Cl^-/H_2O$ System at a Pu Concentration of 1×10^{-4} M and a Total Ionic Strength of 0.1 M. The Calculation Includes Pu Hydrolysis Complexes But Excludes Solid Pu Phases. Reprinted from Altmaier et al. (2019), Copyright 2019, with Permission from the American Nuclear Society.

oxalate complexes; this ligand has been extensively used to separate and recover Pu(III) and Pu(IV) during purification of Pu and reprocessing of spent fuel. Thermal decomposition of oxalate complexes is one of the main methods used for the preparation of PuO_2 (Haire, 2019). Strong and sparingly soluble complexes are also formed by Pu with phosphates. In nitrate media, Pu forms complexes with TBP in its +4 oxidation state. The resulting $Pu(NO_3)_4(TBP)_2$ complexes can be readily recovered in organic solvents by liquid–liquid extraction and can be readily back-extracted from the organic phase by reduction of Pu(IV) to Pu(III), which does not form stable complexes with TBP (Clark et al., 2010). This has been extensively used for the separation of Pu (along with U) during the reprocessing of spent fuel, for example, in the PUREX process.

Plutonium forms two main stoichiometric oxides, PuO_2, where it is in its +4 oxidation state, and the sesquioxide Pu_2O_3, where Pu is in its +3 oxidation state. Non-stoichiometric oxides in between PuO_2 and Pu_2O_3 also exist in the form of PuO_{2-x} or Pu_2O_{3+y} (Haire, 2019; Hou & Ding, 2009). PuO_2 is difficult to dissolve; a mixture of boiling 16 M HNO_3 and

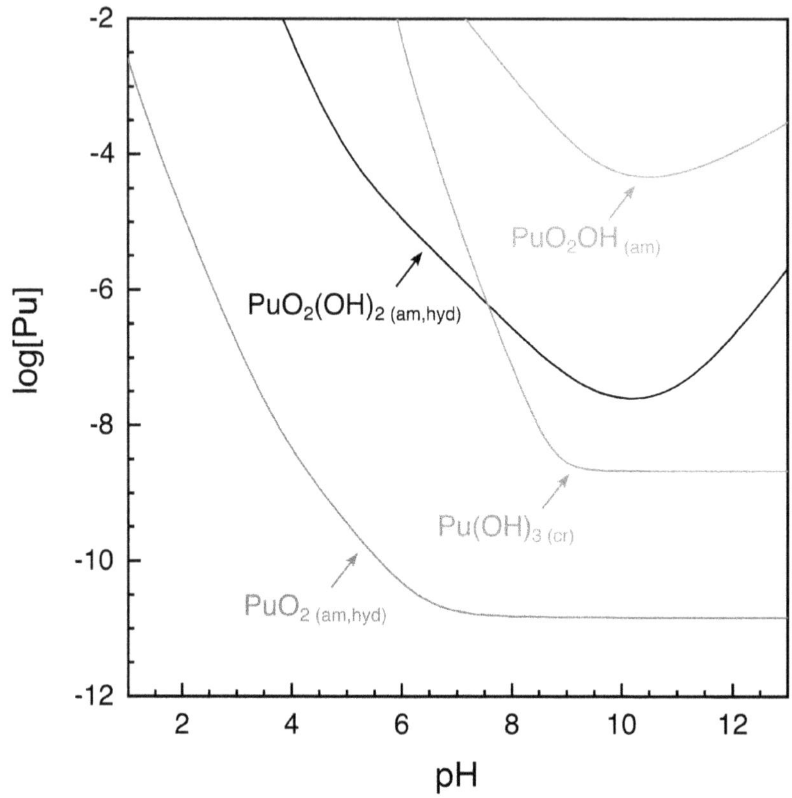

FIGURE 7.8 Solubility Curves for Binary Hydroxide Solid Phases of Pu(III)-Pu(VI), Calculated for the System $H^+/Na^+/H_2O/OH^-$ with a 0.1 M Total Ionic Strength (Reprinted from Altmaier et al. (2019), Copyright 2019, with Permission from American Nuclear Society).

1 M HF is generally used for its dissolution (Clark et al., 2010). In usual UO_2 fuels, PuO_2 is included as a solid solution in the UO_2 matrix after burn-up and can be released during its corrosion (Ewing, 2015; Konings et al., 2011). The solid solution of PuO_2 with UO_2 is usually used as fuel in the form of (U, Pu)O_2 mixed-oxide (MOX) fuels, with a typical weight content of 5% to 10% Pu for light water reactor (Konings et al., 2011).

7.3.1.3 Americium

Americium (Am) is the seventh element of the actinide series ($Z = 95$). Americium exists in oxidation states + 2 to +6; only oxidation states from +3 to +6 are known to exist in aqueous solution, with +3 being the most common and the most stable. In aqueous solution, Am exists as the ions Am^{3+}, Am^{4+}, AmO_2^+, and AmO_2^{2+} for the oxidation states +3, +4, +5, and +6, respectively. Am is especially stable in the +3 oxidation state; the oxidation states from +4 to +6 are rather complicated to obtain and stabilize (Runde & Mincher, 2011; Runde & Schulz, 2011). Am^{4+} is quickly reduced by water to Am^{3+} and is only stable in the presence of strong complexants. AmO_2^+ disproportionates to Am^{3+} and AmO_2^{2+} in strong acids and is only stable in dilute acid (near-neutral) to alkaline conditions. AmO_2^{2+} only exists in alkaline and strong oxidizing conditions. In addition, the oxidation states +4 to +6 are easily reduced under the effect of their own α-radiation. As a result, the chemistry of Am is better characterized in its Am^{3+} ion form. It behaves similarly to other +3 actinides and lanthanides, as a hard Lewis acid that forms strong complexes with hard Lewis-base ligands. It interacts with inorganic oxo-anions or fluorides to form complexes in the following order of increasing strength (see Table 7.4):

$$SiO(OH)_3^- > CO_3^{2-} > OH^- > F^- > SO_4^{2-} > H_2PO_4^- > NO_3^-.$$

Am^{3+} starts to hydrolyze when near-neutral conditions are reached (pH 5–6), forming sparingly soluble hydroxides. Am^{3+} forms stronger complexes with carbonates than with hydroxides, which increases Am solubility in alkaline media. Organic oxygen-donor ligands also form strong complexes with Am^{3+}. Like Pu, Am mainly forms a dioxide, AmO_2, and a sesquioxide, Am_2O_3, and is typically included as AmO_2 in the UO_2 fuel matrix after its burn-up (Geckeis et al., 2019).

7.3.2 ^{99}Tc

Technetium (Tc, $Z = 43$) is a transition metal in group 7 of the periodic table. All of its isotopes, with mass numbers ranging from 85 to 121, are radioactive (Browne & Tuli, 2017). The most important isotope in the context of nuclear waste management is the long-lived isotope ^{99}Tc with a half-life of 2.11×10^5 years (Poineau et al., 2013). Due to its long half-life, ^{99}Tc is expected to be the primary radiation emitter in radioactive waste between 10^4 and 10^6 years after disposal.

Technetium is considered the chemical homologue of rhenium. It dissolves in nitric acid but is insoluble in hydrochloric acid, regardless of concentration, or sulfuric acid. Technetium exhibits a wide range of oxidation states, from +7 to −1, and is prone to disproportionate reactions (oxidation states between +4 and +7 usually disproportionate to corresponding mixtures of Tc(IV) and Tc(VII)) (Darab & Smith, 1996; Spitsyn et al., 1977). In the absence of complexing agents other than oxygen and water, the +4 and +7 oxidation states are the most stable. In nitric acid solutions, technetium stability follows the order: Tc(VII) > Tc(IV) > Tc(V) > Tc(VI) (Popova & others, 2003).

The heptavalent form, Tc(VII), is thermodynamically stable over broad ranges of pH and potentials. Under oxic conditions, technetium forms the pertechnetate oxo-anion $Tc^{VII}O_4^-$, which is highly soluble and predicted to be one of the most mobile radionuclides in the environment due to its weak interaction with mineral surfaces (Bianchetti et al., 2023; Meena & Arai, 2017; Morris et al., 2008). Under reducing conditions, Tc(IV) is favored. In alkaline, neutral, and weakly acid solutions, Tc(IV) exists in the form of sparingly soluble hydrous oxides TcO_2 (solubility limit is 10^{-9} mol/L at ~pH 7) while $TcO(OH)^+$ or TcO_2^{2+} are characteristic of more concentrated acidic solutions (Chotkowski & Czerwiński, 2021; Warwick et al., 2007). The thermodynamic data for technetium and its compounds were reviewed in Rard (2005).

Two technetium oxides of technetium have been identified, TcO_2 and Tc_2O_7. TcO_2 sublimes under vacuum at around 1300 K and undergoes disproportionation to metallic Tc and Tc_2O_7 above 1400 K, whereas the melting point of Tc_2O_7 is a 392.5 K. The volatility of technetium is a significant challenge in the nuclear fuel cycle, particularly during waste treatment and vitrification processes.

In the PUREX process, technetium is found as an insoluble metal phase together with molybdenum, ruthenium, rhodium, and palladium after nitric acid dissolution of the uranium oxide fuel. Technetium is also present as a solution of pertechnetate species. Technetium can be extracted with TBP into the organic phase as the adduct $UO_2(NO_3)(TcO_4)(TBP)_2$. At high nitric acid concentrations, volatile technetium species such as TcO_2 and $HTcO_4$ may form. This volatility is problematic during glass waste formation process, where Tc(VII) species tend to volatilize, and Tc(IV) can disproportionate during glass formation. Upwards of 70% of the Tc may volatilize during the vitrification (Lawler et al., 2018).

7.3.3 ^{79}Se

Selenium (Se, $Z = 34$) is a chalcogen. Its chemical properties are similar to sulfur, which is explained by their equivalent electron configurations. Elemental selenium can be reduced to the −2 (selenides, Se^{2-}) oxidation state or oxidized to the +4 (selenites, SeO_3^{2-}) or +6 (selenates, SeO_4^{2-}) oxidation states (see Figure 7.9). Hydrogen selenide and other selenium compounds are extremely toxic (Perrone et al., 2015). Twenty-four radioactive isotopes of selenium have been identified. Selenium-79 (^{79}Se) is a beta-emitter produced by the fission of ^{235}U in nuclear reactors. It is of interest in the management of nuclear waste repositories due to its long radioactive period of 2.95×10^5 years (Brown et al., 2018) and its potential to migrate from the repository to the environment. Non-volatile ^{79}Se in spent fuel is not separated during reprocessing and remains in residual high-level liquid waste, which is then vitrified. ^{79}Se could also be found in low-level wastes, mainly in stainless steel. In this case, ^{79}Se is a neutron activation product of trace amounts of selenium-78 in reactor steel (Hummel, 2017).

Then, the speciation of selenium in aqueous solutions is complicated because selenium is a redox-sensitive element with four stable oxidation states. In pure water, it presents as an

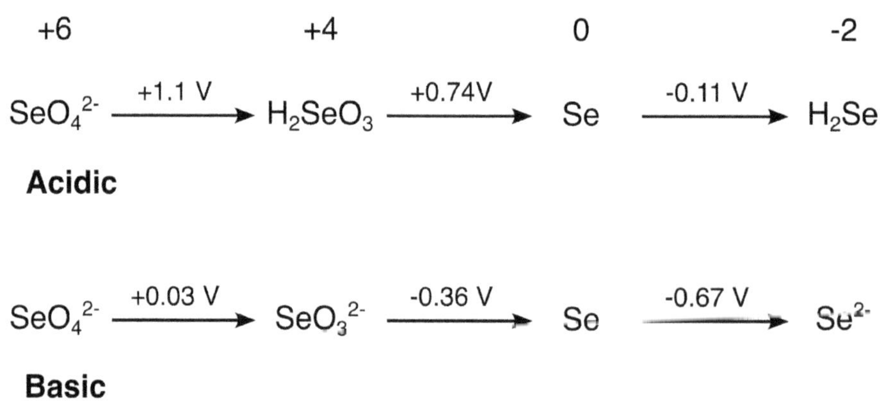

FIGURE 7.9 Latimer Diagram of Main Species of Se in Aqueous Solutions. Electrode Potentials Were Taken from Bard et al. (1985).

anionic species, and in acidic media, it presents as a neutral-charged acid (see Figure 7.9). In 2005, a review of published experimental data on the chemical thermodynamics of selenium was published (Olin et al., 2005). However, Hummel (2017) later concluded that the speciation of the selenium–water system is incomplete, particularly under near-neutral conditions.

Recent results on the chemical state of ^{79}Se in spent nuclear fuel indicate that substantial amounts of Se are homogeneously distributed in UO_2 spent fuel (Curti et al., 2015). Se(-II) substitutes for O(-II) in the expanded UO_2 lattice. This form explains high insolubility and low release via UO_2 matrix dissolution. In vitrified radioactive waste, selenium occurs as SeO_3^{2-}. This was confirmed through the study of an actual radioactive high-level glass fragment (Curti et al., 2014). Depending on the pH of pore water, selenium could be released as SeO_3^{2-} or $HSeO_3^-$ by glass corrosion and then reduced to Se(aq) and further reduced to HSe^-.

7.3.4 ^{129}I

Iodine (I, $Z = 53$) is a halogen element in group 17 of the periodic table. There are only two isotopes of iodine present in nature, ^{127}I and ^{129}I (Hou & Ding, 2009). ^{127}I is the only stable isotope of iodine, whereas ^{129}I is a radioisotope, and it is the longest-lived among the major fission products (see Table 7.3).

Iodine exhibits five oxidation states: −1, 0, +1, +5, and +7, and the associated standard redox potential among different species is shown in Figure 7.10. A notable feature of iodine redox chemistry is the disproportionation of IO^- to IO_3^- and I^- under alkaline conditions.

In simple aqueous systems without complexing agents, I(−I) has a wide stability field, which dominates under reducing conditions, whereas I(V) (IO_3^- and HIO_3) is stable only under oxidizing conditions. At even higher redox potential, I(VII) species are predominant in different pH ranges as HIO_4, IO_4^-, $H_2IO_5^{2-}$, and IO_5^{3-} (Fuge & Johnson, 2015). In neutral to acidic solutions, molecular iodine I_2 may also occur under oxidizing conditions, though

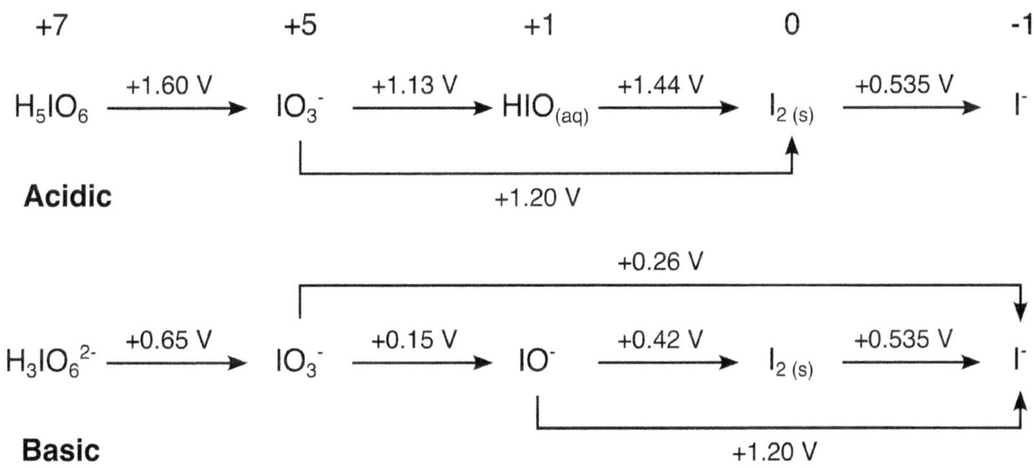

FIGURE 7.10 Latimer Diagram of Main Species of I in Aqueous Solutions. Data Obtained from Bard et al. (1985).

its solubility in water is relatively low, that is, 0.34 g/kg (Smith & Smith, 2005). Molecular iodine can react with iodide, forming more soluble triiodide species (Kaiho, 2014).

Under environmental conditions, iodine is a redox-sensitive element that forms a wide variety of inorganic and organic compounds. The most common inorganic forms are I^-, HOI, I_2, and IO_3^-, while its biophilic nature also enables incorporation into organic molecules such as alkyl iodides, proteins, polyphenols, and humic substances (Hou et al., 2009).

From a coordination chemistry perspective, iodide behaves as a soft base, forming only weak complexes with hard cations such as lanthanides and actinides. It readily forms highly soluble salts with the major and minor constituents of ground and surface waters, including alkali and alkaline earth elements, as well as transition metals such as Fe, Mn, Ni, and Al. In contrast, iodide interacts strongly with soft cations such as Ag, Cu, Hg, Pb, and Bi, yielding sparingly soluble phases, such as CuI and PbI_2, and highly insoluble compounds such as AgI, HgI_2, and Hg_2I_2 (Hummel, 2017). Solubility product constants for iodide and iodate salts with various cations have been summarized by Asmussen et al. (2020), highlighting their role in controlling iodine solubility in groundwater.

In the nuclear fuel cycle, iodine chemistry also plays a role in solvent extraction processes (Asmussen et al., 2020). Molecular iodine can form $TBP{\cdot}I_2$ complexes with tributyl phosphate, and the equilibrium constants were determined at different temperatures (Tsubomura & Kliegman, 1960). However, such complexes are unstable under irradiation and can decompose through radiolysis during PUREX operations (Nowak, 1988).

7.4 CONCLUSION

This chapter has examined the relative importance of radionuclides in HLW by considering three main criteria: their inventory, biological toxicity, and potential for release and mobility. From the perspective of both spent fuel reprocessing and long-term safety assessment, isotopes of uranium, plutonium, and americium, together with long-lived fission products, such as ^{79}Se, ^{99}Tc, and ^{129}I, emerge as key radionuclides requiring particular attention.

The fundamental chemistry of selected radionuclides highlights the contrasting behaviors that govern HLW management. Actinide chemistry is controlled by the filling and contraction of the 5f orbitals, which yield diverse oxidation states for the early actinides and predominantly trivalent chemistry for the heavier ones.

Uranium is dominated by the redox couple U(IV)/U(VI), with UO_2 providing stability under reducing conditions and the uranyl ion (UO_2^{2+}) forming highly soluble and mobile complexes, particularly with carbonates, under oxidizing conditions. Plutonium shows the richest redox chemistry, with oxidation states from +III to +VI co-existing in solution and a strong tendency toward hydrolysis, colloid formation, and complexation, all of which complicate its environmental behavior. Americium, in contrast, is dominated by the trivalent state Am^{3+}, which is chemically similar to lanthanides, and forms sparingly soluble hydroxides but also stable complexes with carbonate and other oxygen donors.

Among long-lived fission products, technetium is dominated by Tc(VII), which forms the highly mobile pertechnetate ion (TcO_4^-) under oxic conditions, and Tc(IV), which precipitates as sparingly soluble oxides under reducing conditions. Its volatility and redox sensitivity complicate immobilization during vitrification. Selenium is redox-sensitive, existing in several oxidation states (–II, +IV, +VI). In spent fuel, Se(–II) substitutes for oxygen in UO_2, limiting release, whereas in glass it occurs primarily as soluble selenite (SeO_3^{2-}). Incomplete thermodynamic data, especially under near-neutral conditions, leaves uncertainties in long-term mobility predictions. Iodine, in contrast, is dominated environmentally by iodide (I^-) and iodate (IO_3^-), both highly soluble and weakly sorbing. Its poor incorporation into glass, volatility during reprocessing, and strong groundwater mobility make ^{129}I a critical radionuclide for HLW management.

Despite intensive studies, significant gaps remain. Thermodynamic data for several redox-sensitive radionuclides under repository-relevant near-neutral and reducing conditions are still incomplete, and the evolution of their chemistry under irradiation and during prolonged interaction with groundwater and host rock requires further investigation.

REFERENCES

Altmaier, M., Gaona, X., Fanghänel, T., 2013. Recent advances in aqueous actinide chemistry and thermodynamics. *Chem. Rev.* 113, 901–943.

Altmaier, M., Gaona, X., Fellhaier, D., Clark, D.L., Runde, W.H., Hobart, D.E., 2019. Aqueous solution and coordination chemistry of plutonium. In: Clark, D.L., Geeson, D.A., Hanrahan, R.J. Jr. (Eds.), *Plutonium Handbook*. Vol. 3. American Nuclear Society, pp. 1543–1726.

Asmussen, R.M., Levitskaia, T.G., Bottenus, C.L., Fountain, M.S., 2020. *Iodine Speciation Basis and Gap Analysis for Hanford Tank Farm Inventory and during Processing*. PNNL-30105 Rev. 0; RPT-OSIF-012 Rev. 0. Pacific Northwest National Laboratory, Richland, WA.

Baker, R.J., 2022. The chemistry of the actinides. In: *The Lanthanides and Actinides*. World Scientific (Europe), pp. 37–77.

Bard, A.J., Parsons, R., Jordan, J., 1985. *Standard Potentials in Aqueous Solution*. Taylor & Francis Group.

Barney, G.S., Wood, B.J., 1980. *Identification of Key Radionuclides in a Nuclear Waste Repository in Basalt. RHO-BWI-ST-9*. Rockwell Hanford Operations, Richland, WA.

Bianchetti, E., Oliveira, A.F., Scheinost, A.C., Di Valentin, C., Seifert, G., 2023. Chemistry of the interaction and retention of Tc VII and Tc IV species at the Fe3O4 (001) surface. *J. Phys. Chem. C* 127, 7674–7682.

Brown, D.A., et al., 2018. ENDF/B-VIII.0: The 8th major release of the nuclear reaction data library with CIELO-project cross sections, new standards and thermal scattering data. *Nucl. Data Sheets* 148, 1–142.

Browne, E., Tuli, J.K., 2017. Nuclear data sheets for A = 99. *Nucl. Data Sheets* 145, 25–340.

Bruno, J., Ewing, R.C., 2006. Spent nuclear fuel. *Elements* 2, 343–349.

Choppin, G., Liljenzin, J.-O., Rydberg, J., Ekberg, C., 2013. *Radiochemistry and Nuclear Chemistry*. 4th ed. Elsevier.

Choppin, G.R., Jensen, M.P., 2010. Actinides in solution: Complexation and kinetics. In: Morss, L.R., Edelstein, N.M., Fuger, J. (Eds.), *The Chemistry of the Actinide and Transactinide Elements*. Springer Netherlands, pp. 2524–2621.

Choppin, G.R., Rao, L.F., 1984. Complexation of pentavalent and hexavalent actinides by fluoride. *Radiochim. Acta* 37, 143–146.

Chotkowski, M., Czerwiński, A., 2021. *Electrochemistry of Technetium*. Springer International Publishing.

Clark, D.L., Hecker, S.S., Jarvinen, G.D., Neu, M.P., 2010. Plutonium. In: Morss, L.R., Edelstein, N.M., Fuger, J. (Eds.), *The Chemistry of the Actinide and Transactinide Elements*. Springer Netherlands, pp. 813–1264.

Council of the European Union, 2011. Council Directive 2011/70/Euratom of 19 July 2011 establishing a Community framework for the responsible and safe management of spent fuel and radioactive waste. *Off. J. Eur. Union L* 199, 48–56.

Curti, E., Froideval-Zumbiehl, A., Günther-Leopold, I., Martin, M., Bullemer, A., Linder, H.P., Borca, C.N., Grolimund, D., 2014. Selenium redox speciation and coordination in high-burnup UO2 fuel: Consequences for the release of 79Se in a deep underground repository. *J. Nucl. Mater.* 453, 98–106.

Curti, E., Puranen, A., Grolimund, D., Jädernas, D., Sheptyakov, D., Mesbah, A., 2015. Characterization of selenium in UO2 spent nuclear fuel by micro X-ray absorption spectroscopy and its thermodynamic stability. *Environ. Sci. Process. Impacts.* 17, 1760–1768.

Darab, J.G., Smith, P.A., 1996. Chemistry of technetium and rhenium species during low-level radioactive waste vitrification. *Chem. Mater.* 8, 1004–1021.

Edelstein, N.M., Fuger, J., Katz, J.J., Morss, L.R., 2010. Summary and comparison of properties of the actinide and transactinide elements. In: Morss, L.R., Edelstein, N.M., Fuger, J. (Eds.), *The Chemistry of the Actinide and Transactinide Elements*. Springer Netherlands, pp. 1753–1835.

Ewing, R.C., 2015. Long-term storage of spent nuclear fuel. *Nat. Mater.* 14, 252–257.

Feiveson, H., Mian, Z., Ramana, M.V., von Hippel, F., 2011. *Managing Spent Fuel from Nuclear Power Reactors: Experience and Lessons from around the World.* International Panel on Fissile Materials.

Fuge, R., Johnson, C.C., 2015. Iodine and human health, the role of environmental geochemistry and diet, a review. *Appl. Geochem.* 63, 282–302.

Geckeis, H., Zavarin, M., Salbu, B., Lind, O.C., Skipperud, L., 2019. Environmental chemistry of plutonium. In: Clark, D.L., Geeson, D.A., Hanrahan, R.J. Jr. (Eds.), *Plutonium Handbook*. Vol. 4. American Nuclear Society, pp. 1979–2118.

Gin, S., Jollivet, P., Tribet, M., Peuget, S., Schuller, S., 2017. Radionuclides containment in nuclear glasses: An overview. *Radiochim. Acta* 105, 927–959.

Grambow, B., 2008. Mobile fission and activation products in nuclear waste disposal. *J. Contam. Hydrol.* 102, 180–186.

Grenthe, I., Drożdżyński, J., Fujino, T., Buck, E.C., Albrecht-Schmitt, T.E., Wolf, S.F., 2010. Uranium. In: Morss, L.R., Edelstein, N.M., Fuger, J. (Eds.), *The Chemistry of the Actinide and Transactinide Elements*. Springer Netherlands, pp. 253–698.

Grenthe, I., Plyasunov, A.V., Runde, W.H., Grambow, B., Konings, R.J.M., Smith, A.L., Moore, E.E., Martinez, J.S., Costa, D., 2021. *Second Update on the Chemical Thermodynamics of Uranium, Neptunium, Plutonium, Americium and Technetium*. Vol. 14. OECD.

Haire, R.G., 2019. Plutonium oxides. In: Clark, D.L., Geeson, D.A., Hanrahan, R.J. Jr. (Eds.), *Plutonium Handbook*. Vol. 3. American Nuclear Society, pp. 1453–1490.

Herbst, R.S., Baron, P., Nilsson, M., 2011. Standard and advanced separation: PUREX processes for nuclear fuel reprocessing. In: *Advanced Separation Techniques for Nuclear Fuel Reprocessing and Radioactive Waste Treatment*. Elsevier, pp. 141–175.

Hou, X., Ding, W., 2009. Isotopes of iodine in thyroid and urine. In: *Comprehensive Handbook of Iodine*. Elsevier, pp. 437–448.

Hou, X., Hansen, V., Aldahan, A., Possnert, G., Lind, O.C., Lujaniene, G., 2009. A review on speciation of iodine-129 in the environmental and biological samples. *Anal. Chim. Acta* 632, 181–196.

Hummel, W., 2017. *Chemistry of Selected Dose-Relevant Radionuclides*. Nagra Technical Report NTB 17-05. Nagra, Wettingen, Switzerland.

IAEA, 2005. *Country Nuclear Fuel Cycle Profiles. 2nd ed. Technical Reports Series No. 425.* International Atomic Energy Agency, Vienna.

IAEA, 2009. *Classification of Radioactive Waste. General Safety Guide No. GSG-1*. International Atomic Energy Agency, Vienna.

Jarzemba, M.S., Pickett, D.A., 1995. *An Evaluation of the Important Radionuclides for Performance Assessment*. Center for Nuclear Waste Regulatory Analyses, San Antonio, TX.

Kaiho, T., 2014. Physical properties of iodine. In: *Iodine Chemistry and Applications*. Wiley, pp. 7–14.

Konings, R.J.M., Wiss, T., Guéneau, C., 2011. Nuclear fuels. In: Morss, L.R., Edelstein, N.M., Fuger, J. (Eds.), *The Chemistry of the Actinide and Transactinide Elements*. Springer Netherlands, pp. 3665–3811.

Lawler, K.V., Childs, B.C., Czerwinski, K.R., Sattelberger, A.P., Poineau, F., Forster, P.M., 2018. Unraveling the mystery of "tech red"—a volatile technetium oxide. *Chem. Commun.* 54, 1261–1264.

Meena, A.H., Arai, Y., 2017. Environmental geochemistry of technetium. *Environ. Chem. Lett.* 15, 241–263.

Meleshyn, A., Noseck, U., 2012. *Radionuclide Inventory of Vitrified Waste after Spent Nuclear Fuel Reprocessing at La Hague: Basic Issues and Current State in Germany*. GRS-294. Gesellschaft für Anlagen- und Reaktorsicherheit, Cologne.

Morris, K., Livens, F.R., Charnock, J.M., Burke, I.T., McBeth, J.M., Begg, J.D.C., Boothman, C., Lloyd, J.R., 2008. An X-ray absorption study of the fate of technetium in reduced and reoxidised sediments and mineral phases. *Appl. Geochem.* 23, 603–617.

Nowak, Z., 1988. The influence of iodine on the radiolysis of purex solvent. *J. Radioanal. Nucl. Chem. Lett.* 127, 309–317.

Olin, Å., Noläng, B., Öhman, L.-O., Osadchii, E.G., Rosén, E., 2005. *Chemical Thermodynamics of Selenium*, Vol. 7. Elsevier for OECD Nuclear Energy Agency.

Perrone, D., Monteiro, M., Nunes, J.C., 2015. Selenium: Chemistry, analysis, function and effects. In: Preedy, V.R. (Ed.), *Selenium: Chemistry, Analysis, Function and Effects*. Royal Society of Chemistry, pp. 3–15.

Poineau, F., Mausolf, E., Jarvinen, G.D., Sattelberger, A.P., Czerwinski, K.R., 2013. Technetium chemistry in the fuel cycle: Combining basic and applied studies. *Inorg. Chem.* 52, 3573–3578.

Popova, N.N., et al., 2003. Technetium: Behaviour during reprocessing of spent nuclear fuel and in environmental objects. *Russ. Chem. Rev.* 72, 101–121.

Rard, J.A., 2005. Current status of the thermodynamic data for technetium and its compounds and aqueous species. J. Nucl. Radiochem. Sci. 6, 197–204.

Runde, W.H., Mincher, B.J., 2011. Higher oxidation states of americium: Preparation, characterization and use for separations. *Chem. Rev.* 111, 5723–5741.

Runde, W.H., Schulz, W.W., 2011. Americium. In: Morss, L.R., Edelstein, N.M., Fuger, J. (Eds.), *The Chemistry of the Actinide and Transactinide Elements*. Springer Netherlands, pp. 1265–1395.

Shannon, R.D., 1976. Revised effective ionic radii and systematic studies of interatomic distances in halides and chalcogenides. *Acta Crystallogr. Sect. A* 32, 751–767.

Shoesmith, D.W., 2000. Fuel corrosion processes under waste disposal conditions. *J. Nucl. Mater.* 282, 1–31.

Singh, B., 2016. Nuclear data sheets for A = 79. *Nucl. Data Sheets* 135, 193–382.

Smith, M.B., Smith, M.B., 2005. Iodine. In: *Encyclopedia of Reagents for Organic Synthesis*. John Wiley & Sons, Ltd.

Spitsyn, V.I., Kuzina, A.F., Oblova, A.A., 1977. Present state of the chemistry of technetium. *Russ. Chem. Rev.* 46, 1030–1039.

Tsubomura, H., Kliegman, J.M., 1960. Molecular complexes and their spectra. XI. The interaction of iodide with tri-n-butyl phosphate. *J. Am. Chem. Soc.* 82, 1314–1317.

Virot, M., Dumas, T., Cot-Auriol, M., Moisy, P., Nikitenko, S.I., 2022. Synthesis and multi-scale properties of PuO2 nanoparticles: Recent advances and open questions. *Nanoscale Adv.* 4, 4938–4971.

Walther, C., Denecke, M.A., 2013. Actinide colloids and particles of environmental concern. *Chem. Rev.* 113, 995–1015.

Warwick, P., Aldridge, S., Evans, N., Vines, S., 2007. The solubility of technetium(IV) at high pH. *Radiochim. Acta* 95, 709–716.

WHO, 2011. *Guidelines for Drinking-Water Quality*. 4th ed. World Health Organization, Geneva.

Xu, Z., Kazimi, M.S., Driscoll, M.J., 2005. Impact of high burnup on PWR spent fuel characteristics. *Nucl. Sci. Eng.* 151, 261–273.

PART III

Nuclear Waste Treatment Technologies

CHAPTER 8

Radioactive Waste Immobilization in Cementitious Materials

Qiu Li

8.1 INTRODUCTION

The safe and secure management of radioactive waste constitutes a fundamental prerequisite for the sustainable utilization of nuclear energy and the beneficial applications of radioisotopes in medicine, industry, and research. Radioactive wastes, generated from activities such as nuclear power generation, fuel reprocessing, medical procedures, and research facilities, present unique long-term hazards due to the ionizing radiation emitted during the decay of their constituent radionuclides. These hazards can persist for periods ranging from decades to millennia, necessitating management strategies capable of providing robust isolation from the biosphere over relevant timescales. The primary objective of radioactive waste management is unequivocally the protection of human health and the environment, both in the present and for future generations, through effective containment and isolation of radioactive materials. This imperative is underscored by international consensus and safety standards (IAEA, 2018).

The global inventory of radioactive waste continues to grow, driven by ongoing nuclear operations, the decommissioning of aging facilities, and the remediation of legacy waste sites. Nuclear power plants (NPPs), for instance, generate significant volumes of low- and intermediate-level waste (LLW/ILW) during routine operation, maintenance, and eventual dismantling. Improper management of this waste carries substantial risks, including potential radiation exposure to workers and the public, environmental contamination of vital soil and water resources, and imposition of long-term societal burdens (IAEA, 2018). Consequently, the development and implementation of safe, reliable, and economically viable waste treatment and disposal technologies transcend a mere technical challenge; they represent a critical societal responsibility and are a pivotal factor influencing public acceptance of nuclear technologies. Within this context, immobilization—the process of

DOI: 10.1201/9781003665793-11

converting waste into a stable, solid form that minimizes the potential for radionuclide release—stands as a cornerstone of most disposal strategies for solidified waste (Ojovan & Lee, 2014).

Cementitious materials, predominantly based on Portland cement and its variants, have emerged as one of the most widely used and extensively studied matrices globally for the immobilization of LLW/ILW. Their prominence arises from a compelling combination of advantageous properties. First, they offer significant chemical stability; key hydration products, particularly calcium silicate hydrate (C–S–H), the primary binding phase; are stable over geological timescales under repository-relevant conditions, forming a durable host matrix (Glasser, 2011). Second, the inherent high alkalinity ($pH > 12.5$) of the cement pore solution promotes low solubility and precipitation for many critical radionuclides, such as actinides and numerous transition metals, while also providing a potent chemical barrier (Atkins & Glasser, 1992). Third, the setting and hardening process physically encapsulates waste particles and debris within a monolithic, low-permeability solid, effectively reducing the surface area exposed to potential leaching agents. Furthermore, cementitious waste forms develop adequate compressive strength to withstand handling, transportation, and disposal loads, ensuring physical integrity over time. Processing simplicity and cost-effectiveness are also major advantages; cementation technologies like in-drum mixing or grouting are relatively simple, mature, and significantly less expensive than high-temperature processes such as vitrification, making them highly suitable for managing large waste volumes typical of LLW/ILW. Finally, cementitious matrices generally exhibit good tolerance to the radiation fields associated with LLW/ILW streams (Ojovan et al., 2011).

While cement-based immobilization offers substantial advantages, it is not a universal solution and possesses inherent limitations that must be carefully evaluated during waste form design and performance assessment (PA). Key advantages include its status as a proven technology with extensive global industrial experience, relatively low processing temperatures, which avoids volatilization issues, tolerance to variations in waste composition and water content, and the ability to incorporate various functional additives like pozzolans or clays to enhance specific properties. Cement systems provide both physical and chemical containment mechanisms and operate within an established regulatory acceptance framework. However, limitations include limited compatibility with certain wastes, such as highly soluble radionuclides like cesium-135/137 and iodine-129 due to their low sorption and potential for rapid leaching, or organic materials, specific metals like aluminum or zinc in large quantities, and strong complexing agents that can interfere with setting or degrade performance (Serne et al., 2008). Volume increase, or inflation, is another concern, as mixing waste with cement and water generally results in a larger final waste form volume compared to the original waste, impacting disposal costs. Long-term performance can also be affected by susceptibility to degradation processes such as carbonation (which reduces pH), sulfate attack, freeze–thaw cycling, and potential radiation-induced changes in microstructure and phase stability. Cementation is generally unsuitable for wastes with high thermal output, such as some ILW or high-level waste (HLW), without significant dilution, as the generated heat can cause cracking and degradation. Finally, uncertainties regarding the long-term stability of chemical bonds formed between radionuclides and cement phases under evolving repository conditions necessitate careful assessment (Bernard et al., 2020). Understanding these advantages and limitations is paramount for selecting appropriate waste streams for cementation, optimizing waste form formulations, and realistically predicting long-term performance within the safety case for disposal. The subsequent sections of this chapter delve into the scientific principles, engineering practices, and ongoing developments underpinning the use of cementitious materials for radioactive waste immobilization.

8.2 FUNDAMENTALS OF CEMENTITIOUS MATERIALS FOR WASTE IMMOBILIZATION

The effectiveness of cementitious materials in immobilizing radioactive waste stems from their complex chemistry, evolving microstructure, and the resulting physicochemical environment they create. Understanding these fundamental aspects is paramount for designing robust waste forms and predicting their long-term performance in a disposal environment.

8.2.1 Chemistry and Microstructure of Cementitious Materials

Cementitious materials used in waste immobilization encompass a range of binders, each with distinct chemistries and microstructures that dictate their immobilization potential. The hydration of Ordinary Portland Cement (OPC), primarily involving tricalcium silicate (C_3S) and dicalcium silicate (C_2S), produces a complex, multi-phase microstructure dominated by several key hydration products (Taylor, 1997). The primary binding phase, constituting 50–60% by volume, is calcium silicate hydrate (C–S–H), a nano-crystalline or amorphous with a variable Ca/Si ratio, typically between 1.2 and 2.0. This phase provides the main mechanical strength and offers a high surface area for chemical sorption, with its structure often described as resembling defective tobermorite or jennite. Another major crystalline phase, comprising 20–25% of the hydrated paste, is calcium hydroxide (CH or portlandite), which is critical for providing and buffering the high alkalinity (pH ~12.5) of the pore solution, an environment crucial for many immobilization mechanisms. Furthermore, the reaction of aluminate (C_3A) with sulfate leads to the formation of calcium sulfoaluminates (CSAs), including ettringite (AFt) and monosulfate (AFm). These phases, particularly the layered AFm structures, play a significant role in anion immobilization through their capacity for ion exchange, allowing for the incorporation of various anions such as OH^-, CO_3^{2-}, Cl^-, and I^- into their interlayers (Glasser, 2011).

To enhance specific properties for waste immobilization, OPC is often partially replaced by supplementary cementitious materials (SCMs) to form blended or composite cements. One common SCM is blast furnace slag (BFS), a latent hydraulic material rich in CaO, SiO_2, and Al_2O_3. When activated by the alkalinity provided by OPC, BFS reacts to form additional C–S–H with a lower Ca/Si ratio while concurrently reducing the CH content. This reaction results in a denser, less permeable microstructure and can enhance the chemical retention of certain radionuclides (Lothenbach et al., 2011). Another widely used SCM is pulverized fuel ash (PFA, or fly ash), a pozzolanic material. Its reaction consumes Portlandite to form additional C–S–H and calcium aluminate silicate hydrate (C–A–S–H) phases, which refines the pore structure and reduces the overall permeability of the cementitious matrix. Beyond traditional blended systems, alkali-activated materials (AAMs) or geopolymers represent an alternative class of binders formed by reacting an aluminosilicate precursor like metakaolin (MK), fly ash, or slag with a concentrated alkaline activator such as NaOH, KOH, or sodium silicate. The reaction products are distinct from OPC hydrates; low-calcium systems (e.g., fly ash-based) primarily form amorphous to semi-crystalline sodium or potassium aluminosilicate hydrate (N-A-S-H or K-A-S-H) gels with a highly cross-linked, zeolite-like framework, while high-calcium systems (e.g., slag-based) form a calcium aluminosilicate hydrate (C–A–S–H) gel, which can be similar to C–S–H but with significant Al substitution and a lower Ca/Si ratio, and sometimes hydrotalcite-like phases (Provis & Bernal, 2014). AAMs often exhibit very low permeability, good chemical resistance, and potentially higher thermal stability than OPC, though their generally lower pore solution pH (10–13.5, depending on formulation) can distinctly impact radionuclide solubility and sorption behavior.

8.2.2 Key Properties Affecting Immobilization

The immobilization performance of a cementitious waste form hinges on several interconnected properties. First, the porosity and pore structure directly control the permeability and diffusivity of the matrix, which are the primary transport mechanisms for radionuclide leaching. The volume fraction, size distribution, and connectivity of pores, particularly the capillary pores, are paramount. Lower capillary porosity and finer, less connected pores significantly reduce the rate of contaminant release, a principle that guides mix design optimization. Techniques such as mercury intrusion porosimetry (MIP) are used for characterization (Scrivener et al., 2016). Second, the high alkalinity ($pH > 12.5$ in fresh OPC systems) generated primarily by the dissolution of portlandite is a cornerstone property. This hyperalkaline environment promotes the precipitation or maintains low solubility for many cationic radionuclides (e.g., U(VI), Pu(IV), Am(III)) as hydroxides, oxides, or carbonates; it can stabilize redox-sensitive elements in less soluble oxidation states (e.g., facilitating the reduction of Tc(VII) to Tc(IV)); and it ensures that the surface of the primary sorption phase, C–S–H, is negatively charged, enabling strong sorption of cations such as Cs^+ and Sr^{2+}. It is noted that blended cements and AAMs can have a lower equilibrium pH (9–13.5), which directly impacts their solubility and sorption behaviors.

Third, the long-term durability of the waste form is essential for its intended function over geological timescales. Key degradation mechanisms relevant to disposal include leaching, where the dissolution of solid phases (especially CH) and the diffusion of ions out of the matrix gradually reduce the internal pH, potentially destabilizing immobilization mechanisms. Carbonation, the reaction of atmospheric or microbial CO_2 with CH and C–S–H to form calcite, consumes hydroxyl ions and lowers the pH to ~8–9, which can dramatically increase the solubility and mobility of many radionuclides previously immobilized at high pH (e.g., actinides) and can destabilize AFm phases (Bernard et al., 2020). Sulfate attack, involving external sulfate ions reacting with AFm and aluminate hydrates to form expansive secondary ettringite or gypsum, can cause cracking and a loss of physical integrity. Furthermore, radiation effects must be considered, as radiolysis of pore water can produce oxidizing and reducing species that may alter radionuclide oxidation states and solubility, while gamma radiation can cause point defects and amorphization in crystalline phases, though C–S–H is relatively resistant (Ojovan et al., 2011).

Finally, the intrinsic chemical binding capacity of the cement phases determines their ability to incorporate or strongly interact with radionuclides. This encompasses several processes: precipitation, where discrete radionuclide-bearing phases (e.g., uranium oxides, technetium sulfides) or solid solutions form; sorption, involving surface complexation and ion exchange onto hydrated phases such as C–S–H (which has a high capacity for cations like Cs^+, Sr^{2+}) and AFm phases (allowing anion exchange for I^-, TcO_4^-), processes that are highly dependent on pH and ionic strength (Tits et al., 2006); and incorporation, where radionuclides substitute into the crystal structure of hydration products (e.g., Ra^{2+} for Ca^{2+} in CH; Sr^{2+} for Ca^{2+} in C–S–H).

8.2.3 Mechanisms of Contaminant Immobilization

Radionuclides are retained within the cementitious matrix through a synergistic combination of physical and chemical mechanisms. Physical encapsulation serves as the first line of defense, where the hardened cement paste acts as a monolithic physical barrier. It encapsulates waste particles, debris, or pre-formed radionuclide phases within its solid matrix, thereby drastically reducing the exposed surface area available for leaching. Furthermore, it severely retards the transport of any dissolved contaminants by forcing them to diffuse through the intricate, low-permeability pore network. The efficacy of this mechanism is wholly dependent on the waste form maintaining its integrity—being free from large cracks—and possessing low permeability and diffusivity.

Chemical binding provides a more specific and often stronger retention through direct chemical interactions between radionuclides and the cement phases. This includes precipitation, where radionuclides form low-solubility solid phases either directly (e.g., $UO_2{\cdot}xH_2O$, TcO_2) or via co-precipitation, where they are incorporated into the structure of common hydration products (e.g., Sr^{2+} in C–S–H, I^- in AFm phases). Sorption is another critical chemical pathway, encompassing both cation exchange, where ions like Cs^+ and Sr^{2+} are reversibly exchanged for Ca^{2+} on the negatively charged C–S–H surface, and surface complexation, the formation of strong inner-sphere complexes between radionuclide ions and surface functional groups (e.g., ≡Si-O⁻) on C–S–H or other hydrates, a process that is particularly important for multi-valent cations and anions (Iler, 1979; Tits et al., 2006). Ion exchange, primarily occurring in the interlayer spaces of layered AFm phases (e.g., hydrocalumite) or intentionally added clay minerals, is a key mechanism for anion immobilization, allowing species like I^-, TcO_4^-, and SeO_4^{2-} to exchange with structural anions.

A particularly powerful mechanism for certain problematic radionuclides is redox stabilization. This leverages the highly reducing environment generated by specific additives (e.g., BFS containing Fe^{2+}/S^{2-}, zero-valent iron) or inherent in certain cement formulations (e.g., low-pH slag cements) to stabilize redox-sensitive elements in lower oxidation states that form vastly less soluble compounds. Paramount examples include the reduction of highly mobile pertechnetate (TcO_4^-, Tc(VII)) to insoluble Tc(IV) oxides ($TcO_2{\cdot}nH_2O$), and the reduction of soluble selenate (SeO_4^{2-}, Se(VI)) to less soluble selenite (SeO_3^{2-}, Se(IV)) or elemental Se(0). The reduction of soluble U(VI) (uranyl) to less soluble U(IV) (uraninite) and Np(V) to Np(IV) are also key processes enabled by this mechanism. The relative importance of each mechanism is not fixed; it depends profoundly on the specific radionuclide's chemistry, its concentration, the overall waste form composition (cement type, additives, and waste stream chemistry), and the evolving conditions over time (pH, Eh, temperature, and leaching). Consequently, a successful and resilient waste form design intentionally leverages multiple complementary mechanisms to ensure robust immobilization.

8.3 TYPES OF RADIOACTIVE WASTE SUITABLE FOR CEMENTATION

Cementation is a mature and widely deployed technology, but its applicability is not universal across all radioactive waste categories. Its effectiveness is primarily demonstrated for LLW and a significant portion of ILW, where the thermal output and radionuclide concentrations are compatible with the physical and chemical limitations of cementitious matrices. Understanding the characteristics of suitable waste streams and the boundaries of applicability is critical for safe and effective implementation.

8.3.1 Low- and Intermediate-Level Waste (LLW/ILW): Characteristics and Examples

LLW and ILW are defined based on their radioactivity content and thermal output, with classifications varying slightly between national regulatory bodies such as the International Atomic Energy Agency (IAEA), the U.S. Nuclear Regulatory Commission (NRC), and the French National Radioactive Waste Management Agency (ANDRA) (IAEA, 2018; NRC, 2021; ANDRA, 2020). In general, LLW contains relatively small amounts of mostly short-lived radionuclides, with low radiation levels and negligible thermal output. This category often constitutes the largest volume of radioactive waste from nuclear operations. Examples of radioactive waste types that are highly suitable for cementation include operational wastes such as slightly contaminated solid materials from NPPs and research facilities (e.g., paper, plastics, cloth, gloves, tools, soil, rubble, and lightly contaminated metals), which are often compacted or incinerated, with the resulting ashes then cemented before immobilization. Decommissioning wastes, comprising bulk materials from the dismantling of nuclear facilities (e.g., concrete, steel, piping, insulation) with low levels of activation or surface contamination, are also prime candidates, though size reduction may be needed before encapsulation. Liquid wastes, particularly low-activity aqueous streams such as laundry waste, floor drains, and chemical lab drains that are often concentrated by evaporation, represent another key stream. The resulting evaporator concentrates—brines containing soluble salts like $NaNO_3$, Na_2SO_4, Na_3BO_3 and radionuclides such as ^{137}Cs, ^{90}Sr, ^{60}Co, ^{3}H, and ^{14}C—are especially suitable for cementation, often requiring tailored binders (e.g., blended cements) and additives to manage high salt content and enhance radionuclide retention (Atkins & Glasser, 1992; IAEA, 1996). Furthermore, spent organic ion-exchange resins and filter media used in reactor coolant purification, fuel pool water treatment, and liquid waste processing can be immobilized via cementation, though they pose specific challenges such as organic degradation, gas generation, and potential swelling, necessitating careful formulation with high water-to-cement ratios, pozzolans, and special additives (Ojovan & Lee, 2014; Plecas et al., 2005).

ILW contains higher concentrations of radionuclides, including long-lived alpha emitters (e.g., actinides), which require greater containment and isolation than LLW. Its thermal output may be significant but is generally low enough (typically < ~2 kW/m^3) to be managed in cementitious waste forms, often with specific design considerations like larger containers, lower waste loading, or enhanced thermal conductivity additives. Examples suitable for cementation include chemical reactor wastes, such as sludges and solids from chemical treatment processes (e.g., precipitation, flocculation) used to decontaminate liquid effluents, which often contain metal hydroxides/oxides co-precipitated with radionuclides like Ni, Co, Cs, Sr, and actinides. Fuel cladding hulls—metallic residues (e.g., Zircaloy, stainless steel) from reprocessed spent nuclear fuel—are typically fragmented and cemented into drums or boxes; while the metal itself is inert, surface contamination from activation and fission products must be immobilized. Activated components, such as metal parts from reactor cores or control rods that have become radioactive through neutron activation, are encapsulated after decontamination and size reduction. Concentrated liquid wastes, including higher activity evaporator concentrates or raffinates from reprocessing (after removal of most actinides and Cs/Sr in HLW vitrification), may contain significant concentrations of problematic radionuclides like ^{99}Tc, ^{129}I, and ^{14}C, requiring optimized formulations. Finally, spent precipitants and adsorbents, like nickel ferrocyanide for Cs removal or titanate/silicate compounds for actinide/strontium removal, can be effectively encapsulated in cement, leveraging both physical encapsulation and potential chemical binding (Serne et al., 2008).

8.3.2 Challenges with High-Level Waste (HLW)

Cementitious materials are generally unsuitable for the direct immobilization of HLW due to several fundamental limitations. The high thermal output of HLW, primarily from reprocessed spent fuel or spent fuel itself, generates significant heat (>> 2 kW/m^3) from the decay of fission products and actinides. This heat can cause excessive temperature rises within the cement matrix (>100°C), leading to dehydration of key hydrate phases like C–S–H, ettringite, and AFm; loss of bound water; and severe cracking; it can also accelerate degradation processes such as leaching and phase transformations, cause pressurization from steam generation, and make it difficult to maintain the required high pH environment. Vitrification (glass) is the standard technology for HLW due to its superior thermal stability and durability (Ojovan & Lee, 2014). Additionally, while cement is tolerant to LLW/ILW radiation fields, the intense gamma and neutron flux from fresh HLW can cause significant radiolytic decomposition of water in the pore structure, generating gases (H_2, O_2) and reactive species that damage the matrix and alter radionuclide chemistry, with cumulative dose effects on long-term phase stability also being a concern (Ojovan et al., 2011). Moreover, HLW contains high concentrations of long-lived, highly mobile radionuclides like ^{129}I, ^{99}Tc, ^{135}Cs, and actinides (Np, Pu, Am, Cm), and standard cement formulations struggle to provide the very high long-term retention required for HLW disposal safety cases, especially under evolving chemical conditions such as pH drop from carbonation, whereas vitrified waste forms offer superior chemical durability for these elements.

8.3.3 Compatibility with Other Waste Forms

Cementation is often used to encapsulate or co-immobilize various solid or semi-solid waste forms, but compatibility must be carefully assessed for each type. For ion-exchange resins, challenges include high organic content, potential for biodegradation and gas generation (H_2, CH_4, CO_2, NH_3), swelling upon water absorption, potential chemical interactions such as release of sulfonic acid groups lowering pH, and radiolytic degradation, with high water demand often leading to high porosity. Mitigation strategies involve pre-treatment, which includes drying or washing, use of high-fluidity formulations with superplasticizers, incorporation of pozzolans like fly ash or silica fume (SF) to densify the matrix and sorb organics/radionuclides, addition of chemical stabilizers for iodine, limiting waste loading, and using specialized binders such as CSA cement for faster setting and reduced swelling potential (Plecas et al., 2005; IAEA, 1996). For ashes from incineration, challenges include high alkalinity or acidity depending on the waste incinerated, the presence of soluble salts such as chlorides and sulfates, potential for heavy metals (Pb, Cd, Zn) and volatilized radionuclides such as Cs and Tc, and fine particle size affecting workability. Mitigation requires thorough characterization, blending with other wastes, use of additives such as phosphates for metal stabilization or clays for sorption, and formulations resistant to leaching and degradation in high-salt environments. Sludges pose challenges due to high water content diluting cement and increasing porosity, variable and complex chemistry, including acidity, organics, sulfates, and heavy metals, potential for gas generation, and heterogeneity. Strategies include pre-treatment such as dewatering, drying, or pH adjustment, thorough characterization, formulation optimization with binders and additives such as bentonite for water retention/sorption, and ensuring good mixing homogeneity, with grout injection often used for in situ stabilization. For metals, challenges include corrosion of reactive metals such as Al, Zn, and Mg in the high-pH cement pore solution, which generates hydrogen gas (H_2). This gas can cause pressurization, cracking, and flammability, while ferrous metals corrode slowly but contribute to long-term gas generation. Mitigation involves limiting the content of reactive metals, pre-treatment such as passivation or coating, use of additives that consume hydroxide or form protective layers (e.g., lithium salts, phosphates), ensuring adequate free volume in containers for gas accumulation and monitoring (NEA, 2012). Finally, boric acid and borates, common neutron absorbers in NPPs present in evaporator concentrates, retard cement hydration by adsorbing onto C–S–H and inhibiting nucleation/growth, with high concentrations severely compromising setting and strength; mitigation strategies include use of borate-resistant cements such as calcium aluminate cements, addition of excess lime or lithium salts to counteract retardation, pre-treatment to remove or fix boron (e.g., precipitation as colemanite), and careful control of waste loading and water content (Glasser, 2011).

A key consideration across all waste types is that thorough waste characterization—physical, chemical, and radiological—is paramount before cementation. Incompatible components can lead to poor waste form properties such as low strength, high porosity, and cracking, as well as gas generation, excessive heat, or inadequate radionuclide retention, jeopardizing long-term safety. Cementation is a versatile tool, but it is not a panacea; its application must be based on rigorous waste-stream-specific assessment and formulation development.

8.4 DESIGN AND OPTIMIZATION OF CEMENTITIOUS WASTE FORMS

The successful immobilization of radioactive waste in cementitious matrices hinges on the deliberate design and optimization of the waste form. This involves selecting appropriate binders, incorporating functional additives, and tailoring formulations to address the specific challenges posed by the waste stream and its radionuclide inventory. The goal is to achieve a monolithic, durable solid with minimized leachability, adequate mechanical strength, and long-term stability under repository conditions.

8.4.1 Selection Criteria for Cement Types

The choice of cement binder is the foundational decision, driven by waste chemistry, processing requirements, and desired long-term performance. OPC is ubiquitous and well-understood, offering a high initial pH (>12.5) beneficial for many cations, good workability, and established standards. However, it is susceptible to certain chemical attacks like sulfate and acid, and its high pH can be detrimental to some anions such as I^- and TcO_4^-. It also has limited compatibility with organics, boron, and reactive metals, making it best for general LLW solidification such as debris, soils, rubble, and evaporator concentrates with moderate salt content. It is often used as a base for blended systems (Glasser, 2011). BFS-blended cements, such as OPC/BFS, offer lower heat of hydration, reduced permeability and porosity due to denser C-(A)-S-H formation, lower equilibrium pH (~12–12.5), which is potentially better for anion retention. They also provide better resistance to sulfate attack, higher long-term strength, potential for redox control via Fe^{2+}/S^{2-}, and reduced CH content. However, they suffer from slower setting and strength development, sensitivity to curing temperature, and potentially higher gas generation if sulfides oxidize. These characteristics make them suitable for liquid wastes such as evaporator concentrates and some ILW streams, wastes containing sulfates, and scenarios where lower long-term pH or redox control is needed for radionuclides such as Tc and Se. They are also used for encapsulation of resins to improve matrix density (Lothenbach et al., 2011; Glasser, 2011). Pulverized fuel ash (PFA or fly ash)-blended cements, such as OPC/PFA, provide reduced heat of hydration, lower permeability through refined pore structure, improved workability, reduced CH content via pozzolanic reaction, good acid resistance compared to OPC, and cost-effective use of an industrial by-product. However, they exhibit slower early strength development, potential variability in ash composition, and may increase water demand if carbon content is high, rendering them best for general LLW/ILW solidification, wastes with high water content, acidic wastes after neutralization, and encapsulation of resins and ashes for improved sorption capacity.

CSA cements offer rapid setting and strength development, high early strength, and a lower pH (10.5–11.5) due to absence of CH, as they form ettringite and strätlingite. They also provide reduced expansion with reactive metals such as Al, good dimensional stability, and potential for anion binding in AFm phases, though they come with higher cost than OPC, sensitivity to mix water and curing conditions, potential for rapid setting requiring admixtures, and a less extensive long-term performance database. These characteristics make them suitable for encapsulation of ion-exchange resins to control swelling and achieve faster setting, wastes containing reactive aluminum, wastes where lower pH is beneficial for I^- immobilization in AFm phases, and rapid turnaround applications such as emergency waste treatment (Zhang & Glasser, 2002; Bernal et al., 2011). AAMs or geopolymers provide very low permeability, high early strength, good chemical resistance to acids and sulfates, good thermal stability, potential for lower CO_2 footprint, tunable chemistry with a pH range of ~10–13.5 depending on activator and precursor, and good immobilization potential for certain cations like Cs^+ and Sr^{2+} and anions via structural incorporation or surface sites. However, they involve complex chemistry, sensitivity to precursor composition and activator concentration, potential for efflorescence, less predictable long-term phase evolution compared to OPC, developing regulatory acceptance, and safety concerns in handling concentrated alkalis. These factors make them best suited for niche applications requiring high chemical resistance, immobilization of specific problematic wastes such as Al-containing wastes and some ashes, development of lower-pH waste forms, and encapsulation where high strength and low permeability are critical, Research is ongoing with regard to Tc, I, and actinides (Provis & Bernal, 2014; Bernal et al., 2011). Key selection factors include waste composition (pH, organics, salts, and metals), radionuclide inventory (types, concentrations, and oxidation states), required setting time and strength development, thermal constraints, long-term performance targets such as leach resistance and pH stability, processability for pumping and mixing, as well as cost and availability.

8.4.2 Additives and Admixtures for Enhanced Performance

Additives are incorporated to modify the cement paste properties, enhance radionuclide retention, or mitigate adverse waste–cement interactions. Pozzolans such as fly ash (PFA), SF, and MK improve microstructure by reducing porosity and permeability, consume CH, provide additional sorption sites, and enhance physical encapsulation, chemical durability, and sorption capacity while reducing leach rates and managing heat generation, with SF being particularly effective for drastic porosity reduction and MK enhancing anion sorption capacity (Tits et al., 2006). Clay minerals such as bentonite (montmorillonite), with high swelling capacity and cation exchange capacity (CEC), are used to absorb excess water, reduce bleed, increase viscosity, and enhance sorption of cations such as Cs^+, Sr^{2+}, and actinides. They also act as a permeability barrier within the waste form or as a backfill. Zeolites such as clinoptilolite and chabazite, with high CEC and specific affinity for certain cations like Cs^+ and Sr^{2+}, are incorporated to selectively trap mobile radionuclides within their crystalline structures This approach was proven effective in Fukushima wastewater treatment and subsequent waste solidification (NEA, 2020; Osmanlioglu, 2002).

pH buffers and reducing agents are crucial for controlling the long-term chemical environment. pH buffers such as hydrotalcite-like layered double hydroxides (LDHs) and calcium silicate hydrates with low Ca/Si ratios are designed to maintain a controlled, often lower pH (around 10–11) over the long term. This helps improve the stability of radionuclides like U(VI) and Sn(II), which may re-dissolve if standard OPC carbonates and pH drops sharply Hydrotalcite and its analogues also offer high anion exchange capacity for I^-, TcO_4^-, and SeO_4^{2-} (Shi & Fernández-Jiménez, 2006; Bernal et al., 2011). Reducing agents are added to stabilize redox-sensitive radionuclides in lower, less soluble oxidation states.

Zero-valent iron (ZVI) is a powerful reductant for Tc(VII)→Tc(IV), U(VI)→U(IV), and Cr(VI)→Cr(III) and can be encapsulated within the cement matrix. BFS provides inherent reducing capacity via Fe^{2+}/S^{2-} in slag-rich cements; titanium dioxide under UV or radiation can facilitate photocatalytic reduction of species like TcO_4^-; and sulfide salts like FeS and Na_2S enable direct precipitation of low-solubility sulfides such as TcS_2 and SnS.

Other functional additives include superplasticizers, set controllers, antifoaming agents, and corrosion inhibitors. Superplasticizers, such as polycarboxylate ethers (PCEs) and sulfonated naphthalene/formaldehyde condensates, reduce water demand and allow lower water-to-cement (W/C) ratios, thereby improving strength and reducing permeability while maintaining workability. Set controllers, including accelerators such as Li_2CO_3 and retarders such as sucrose or citric acid, help adjust setting time to meet processing requirements. Antifoaming agents suppress air entrainment during mixing and improve density and homogeneity, whereas corrosion inhibitors such as $LiNO_3$ help passivate reactive metals in waste stream.

8.4.3 Formulation Strategies for Specific Radionuclides

Optimization often requires tailoring the formulation to target the immobilization mechanisms of key mobile or long-lived radionuclides. For Cesium-137/135 ($^{137}Cs/^{135}Cs$), which has high solubility and weak sorption on C–S–H at high pH due to competition with K^+/Na^+, strategies include using additives with high Cs^+ selectivity like zeolites (e.g., clinoptilolite, chabazite), clay minerals (illite, vermiculite), or hexacyanoferrates (e.g., copper or nickel ferrocyanide) often pre-loaded on substrates; minimizing W/C ratio with superplasticizers and incorporating pozzolans like SF and FA to densify the matrix and reduce diffusion pathways; and cautiously using blended cements (BFS, PFA) or CSA cements with lower equilibrium pH (10–12) to potentially increase Cs^+ sorption on C–S–H (Tits et al., 2006; Osmanlioglu, 2002).

Strontium-90 (^{90}Sr) behaves similarly to Ca^{2+} and is relatively well retained in OPC systems through sorption onto C–S–H, surface complexation, ion exchange, and possible incorporation into C–S–H by substitution for Ca^{2+}. Strategies to improve ^{90}Sr immobilization include increasing the C–S–H surface area with pozzolans, forming low-Ca/Si C–S–H in blended or silica-fume-containing cements, promoting incorporation into CH or AFm phases, precipitating $SrSO_4$ (celestine) through sulfate addition where appropriate, and reducing porosity and permeability.

For iodine-129 (^{129}I), an anionic, very mobile, and long-lived radionuclide with negligible sorption on C–S–H at high pH and volatility during processing, strategies involve anion exchange using additives with high I^- affinity such as AFm phases like hydrocalumite (requiring careful formulation for stability), hydrotalcite (LDHs), silver-impregnated materials (e.g., silver-exchanged zeolites, Ag-mordenite), copper-coated materials, or bismuth oxyhydroxides, often pre-loaded onto substrates then encapsulated. Another strategy involves using low-pH matrices like CSA cements or carefully designed AAMs (pH ~10.5–11.5), where I^- sorption on C–S–H or C–A–S–H may be slightly enhanced compared to OPC pH>12.5, with hydrotalcite helping to buffer in this range. Additional approaches include precipitation by incorporating heavy metals like Pb or Bi to form low-solubility iodides (e.g., PbI_2, BiOI), although this may pose processing challenges. Finally, synthesizing mineral analogs like iodoapatite ($Pb_{10}(VO_4)_6I_2$) or other durable iodide minerals within the matrix has been explored, although it remains largely at the research stage (Serne et al., 2008).

For technetium-99 (^{99}Tc), which is highly mobile as the pertechnetate anion (TcO_4^-) under oxidizing conditions, exhibits weak sorption and very long lifetime, the primary strategy is reduction and precipitation to convert soluble Tc(VII) into insoluble Tc(IV) (e.g., $TcO_2{\cdot}nH_2O$) using reducing additives like BFS (Fe^{2+}/S^{2-}), zero-valent iron (ZVI), tin(II) compounds ($SnCl_2$), ferrous sulfides (FeS), or titanium(III) compounds, while maintaining reducing conditions over the long term. AAMs based on slag may provide a reducing environment. Anion exchange using silver-functionalized materials or hydrotalcites can be partially effective, but reduction is preferred. In addition, sulfide precipitation can form TcS_2 under sulfidic conditions if reduced, with sulfides provided by BFS or added sulfides.

Uranium, including radionuclides in the ^{238}U decay series and ^{235}U, has complex chemistry and can occur in multiple oxidation states, mainly U(IV) and U(VI). Under high-pH conditions, U(VI) occurs as the uranyl ion (UO_2^{2+}) and can form soluble carbonate complexes, although it may initially precipitate as uranyl hydroxides or calcium uranates. In contrast, U(IV) is relatively insoluble; therefore, long-term immobilization strategies often involve reducing U(VI) to U(IV) using reductants such as blast furnace slag (BFS), zero-valent iron (ZVI), or Sn^{2+}. Additional strategies include maintaining high pH while limiting carbonate availability through low-permeability formulations, pozzolans, and controlled carbonate sources in wastes or additives. Uranium immobilization can also be enhanced by sorption onto C–S–H and hydrotalcite, incorporation into AFm or calcium silicate hydrate phases, and precipitation as low-solubility uranyl phosphate minerals, such as autunite-group minerals, through phosphate additives (Tits et al., 2011). Phosphate cements, such as magnesium potassium phosphate (MKP) and calcium phosphate cements, may also provide high retention of uranium and other actinides through precipitation, although their long-term durability under wet conditions remains a concern.

Formulation optimization is an iterative process involving waste characterization, laboratory-scale testing for workability, setting, strength, and leachability (see Section 8.5), and often computational modeling, while balancing waste loading to minimize final volume, processing requirements, short-term properties, and predicted long-term performance.

8.5 PERFORMANCE ASSESSMENT AND DURABILITY TESTING

Ensuring the long-term safety of cemented radioactive waste forms requires rigorous assessment of their performance over geological timescales relevant to disposal (hundreds to hundreds of thousands of years). This involves understanding potential degradation pathways, developing predictive models based on experimental data, and evaluating performance under anticipated repository conditions. PA integrates laboratory testing, modeling, and natural/industrial analogue studies to build a safety case demonstrating that radionuclide release rates will remain below regulatory thresholds.

8.5.1 Short- and Long-Term Degradation Mechanisms

Cementitious waste forms are not inert; they evolve chemically and physically over time due to interactions with the surrounding environment (groundwater, gases, or microbes) and internal processes (continued hydration, phase transformations, or radiation effects). The most critical degradation process for radionuclide release is leaching, a chemical degradation mechanism involving the dissolution of solid phases and the diffusion and transport of ions out of the matrix. The process begins when pore water, initially saturated with cement ions (Ca^{2+}, Na^+, K^+, OH^-, SO_4^{2-}), diffuses outwards upon contact with external water such as groundwater, creating concentration gradients that drive the dissolution of soluble phases like Portlandite ($Ca(OH)_2$) and the decalcification of C–S–H (reducing its Ca/Si ratio). Alkalis (Na^+, K^+) are rapidly leached initially, and the pH gradually decreases from >12.5 toward neutrality over very long timescales, with radionuclides released either by dissolution of their host phases or by desorption and diffusion. Standardized leaching tests are essential tools for quantifying release rates and understanding these mechanisms. Key tests include ANS 16.1 (ANSI/ANS-16.1-2019) (ANS, 2019), which measures the Leachability Index (LI) to determine how effectively the matrix retards diffusion, with LI > 6 often a target for good performance, based on sequential leaching in deionized water or simulated groundwater; ASTM C1308 (ASTM C1308–21) (ASTM, 2021), which provides a more detailed characterization of diffusive release and allows calculation of effective diffusion coefficients (D_e) for specific constituents; and other tests such as tank leaching tests (e.g., NEN 7345, EA NEN 7375), pH-stat tests to study dissolution at controlled pH, and tests with site-specific groundwater compositions, with long-term (>1 year) static or dynamic tests offering more realistic but time-consuming data (Atkins & Glasser, 1992; Plecas & Peric, 2006).

Carbonation is another significant degradation mechanism, involving the reaction of atmospheric or biogenic CO_2 with the alkaline cement hydrates. CO_2 dissolves in pore water forming carbonic acid (H_2CO_3), which reacts primarily with Portlandite ($Ca(OH)_2 + CO_2 \rightarrow CaCO_3 + H_2O$) and subsequently with C–S–H (decarbonation: C–S–H + $CO_2 \rightarrow CaCO_3$ + SiO_2 gel + H_2O), consuming OH^- ions and significantly lowering the pore solution pH to ~8–9 in fully carbonated zones. This pH drop can dramatically increase the solubility and mobility of many radionuclides previously immobilized under high pH conditions, such as U(VI), Pu(IV), Am(III), Sn(II), Ni(II), and Co(II), destabilize AFm phases by converting monosulfate to calcite and gypsum, and cause shrinkage and microcracking, with carbonation front penetration being diffusion-controlled and dependent on permeability, CO_2 concentration, humidity, and temperature (Bernard et al., 2020; Johannesson & Utgenannt, 2001).

Sulfate attack occurs when external sulfate ions (SO_4^{2-}) from groundwater or internal sources diffuse into the matrix and react with monosulfate (AFm) and aluminate hydrates to form expansive ettringite (AFt): "$3CaO{\cdot}Al_2O_3{\cdot}CaSO_4{\cdot}12H_2O + 2CaSO_4 + 2OH^- + 20H_2O \rightarrow 3CaO{\cdot}Al_2O_3{\cdot}3CaSO_4{\cdot}32H_2O$," with sulfates also reacting with CH to form gypsum ($CaSO_4{\cdot}2H_2O$). The formation of ettringite and gypsum crystals within confined pores generates internal stresses, leading to expansion, cracking, spalling, and loss of mechanical integrity, which increases permeability and accelerates leaching. Blended cements, especially with slag or fly ash, generally offer better sulfate resistance than pure OPC due to lower reactive alumina content and denser microstructure (Glasser et al., 1999).

Radiation damage involves ionizing radiation (gamma, beta) interacting with the waste form, primarily causing radiolysis of pore water: "$H_2O \rightarrow \cdot OH, H\cdot, H_2O_2, H_2, e^-{}_{a}q$," generating reactive species (oxidizing: $\cdot OH$, H_2O_2; reducing: $H\cdot$, $e^-{}_{a}q$, H_2) and gases (H_2, O_2). Gamma radiation can also cause ionization and atomic displacement (radiolytic defects) in crystalline phases like CH and AFm/AFt, potentially leading to amorphization at very high doses, while alpha radiation from actinides causes more severe localized damage due to high linear energy transfer (LET), displacing thousands of atoms per alpha particle and causing significant structural damage ("alpha-recoil damage") in host phases. The impact includes alteration of radionuclide oxidation states (e.g., oxidizing Tc(IV) to mobile Tc(VII), reducing Np(V) to Np(IV)), affecting solubility, causing gas pressure buildup (H_2, O_2) potentially leading to cracking, and increasing the solubility of crystalline phases due to radiolytic defects, though C–S–H is relatively radiation-resistant. The overall impact depends on dose rate, total accumulated dose, waste form composition, and temperature, with cementitious matrices generally suitable for LLW/ILW dose rates ($< 10^6$ Gy/h) and total doses ($< 10^8$–10^9 Gy) (Ojovan et al., 2011).

8.5.2 Accelerated Aging Tests and Modeling Approaches

Given the impracticality of testing over millennia, accelerated aging tests and predictive modeling are essential. Accelerated aging tests include elevated temperature, the most common method, which increases reaction kinetics like leaching, carbonation, and hydration according to the Arrhenius law, with temperatures typically from 40°C to 80°C (higher temperatures risk altering degradation mechanisms), used extensively in leaching studies and phase stability research; enhanced driving forces using more aggressive leachants like lower pH, higher ionic strength, or complexing agents such as EDTA or citrate, higher CO_2 concentrations for carbonation, or higher sulfate concentrations for sulfate attack, though interpretation requires caution to avoid unrealistic mechanisms; radiation exposure by irradiating samples with gamma sources like ^{60}Co or incorporating alpha emitters like ^{238}Pu doped samples to study radiation effects on microstructure, phase stability, gas generation, and radionuclide retention; and combined stresses subjecting samples to simultaneous stresses like temperature + radiation or leaching + carbonation to study synergistic effects, which are critical for long-term performance but complex to predict (Bernard et al., 2020; Plecas & Pcric, 2006).

Modeling approaches encompass empirical/parametric models that fit experimental leaching data to equations derived from simplified transport models, used to calculate effective D_e and LI). They also include thermodynamic equilibrium modeling, using tools like Gibbs Energy Minimization Software (GEMS), PHREEQC, or Gibbs Energy Minimization Selektor (GEM-Selektor) to predict stable mineral phases controlling radionuclide solubility under specific chemical conditions such as pH, Eh, and major ion concentrations. This approach is essential for understanding chemical binding and solubility limits over the evolving chemical environment as pH drops due to carbonation, and it requires robust thermodynamic databases for cement phases and radionuclides (Lothenbach et al., 2019). Another approach involves reactive transport modeling (RTM), using tools like Crunch, MIN3P, or HYTEC. These tools integrate chemical reactions—like dissolution, precipitation, sorption, and redox—with physical transport processes—like advection, diffusion, and dispersion—through the porous cement matrix and surrounding geosphere, representing the most comprehensive approach for simulating long-term evolution and radionuclide release under repository conditions, although they are computationally intensive and require extensive input data (Steefel et al., 2015). Finally, microstructure evolution models simulate changes in porosity, permeability, and phase assemblage over time due to degradation, and they provide input parameters for RTM.

8.5.3 Role of Environmental Conditions

The repository environment significantly influences degradation rates and mechanisms. Humidity and moisture are critical as water is the primary vector for degradation, like leaching, carbonation, sulfate attack, and microbial activity. Saturated conditions favor diffusion-controlled leaching but limit carbonation, while unsaturated conditions allow gas transport but may concentrate aggressive ions in pore water, with cyclic wetting/drying exacerbating physical damage through cracking and salt crystallization. Temperature affects all chemical reactions and transport processes, with elevated temperatures accelerating degradation; repository temperatures depend on waste thermal output, disposal depth, and geothermal gradient, and temperature also influences phase stability, such as the stability of ettringite below ~70°C, and hydration kinetics. Groundwater composition—including pH, Eh, ionic strength, and major ions like Na^+, K^+, Ca^{2+}, Mg^{2+}, Cl^-, SO_4^{2-}, and HCO_3^-/CO_3^{2-}—controls dissolution/precipitation equilibria, ion exchange, complexation, and corrosion, with seawater intrusion or brine contact posing specific challenges due to high ionic strength and Mg^{2+} leading to brucite formation and loss of cohesion.

The gaseous environment, particularly CO_2 concentration, drives carbonation, while oxygen (O_2) availability influences redox conditions, affecting radionuclide speciation for elements like Tc, U, Np, and Pu, as well as the corrosion of encapsulated metals. In addition, hydrogen (H_2) buildup from radiolysis or metal corrosion must be managed. Microbial activity from bacteria and archaea in the host rock or introduced with the waste can influence degradation through several mechanisms. These include biofilm formation, which can physically clog pores or enhance transport along interfaces. Another mechanism is acid production, where sulfur-oxidizing bacteria (SOB) produce sulfuric acid and nitrifying bacteria produce nitric acid, thereby accelerating leaching. Redox transformations also play an important role. For example, sulfate-reducing bacteria (SRB) reduce SO_4^{2-} to H_2S, potentially forming sulfide minerals with radionuclides like TcS_2 or SnS, while also contributing to sulfide corrosion of metals. In addition, iron-reducing bacteria can alter Fe(III) phases, and methanogens may consume H_2. Finally, microbial metabolism can generate gases, producing CO_2 and CH_4, which may contribute to gas buildup and potentially alter pH/Eh (NEA, 2012; Stroes-Gascoyne et al., 2010). Mechanical stress from ground movement, seismic events, or overburden pressure can cause cracking, compromising physical containment.

PA integration involves feeding data from laboratory tests under controlled and accelerated conditions, informed by understanding of fundamental mechanisms and environmental factors, into integrated PA models. These models simulate the coupled thermal, hydrological, chemical, and mechanical (THCM) evolution of the waste form, near-field engineered barriers, and far-field host rock over geological timescales to predict radionuclide release and transport to the biosphere, forming the basis of the repository safety case.

8.6 EMERGING TRENDS AND FUTURE CHALLENGES

The field of radioactive waste immobilization in cementitious materials is dynamic, driven by the need for improved performance, enhanced safety, greater sustainability, and solutions for increasingly complex waste streams. While traditional Portland cement-based systems remain dominant, research and development are actively exploring novel materials, hybrid approaches, and strategies to address persistent limitations. This section highlights key emerging trends and the significant challenges that lie ahead.

8.6.1 Advanced Cementitious Materials

Moving beyond conventional OPC and blended systems, several advanced binders offer potential advantages for specific waste immobilization challenges. Geopolymers and low-pH alkali-activated materials (AAMs) are a major focus in the development of advanced cementitious waste forms. Their chemistry can be tailored by selecting suitable precursors, such as fly ash, slag, and metakaolin (MK), and activators, such as silicate and hydroxide solutions, to achieve specific performance targets. These materials can provide a lower equilibrium pH of approximately 10.5–12.5, which may benefit the immobilization of anions such as I^- and TcO_4^-, whose sorption decreases at very high pH, and may help reduce the solubility gap of amphoteric radionuclides such as U(VI) and Sn(II) during long-term carbonation. Geopolymers and low-pH AAMs may also offer improved chemical resistance to acid, sulfate, and saline-water attack compared with OPC, tailored pore structures with low permeability and high specific surface area, and enhanced actinide retention through high-Ca slag-based or phosphate-modified formulations that promote precipitation or structural binding (Bernal et al., 2011; Shi & Fernández-Jiménez, 2006; Neeway et al., 2016). However, challenges include long-term phase stability, such as zeolite crystallization from gels, potential for efflorescence, understanding and controlling radionuclide speciation and retention mechanisms within the complex gel structure, regulatory acceptance based on a less extensive database than OPC, handling safety of concentrated alkaline activators, and variability in precursor materials (Provis, 2018; Chindaprasirt & Cao, 2021).

Magnesium phosphate cements (MPCs), also known as chemically bonded phosphate ceramics (CBPCs), are produced by an acid–base reaction between magnesium oxide (MgO) and a soluble phosphate source, such as KH_2PO_4 or $NH_4H_2PO_4$; this reaction forms phosphate hydrate phases and results in a rapid-setting, near-neutral-pH binder. They offer rapid setting and high early strength, making them useful for rapid-turnaround applications and for encapsulating reactive wastes. MPCs also provide a near-neutral pH of 7–9, which avoids the high-pH plume associated with ordinary Portland cement (OPC), may simplify safety cases, and can improve compatibility with some host rocks and clays. This pH range is advantageous for immobilizing metals that are prone to corrosion under highly alkaline conditions and radionuclides such as U(VI), whose phosphate phases have low solubility near neutral pH. In addition, MPCs can promote the direct formation of highly insoluble actinide phosphate phases, such as autunite-group uranium minerals, within the matrix (Gardner et al., 2020). However, challenges remain, including long-term durability under saturated conditions, possible phase transformations and strength loss, sensitivity to water content and mixing conditions, potential ammonia release when ammonium phosphate is used, higher cost than OPC, limited large-scale experience in radioactive waste processing, and the need to better understand long-term radionuclide retention mechanisms beyond precipitation (Singh et al., 2019). Additionally, calcium aluminate phosphate cements are emerging systems combining aspects of calcium aluminate and phosphate chemistries, potentially offering rapid setting, high early strength, and good chemical resistance with a more controlled pH profile than OPC, though research remains at an early stage for nuclear waste applications (Gougar & Scheetz, 2015).

8.6.2 Hybrid Systems and Encapsulation Technologies

Recognizing that no single matrix is optimal for all wastes or radionuclides, hybrid approaches are gaining traction. Cement-composite coatings involve applying functionalized coatings to cement waste forms to enhance performance. These include permeability barriers, where thin layers of polymers, like epoxy or polyurethane, or geopolymer, are applied to the waste form surface to drastically reduce water ingress and leaching rates, thereby acting as an additional engineered barrier. Another approach involves using chemical barriers with coatings containing specific sorbents like Ag-functionalized materials for I^-, LDHs for anions, or zeolites for Cs^+ to intercept radionuclides diffusing out of the waste form (Vance & Perera, 2018). Challenges include ensuring long-term adhesion under repository conditions like temperature, humidity, and radiation, verifying coating integrity and performance over millennia, and addressing the cost and complexity of application. Multi-barrier encapsulation combines cementation with other immobilization techniques, such as pre-treatment/pre-immobilization, where waste particles are pre-treated (e.g., spent resins are stabilized with phenol-formaldehyde, TcO_4^- is reduced and precipitated on ZVI, I^- is captured on Ag-mordenite) before encapsulation within a cement matrix. thereby leveraging specific immobilization mechanisms together with the physical encapsulation/chemical environment of cement. Another approach involves structured waste forms with layered designs, where different zones have tailored compositions. For instance, a reducing slag-rich core may be used for Tc immobilization, surrounded by a low-permeability OPC/PFA shell (Ojovan & Lee, 2014). Challenges involve complexity of design and fabrication, potential for interface weaknesses, and demonstrating long-term synergistic performance of the barriers.

8.6.3 Addressing High-Heat and Long-Lived Radionuclides

The fundamental limitations of cement for HLW immobilization remain, but research explores niche applications and mitigations. Immobilization of partitioned long-lived fission products (LLFPs) and minor actinides (MAs) from advanced fuel cycles aims to separate LLFPs like ^{129}I, ^{99}Tc, and ^{135}Cs and MAs like Np, Am, and Cm from HLW for dedicated transmutation or tailored immobilization. Cement could potentially immobilize separated LLFP streams or specific MA-bearing streams if thermal output is managed through low concentration, dilution, or specialized matrices like MPCs for actinide phosphates. This approach would require highly optimized formulations targeting specific radionuclides and demonstration of long-term performance under expected repository conditions (Glasser, 2011; NEA, 2006). Mitigating heat effects for incorporating higher-activity wastes includes strategies like enhanced thermal conductivity by adding conductive fillers such as graphite or steel fibers to cement to improve heat dissipation and reduce peak temperatures within the waste form; heat-resistant matrices utilizing geopolymers or specialized refractory cements with higher dehydration temperatures; and dilution through lower waste loading, though this increases final waste volume and is generally only feasible for ILW approaching HLW thermal limits (Glasser, 2011). Challenges include trade-offs in cost, complexity, and potential impact on other properties like leaching, with direct cementation of true HLW remaining impractical.

8.7 CONCLUSION AND OUTLOOK

Cementitious materials have established themselves as indispensable matrices for the safe immobilization of LLW/ILW. This chapter has synthesized the scientific principles, engineering practices, and evolving landscape of this critical technology.

Cement-based systems provide a robust immobilization mechanism through a multifaceted barrier encompassing physical encapsulation, chemical binding via precipitation, sorption, ion exchange, and redox stabilization. The inherent high alkalinity ($pH > 12.5$ for OPC) effectively precipitates many cationic radionuclides such as actinides, Sn, and Ni, and promotes sorption of species like Cs^+ and Sr^{2+} on C–S–H (Atkins & Glasser, 1992; Glasser, 2011). The material versatility is demonstrated by a range of cementitious binders, including OPC, blended cements like OPC/BFS and OPC/PFA, CSA cement, and emerging AAMs/geopolymers, which offer solutions tailored to waste chemistry, such as CSA for resins and AAMs for lower pH and anion retention (Ojovan & Lee, 2014; Provis, 2018). Performance optimization is critical through formulation design, where functional additives like zeolites for Cs^+, bentonite for rheology and sorption, hydrotalcite for pH buffering and anion exchange, and zero-valent iron for Tc(VII) reduction significantly enhance retention of problematic radionuclides, including Cs, I, Tc, and U (Serne et al., 2014; Singh et al., 2015).

Proven industrial application is evidenced by cementation's maturity, deployed globally via container filling (in-drum mixing), in situ stabilization, and historical bulk grouting, with case studies like Fukushima ALPS secondary wastes (NEA, 2020; TEPCO, 2021), Sellafield encapsulation (NDA, 2021; Thomson, 2011), and Hanford tank residuals (DOE, 2014) demonstrating adaptability across diverse waste streams and scales. A robust PA framework exists, with well-studied degradation mechanisms like leaching, carbonation, sulfate attack, and radiation damage. This framework is supported by standardized tests, such as ANSI/ANS-16.1 and ASTM C1308, as well as advanced modeling approaches, including thermodynamic equilibrium and reactive transport modeling, which underpin long-term safety cases. These assessments also consider environmental factors like humidity, temperature, and microbes (Lothenbach et al., 2019; Steefel et al., 2015). The evolving landscape is marked by advanced materials like tailored geopolymers and magnesium phosphate cements (MPCs); hybrid systems such as composite coatings and macro-encapsulation; and strategies for sustainability, including low-CO_2 binders like LC^3 and circular economy principles, which represent active research frontiers aimed at addressing limitations and future challenges (Scrivener et al., 2018; Wagh, 2016).

Problematic species require robust and cost-effective immobilization strategies that can provide long-term retention acceptable to regulators. For highly mobile anions, such as I^- and TcO_4^-, future work should move beyond simple encapsulation and focus on durable mineral phases, such as iodoapatite, or stable getters that remain effective as cementitious matrices age. For volatile and organic species, such as ^{14}C, ^{3}H, and organic iodine, improved matrix-trapping mechanisms and pre-treatment methods are needed. Redox-control strategies should also be developed to maintain reducing conditions and limit the mobility of redox-sensitive elements such as Tc, U, Np, and Pu over millennial timescales (Serne et al., 2014; Singh et al., 2015). Advanced materials and systems require establishing comprehensive long-term performance databases for geopolymers/AAMs and MPCs under repository-relevant conditions like saturation, radiation, and temperature; optimizing hybrid systems such as cement–polymer coatings and cement composites with engineered nano-materials for enhanced barrier properties and adhesion durability; and developing and validating "smart" waste forms with self-sealing or self-healing capabilities upon crack formation or pH drop (Provis, 2018; Vance & Perera, 2018).

Addressing high-heat and complex wastes involves refining thermal management strategies like optimized high-conductivity additives and engineered container designs for higher-activity ILW approaching thermal limits (~2 kW/m^3); and developing effective formulations for organic-rich wastes, extreme pH wastes, and heterogeneous debris streams that challenge conventional cementation (Glasser, 2011; NEA, 2006). Future work should address both sustainability and regulatory acceptance. This includes rigorously assessing the long-term immobilization performance and durability of low-CO_2 cementitious materials, such as high-SCM blends, calcined-clay cements, and carbonatable cements, for radioactive waste applications. It also requires developing performance-based regulatory pathways that support the safe deployment of validated novel waste forms and technologies, rather than relying solely on prescriptive standards. Finally, life cycle assessment (LCA) methodologies tailored to radioactive waste immobilization should be advanced so that long-term safety performance is considered alongside environmental impacts (NEA, 2019; Scrivener et al., 2018).

Cementitious materials will continue to be a cornerstone of radioactive waste management for LLW/ILW, underpinned by decades of research, industrial experience, and regulatory acceptance. The future lies not in replacing cement, but in strategically evolving it. Success requires a concerted effort: advancing fundamental understanding of long-term material–radiolysis–contaminant interactions; developing and rigorously validating next-generation materials and hybrid systems; optimizing for sustainability; and fostering collaboration between researchers, engineers, regulators, and the public. Addressing the identified knowledge gaps is paramount to ensuring the safe, effective, and environmentally responsible immobilization of radioactive wastes for generations to come.

REFERENCES

ANDRA, 2020. *National Inventory of Radioactive Materials and Waste*. Agence nationale pour la gestion des déchets radioactifs, Châtenay-Malabry, France.

ANS, 2019. *Measurement of the Leachability of Solidified Low-Level Radioactive Wastes by a Short-Term Test Procedure*. ANSI/ANS-16.1-2019. American Nuclear Society, La Grange Park, IL.

ASTM, 2021. Standard Test Method for Accelerated Leach Test for Measuring Contaminant Releases From Solidified Waste. ASTM C1308-21. ASTM International, West Conshohocken, PA.

Atkins, M., Glasser, F.P., 1992. Application of Portland cement-based materials to radioactive waste immobilization. *Waste Manag.* 12, 105–131.

Bernal, S.A., Provis, J.L., Rose, V., Mejía de Gutiérrez, R., 2011. Evolution of binder structure in sodium silicate-activated slag-metakaolin blends. *Cement and Concrete Composites* 33(1), 46–54.

Bernard, E., Lothenbach, B., Rentsch, D., Pochard, I., Dauzères, A., Nied, D., 2020. Long-term performance of cement paste during combined calcium leaching–sulfate attack: Kinetics and size effect. *Cem. Concr. Res.* 131, 106017.

Chindaprasirt, P., Cao, T., 2021. Reuse of recycled aggregate in alkali-activated concrete. In: *New Trends in Eco-efficient and Recycled Concrete*. Woodhead Publishing, pp. 465–487.

DOE (U.S. Department of Energy), 2014. *Hanford Site Tank Closure and Waste Management Environmental Impact Statement* (Final EIS). DOE.

Gardner, L.J., Bernal, S.A., Walling, S.A., Corkhill, C.L., Provis, J.L., Hyatt, N.C., 2020. Characterisation of magnesium potassium phosphate cements blended with fly ash and ground granulated blast furnace slag. *Cem. Concr. Res.* 132, 106041.

Glasser, F.P., 2011. Application of inorganic cements to the conditioning and immobilisation of radioactive wastes. In: *Handbook of Advanced Radioactive Waste Conditioning Technologies*. Woodhead Publishing, pp. 67–135.

Glasser, F.P., Marchand, J., Samson, E., 1999. Durability of concrete—Degradation phenomena involving detrimental chemical reactions. *Cem. Concr. Res.* 29, 1145–1153.

Gougar, M.L.D., Scheetz, B.E., 2015. *Calcium Aluminate Phosphate Cement for Low-Level Radioactive Waste Immobilization*. MRS Proceedings 1744. Materials Research Society.

IAEA, 1996. *Processing of Nuclear Power Plant Waste Streams Containing Boric Acid. IAEA-TECDOC-911*. International Atomic Energy Agency, Vienna.

IAEA, 2018. *Status and Trends in Spent Fuel and Radioactive Waste Management*. IAEA Nuclear Energy Series No. NW-T-1.14. International Atomic Energy Agency, Vienna.

Iler, R.K., 1979. *The Chemistry of Silica: Solubility, Polymerization, Colloid and Surface Properties, and Biochemistry*. John Wiley & Sons.

Johannesson, B., Utgenannt, P., 2001. Microstructural changes caused by carbonation of cement mortar. *Cem. Concr. Res.* 31, 925–931.

Lothenbach, B., Kulik, D.A., Matschei, T., Balonis, M., Baquerizo, L., Dilnesa, B., Wadsö, L., 2019. Cemdata18: A chemical thermodynamic database for hydrated Portland cements and alkali-activated materials. *Cem. Concr. Res.* 115, 472–506.

Lothenbach, B., Scrivener, K., Hooton, R.D., 2011. Supplementary cementitious materials. *Cem. Concr. Res.* 41, 1244–1256.

NEA, 2006. *Advanced Nuclear Fuel Cycles and Radioactive Waste Management*. OECD Publishing, Paris.

NEA (Nuclear Energy Agency), 2012. *Gas Generation and Release from Radioactive Waste Repositories*. NEA No. 6892. OECD Nuclear Energy Agency, Paris.

NEA (Nuclear Energy Agency), 2019. *Reducing the Carbon Footprint of the Nuclear Fuel Cycle: Opportunities and Challenges*. OECD Nuclear Energy Agency, Paris.

NEA (Nuclear Energy Agency), 2020. *Radioactive Waste Management Programmes in OECD/NEA Member Countries: Japan*. OECD Nuclear Energy Agency, Paris.

Neeway, J.J., Qafoku, N.P., Serne, R.J., Brown, C.F., Valenta, M.M., 2016. Geopolymer as a potential waste form for Hanford secondary waste: A preliminary study. *Environ. Prog. Sustain. Energy* 35, 478–485.

NRC, 2021. *10 CFR Part 61—Licensing Requirements for Land Disposal of Radioactive Waste*. U.S. Nuclear Regulatory Commission.

Nuclear Decommissioning Authority (NDA), 2021. *Sellafield Plan: 2021 Update*. NDA.

Ojovan, M.I., Lee, W.E., 2014. *An Introduction to Nuclear Waste Immobilisation*. 2nd ed. Elsevier.

Ojovan, M.I., Pomet, M.J., Lee, W.E., 2011. Radiation effects in cementitious materials for immobilization. In: *Handbook of Advanced Radioactive Waste Conditioning Technologies*. Woodhead Publishing, pp. 136–159.

Osmanlioglu, A.E., 2002. Immobilization of radioactive waste by cementation with purified kaolin clay. *Waste Manag.* 22, 481–483.

Peuget, S., Cachia, J.N., Jégou, C., Deschanels, X., Roudil, D., Broudic, V., Bart, J.M., 2014. Effect of alpha decay on the properties of borosilicate glass for waste confinement. *J. Nucl. Mater.* 354, 115–121.

Plecas, I., Dimovic, S., Orta, M.M., 2005. Immobilization of spent ion exchange resins in cement mortar. *J. Serb. Chem. Soc.* 70, 1195–1202.

Plecas, I., Peric, A., 2006. Leaching behavior of ^{60}Co and ^{137}Cs from spent ion exchange resins in cement–bentonite clay matrix. *J. Nucl. Mater.* 355, 203–207.

Provis, J.L., 2018. Alkali-activated materials. *Cem. Concr. Res.* 114, 40–48.

Provis, J.L., Bernal, S.A., 2014. Geopolymers and related alkali-activated materials. *Annu. Rev. Mater. Res.* 44, 299–327.

Scrivener, K.L., John, V.M., Gartner, E.M., 2018. Eco-efficient cements: Potential economically viable solutions for a low-CO_2 cement-based materials industry. *Cem. Concr. Res.* 114, 2–26.

Scrivener, K.L., Juilland, P., Monteiro, P.J.M., 2016. Advances in understanding hydration of Portland cement. *Cem. Concr. Res.* 78, 38–56.

Serne, R.J., Cantrell, K.J., Lindenmeier, C.W., 2008. *Review of Potential Technologies for Immobilization of Iodine-129 and Technetium-99*. Pacific Northwest National Laboratory, Richland, WA.

Serne, R.J., Westsik, J.H., Lindberg, M.J., 2014. *Technetium and Iodine Issues Associated with Hanford Tank Waste Residuals and Grout Disposal Packages*. Pacific Northwest National Laboratory, Richland, WA.

Shi, C., Fernández-Jiménez, A., 2006. Stabilization/solidification of hazardous and radioactive wastes with alkali-activated cements. *J. Hazard. Mater.* 137, 1656–1663.

Singh, D., Wagh, A.S., Tlustochowicz, M., Jeong, S.Y., 2015. Phosphate ceramic solidification and stabilization of Cs and Sr laden fly ash. *J. Nucl. Mater.* 467, 888–896.

Singh, D., Wagh, A.S., Tlustochowicz, M., Jeong, S.Y., 2019. Phosphate ceramic solidification and stabilization of radioactive and hazardous wastes. In: *Handbook of Advanced Radioactive Waste Conditioning Technologies*. 2nd ed. Woodhead Publishing, pp. 289–324.

Steefel, C.I., Appelo, C.A.J., Arora, B., Jacques, D., Kalbacher, T., Kolditz, O., Yabusaki, S.B., 2015. Reactive transport codes for subsurface environmental simulation. *Comput. Geosci.* 19, 445–478.

Stroes-Gascoyne, S., Hamon, C.J., Maak, P., Russell, S., 2010. The effects of the physical and chemical properties of highly compacted smectitic clay (bentonite) on the culturability of indigenous microorganisms. *Appl. Clay Sci.* 47, 155–162.

Taylor, H.F.W., 1997. *Cement Chemistry*. 2nd ed. Thomas Telford Publishing.

Tits, J., Gaona, X., Mace, N., 2011. Uptake of uranium(VI) by calcium silicate hydrate phases. *Appl. Geochem.* 26, 151–161.

Tits, J., Wieland, E., Bradbury, M.H., Eckert, P., 2006. Strontium binding by calcium silicate hydrates. *J. Colloid Interface Sci.* 300, 78–87.

Tokyo Electric Power Company Holdings, Inc. (TEPCO), 2021. *Mid-and-Long-Term Roadmap Towards the Decommissioning of TEPCO's Fukushima Daiichi Nuclear Power Station Units 1–4.* Tokyo Electric Power Company Holdings, Inc., Tokyo.

Vance, E.R., Perera, D.S., 2018. Developments in glass, ceramic, and other waste forms. In: *Handbook of Advanced Radioactive Waste Conditioning Technologies.* 2nd ed. Woodhead Publishing, pp. 325–360.

Wagh, A.S., 2016. *Chemically Bonded Phosphate Ceramics.* 2nd ed. Elsevier.

Zhang, L., Glasser, F.P., 2002. Investigation of the microstructure and carbonation of CSA-based concretes removed from service. *Cem. Concr. Res.* 32, 1235–1243.

CHAPTER 9

Radioactive Waste Immobilization in Glass Composite Material

Shi-Kuan Sun and Sheng-Heng Tan

9.1 INTRODUCTION

The immobilization of high-level radioactive waste is essential to ensure the long-term safety of nuclear energy. High-level waste (HLW) from reprocessing, together with intermediate-level waste (ILW), contains fission products, minor actinides, and long-lived radionuclides that should be stabilized for a geological timescale. The target of waste forms is to immobilize these elements in durable solids with low leaching rate, radiation tolerance, and stability in a deep repository environment (Thorpe et al., 2021).

Pure glass materials, especially borosilicate glass, are an established system as a waste form. The disordered networks of SiO_4, BO_3/BO_4, and AlO_4 units can host diverse radionuclides with good chemical flexibility and industrial scalability, as demonstrated in vitrification facilities worldwide (McCloy et al., 2024). Phosphate glass, though less widely deployed, offers lower melting temperature and higher solubility for problematic species such as sulfates and halides, making it attractive for challenging waste chemistry (Yang et al., 2017).

However, glass materials still face limits: some elements (Mo, S, Cl) induce phase separation; volatile nuclides (Cs, Tc, I) may be released during melting, and high waste loading can compromise durability (Goel et al., 2019). These motivate the development of glass-composite materials (GCMs), or glass–ceramics, where crystalline phases are embedded in a glass matrix. Such crystalline phases, for instance zirconolite, pyrochlore, apatite, pollucite, selectively immobilize actinides and/or rare-earth elements while the glass accommodates the remainder (Ojovan et al., 2021).

This chapter reviews both pure glass and glass-composite waste forms, where the structures, processing, and performance are introduced, outlining how their complementary strengths contribute to safe and effective nuclear waste immobilization. Furthermore,

DOI: 10.1201/9781003665793-12

advances in data-driven and multi-scale modeling are also introduced to enable predictive glass design and optimize composition and durability efficiently.

9.2 PURE GLASS

The pure glass system was initially developed as one of the earliest and most fundamental approaches for the immobilization of high-level radioactive waste. Among the numerous compositions explored, borosilicate and phosphate glasses have emerged as the two principal families for HLW vitrification. These materials differ fundamentally in their network-forming chemistry, which governs their thermal behavior, radionuclide solubility, and long-term stability under repository conditions. The following sections provide an overview of their compositional characteristics, immobilization mechanisms, processing routes, and performance, highlighting both established practices and emerging challenges in the field.

9.2.1 Types of Glass Matrices

Pure glass systems developed for the immobilization of high-level radioactive waste are commonly grouped into two main categories: borosilicate glass and phosphate glass. These categories are distinguished by the fundamental network-forming chemistry, which controls their structural adaptability, thermal behavior, and long-term stability. Over the past four decades, borosilicate glass has become the reference material for large-scale vitrification, whereas phosphate-based glass continues to attract attention as tailored solutions for specific, compositionally challenging waste streams (McCloy et al., 2024).

The structural framework of borosilicate glass is based on silica (SiO_2), modified with boron oxide (B_2O_3), alkali oxides such as Na_2O and Li_2O, alkaline-earth oxides such as CaO and MgO, and minor amounts of Al_2O_3. The addition of B_2O_3 lowers the melting temperature, improves melt fluidity, and enhances the solubility of a broad range of waste constituents. Meanwhile, the inclusion of Al_2O_3 contributes to network reinforcement by generating AlO_4 tetrahedra, charge-compensated by alkali or alkaline-earth cations. These combined effects suppress hydrolysis and delay the progression of alteration; therefore, the overall durability can be enhanced (Thorpe et al., 2021). Borosilicate glass has been employed at nearly all major vitrification plants worldwide, including La Hague in France, Sellafield in the United Kingdom, and the Defense Waste Processing Facility (DWPF) in the United States (McCloy et al., 2024). Its broad deployment reflects the robustness of the composition, its versatility across different waste streams, and its demonstrated capacity to immobilize a wide spectrum of radionuclides ranging from fission products such as Cs and Sr to minor actinides such as Np and Am.

With respect to performance, borosilicate glass typically accommodates waste loadings between 18 and 35 wt %, depending on the complexity of waste streams and glass formulation. Processing properties, such as viscosity and thermal expansion, can be tailored to match the requirements of industrial melters. Long-term stability has been extensively confirmed by standardized tests, including the product consistency test (PCT), which routinely reports normalized leach rates of less than 10^{-3}–10^{-2} $g·m^{-2}·d^{-1}$.

Phosphate glass is constructed from networks of PO_4 tetrahedra joined by P–O–P bonds. While these bonds provide structural flexibility, they are more susceptible to hydrolysis than Si–O bonds, resulting in comparatively lower intrinsic durability. To address this issue, modified systems, such as iron phosphate and aluminophosphate glasses, have been developed. In these variants, P–O–P linkages are replaced by stronger P–O–Fe or P–O–Al bonds, significantly improving resistance to hydrolysis while maintaining high waste-loading potential (Sengupta, 2012).

A notable processing advantage of phosphate glass is its low melting temperature, which generally lies between 850°C and 1000°C. This feature offers practical benefits: reduced volatilization of sensitive radionuclides, diminished refractory corrosion, and improved control over redox conditions. The composition of phosphate is also highly effective at dissolving problematic species such as sulphates, halides, and MoO_3, which present serious challenges in silicate glass systems (Kofuji et al., 2012). Nevertheless, these benefits are offset by serious drawbacks. The chemical durability of phosphate glass remains inferior to that of borosilicates, owing to the hydrolytic weakness of P–O–P bonds, which accelerates dissolution under groundwater conditions relevant to repository scenarios. Moreover, the industrial application of phosphate vitrification is limited, with most demonstrations restricted to laboratory or pilot-scale facilities. Consequently, phosphate glass has so far been adopted only in specialized applications rather than as the mainstream waste form (Sengupta, 2012).

9.2.2 Mechanisms of Immobilization in Glass

The capacity of glass to immobilize radionuclides originates fundamentally from its disordered atomic structure. Unlike crystalline structure, which is defined by rigid and periodic lattices, glass consists of metastable, three-dimensional networks with significant structural flexibility. This flexibility allows glass to accommodate a wide range of cations and anions, making it a highly versatile matrix for nuclear waste immobilization (Pilania & Dube, 2023). In these systems, the network structure is primarily built by glass-forming elements such as silicon, boron, phosphorus, and aluminum. These atoms generally occur in tetrahedral coordination, forming building units such as SiO_4, BO_3/BO_4, PO_4, or AlO_4. The way these tetrahedra are linked determines the degree of polymerization of the glass network, which in turn strongly influences its long-term chemical durability and resistance to alteration (Thorpe et al., 2021; Zubekhina et al., 2020). Alongside these formers, network modifiers including sodium, potassium, calcium, and magnesium play essential roles. By disrupting bridging oxygen bonds, they generate non-bridging oxygens (NBOs), which act as flexible charge-balancing sites. These NBO-rich domains tend to host larger, mobile radionuclides such as Cs^+ and Sr^{2+}, which often localize in the more depolymerized regions of the glass network (Pilania & Dube, 2023).

Certain elements can also function as intermediates. Aluminum is the most notable example, as it can serve either as a network former or as a charge-balancing cation, depending on its coordination environment and glass composition. When present as AlO_4 tetrahedra, aluminum reduces the concentration of NBOs. Consequently, network connectivity can be enhanced, and chemical durability would be improved. This versatile behavior gives

Al a significant stabilizing role in many HLW glass formulations. The pathways by which radionuclides are accommodated in glass networks are diverse. Monovalent and divalent fission products, such as Cs^+ and Sr^{2+}, are typically located within depolymerized regions near NBOs, where charge neutrality can be maintained. Actinides and lanthanides, being larger and often multi-valent, can be incorporated either by substituting for modifier cations or by occupying interstitial positions in less densely packed domains of the network. In contrast, anionic species such as MoO_4^{2-} and SO_4^{2-} exhibit poor solubility in silicate-based glass, leading to the formation of "yellow phase" (see Figure 9.1; Sengupta, 2012). The limited compatibility often results in phase separation or the formation of crystalline secondary phases, which compromise the homogeneity and durability of the final waste form (Pilania & Dube, 2023; Sengupta, 2012).

Borosilicate glass displays structure–property relationships that are especially sensitive to the roles of boron and aluminum. Recent studies suggest that when boron content exceeds a certain threshold in borosilicate glasses, the formation of B–O–Si and B–O–B linkages becomes more prevalent, and these linkages tend to be more vulnerable to hydrolysis than pure silicate bonds. This structural feature increases both the initial and residual dissolution rates of the glass, particularly under radiation conditions (see Figure 9.2, Gin et al., 2024). In contrast, aluminum contributes to network stabilization. By forming AlO_4 units integrated into the network, it strengthens connectivity, suppresses hydrolytic attack, and enhances the passivation of alteration layers. Together, these processes result in reduced long-term dissolution and improved overall durability of borosilicate glass.

Phosphate glass immobilizes radionuclides through distinct structural mechanisms. The ability to accommodate radionuclides depends strongly on the capacity of phosphate tetrahedra (PO_4) to form cation–phosphate complexes. Multi-valent cations such as actinides and rare-earth elements are stabilized within the network via P–O–M bonds, where M

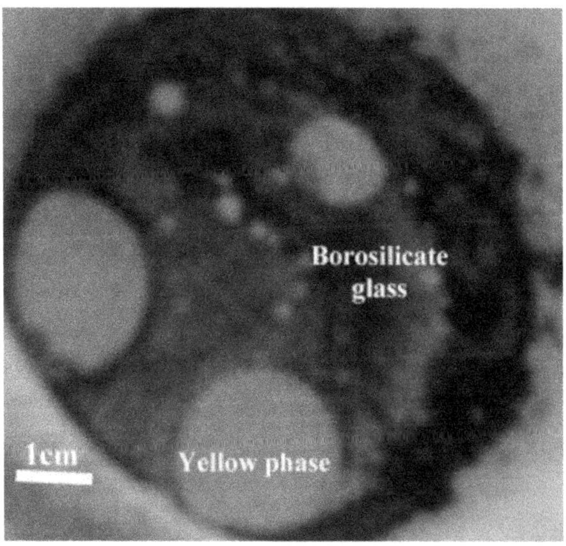

FIGURE 9.1 Formation of Yellow Phase within Borosilicate Glass Matrix (Sengupta, 2012).

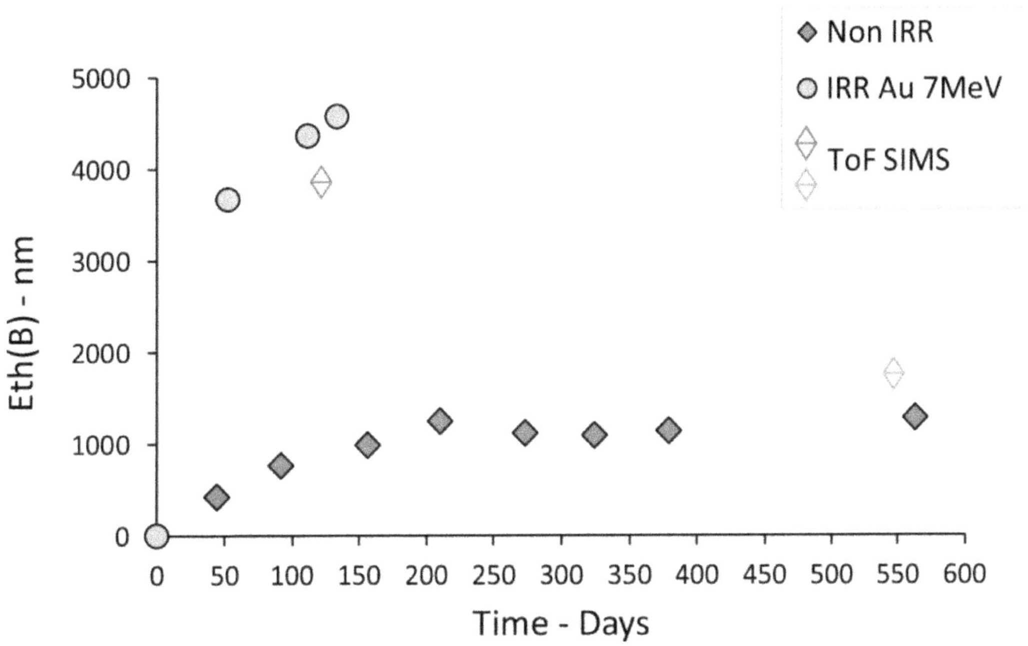

FIGURE 9.2 Time Evolution of Altered Glass Thickness for the Static Experiment Conducted in Silica-Saturated Solution at 90°C and pH 9. Gel Thickness Obtained by time-of-flight secondary ion mass spectrometry (ToF-SIMS) Is Shown for Comparison (Gin et al., 2024).

denotes a high-valent metal. While this mechanism allows phosphate glass to incorporate relatively high loading of actinides, it also reduces network connectivity, leaving the glass more vulnerable to hydrolytic attack and diminishing long-term durability (Sengupta, 2012). To mitigate these issues, modified phosphate formulations have been developed. Iron phosphate and aluminophosphate systems, in particular, substitute hydrolytically unstable P–O–P linkages with stronger P–O–Fe or P–O–Al bonds. This adjustment significantly enhances resistance to dissolution, offering phosphate-based glass improved prospects for the immobilization of complex waste streams (Brow et al., 2020).

9.2.3 Processing of Pure Glass Waste Forms

The vitrification of high-level radioactive waste in borosilicate glass is the most widely deployed industrial strategy, practiced for decades in France, the United States, Russia, the United Kingdom, and Japan (Marra et al., 1998; Masson, 2001). The process typically begins with preparation of calcined (or otherwise conditioned) waste, which is homogenized with a tailored glass frit; the mixture is then evaporated to target solids, melted at ~1100–1200°C, and poured into stainless steel canisters to cool as monoliths suitable for storage and disposal (Fellinger, 1999; Marra & Harbour, 2004).

Two industrial routes are in common use. In the one-stage, slurry-fed process (as at the U.S. DWPF), waste slurries are acid-adjusted, mixed with frit, concentrated, and fed directly to a joule-heated melter for co-melting. By contrast, the two-stage French/UK route (La Hague/Sellafield) employs an upstream rotary calciner to convert nitrate feeds into an oxide calcine that is then blended with frit and vitrified, improving control over volatile species and melter stability (Masson, 2001; Chauvin et al., 2010). The reliability of the process largely depends on melter technology. The joule-heated ceramic melter (JHCM) serves as the primary approach of vitrification technology, which delivers electrical resistance heating directly to the melt and has accumulated millions of operating hours at DWPF and La Hague (Marra et al., 1998). More recently, cold crucible induction melters (CCIMs) have gained attention: induction heating in a water-cooled metallic shell enables higher operating temperature (>1300°C), lower refractory corrosion, and potential for higher waste loading; these advantages have been demonstrated at pilot and industrial scales in Europe and Russia (Amoroso et al., 2021; Pinet et al., 2019).

Phosphate glasses are generally processed at lower temperatures (~850–1000°C), which reduces volatilization of sensitive species and limits refractory corrosion, while retaining high solubility for problematic components such as sulphates, halides, and Mo-rich streams. A critical consideration is redox control during processing: maintaining Fe^{3+} over Fe^{2+} is essential for durability and to suppress unwanted crystallization during cooling. Despite successful pilot and plant-scale demonstrations, exclusive, long-term reliance on phosphate glass remains limited due to comparatively lower intrinsic aqueous durability relative to borosilicates.

9.2.4 Performance and Recent Advances in Pure Glass

The performance of pure glass systems for radioactive waste immobilization is evaluated primarily in terms of chemical durability, radiation stability, and their ability to accommodate diverse waste compositions without significant phase separation or uncontrolled crystallization. Among the various formulations, borosilicate glass has become the international reference material, while phosphate glass remains of interest for specialized applications despite its more limited industrial deployment.

A principal advantage of borosilicate glass is its high chemical durability. Standardized leach protocols, such as the PCT, routinely report normalized release rates below 10^{-5} $g{\cdot}cm^{-2}{\cdot}d^{-1}$, confirming their strong resistance to aqueous alteration over extended timescales. Durability is, however, strongly composition-dependent. The incorporation of aluminum enhances stability by generating AlO_4 tetrahedra, which reduces the proportion of NBOs and, in turn, delays the destabilization of alteration gels. In contrast, excessive boron contents increase the fraction of hydrolytically weak B–O–B and B–O–Si linkages, accelerating glass dissolution in both near-field and repository environments (Zubekhina et al., 2020). Taron confirms that co-optimization of higher Al_2O_3 and lower B_2O_3 contents markedly enhances borosilicate durability, reducing aqueous leaching rates by more than an order of magnitude through suppression of weak B–O–B linkages and reinforcement of the aluminosilicate network (see Table 9.1, Taron et al., 2025).

Borosilicate glass also demonstrates notable radiation tolerance. Under irradiation up to 10^8 rads from ^{60}Co γ-sources, it is reported that the borosilicate glass darkened considerably, but only a slight densification was measured; the distributions of strength, the slow-crack growth exponent (N), elastic modulus, and fracture toughness all remained

TABLE 9.1 Initial Dissolution Rates ($g \cdot m^{-2} \cdot d^{-1}$) of Si, B, Al, and Na Measured from Solution Analyses for Borosilicate Glass Compositions with Varying Al and B Contents

Sample	Si ($g \cdot m^{-2} \cdot d^{-1}$)	B ($g \cdot m^{-2} \cdot d^{-1}$)	Al ($g \cdot m^{-2} \cdot d^{-1}$)	Na ($g \cdot m^{-2} \cdot d^{-1}$)
B6A4	2.4 ± 0.6	2.3 ± 0.4	3.0 ± 1.0	3.5 ± 1.9
B11A3	10.3 ± 2.6	9.8 ± 2.8	10.3 ± 2.1	10.3 ± 3.5
B14A3	32.6 ± 6.3	28.7 ± 3.2	33.4 ± 3.5	27.6 ± 4.8
B11A6	10.0 ± 2.0	9.2 ± 2.7	10.1 ± 2.0	9.2 ± 3.1
B10A2	5.1 ± 0.6	4.9 ± 0.8	6.3 ± 1.0	7.2 ± 3.3
B10A1	15.5 ± 2.6	14.3 ± 2.8	11.9 ± 2.5	17.6 ± 3.6

(Taron et al., 2025).

essentially unchanged. In other words, γ-radiation below 10^8 rads does not significantly affect the mechanical strength of glass (Zdanievski et al., 1983). Nevertheless, irradiation can indirectly accelerate alteration processes by modifying solution chemistry, for instance, through pH shifts or the generation of reactive radiolytic species. Such feedback mechanisms highlight the importance of coupling structural irradiation studies with corrosion testing under repository-relevant conditions.

Despite these strengths, borosilicate glass is subject to several limitations. Certain fission products and process-derived elements, notably molybdenum, sulphur, and chlorine, exhibit low solubility in the melt. Their accumulation frequently leads to the segregation of alkali-molybdate or sulphate phases, commonly referred to as the "yellow phase," which compromises waste forms' homogeneity and long-term durability (Vienna et al., 2002). Another limitation is derived from uncontrolled crystallization during cooling, which redistributes radionuclides and reduces chemical stability. Furthermore, volatility losses of species such as cesium and technetium during high-temperature melting remain an operational concern, as these elements are both difficult to retain in the glass and hazardous if released (Kim & Kruger, 2018).

Phosphate glass generally exhibits lower chemical durability relative to borosilicates, owing to the hydrolytic vulnerability of P–O–P bonds, which are readily attacked in aqueous solutions. Consequently, pure phosphate glass often dissolves at rates an order of magnitude higher than borosilicate counterparts under comparable conditions (Zubekhina et al., 2020). Nonetheless, significant improvements can be achieved through compositional modification. Iron phosphate and aluminophosphate formulations replace weak P–O–P linkages with more robust P–O–Fe or P–O–Al bonds, which markedly improve resistance to corrosion while retaining the ability to accommodate high/various waste loadings (see Figure 9.3; Joshi et al., 2021).

The radiation stability of phosphate glass is generally adequate, though less robust than that of borosilicates. While these networks can accommodate damage from α and β decay without catastrophic collapse, they tend to undergo more pronounced structural modification under prolonged self-irradiation (Petit, 2020). Processing factors also influence long-term stability. In particular, the Fe^{2+}/Fe^{3+} ratio must be carefully managed: an excess of Fe^{2+} promotes crystallization during cooling, undermining durability and radionuclide homogeneity. These sensitivities, coupled with lower intrinsic stability in groundwater environments, have constrained phosphate glass to pilot-scale demonstration rather than full industrial adoption (Dube et al., 2020).

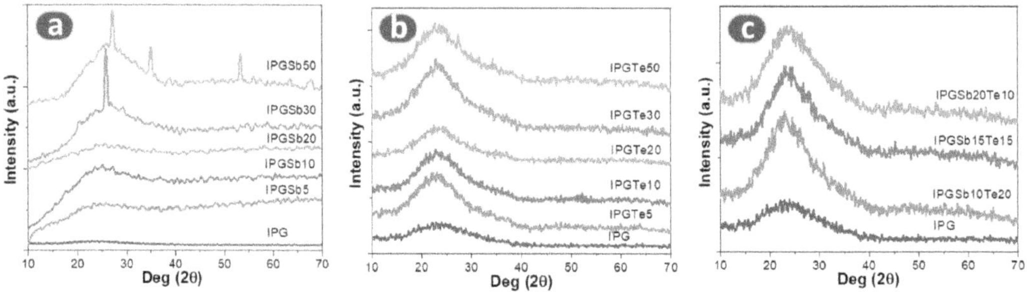

FIGURE 9.3 X-Ray Diffraction (XRD) Pattern of (a) Sb-loaded; (b) Te-loaded; and (c) Sb/Te Co-Loaded Iron Phosphate Glass Samples (Joshi et al., 2021).

9.3 GLASS-COMPOSITE WASTE FORM

Pure glass waste forms are the global standard for radioactive waste immobilization owing to their durability, processing simplicity, and ability to incorporate diverse radionuclides (Vienna et al., 2013). Yet, challenges remain, including limited solubility of Mo, S, Cl, and halides, volatilization of Cs, Tc, and I, and issues of devitrification, radiation effects, and redox control at high waste loadings (Riley et al., 2014; Gong et al., 2018; Löwy et al., 2025). To mitigate these issues, glass-composite waste forms have been developed. In such hybrid systems, problematic radionuclides are selectively partitioned into durable crystalline phases, while the surrounding glass immobilizes the remainder. This dual strategy alleviates solubility constraints, enables higher overall waste loading, and improves durability by exploiting the superior corrosion resistance and radiation tolerance of crystalline hosts. Glass-composite waste forms constitute hybrid immobilization strategies that integrate the chemical versatility of glass with the structural stability of crystalline phases. The terms *glass-composite* and *glass ceramics* are sometimes used interchangeably (Ojovan et al., 2021). As emphasized by McCloy and Goel (McCloy & Goel, 2017), glass composites are externally assembled systems in which pre-existing crystalline phases are mixed with a glass matrix prior to consolidation, in order to produce crystalline inclusions dispersed within the glass. In contrast, glass–ceramics are internally evolved structures formed by controlled crystallization of homogeneous melt, resulting in the interpenetrating network of glass and crystal domains. This structural integration highlights glass–ceramics as the more advanced evolution of glass-composite waste forms. Therefore, in this section, the discussion will primarily introduce glass–ceramic waste forms as representative systems, focusing on the principles, synthesis strategy, and performance advantage.

9.3.1 Design Strategy and Crystalline Phase Selection

In the development of glass–ceramic or composite waste forms, the selection of crystalline host phases is one of the most critical design choices. The ideal crystalline phase is expected to simultaneously satisfy several requirements: high stability under repository-relevant conditions and radiation damage and the capacity to incorporate key radionuclides, including actinides, lanthanides, molybdenum, and cesium, into its lattice without promoting segregation into undesirable secondary phases (Ewing, 1999).

Zirconolite ($CaZrTi_2O_7$) has historically served as a typical host and remains among the most extensively studied phases. Its appeal lies in its structural flexibility, which allows substitution of cations across a wide range of valence states and ionic radii, while maintaining resistance to both corrosion and radiation-induced amorphization. Experimental studies have demonstrated that surrogate actinides such as neodymium (Nd) and cerium (Ce) can be effectively incorporated into the zirconolite lattice through direct or coupled substitution mechanisms that preserve charge balance. This versatility explains the frequent inclusion of zirconolite in advanced glass–ceramic waste form designs (Thornber, 2018). Furthermore, it is reported that zirconolite-based glass–ceramics can accommodate chloride contamination through selective partitioning of Cl^- into the albite glass phase up to a solubility limit of approximately 1.0 ± 0.1 wt%, without disrupting Ce or Pu surrogate incorporation (see Figure 9.4, Thornber et al., 2020).

Derived from Synroc systems, several other titanate-based phases provide complementary functions. Pyrochlore ($A_2B_2O_7$) is valued for its ability to accommodate diverse cations and for its natural analogues, which suggest geological-scale durability (Ewing, 1999). Perovskite structures (ABO_3) offer stable frameworks capable of incorporating multi-valent cations, while hollandite (e.g., $BaAl_2Ti_6O_{16}$) is particularly important for immobilizing large monovalent ions such as Cs^+, facilitated by tunnel-like sites within its structure. For cesium immobilization, pollucite ($CsAlSi_2O_6$) has also been investigated, both as a single-phase material and as a constituent of glass–ceramic composites. Pollucite's aluminosilicate framework provides a durable host for Cs, and recent studies have examined pollucite-based composites derived from Cs-loaded IONSIV® crystalline silicotitanate (CST) ion-exchange material (Bahmanrokh et al., 2025).

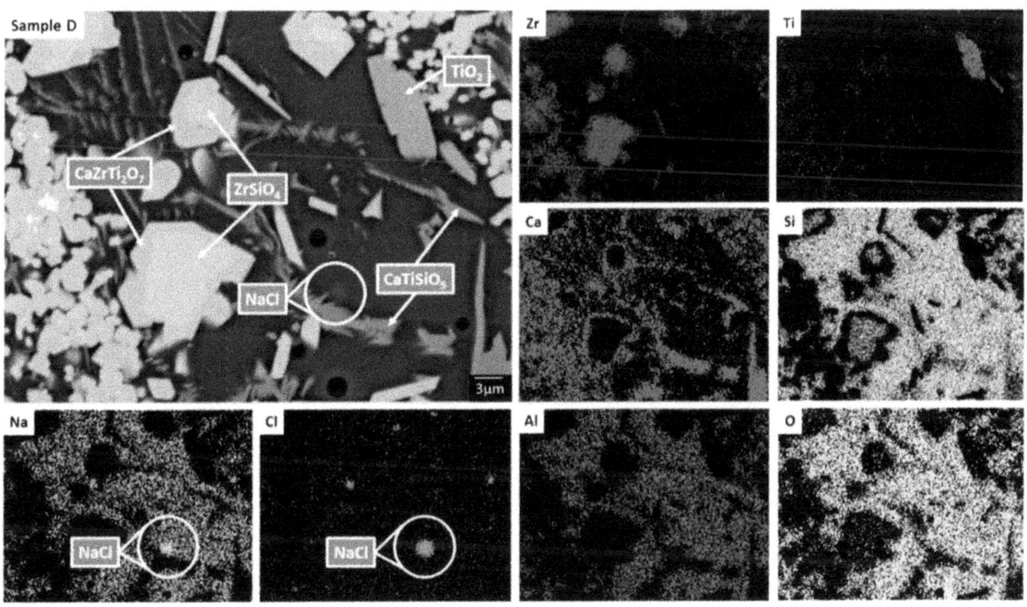

FIGURE 9.4 Backscattered Electron Micrograph and EDX Elemental Maps Showing the Microstructure of Zirconolite Glass–Ceramic Containing 1.7 wt % Cl (Thornber et al., 2020).

Apatite and oxyapatite phases are also promising for the immobilization of rare-earth and minor actinides. Their phosphate-silicate frameworks can accommodate trivalent and multi-valent cations on substitutional sites, making them suitable for use in composites where glass provides a flexible matrix and the crystalline phase secures chemically challenging elements (Crum et al., 2010; Donald, 2010).

Compatibility between the selected crystalline phases and the surrounding glass matrix is a key consideration in waste form design. Borosilicate glass remains the dominant matrix owing to its well-established durability, compositional flexibility, and industrial maturity. In such systems, crystalline phases must remain stable in contact with the melt, avoiding dissolution back into the glass, thermal-expansion mismatch, and weak glass–crystal interfaces. Phosphate glasses, on the other hand, offer superior solubility for problematic species. However, phosphate matrices are less hydrolytically durable, meaning that crystalline hosts in these systems must compensate for, rather than amplify, their inherent weaknesses.

Both thermodynamic and kinetic factors determine whether a crystalline host will perform as intended. Thermodynamically, the redox equilibria of multi-valent species (e.g., Ce^{3+}/Ce^{4+}, U^{3+}/U^{4+}) strongly affect whether radionuclides partition into target crystalline lattices or segregate into secondary oxides or residual glass. Unintended phase changes under repository conditions, such as oxidation or hydration, should also be anticipated. From a kinetic perspective, diffusion rates within residual glass and across interfaces control the efficiency of radionuclide incorporation. Slow diffusion may leave radionuclides stranded in the glass, reducing immobilization efficiency, while the composition of the residual glass after crystallization can impose further constraints. Glasses depleted or enriched in certain network formers or modifiers may become more fragile, susceptible to devitrification, or less durable in aqueous environment. In some designs, supplementary "getter" or buffer phases are introduced. For instance, minor additions of calcium or zinc can stabilize molybdenum as $CaMoO_4$ or $ZnMoO_4$, which reduces the burden on the glass to retain Mo in solution and prevents deleterious phase separation. Such targeted strategies expand the compositional design space and enable higher overall waste loading (Guo et al., 2025; Short et al., 2005).

9.3.2 Processing Routes

A wide range of processing routes has been developed to fabricate glass–ceramic waste forms, aiming to control microstructure, densification, and radionuclide partitioning. One of the most established approaches is the melt-and-crystallize method, in which a homogeneous parent glass, often incorporating nucleating agents such as TiO_2 or ZrO_2, is subjected to a controlled thermal schedule comprising nucleation and growth stages. This approach has been widely applied to systems such as zirconolite- and apatite-based glass–ceramics, where internal crystallization is promoted by titania or zirconia additions that drive in situ nucleation within the glass matrix (Thornber, 2018). Surface-initiated crystallization represents a variant of this route, particularly evident in cordierite systems, where nucleation is concentrated at the glass surface, and directional growth is achieved through optimized heating profiles combined with nucleating additives (Yu et al., 2021).

Beyond melt crystallization, powder-based consolidation techniques offer greater independence in tailoring phase identity and microstructure. Hot pressing and hot isostatic pressing (HIP) allow glass and crystalline powders to be co-sintered at lower temperature, which enables the incorporation of metastable or pre-synthesized crystalline phases. These approaches mitigate the volatilization (e.g., Cs) and corrosion problems often associated with high-temperature melting. In particular, the HIP route effectively suppresses Cs volatilization and secondary phase formation by achieving full densification at ~950°C under 100 MPa Ar, resulting in uniform encapsulation of Cs-bearing chabazite glass ceramics with minimal interfacial porosity. This leads to the enhancement of the retention of volatile radionuclides and the overall chemical durability of the waste form (see Figure 9.5, Gardner et al., 2021). As emphasized in comprehensive reviews of nuclear waste glass processing and glass–ceramics (Yudintsev et al., 2024), powder-based consolidation represents a natural extension of established ceramic processing to glass–ceramic composites. Recent surveys of SPS fundamentals and applications in glass/ceramic systems further document its ability to densify glass–ceramics rapidly while limiting grain coarsening (Babu et al., 2020).

Cold sintering has more recently emerged as a promising low-temperature alternative for composite densification. This method uses a transient liquid phase, typically water or another solvent, combined with uniaxial pressure at relatively mild temperatures to promote particle rearrangement and densification, followed by gentle heating to complete diffusion. Foundational studies demonstrate the underlying mechanisms and process envelopes that enable densification of ceramic and glass–ceramic systems at temperature orders of magnitude below conventional sintering (Guo et al., 2016). The reduced thermal stress and suppression of volatilization make cold sintering particularly attractive for radionuclides that are volatile or heat-sensitive.

9.3.3 Performance Characteristics

The hybrid structure of glass–ceramics and glass composites often yields performance advantages over either monolithic glass or single-phase ceramics, especially for difficult-to-immobilize radionuclides. Crystalline phases typically capture elements poorly retained by

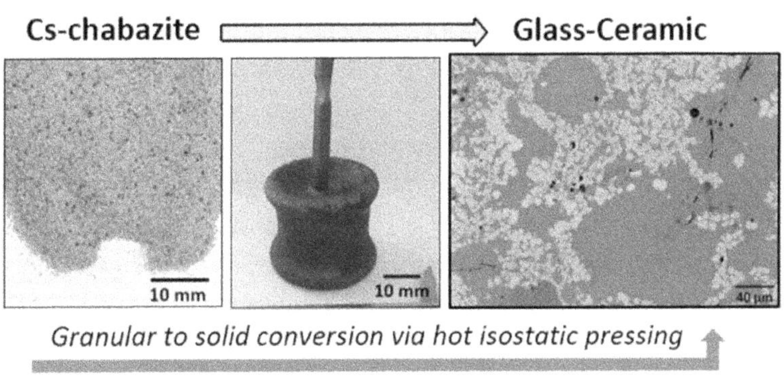

FIGURE 9.5 Conversion of Cs-Chabazite Precursor into a Dense Glass–Ceramic Waste Form via HIP (Gardner et al., 2021).

glass, for example, Cs in pollucite-like structures, Mo in powellite-type phases, or actinides in zirconolite, thereby reducing the burden on the glass matrix. Interfaces between glass and crystals can further act as corrosion barriers, slowing the advance of alteration fronts (McCloy & Goel, 2017; Neeway et al., 2019).

The behavior of glass–crystal interfaces is pivotal. Well-bonded interfaces can promote passivation, suppress ion exchange, and reduce leach propagation. Conversely, poorly bonded or cracked interfaces may serve as fast diffusion pathways, eroding performance benefits. However, Neeway et al. (2019) observed that when powellite and other crystalline phases are encapsulated within a glass matrix, the dissolution behavior can become locally accelerated, particularly for Mo, due to interfacial environments with elevated ionic strength or pH, highlighting that imperfect glass–crystal interfaces may increase leaching rates and partially offset the barrier effect. On the other hand, the normalized leaching rates (NRi) of key elements such as Ca, Zr, and Ti for zirconolite glass–ceramics were on the order of 10^{-5}–10^{-6} g m^{-2} d^{-1} after 28 days in deionized water at 90°C, and these rates are comparable to or lower than those of durable borosilicate glasses, demonstrating the excellent chemical durability of zirconolite-based glass ceramics (Yudintsev et al., 2024).

Radiation stability in composites presents a more nuanced picture. While glasses remain resistant to amorphization from α, β, and γ irradiation, they are metastable and can evolve through phase separation or network rearrangement. Crystalline phases, although susceptible to disorder or defect accumulation under heavy-ion or α-recoil damage, often retain lattice-bound radionuclides even after damage; moreover, interfaces can act as sinks for radiation-induced defects and gases, mitigating microcrack formation (Xu et al., 2022; Gin et al., 2024). Mechanically, composites leverage the complementary properties of glass and crystals. The glassy phase provides compliance, crack blunting, and strain accommodation, while crystalline domains contribute stiffness and thermal stability. Where thermal expansion is well matched, the composite resists cracking under thermal cycling or repository-induced stresses, preserving structural integrity (Yudintsev et al., 2024).

A number of glass–ceramic systems have been demonstrated at laboratory and pilot scales, underscoring both the flexibility and the potential of composite waste forms. Among these, zirconolite-based glass–ceramics for actinide-rich waste streams remain the most extensively investigated. In particular, neodymium (Nd) and cerium (Ce), commonly employed as surrogates for minor actinides and plutonium, have been incorporated with high efficiency into zirconolite phases within glass–ceramic matrices; synthesis–structure relationships and actinide surrogate uptake in zirconolite-bearing glass–ceramics have been summarized in recent reviews and experimental studies (Zhu et al., 2020). Complementary work shows that aliovalent substitutions and processing atmosphere can redirect cation partitioning: under reducing conditions, $Ce^{4+} \rightarrow Ce^{3+}$ reduction increases the potential of perovskite formation, illustrating the need for strict redox control during zirconolite glass–ceramic processing. In the United Kingdom, zirconolite glass–ceramics and related titanate assemblages have also been evaluated for plutonium disposition. HIP has been used to densify Pu-bearing ceramic and glass–ceramic bodies without forming deleterious second phases, yielding microstructures comparable to those of conventionally sintered analogues (Thornber, 2018).

Pollucite-based glass–ceramics are emerging as a particularly promising option for cesium immobilization. A recent scoping study coupling Cs-loaded IONSIV with tailored glass formers demonstrated that cesium partitions predominantly to crystalline pollucite while a residual glass provides cohesion and processing flexibility; notably, the system achieved high waste loadings and exhibited favorable aqueous durability under ASTM C1285 testing (Bahmanrokh et al., 2024). Given the volatility of Cs in borosilicate vitrification, the pollucite-glass–ceramic route represents a robust pathway for high-Cs immobilization. Oxyapatite-containing glass–ceramics have also been investigated for selective uptake of rare-earth and actinide surrogates. Their silico-phosphate frameworks can stabilize trivalent cations, and heat-treatment schedules can be tuned to co-crystallize oxyapatite with accessory phases (e.g., powellite) that help accommodate problematic species; partitioning trends and corrosion behavior in rare-earth-rich aluminoborosilicate systems provide design guidance for such composites (Quintas et al., 2009).

Beyond strictly nuclear applications, advances in structural and functional glass–ceramics provide transferable insights. Cordierite glass–ceramics, widely used for electronic substrates, are canonical examples of surface-initiated crystallization controlled via nucleation agents and thermal profiles (Synkiewicz-Musialska et al., 2021). Similarly, oxyfluoride glass–ceramics offer a mature platform for controlled nano-crystallization with long-term chemical stability and moisture tolerance, processing lessons that translate to nuclear glass–ceramic design where interfacial control and residual-glass continuity are critical (Dejneka, 1998). Together, these examples supply microstructural control strategies, fine phase dispersion, tuned nucleation/growth windows, and interface engineering, which can be adapted to engineer durable, high-loading nuclear glass–ceramic waste forms.

Looking forward, promising directions include: (i) the design of crystalline hosts specifically tailored for partitioned waste streams from advanced fuel cycles (e.g., minor actinides or transuranics); (ii) the application of multi-scale modeling, from atomistic simulations to continuum models, to predict microstructural evolution and degradation; (iii) advanced in situ characterization techniques, such as synchrotron X-ray tomography, high-resolution transmission electron microscopy (TEM), and atom-probe tomography, to probe interface chemistry, microcrack development, and radiation effects; (iv) coupling with partitioning and separation technologies, enabling waste streams to feed directly into optimized composite designs (e.g., Mo-rich streams stabilized by powellite-zirconolite glass–ceramics, Cs-rich streams immobilized via pollucite-based composites); and (v) long-term testing under repository analogues, including accelerated corrosion and radiation experiments, to validate predictive models and licensing cases.

9.4 MODELING AND DATA-DRIVEN APPROACHES FOR GLASS WASTE FORM DESIGN

The development of predictive and data-driven modeling has transformed the design of glass waste forms into a quantitative science. Traditional formulation approaches, based on phase diagrams, trial melts, and iterative testing, are increasingly complemented by machine learning (ML), regression frameworks, and multi-scale simulation tools that enable rapid screening, uncertainty-aware optimization, and mechanistic interpretation.

These modeling paradigms collectively shorten development cycles and enhance confidence in performance predictions under repository conditions.

9.4.1 Predictive Modeling of Crystallization and Durability

Predicting crystallization during glass cooling remains one of the central challenges in vitrification. Undesired crystalline phases, such as nepheline ($NaAlSiO_4$), consume network-forming Al and Si, weakening the amorphous network and accelerating dissolution in groundwater. Empirical phase rules are inadequate for multi-component feeds that may include more than 30 oxides. Recent ML classifiers, such as support vector machines, random forests, and neural networks, trained on composition–outcome datasets, achieve ≈90 % accuracy for nepheline formation prediction (Lu et al., 2021; Gunnell et al., 2022). In addition to high predictive accuracy, Lu et al. (2021) demonstrated that Si/Al and Na/Al ratios exert the strongest control over nepheline crystallization propensity, providing a quantitative basis for compositional tuning to suppress undesired phase formation during vitrification (see Lu et al., 2021). Unlike binary boundaries, these models return probabilistic crystallization maps across high-dimensional composition space, supporting risk-based control of melter feeds. Apart from nepheline, analogous classifiers are being developed to predict other crystallization or phase-separation risks. Embedding such ML models in process control (as proposed by Lu et al., 2024b; see Figure 9.6) would enable real-time feed adjustments. Traditional constraints (McCloy & Vienna, 2010) have been used in an offline manner to avoid regimes prone to nepheline formation.

Regression-type models complement classification by predicting continuous properties. Mixture models (e.g. polynomial or quadratic mixture) are often used to capture non-linear relationships between composition and properties, such as viscosity, liquidus

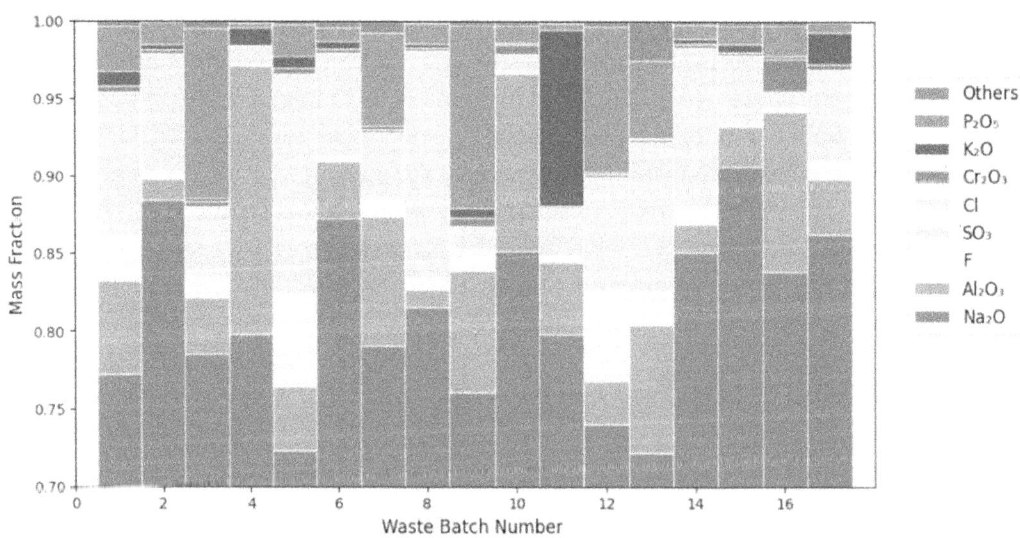

FIGURE 9.6 Major Waste Component Concentrations (Mass Fraction) as a Function of Batch Number (Lu et al., 2024b).

temperature, and leachability, frequently incorporating Vogel–Fulcher–Tammann or Arrhenius temperature dependencies. These models, when trained on curated databases, can reproduce processing and performance metrics with useful accuracy (Lu et al., 2024a). Multi-dimensional analyses quantify trade-offs, e.g., Al additions form AlO_4 units that enhance durability, while excess B promotes B–O–B and B–O–Si linkages that hydrolyze readily (Du et al., 2021). In combination, ML classifiers for crystallization and mixture-based regressions for thermophysical and durability properties now underpin adaptive, data-driven design of vitrified waste forms.

9.4.2 Property Databases and Mixture Model Frameworks

Reliable modeling depends on comprehensive, high-quality datasets. Institutions such as Pacific Northwest National Laboratory (PNNL) have assembled extensive databases covering composition, viscosity, liquidus temperature, thermal expansion, and PCT leach rates, spanning thousands of borosilicate and phosphate formulations (Lu et al., 2024a). These repositories ensure continuity between decades of laboratory work and modern ML pipelines, providing the empirical foundation for safety-case-defensible predictions.

Composition is typically represented via the network-former/intermediate/modifier framework, which fits naturally into partial quadratic mixture (PQM) models incorporating oxide-interaction terms (Heredia-Langner et al., 2022). Temperature-dependent variants unify viscosity and durability fits across melt and repository conditions (Fluegel, 2007). The appeal of such mixture models lies in their interpretability and computational efficiency: coefficients link directly to physical interactions and can be re-fitted as new data accumulate. Nonetheless, data coverage remains uneven. Classic silicate-rich systems are densely sampled, whereas halide- or sulfate-bearing compositions are sparse, limiting predictive reliability in those domains. To mitigate extrapolation risk, practitioners define explicit "validity domains," composition bounds within which model errors are quantified, and treat external predictions as high-uncertainty (Hrma, 2008). Cross-validation, independent test sets, and targeted experiments at domain edges strengthen model defensibility. Newly measured viscosity or PCT data are periodically integrated into centralized repositories, allowing re-training that enhances both global accuracy and local fidelity. Open, machine-readable data standards and consistent analytical reporting promote reproducibility and inter-laboratory integration. When mixture-model regressions and ML classifiers draw from such continually refreshed datasets, they provide fast, trustworthy guidance for feed adjustment and formulation screening (Heredia-Langner et al., 2022; Crum et al., 2023).

9.4.3 Uncertainty Quantification and Model Validation

Given the complexity and non-linearity of multi-oxide glass systems, explicit uncertainty quantification (UQ) is critical. Measurement variability across laboratories, compositional and analytical errors, and non-linear oxide–oxide couplings can amplify uncertainties in model predictions. In the absence of quantified confidence intervals, outputs from ML or regression models may misleadingly be perceived as exact rather than probabilistic estimations (Lu et al., 2024b; Schmidt et al., 2019; Tavazza et al., 2021). Common UQ practices include k-fold cross-validation and bootstrap resampling to estimate variability in model performance, alongside validation against independent (held-out) datasets as a measure of generalizability (Ward et al., 2016). Bayesian methods further enrich this framework: Gaussian process regression (GPR) and related probabilistic regressors yield predicted means with uncertainty intervals (or credible bands). These calibrated uncertainty bounds can be propagated through downstream process models— for example, melt viscosity or crystallization risk models—to define risk-aware operating windows (Zuo et al., 2021).

Such probabilistic predictions are especially valuable in contexts such as regulatory safety assessments, where explicit safety margins and worst-case scenarios must be considered. Model reliability also depends on how compositional features are represented. Featurization choices—which normalize by modifier content, group oxides by structural role, or adopt elemental-descriptor sets such as MAGPIE—can alter both accuracy and interpretability. Best-practice guidelines from materials informatics emphasize systematic descriptor testing, scaling, and sensitivity analyses to eliminate spurious correlations and improve stability (Ward et al., 2016; Ghiringhelli et al., 2015). Ultimately, model validation is iterative. Targeted experiments near edge compositions—where crystallization or phase separation is likely—provide the most stringent tests of model fidelity and expand the empirical basis for re-training. Data-driven predictions gain greater depth when coupled with physics-based simulations that resolve atomic and process mechanisms. Molecular dynamics (MD) studies, supported by improved interatomic potentials and validated through in situ spectroscopy, reveal how local motifs such as $BO_3 \leftrightarrow BO_4$ conversion and Fe^{2+}/Fe^{3+} redox regulate network connectivity and, by extension, glass durability (Le Losq et al., 2021; Brow et al., 2020; Tasheva et al., 2023). While Tasheva et al. (2023) reported that high Fe_2O_3 concentrations in silicate glasses promote Fe–O–Si bonding and network depolymerization, Brow et al. (Brow et al., 2020) showed that balanced Fe redox states can instead strengthen phosphate networks through P–O–Fe linkages. Together, these insights illustrate the dual structural and chemical roles of iron in controlling connectivity and corrosion resistance across glass families.

Translating such atomic-scale descriptors, including non-bridging-oxygen fraction, short-range-order parameters, and bond-length or ring-size distributions, into ML feature spaces enables hybrid models that combine mechanistic fidelity with statistical efficiency (Liu et al., 2019). Liu et al. (2019) demonstrated that including molecular-topological descriptors, such as network connectivity and ring-size distributions, allows ML models to predict dissolution rates spanning five orders of magnitude with high fidelity (as seen in Figure 9.7), linking atomic structure to macroscopic reactivity in silicate glasses. More broadly, physics-aware learning frameworks for disordered materials are exposing structure–property relationships that conventional empirical regressions overlook (Jung et al., 2024).

At larger scales, continuum modeling translates composition–structure relationships into process-level predictions. Property regressions for viscosity, liquidus temperature, and crystallization tendency are coupled with computational fluid dynamics (CFD) simulations of cold-crucible and joule-heated melters to predict convection patterns, temperature distributions, and feed homogenization behavior (Bojarevics et al., 2003; Rigby

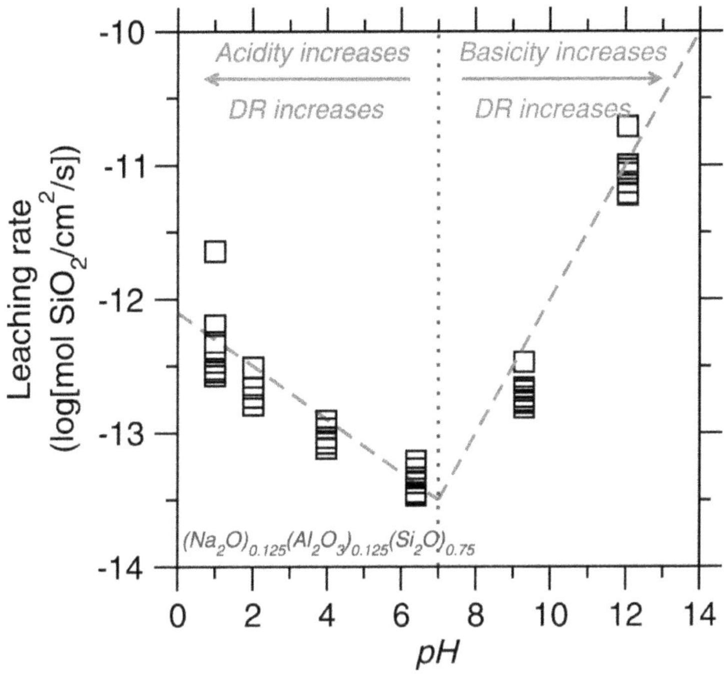

FIGURE 9.7 Comparison of Predicted and Experimentally Measured Dissolution Rates for Silicate Glasses using Topology-Informed Machine Learning Models (Liu et al., 2019).

et al., 2023). Such CFD studies provide insight into melt stirring, cold-cap dynamics, and bubble transport, which influence the efficiency and stability of vitrification. In parallel, finite-element thermo-mechanical models are applied to simulate canister cooling, using temperature-dependent thermal and rheological properties to estimate residual stress development, potential cracking, and the onset of crystallization during cooldown (Wescott & Slate, 1982). Although early, the work of Wescott and Slate remains foundational for quantifying thermal gradients and stress buildup in large HLW canisters, and its methodology continues to inform modern modeling efforts that integrate viscoelastic constitutive laws and heat-transfer data from contemporary systems (Wescott & Slate, 1982). Together, these process-scale simulations enable the propagation of uncertainty from composition–property models (e.g., Bayesian intervals on viscosity or liquidus temperature) into engineering design and operational constraints, thereby linking UQ directly to melter performance optimization and repository-relevant safety margins.

9.5 CONCLUSION

This chapter highlights the complementary strengths of glass and glass-composite waste forms for immobilizing high-level radioactive waste. Borosilicate glass remains the industrial benchmark, offering proven durability and scalability, while phosphate glass provides specialized advantages for sulphate- and halide-rich wastes. Their limitations, such as phase separation, volatilization, and restricted solubility, have motivated glass–ceramic strategies, where crystalline phases such as zirconolite, pollucite, and apatite selectively immobilize challenging radionuclides. The synergy between crystalline hosts and residual glass improves durability, enhances radionuclide partitioning, and increases waste loadings. Advances in processing, from cold crucible melters to powder-based consolidation, expand fabrication routes, while predictive modeling, ML, and property databases provide new tools for composition optimization. UQ and multi-scale modeling reinforce confidence in PA, bridging laboratory studies with repository timescale. Looking forward, the integration of experimental innovation with data-driven design and validation will be essential for scaling promising concepts to industrial practice, ensuring that next-generation waste forms meet the requirements for long-term nuclear safety.

REFERENCES

Amoroso, J., Addleman, R., Marra, S., Smith, G.L., 2021. *Appropriateness and Readiness of Cold Crucible Vitrification for U.S. DOE HLW and LAW*. SRNL Report SRNL-STI-2021-00342. Savannah River National Laboratory, Aiken, SC.

Babu, S., Galusek, D., Durán, A., Pascual, M., 2020. Glass-ceramics processed by spark plasma sintering (SPS) for optical applications. *Appl. Sci.* 10(8), 2791.

Bahmanrokh, G., Degeling, J., Dayal, P., Peristyy, A., Mcrae-Black, A., Farzana, R., Koshy, P., Gregg, D., 2025. Impact of Cs loading on pollucite formation in glass–ceramics for IONSIV immobilization. *MRS Advances* 10, 1825–1832.

Bahmanrokh, G., Whitelock, E., Dayal, P., Peristyy, A., McRae-Black, A., Farzana, R., Koshy, P., Gregg, D., 2024. Candidate glass–ceramic wasteforms for the immobilisation of Cs-loaded IONSIV® wastes: A scoping study. *MRS Adv.* 9, 420–425.

Bojarevics, V., Djambazov, G., Harding, R.A., Pericleous, K., Wickins, M., 2003. Investigation of the cold crucible melting process: Experimental and numerical study. *Magnetohydrodynamics* 39(4), 395–402.

Brow, R.K., Schue, L., Chitambar, S., Selvaraj, M., 2020. Iron polyphosphate glasses for waste immobilization. *Int. J. Appl. Glass Sci.* 11(4), 661–675.

Chauvin, E., Gasperini, M., Gordon, A., 2010. La Hague continuous improvement program to go beyond. In: *Proceedings of Waste Management Symposia 2010 (WM2010), Paper 10128*. Waste Management Symposia, Phoenix, AZ.

Crum, J.V., Turo, L., Riley, B.J., Tang, M., Kossoy, A., Sickafus, K., 2010. *Glass Ceramic Waste Forms for Combined Cs + Ln + TM Fission Products Waste Streams. PNNL-19780*. Pacific Northwest National Laboratory, Richland, WA.

Crum, J.V., Vienna, J.D., Riley, B.J., Silverstein, J.A., Kissinger, R.M., Neeway, J.J., George, J.L., Bellofatto, D.L., Hall, M.A., 2023. Formulation and testing of a high-tin borosilicate nuclear waste glass for in-can melting. *J. Nucl. Mater.* 585, 154643.

Dejneka, M.J., 1998. The luminescence and structure of novel transparent oxyfluoride glass-ceramics. *J. Non-Cryst. Solids* 239, 149–155.

Donald, I.W., 2010. *Waste Immobilization in Glass and Ceramic Based Hosts: Radioactive, Toxic and Hazardous Wastes*. Wiley, Chichester.

Du, J., Lu, X., Gin, S., Delaye, J.-M., Deng, L., Taron, M., Bisbrouck, N., Bauchy, M., Vienna, J.D., 2021. Predicting the dissolution rate of borosilicate glasses using QSPR analysis based on molecular dynamics simulations. *J. Am. Ceram. Soc.* 104(9), 4445–4458.

Dube, C.L., Stennett, M.C., Ananthanarayanan, A., Hyatt, N.C., 2020. Radiation stability study on cerium loaded iron phosphate glasses. *J. Radioanal. Nucl. Chem.* 323(3), 1381–1386.

Ewing, R.C., 1999. Nuclear waste forms for actinides. *Proc. Natl. Acad. Sci. USA* 96(7), 3432–3439.

Fellinger, T.L., Bibler, N.E., Harbour, J.R., Jantzen, C.M., 1999. Demonstration of the Defense Waste Processing Facility (DWPF) flowsheet in the Shielded Cells Facility. *MRS Proc.* 556, 199–206.

Fluegel, A., 2007. Glass viscosity calculation based on a global statistical modeling approach. *Glass Technol. Eur. J. Glass Sci. Technol. A* 48(1), 13–30.

Gardner, L.J., Walling, S.A., Corkhill, C.L., Hyatt, N.C., 2021. Thermal treatment of Cs-exchanged chabazite by hot isostatic pressing to support decommissioning of Fukushima Daiichi Nuclear Power Plant. *Journal of Hazardous Materials* 413, 125250.

Ghiringhelli, L.M., Vybiral, J., Levchenko, S.V., Draxl, C., Scheffler, M., 2015. Big data of materials science: Critical role of the descriptor. *Phys. Rev. Lett.* 114(10), 105503.

Gin, S., Taron, M., Arena, H., Delaye, J.-M., et al., 2024. Effect of structural disorder induced by external irradiation on the alteration of a four-oxide borosilicate glass. *NPJ Mater. Degrad.* 8, 64.

Goel, A., Pokorny, R., Hrma, P., Kruger, A.A., 2019. Challenges with vitrification of Hanford High-Level Waste (HLW) to borosilicate glass—An overview. *J. Non-Cryst. Solids X* 1, 100033.

Gong, Y., Xu, J., Buchanan, R.C., 2018. The aqueous corrosion of nuclear waste glasses revisited: Probing the surface and interfacial phenomena. *Corros. Sci.* 143, 65–75.

Gunnell, L.L., Manwaring, K., Lu, X., Reynolds, J., Vienna, J., Hedengren, J., 2022. Machine learning with gradient-based optimization of nuclear waste vitrification with uncertainties and constraints. *Processes* 10(11), 2365.

Guo, J., Guo, H., Baker, A.L., Lanagan, M.T., Kupp, E.R., Messing, G.L., Randall, C.A., 2016. Cold sintering: A paradigm shift for processing and integration of ceramics. *Angew. Chem. Int. Ed.* 55(38), 11457–11461.

Guo, L., Shan, Z., Zhu, Y., Cui, Z., Yang, D., Liu, S., 2025. Increasing the incorporation of MoO3 in borosilicate glass via the substitution of B2O3 with Al2O3. *Ceram. Int.* 51(19, Part B), 28858–28866.

Heredia-Langner, A., Gervasio, V., Cooley, S.K., Lonergan, C.E., Kim, D.-S., Kruger, A.A., Vienna, J.D., 2022. Hanford low-activity waste glass composition-temperature-melt viscosity relationships. *Int. J. Appl. Glass Sci.* 13(4), 514–525.

Hrma, P., 2008. Glass viscosity as a function of temperature and composition: A model based on Adam–Gibbs equation. *J. Non-Cryst. Solids* 354(29), 3389–3399.

Joshi, A.C., Roy, M., Dutta, D.P., Mishra, R.K., Meena, S.S., Kumar, R., Bhattacharyya, D., Alexander, R., Kaushik, C.P., Tyagi, A.K., 2021. Effect on the structure and stability of iron phosphate glass with Sb and Te-ion loading for nuclear waste storage application. *J. Non-Cryst. Solids* 570, 121016.

Jung, G., Biroli, G., Berthier, L., 2024. Dynamic heterogeneity at the experimental glass transition predicted by transferable machine learning. *Phys. Rev. B* 109(6), 064205.

Kim, D., Kruger, A.A., 2018. Volatile species of technetium and rhenium during waste vitrification. *J. Non-Cryst. Solids* 481, 41–50.

Kofuji, H., Yano, T., Myochin, M., Matsuyama, K., Okita, T., Miyamoto, S., 2012. Chemical durability of iron-phosphate glass as the high level waste from pyrochemical reprocessing. *Procedia Chem.* 7, 764–771.

Le Losq, C., Valentine, A.P., Mysen, B.O., Neuville, D.R., 2021. Structure and properties of alkali aluminosilicate glasses and melts: Insights from deep learning. *Geochim. Cosmochim. Acta* 314, 27–54.

Liu, H., Zhang, T., Krishnan, N.M.A., Wang, C., Sant, G., Bauchy, M., 2019. Predicting the dissolution kinetics of silicate glasses by topology-informed machine learning. *NPJ Mater. Degrad.* 3, 32.

Löwy, L., Eret, M., Kloužek, J., Cincibusová, P., Vernerová, M., Kohoutková, M., George, J., Killmer, M., Ferkl, P., Hrma, P., Kruger, A.A., Pokorný, R., 2025. Increasing Tc retention during nuclear waste vitrification: Effect of alumina source on the rate of rhenium incorporation into glass. *J. Nucl. Mater.* 614, 155916.

Lu, X., Sargin, I., Vienna, J.D., 2021. Predicting nepheline precipitation in waste glasses using ternary submixture model and machine learning. *J. Am. Ceram. Soc.* 104(11), 5636–5647.

Lu, X., Vienna, J.D., Du, J., 2024a. Glass formulation and composition optimization with property models: A review. *J. Am. Ceram. Soc.* 107(3), 1603–1624.

Lu, X., Weller, Z.D., Gervasio, V., Vienna, J.D., 2024b. Glass design using machine learning property models with prediction uncertainties: Nuclear waste glass formulation. *J. Non-Cryst. Solids* 631, 122907.

Marra, S.L., Harbour, J.R., 2004. *Characterization of Defense Waste Processing Facility (DWPF) Glass and Deposit Samples from Melter #2.* WSRC-TR-2003-00504, Rev. 0. Westinghouse Savannah River Company, Aiken, SC.

Marra, S.L., Harbour, J.R., Bibler, N.E., 1998. *The Defense Waste Processing Facility: Two Years of Radioactive Operations.* Westinghouse Savannah River Company Report WSRC-MS-98-00050, 1–12.

Masson, H., 2001. *The R7/T7 vitrification at La Hague: 10 years of operation.* IAEA-SM-357/33. International Atomic Energy Agency, Vienna.

McCloy, J.S., Goel, A., 2017. Glass-ceramics for nuclear-waste immobilization. *MRS Bull.* 42(3), 233–240.

McCloy, J.S., Riley, B.J., Dixon Wilkins, M., Evarts, J., Bussey, J., Vienna, J.D., Bingham, P.A., Gregg, D., Ojovan, M.I., Schuller, S., Uruga, K., Perret, D., Regnier, É., Giboire, I., Um, W., Xu, K., Goel, A., Kruger, A.A., 2024. International perspectives on glass waste form development for low-level and intermediate-level radioactive waste. *Mater. Today* 80, 594–618.

McCloy, J.S., Vienna, J.D., 2010. *Glass Composition Constraint Recommendations for Use in Life-Cycle Mission Modeling.* PNNL-19372. Pacific Northwest National Laboratory, Richland, WA.

Neeway, J.J., Asmussen, R.M., McElroy, E.M., Peterson, J.A., Riley, B.J., Crum, J.V., 2019. Kinetics of oxyapatite [Ca2Nd8(SiO4)6O2] and powellite [(Ca, Sr, Ba)MoO4] dissolution in glass-ceramic nuclear waste forms in acidic, neutral, and alkaline conditions. *J. Nucl. Mater.* 515, 227–237.

Ojovan, M.I., Petrov, V.A., Yudintsev, S.V., 2021. Glass crystalline materials as advanced nuclear wasteforms. *Sustainability* 13(8), 4117.

Petit, L., 2020. Radiation effects on phosphate glasses: Review. *Int. J. Appl. Glass Sci.* 11(3), 511–521.

Pilania, R.K., Dube, C.L., 2023. Matrices for radioactive waste immobilization: A review. *Front. Mater.* 10, 1236470.

Pinet, O., Hollebecque, J.-F., Hugon, I., Debono, V., Campayo, L., Vallat, C., Lemaitre, V., 2019. Glass ceramic for the vitrification of high-level waste with a high molybdenum content. *J. Nucl. Mater.* 519, 121–127.

Quintas, A., Caurant, D., Majérus, O., Charpentier, T., Dussossoy, J.-L., 2009. Effect of compositional variations on charge compensation of AlO4 and BO4 entities and on crystallization tendency of a rare-earth-rich aluminoborosilicate glass. *Mater. Res. Bull.* 44(9), 1895–1898.

Rigby, J.C., Dixon, D.R., Kloužek, J., Pokorný, R., Thompson, P.B.J., Scrimshire, A., Kruger, A.A., Bell, A.M.T., Bingham, P.A., 2023. Alternative reductants for foam control during vitrification of high-iron High Level Waste (HLW) feeds. *J. Non-Cryst. Solids* 608, 122240.

Riley, B.J., Schweiger, M.J., Kim, D.-S., Lukens, W.W., Williams, B.D., Iovin, C., Rodriguez, C.P., Overman, N.R., Bowden, M.E., Dixon, D.R., Crum, J.V., McCloy, J.S., Kruger, A.A., 2014. Iodine solubility in a low-activity waste borosilicate glass at 1000 °C. *J. Nucl. Mater.* 452(1–3), 178–188.

Schmidt, J., Marques, M.R.G., Botti, S., et al., 2019. Recent advances and applications of machine learning in solid-state materials science. *NPJ Comput. Mater.* 5, 83.

Sengupta, P., 2012. A review on immobilization of phosphate containing high level nuclear wastes within glass matrix—Present status and future challenges. *J. Hazard. Mater.* 235–236, 17–28.

Short, R.J., Hand, R.J., Hyatt, N.C., Möbus, G., 2005. Environment and oxidation state of molybdenum in simulated high level nuclear waste glass compositions. *J. Nucl. Mater.* 340(2–3), 179–186.

Synkiewicz-Musialska, B., Szwagierczak, D., Kulawik, J., Pałka, N., Piasecki, P., 2021. Structural, thermal and dielectric properties of low dielectric permittivity cordierite-mullite-glass substrates at terahertz frequencies. *Materials* 14(14), 4030.

Taron, M., Gin, S., Delaye, J.M., 2025. Impact of B and Al on the initial and residual dissolution rate of alumino-borosilicate glasses. Part I: Kinetic data. *NPJ Mater. Degrad.* 9(1), 32.

Tasheva, T., Harizanova, R., Mihailova, I., Cherkezova-Zheleva, Z., Paneva, D., Nedkova, M., Rüssel, C., 2023. Structure and redox ratio of soda-lime-silica glasses with high iron oxide concentrations. *Int. J. Appl. Glass Sci.* 14(3), 445–454.

Tavazza, F., DeCost, B., Choudhary, K., 2021. Uncertainty prediction for machine learning models of material properties. *ACS Omega* 6(48), 32431–32440.

Thornber, S.M., 2018. *The Development of Zirconolite Glass-Ceramics for the Disposition of Actinide Wastes*. University of Sheffield (Ph.D. thesis).

Thornber, S.M., Mottram, L.M., Mason, A.R., Thompson, P., Stennett, M.C., Hyatt, N.C., 2020. Solubility, speciation and local environment of chlorine in zirconolite glass–ceramics for the immobilisation of plutonium residues. *RSC Adv.* 10(54), 32497–32510.

Thorpe, C.L., Neeway, J.J., Pearce, C.I., Hand, R.J., Fisher, A.J., Walling, S.A., Hyatt, N.C., Kruger, A.A., Schweiger, M.J., et al., 2021. Forty years of durability assessment of nuclear waste glass by standard methods. *NPJ Mater. Degrad.* 5(1), 61.

Vienna, J.D., Hrma, P.R., Smith, D.E., Schweiger, M.J., 2002. *Sodium Bearing Waste Glass Formulation Status*. PNNL Technical Report PNNL-14050, Pacific Northwest National Laboratory, Richland, WA.

Vienna, J.D., Ryan, J.V., Gin, S., Inagaki, Y., 2013. Current understanding and remaining challenges in modeling long-term degradation of borosilicate nuclear waste glasses. *Int. J. Appl. Glass Sci.* 4, 283–294.

Ward, L., Agrawal, A., Choudhary, A., et al., 2016. A general-purpose machine learning framework for predicting properties of inorganic materials. *NPJ Comput. Mater.* 2, 16028.

Wescott, R.I., Slate, S.C., 1982. Vitrification of high-level wastes: A review of the computer thermal analyses for storage canisters. *Nucl. Eng. Des.* 67(3), 447–458.

Xu, H., Kim, S.Y., Chen, D., Monchoux, J.-P., Voisin, T., Sun, C., Li, J., 2022. Materials genomics search for possible helium-absorbing nano-phases in fusion structural materials. *Adv. Sci.* 9(32), 2203555.

Yang, J.-H., Choi, J., Lee, J.-S., Choi, K.-W., Kim, S.-H., Park, D.-H., 2017. Al2O3-containing silver phosphate glasses as hosting media for radioactive iodine. *J. Nucl. Sci. Technol.* 54(11), 1330–1337.

Yu, W., Cao, S., Wang, J., Zhang, Z., Han, J., Liu, C., Ruan, J., 2021. Crystallization mechanisms of cordierite glass-ceramics with 'surface-center' crystallization behavior. *J. Eur. Ceram. Soc.* 41(13), 6708–6721.

Yudintsev, S.V., Ojovan, M.I., Malkovsky, V.I., 2024. Thermal effects and glass crystallization in composite matrices for immobilization of the rare-earth element–minor actinide fraction of high-level radioactive waste. *J. Compos. Sci.* 8(2), 70.

Zdanievski, W.A., Easler, T.E., Bradt, R.C., 1983. Gamma radiation effects on the strength of a borosilicate glass. *J. Am. Ceram. Soc.* 66(5), 311–313.

Zhu, H., Wang, F., Liao, Q., Zhu, Y., 2020. Synthesis and characterization of zirconolite-sodium borosilicate glass-ceramics for nuclear waste immobilization. *J. Nucl. Mater.* 532, 152026.

Zubekhina, B.Y., Burakov, B.E., Ojovan, M.I., 2020. Surface alteration of borosilicate and phosphate nuclear waste glasses by hydration and irradiation. *Challenges* 11(2), 14.

Zuo, Y., Qin, M., Chen, C., Ye, W., Li, X., Luo, J., Ong, S.P., 2021. Accelerating materials discovery with Bayesian optimization and graph deep learning. *Mater. Today* 51, 126–135.

CHAPTER 10

Radioactive Waste Immobilization in Ceramic and Novel Materials

Xiangbiao Yin and Yi Zhou

10.1 ZIRCON

Zircon ($ZrSiO_4$) crystallizes in the tetragonal system with the space group I4/amd and $Z = 4$. As illustrated in Figure 10.1, its structure consists of chains of alternating ZrO_8 dodecahedra and SiO_4 tetrahedra extending along the c-axis. These chains are interconnected via edge-sharing ZrO_8 polyhedra parallel to the a-axis. Zircon exhibits a notable capacity to incorporate various radionuclides—such as U, Pu, Cm, Np, and Gd—through substitution into the Zr site, forming solid solutions. Prolonged α-decay events may lead to partial or even complete metamictization (amorphization) of the crystal structure. Nevertheless, even in the fully metamict state, zircon retains excellent resistance to hydrothermal corrosion and maintains the ability to securely immobilize radionuclides, demonstrating high chemical durability under demanding environmental conditions.

10.1.1 Immobilization of Radioactive Waste

The incorporation of tetravalent actinides into zircon ($ZrSiO_4$) has been extensively studied, with plutonium-doped zircon receiving particular attention. In 2000, Burakov et al. successfully incorporated ^{239}Pu into the zircon structure via a sol–gel method followed by high-temperature solid-state reaction, achieving a maximum Pu solubility of 7 wt% (Burakov et al., 2000). Using a similar approach, Weber synthesized Pu-doped $ZrSiO_4$ under an argon atmosphere at 1465°C, with a Pu solubility reaching 9.2 wt% (Weber, 1991). In 2004, Hanchar et al. obtained single crystals of Pu-bearing zircon using a flux method and synthesized zircon waste forms after 204 h at 1220°C. Their results indicated that Pu solubility in (Zr, Pu)SiO_4 ranged from 5 to 14 wt%, with no evidence of Pu-rich secondary phases, demonstrating zircon's capacity to incorporate over 10 wt% Pu (Hanchar et al., 2004).

Microwave sintering has also been employed to produce Ce^{4+}-doped zircon waste forms with high synthesis efficiency at 1500°C. When the Ce doping level was below 5 wt%, no

 DOI: 10.1201/9781003665793-13

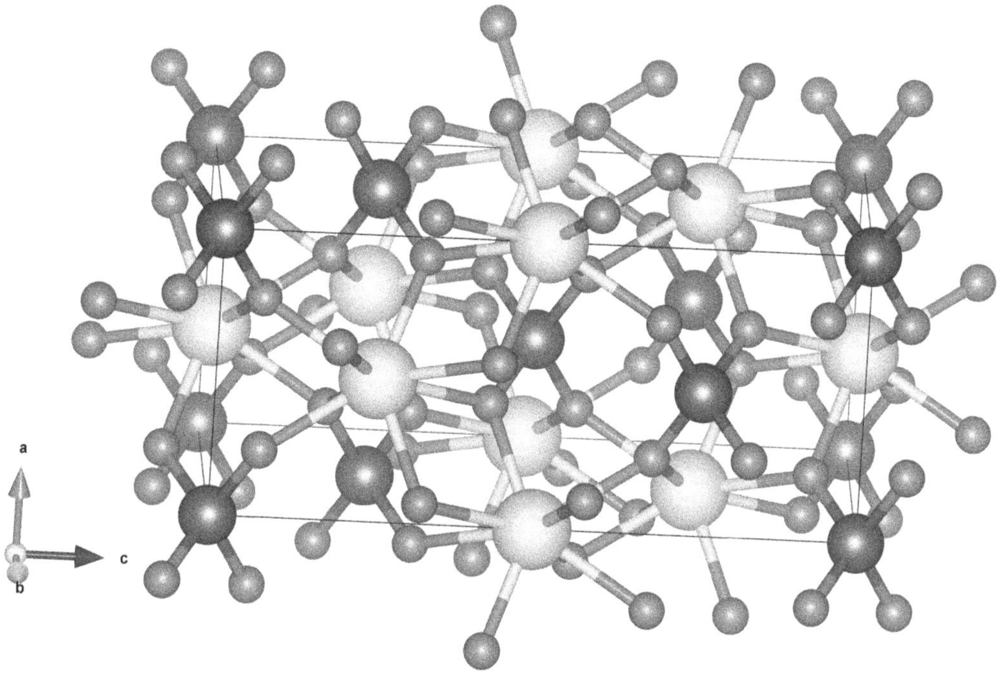

FIGURE 10.1 The Crystal Structure of $ZrSiO_4$.

Ce-bearing secondary phases were observed. Furthermore, based on studies of uranium content in "Chernobyl lavas," Burakov and Pazukhin proposed a strategy for synthesizing highly U-doped (Zr, U)SiO_4, suggesting that zircon can accommodate U concentrations between 6.1 and 12.9 wt% (Burakov & Pazukhin, 1993).

Research on the trivalent actinide incorporation into $ZrSiO_4$ often employs Nd^{3+} and Eu^{3+} as surrogate ions. Under conditions of 1500°C with a 72-hour holding time, Nd^{3+}-substituted zircon waste forms can be successfully synthesized, achieving a solid solution limit of approximately 1 mol% at the Zr site. Similarly, Eu^{3+} can be incorporated into the zircon structure, yielding a single-phase ceramic after annealing at 1500°C for 22 hours, with a maximum solubility of up to 10 wt%.

Zircon also enables co-doping with multiple elements. For instance, a combined solid solution of Nd and Ce has been achieved, resulting in a single-phase $ZrSiO_4$ solid solution after 72 hours at 1550°C with total dopant concentrations up to 4 mol%. Beyond this threshold, a secondary (Nd, Ce)$_2Si_2O_7$ phase forms. Furthermore, studies utilizing sol–gel methods combined with hot pressing have successfully incorporated both Nd and U into the zircon matrix, achieving a high total dopant solubility of up to 20 wt% (Ding et al., 2015).

10.1.2 Sintering Preparation of Zircon Waste Forms

The synthesis of the target zircon phase via solid-state reaction typically requires high temperatures (1400–1500°C) and extended dwell times due to the involvement of SiO_2 as a raw material. For instance, the sol–gel method combined with high-temperature solid-state reaction, as mentioned earlier, necessitates processing at 1500°C for 3 hours to achieve complete phase formation (Burakov et al., 2000). In efforts to reduce sintering temperatures, external field-assisted sintering has shown promise. Notably, in a study on co-immobilization of Nd and U, the combination of sol–gel processing with hot pressing enabled successful reaction sintering at a reduced temperature of 1200°C under a pressure of 20 MPa (Ding et al., 2015).

10.2 PEROVSKITE

The chemical formula of perovskite is generally denoted as ABO_3, which, in the field of ceramic waste forms, commonly refers to $CaTiO_3$. Its structure belongs to the simple cubic lattice with the space group Pm3m. As shown in Figure 10.2, the crystal structure of $CaTiO_3$ consists of a three-dimensional network of corner-sharing TiO_6 octahedra. Calcium ions occupy the body-center positions, where each is coordinated by 12 oxygen ions, forming a cuboctahedron, while titanium ions reside at the cube corners, each surrounded by six oxygen ions in an octahedral coordination.

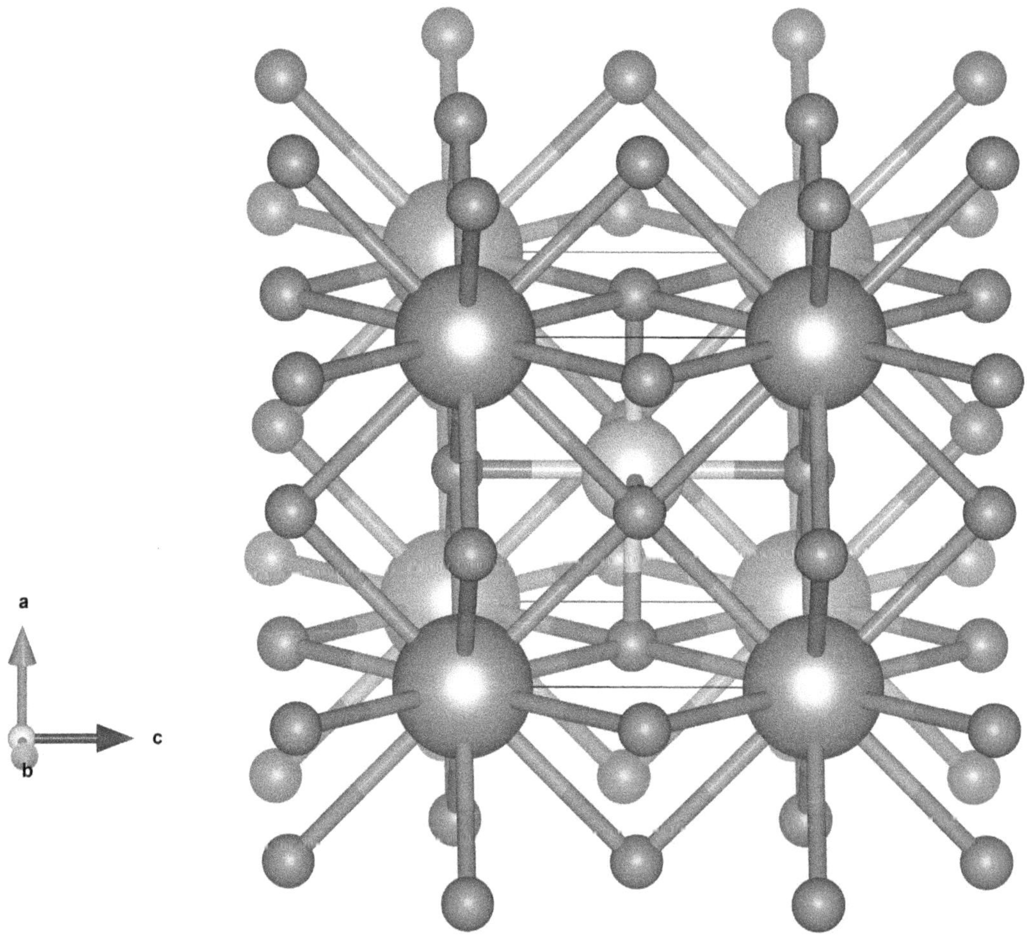

FIGURE 10.2 The Crystal Structure of $CaTiO_3$.

Pu, along with other actinides and rare earth elements, can incorporate into the Ca site of $CaTiO_3$. The stability and potential structural evolution of the doped perovskite are primarily governed by the tolerance factor, $t = (R_{Ca} + R_O)/2^{1/2}(R_{Ti} + R_O)$, and the octahedral factor. These parameters underpin the phase's ability to self-adjust and accommodate structural distortions induced by dissolved radionuclides.

10.2.1 Immobilization of Radioactive Waste

Perovskite-type ceramics have been demonstrated to incorporate tetravalent actinides. Ce^{4+}, as a surrogate for tetravalent actinides, can be dissolved into the $CaTiO_3$ structure with a solid solubility of up to 19.84 wt%. Using $Ca_{0.9}Pu_{0.1}Ti_{1-x}Al_xO_3$ ($x = 0.1$ and 0.2) and $Ca_{0.9}Pu_{0.1}TiO_3$ as target phases, it was found that Pu can be incorporated into the perovskite lattice after hot pressing and annealing. The Al-substituted composition exhibited less sensitivity to the annealing atmosphere, yielding a final product consisting primarily of perovskite with minor PuO_2 (< 2 wt%). In contrast, the unmodified $Ca_{0.9}Pu_{0.1}TiO_3$ contained 6–10% of a pyrochlore secondary phase alongside perovskite. The annealing atmosphere also influences the valence state of incorporated Pu: under oxidizing conditions, Pu is stabilized in the +4 state, while under reducing conditions, it is reduced to +3. The maximum solid solubility of PuO_2 in the perovskite matrix was reported to reach 18.14 wt% (Proust et al., 2019).

Perovskite-type ceramics have also been utilized for the immobilization of ^{90}Sr (Blackburn et al., 2020). A waste form rich in perovskite (approximately 85 wt% perovskite, 10 wt% zirconolite, and 5 wt% rutile) has been synthesized for this purpose. Furthermore, perovskite ceramics demonstrate effective retention of ^{90}Zr, the decay product of ^{90}Sr. When the solid solubility of Zr^{4+} in the perovskite is below 22.23 wt%, a pure $SrTiO_3$ phase is obtained, with a Zr^{4+} leaching rate as low as 1.3×10^{-4} g·m^{-2}·d^{-1} (Maddrell, 2013).

Perovskite ceramics also enable the co-incorporation of Sr with other radionuclides. For instance, Sr and Gd have been successfully co-doped into the perovskite structure, with a total solid solubility reaching 20 mol% (Jung et al., 2016). Zou et al. used ^{88}Sr as a surrogate for ^{90}Sr and Ce^{4+} as a simulant for tetravalent actinides (Clark et al., 2017). By directly reacting $SrCO_3$ and CeO_2, they synthesized a perovskite-type $SrCeO_3$ waste form, achieving high incorporation capacities: 46.17 wt% for $SrCO_3$ and 53.83 wt% for CeO_2.

Studies on the solubility of Cs in perovskite ceramics have also been conducted. Zhao et al. prepared $Ba_{1-x}(La, Cs)_xZrO_3$ ($x = 0, 0.1, 0.2, 0.3, 0.4$) precursor powders via a sol–gel method (Zhao et al., 2020). After sintering at 900°C for 2 hours, Cs was successfully incorporated into the ceramic lattice, with a maximum solid solubility of 9.63 wt% and a leaching rate of 0.957×10^{-3}g·m^{-2}·d^{-1}.

10.2.2 Sintering Preparation of Perovskite Waste Forms

Perovskite ceramics are primarily synthesized via solid-state reaction. In the aforementioned studies on Pu immobilization using perovskite, sintering was typically conducted at 1500°C in air. Alternatively, hot pressing has also been employed under conditions of 1200°C and 20 MPa to fabricate perovskite-based waste forms. Owing to their applications in dielectric ceramics, perovskite materials are also processed using various advanced sintering techniques, including spark plasma sintering (SPS), cold sintering, and flash sintering.

10.3 PYROCHLORE

The pyrochlore mineral phase has the general chemical formula $A_{2-m}B_2O_6(O, OH, F)_{1-n}$, which can be simplified as $A_2B_2O_7$. The A-site accommodates larger cations (e.g., Ca, U, Th, and lanthanides), while the B-site hosts smaller, higher-valent cations (e.g., Nb, Ta, Ti, Zr). Naturally occurring pyrochlore minerals have demonstrated the capacity to incorporate actinides for up to 1.4 billion years.

As illustrated in Figure 10.3, the ideal crystal structure of pyrochlore is face-centered cubic (space group: Fd 3m, $Z = 8$). Under external influences such as irradiation, pressure, or specific sintering conditions, pyrochlore may undergo a structural transition to a defective fluorite-type structure. Currently, the primary radionuclides incorporated into pyrochlore-based waste forms include U, ^{238}Pu, Th, ^{244}Cm, among others.

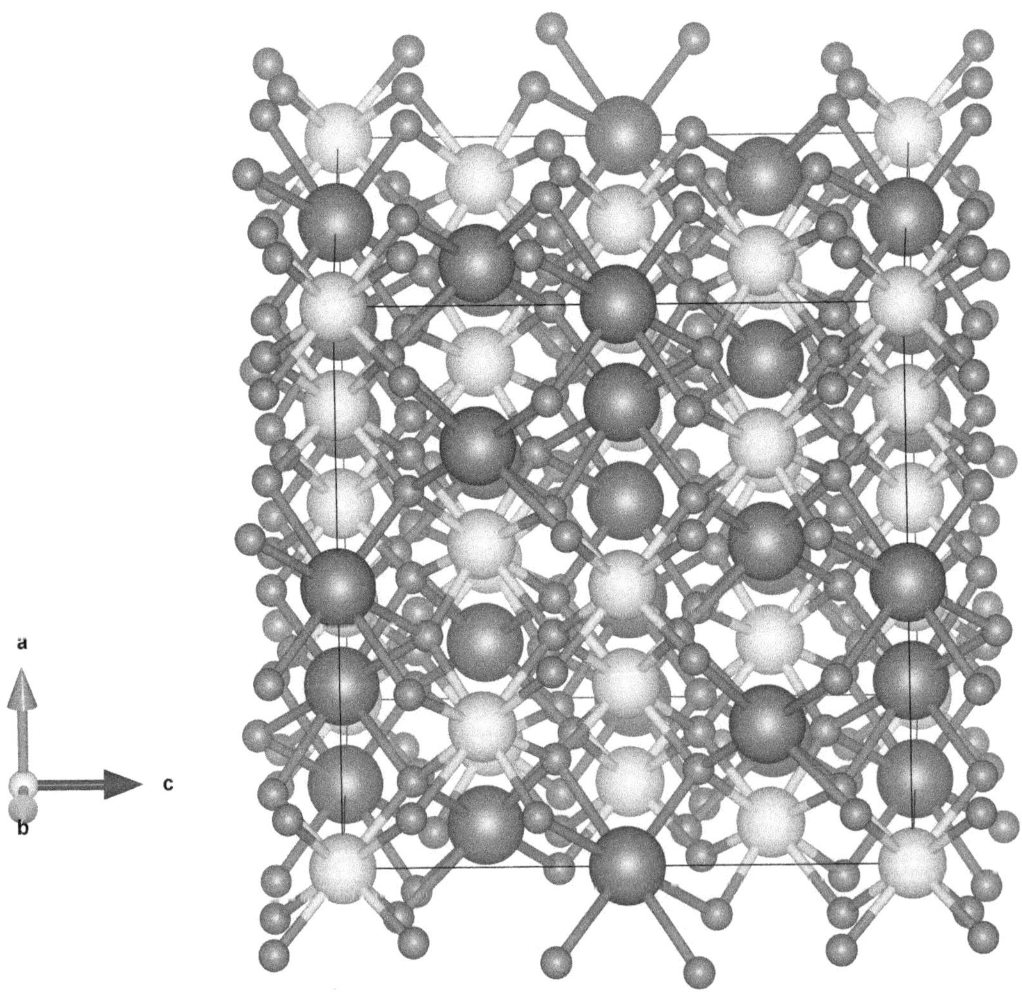

FIGURE 10.3 The Crystal Structure of Pyrochlore.

10.3.1 Immobilization of Radioactive Waste

During the early development of pyrochlore-based waste forms, researchers explored its application for immobilizing tetravalent actinides. Initial attempts to incorporate uranium did not yield a pure pyrochlore phase. Dickson et al. prepared $CaUTi_2O_7$ from stoichiometric mixtures in 1989, but the final product contained impurities such as UO_2 and perovskite (Dickson et al., 1989). Helean et al. synthesized U-bearing pyrochlore with a molar ratio of 2.1Ca: 1.0U: 3.7Ti, achieving a product consisting of 86.31% pyrochlore (Helean et al., 2002). It was not until 2010 that a phase-pure uranium-bearing pyrochlore, with the composition $Ca_{1.25}U_{0.75}Ti_2O_7$, was successfully synthesized (James et al., 2010). This compound has a theoretical UO_2 loading capacity of 46.4 wt%, with uranium present in the +4 oxidation state.

Pyrochlore has also been investigated for thorium immobilization. Studies show that nearly phase-pure $Nd_2Zr_2O_7$ ceramics can be synthesized under a reducing atmosphere at 1650°C with a 30-hour holding time (Nästren et al., 2009). In contrast, studies on Pu incorporation into pyrochlore are relatively limited. $Nd_2Zr_2O_7$-based matrices have been shown to accommodate 5 mol% and 10 mol% of Pu, which exists in the tetravalent state. The substitution of Pu leads to a reduction in symmetry at the A-site and results in relatively low loading capacities (Finkeldei et al., 2020).

Pyrochlore-type ceramic waste forms are also capable of incorporating trivalent actinide nuclides. Studies have shown that Nd^{3+}, as a surrogate for Am^{3+}, can be dissolved into the $Gd_2Zr_2O_7$ ceramic structure. Using a high-temperature solid-state reaction method with three sequential sintering steps at 1473 K (36 h), 1573 K (36 h), and 1673 K (48 h), a continuous $Gd_{2-x}Nd_xZr_2O_7$ ($0 \leq x \leq 2.0$) pyrochlore solid solution was obtained. The formation of a pure phase across the entire composition range is attributed to the suitable ratio of A- and B-site ionic radii in $Gd_{2-x}Nd_xZr_2O_7$. The corresponding solid solubility of the simulated nuclide reached 49.48 wt%, and no order-disorder phase transition was observed in this series of waste forms (Mandal & Tyagi, 2007).

Furthermore, Nd^{3+} (as a surrogate for Pu^{3+} and Am^{3+}) and La^{3+} (as a surrogate for its decay products U^{3+} and Np^{3+}) can be incorporated into zirconate-based pyrochlores with the general formula $(La_xNd_{1-x})_2Zr_2O_7$. All compositions in this series exhibited a single-phase pyrochlore structure. Under high-dose γ-irradiation (total dose: 233.78 kGy), the pyrochlore structure remained largely intact. However, with increasing concentration of decay products, the An-O_{48f} bond tends to undergo rearrangement, leading to disordering of the pyrochlore structure and a consequent reduction in its radiation tolerance (Lawson et al., 2019).

In addition to actinides, pyrochlore has also demonstrated excellent chemical compatibility with other radionuclides. Begg et al. successfully incorporated 30% ^{93}Zr into a $Gd_2Zr_2O_7$ ceramic waste form by sintering at 1600°C for 50 hours (Begg et al., 2001). The presence of Gd, which has a high neutron absorption cross-section, further enhances the waste form's performance.

Studies have also shown that $Gd_2Zr_2O_7$-based ceramics can accommodate up to 11 simulated radionuclides (including La, Pr, Nd, Sm, Eu, Gd, and Ce) while maintaining high sample density. These materials were synthesized via SPS at 1600°C under vacuum with an applied pressure of 30 MPa, achieving full densification in only 3 minutes. When the total doping content of mixed waste simulants was below 40%, a purer pyrochlore phase was obtained (Shu et al., 2019).

10.3.2 Sintering Preparation of Pyrochlore Waste Forms

Pyrochlore ceramic waste forms are predominantly synthesized via conventional solid-state reaction sintering, which involves mixing raw oxide powders, pelletizing, and subsequent high-temperature sintering. This method offers advantages such as scalability and procedural simplicity. However, pyrochlore typically requires high sintering temperatures (generally above 1500°C) to achieve full densification, which limits its applicability for immobilizing radionuclides with high vapor pressure. Additionally, secondary radioactive contamination generated during processing requires careful management.

External field-assisted sintering techniques can effectively reduce sintering temperatures and mitigate the issues mentioned earlier. Xu et al. successfully synthesized single-phase nano-crystalline $Gd_2Zr_2O_7$ ceramics within 1 minute using flash sintering at 1050°C under an electric field of 100 V/cm (Xu et al., 2020). SPS, which couples electric and mechanical fields, significantly enhances densification at reduced temperatures. Meanwhile, microwave sintering enables the densification of $Gd_2Zr_2O_7$ and $Gd_{2-x}Nd_xZr_{2-5x}Ce_{5x}O_7$ pyrochlore systems at temperatures between 1100°C and 1300°C (Zhou et al., 2019). He et al. also employed self-propagating high-temperature synthesis combined with quick pressing (SHS/QP) to fabricate $Gd_{0.59}Nd_{1.39}Ti_{1.33}Zr_{0.69}O_7$ under a pressure of 50 MPa (held for 60 seconds), completing the sintering process within 5 minutes (He et al., 2018). The results indicated that the solid solubility of Nd_2O_3 reached up to 70 mol%.

10.4 ZIRCONOLITE

The zirconolite mineral phase has the general chemical formula $CaZrTi_2O_7$ and adopts a defect fluorite-related structure with multiple polytypes, including monoclinic (2M or 4M) and orthorhombic (3O) forms. Among these, the 2M polytype of $CaZrTi_2O_7$ is the most commonly studied one due to its excellent chemical durability and high stability. It crystallizes in the space group C2/c, with $Z = 8$, and its composition can be expressed as $CaZr_xTi_{3-x}O_7$ ($0.85 \le x \le 1.30$). The crystal structure is illustrated in Figure 10.4. Both the Ca and Zr sites in zirconolite can accommodate a wide range of radionuclides, including actinides and rare earth elements, making it a promising host phase for radioactive waste immobilization.

10.4.1 Immobilization of Radioactive Waste

Extensive research has been conducted on the incorporation of tetravalent actinides into zirconolite-based ceramic waste forms. Studies have shown that $CaZrTi_2O_7$ can accommodate U^{4+} through substitution at the Zr-site, forming solid solutions with the composition $CaU_xZr_{1-x}Ti_2O_7$ ($0 \le x \le 1.0$). With increasing uranium content, the crystal structure of zirconolite transitions from the 2M polytype to the 4M polytype, and eventually to

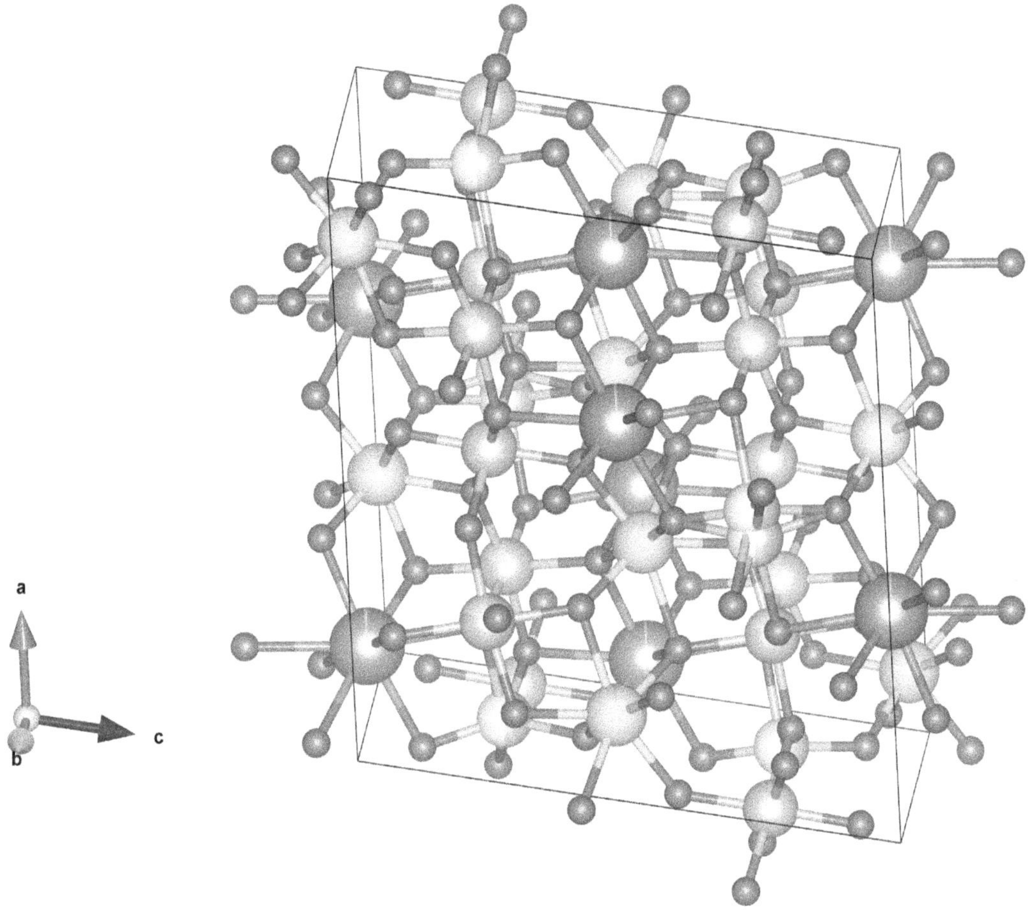

FIGURE 10.4 The Crystal Structure of Zirconolite-2M $CaZrTi_2O_7$.

a pyrochlore-dominated phase. The maximum solubility of U^{4+} in the stable zirconolite phase was found to be 33.43% (Vance et al., 2002).

Zhang et al. synthesized two types of zirconolite-rich waste forms: one co-doped with U^{4+}/Th^{4+} and another solely doped with Pu^{4+} (Zhang et al., 2009). In both cases, the dopants were effectively incorporated into the structure. The incorporation levels reached 4.51 wt% for UO_2 and 4.41 wt% for ThO_2. The normalized leaching rates of Gd, Th, and U were all below $10^{-3}g{\cdot}m^{-2}{\cdot}d^{-1}$ after 28 days. For the Pu-doped sample, the solid solubility of PuO_2 reached 13 wt%, with leaching rates of Pu and Hf below $10^{-3}g{\cdot}m^{-2}{\cdot}d^{-1}$ after 35 days.

As a surrogate for tetravalent actinides such as Pu^{4+} and Np^{4+}, Ce^{4+} has been successfully incorporated into the zirconolite structure. Cerium can occupy either the Ca-site, forming $Ca_{1-2x}Ce_xZrTi_2O_7$, or the Zr-site, yielding $CaZr_{1-x}Ce_xTi_2O_7$ solid solutions, with maximum CeO_2 incorporation levels of 8.87 wt% and 10.84 wt%, respectively (Caurant et al., 2006). Furthermore, the sintering atmosphere significantly influences both the yield of Pu-bearing zirconolite and the valence state of plutonium.

For the simulation of trivalent actinides, zirconolite-based ceramics have been shown to incorporate Nd^{3+} effectively. Jafar et al. synthesized a series of $Ca_{1-x}Zr_{1-}x(Sm/Nd)_{2x}Ti_2O_7$ ($0 \leq x \leq 1.0$) solid solutions via a high-temperature solid-state reaction (Jafar et al., 2014). In these systems, Nd^{3+} and Sm^{3+} primarily occupy Ca-sites, with a minor fraction substituting at Zr-sites. As the concentration of trivalent rare-earth elements increases, the phase evolution progresses from 2M-$CaZrTi_2O_7$ to 4M–4M-$CaZrTi_2O_7$, and finally to $(Sm/Nd)_2Ti_2O_7$ with a pyrochlore structure, accompanied by a gradual reduction and eventual disappearance of the $CaTiO_3$ secondary phase. The maximum solubilities of Nd and Sm within the stable zirconolite phase were determined to be 49.48% and 59.15%, respectively. The presence of a pyrochlore phase in such $CaZrTi_2O_{7\_}A_2Ti_2O_7$ systems enables the accommodation of a wider range of heterovalent cations.

Further studies indicate that lanthanide elements (Ln = La, Nd, Gd, Ho, Yb) can be incorporated into the zirconolite structure via compositions such as $Ca_{1-x}Ln_xZrTi_{2-x}(Al, Fe)_xO_7$ (x = 0.1–0.9). With increasing dopant content, the structure transitions from 2M-$CaZrTi_2O_7$ to 3O-$CaZrTi_2O_7$, allowing for the accommodation of more Ln^{3+} ions and achieving a solid solubility of up to 14.52 mol% Ln_2O_3 (Ji et al., 2020). Additionally, zirconolite has also been investigated for the immobilization of Np and Tc.

10.4.2 Sintering Preparation of Zirconolite Waste Forms

Zirconolite achieves sintering densification at temperatures between 1300°C and 1400°C, which facilitates the co-immobilization of other radionuclides with low vapor pressures at elevated temperatures. The predominant synthesis method remains conventional high-temperature solid-state reaction.

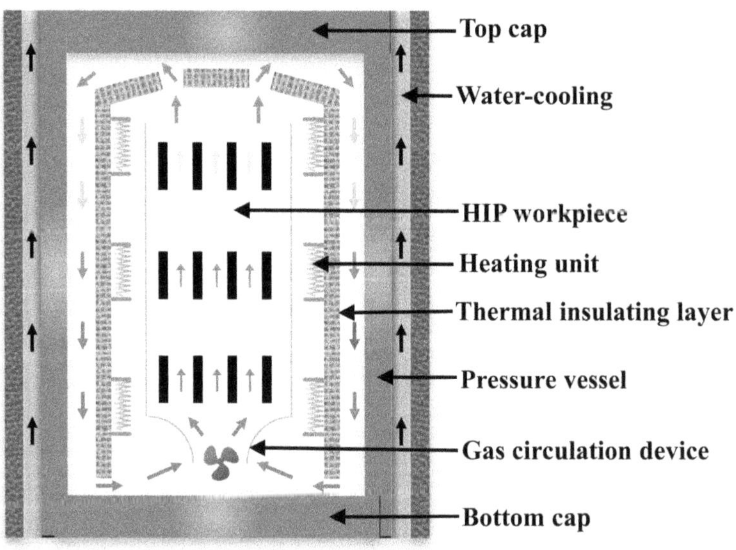

FIGURE 10.5 The Schematic of Hot Isostatic Pressing Technique for the Fabrication of Waste Forms.

Isolated reports on hot pressing emerged in the 1980s. Hot isostatic pressing (HIP) has gained attention for producing waste forms with reduced volume and enhanced integrability within stainless steel canisters, thereby improving handleability and safety during transportation and storage. Major nuclear research institutions in the United States and the United Kingdom have identified HIP as a promising candidate technology for plutonium immobilization. Additionally, reactive SPS has been successfully applied to fabricate $CaZrTi_2O_7$ ceramics. Studies demonstrate that dense $CaZrTi_2O_7$ waste forms can be rapidly synthesized via reactive SPS at 1200°C within just 1 hour, with controlled grain sizes of (1.9 ± 0.6) μm (Sun et al., 2018). A schematic diagram of the HIP process for fabricating nuclear waste forms is shown in Figure 10.5.

10.5 ZIRCONOLITE-BASED GLASS–CERAMICS

Glass–ceramic waste forms are composite materials consisting of both crystalline phases, obtained through controlled annealing and crystallization of molten glass. Radionuclides are either immobilized via isomorphous substitution within the crystalline phases or dispersed within the glass network structure. These waste forms exhibit superior mechanical strength, thermal stability, and chemical durability compared to monolithic glass waste forms. Glass–ceramic immobilization represents a significant advancement in the treatment of high-level radioactive waste (HLW) and is recognized as one of the most promising matrices for the disposal of actinide-rich HLW (Deokattey et al., 2013; Crum et al., 2014; Taurines & Boizot, 2012).

In recent years, researchers in nuclear materials science have conducted extensive studies on glass–ceramics containing stable crystalline phases—such as zirconolite ($CaZrTi_2O_7$), sphene ($CaTiSiO_5$), and powellite ($CaMoO_4$)—focusing on phase composition, microstructure, synthesis methods, and long-term stability, with considerable progress being achieved. Among these, zirconolite is particularly notable for its exceptional stability and high solid solubility for actinides (and lanthanides) at both Ca and Zr sites (Vance et al., 1994). As a result, zirconolite-based glass–ceramics have become one of the most extensively studied matrices for nuclear waste immobilization.

10.5.1 Immobilization of Radioactive Waste

Loiseau et al. employed analytical techniques such as energy dispersive X-ray spectroscopy (EDX) and electron paramagnetic resonance (EPR) to investigate the distribution of simulated actinide nuclides within the calcium aluminosilicate glass–ceramic system (Loiseau et al., 2004). The results indicated a relatively low incorporation of the simulated actinides into the zirconolite phase. Specifically, the molar fractions of neodymium (Nd) and thorium (Th) partitioned into zirconolite were only 23% and 19%, respectively. This low partitioning efficiency may be attributed to the relatively low content of the zirconolite crystalline phase in the glass–ceramic matrix, which accounted for only 9–11% by volume. By increasing the concentrations of ZrO_2 and TiO_2, the volume fraction of the zirconolite phase was elevated to 19%, and the molar fraction of Nd incorporated into zirconolite increased to 43% (Caurant et al., 2006).

Wu et al. successfully synthesized borosilicate glass–ceramics containing zirconolite, as well as co-existing zirconolite and sphene phases, by incorporating varying amounts of CaO, $ZrSiO_4$, and TiO_2 (denoted as CZST, with a molar ratio of 2:1:3) as nucleating agents (Martin et al., 2007). The parent glass was melted at 1150–1200°C, followed by heat treatment at 740–810°C for 2 hours. When the nucleating agent content was 30 wt%, a glass–ceramic containing solely the zirconolite phase was obtained. At 40 wt% nucleating agent, the crystalline phases consisted of both zirconolite and sphene, with the degree of crystallinity increasing with higher heat-treatment temperatures.

France and Australia conducted engineering-scale verification studies on the cold crucible vitrification of calcined HLW powders from the Idaho National Engineering and Environmental Laboratory (INEEL). The waste loading of the resulting waste forms reached 50 wt%. Using a cold crucible with a diameter of 50 cm, melting was carried out at 1260–1300°C, followed by cooling at a rate of 414°C/h when the central temperature of the crucible dropped below 900°C. Microstructural analysis revealed that the primary crystalline phase in the obtained glass–ceramics was zirconolite, accounting for 21 vol% of the solidified waste form (Crum et al., 2014). The cold crucible glass–ceramic process is considered to be the most promising technology to replace the current joule-heated melter vitrification method for HLW, demonstrating significant potential for future application (Taurines & Boizot, 2012).

10.5.2 Sintering Preparation of Zirconolite-Based Glass–Ceramics

Glass–ceramics are typically fabricated using methods such as the melt-quenching process, sintering, and sol–gel techniques.

The melt-quenching process involves melting the raw materials to form a glass, followed by shaping, annealing, and subsequent heat treatment for nucleation and crystallization. A key advantage of this method is its compatibility with various glass-forming techniques, producing materials with high density and uniform composition. However, it generally requires high melting temperatures.

The sintering method entails melting the glass, water-quenching it to form frits, followed by crushing, shaping, and sintering. Due to the increased surface area of the finely quenched particles, crystallization is generally more facile compared to the melt-quenching route. However, sintered materials often contain porosity, which compromises their density compared with melt-quenched samples. This reduced density can significantly impair the leaching resistance of radionuclides in the final waste form.

The sol–gel method is a low-temperature synthesis route wherein raw materials undergo hydrolysis to form a gel, which is then dried, shaped, and sintered. This technique offers advantages such as lower processing temperatures—preventing the volatilization of certain components—and the ability to produce highly pure and homogeneous materials. Nevertheless, it is associated with higher costs, longer production cycles, and more complex processing steps.

Currently, the melt-quenching process is the most widely used method for fabricating zirconolite-based glass–ceramics. A major focus and challenge in current domestic and international research lies in achieving highly crystalline and dense glass–ceramic waste forms with superior performance (Taurines & Boizot, 2012).

10.6 HIGH-ENTROPY CERAMICS

In recent years, the high-entropy concept has been introduced into the field of ceramics, attracting widespread attention. High-entropy ceramics (HECs) refer to inorganic compound solid solutions with one or more Wyckoff sites occupied by multiple principal elements in equimolar or near-equimolar ratios (Oses et al., 2020). HECs exhibit a range of remarkable properties, including excellent phase stability (Williford et al., 1999), favorable mechanical performance (Oses et al., 2020), and high corrosion resistance (Connelly et al., 2011). They also demonstrate outstanding radiation tolerance (Wu et al., 2023), showing broad potential for applications in the nuclear industry. Given the diverse elemental composition of radioactive waste, the extensive design space of HECs enables the incorporation of multiple elements into simple crystal structures, forming a unified solid phase and thereby simplifying waste treatment processes.

The sluggish diffusion effect and severe lattice distortion inherent in HECs contribute significantly to the enhanced leaching resistance and radiation stability of the waste forms. Among them, high-entropy pyrochlores are considered particularly promising for nuclear waste immobilization. However, research in this area remains limited. Yang et al. successfully incorporated uranium into high-entropy rare-earth titanate ceramics with a pyrochlore structure using SPS (Yang et al., 2021). Compared to single-component rare-earth titanate ceramics, the uranium-doped HECs exhibited lower Vickers hardness (8.8 GPa) and higher fracture toughness (crack length: 24.1 μm). The reduction in mechanical properties of the uranium-containing HECs—consistent with the rule of mixtures—can be attributed to the relatively poor mechanical performance of monolithic UO_2 (Vickers hardness ~6.8 GPa, crack length ~31.4 μm).

10.7 PROPERTIES OF CERAMIC-SOLIDIFIED BODIES

Chemical stability is a key performance indicator in evaluating whether a waste form is suitable for immobilizing HLW. The high radioactivity and long half-lives of actinide elements (such as Pu) necessitate extremely low release rates of radionuclides into the environment, that is, low leaching rates. Ceramic waste forms have been widely advocated for the immobilization of plutonium due to their superior performance, which is of great significance for safe long-term storage.

Taking various ceramic systems for immobilizing Pu as an example, differences in the chemical stability of ceramic waste forms can be discussed by comparing their Pu leaching rates. As shown in Table 10.1, pyrochlore-based ceramics incorporating Pu exhibit leaching rates in the range of 10^{-6}–10^{-4}g·m^{-2}·d^{-1}, demonstrating better chemical stability compared to conventional borosilicate glasses (Pu leaching rates: 10^{-4}–10^{-2}g·m^{-2}·d^{-1} (Bates et al., 1995)). Among them, after being subjected to self-irradiation from incorporated Pu, the leaching rate of a pyrochlore ceramic increased from 8.26×10^{-5} to 2.46×10^{-4} g·m^{-2}·d^{-1}, indicating that irradiation-induced amorphization leads to an increase in leaching rate.

Furthermore, representative compositions such as $Gd_2Zr_2O_7$ and $Gd_2Ti_2O_7$ generally show leaching rates between 10^{-6} and 10^{-4}g·m^{-2}·d^{-1}, depending on the temperature and pH conditions of the leaching tests (Lu et al., 2015). In the case of flash-sintered $Gd_2Zr_2O_7$, the samples exhibited leaching resistance comparable to those prepared by conventional

TABLE 10.1 Leaching Rates of Pu in Various Ceramic Waste Forms

Form	Chemical Formula	Experimental Parameters			Leach Rate/	Ref
		T/°C	Days	pH	($g·m^{-2}·d^{-1}$)	
Pyrochlore	$CaPuTi_2O_7$	70	150	5.6	5.50×10^{-6}	(Zhang et al., 2000)
	$CaGd_{0.21}Hf_{0.21}Pu_{0.21}U_{0.37}Ti_2O_7$	70	150	5.6	5.50×10^{-6}	(Zhang et al., 2000)
	(Ca, Gd, Hf, Pu, U)(Ti, Hf)O_7	85	10	2.0	8.26×10^{-5}	(Strachan et al., 2005)
	(Ca, Gd, Hf, Pu, U)(Ti, Hf) O_7-Irradiation	85	10	2.0	2.46×10^{-4}	(Strachan et al., 2005)
	$Ca_{0.78}Gd_{0.19}Pu_{0.19}U_{0.39}Ti_{2.02}Hf_{0.20}O_7$	85	-	2.0	3.67×10^{-4}	(Kholghy et al., 2010)
Perovskite	$Ca_{0.8}Gd_{0.05}Pu_{0.12}Hf_{0.90}Ti_{1.85}Al_{0.15}O_7$	85	-	2.0	5.50×10^{-6}	(Kholghy et al., 2010)
	^{238}Pu-bearing zirconolite	90	7	-	5.50×10^{-6}	(Stefanovsky et al., 2004)
	(Ca, Gd, Pu, U.Hf)(Ti, Al)O_7	90	50	7.0	5.50×10^{-6}	(Hart et al., 1999)
	(Ca, Gd, Pu, U.Hf)(Ti, Al)O_7	90	150	7.0	5.50×10^{-6}	(Hart et al., 1999)
Zirconolite	$CaZrO_3$-Ce	90	32	1.0	5.50×10^{-6}	(Kamizono et al., 1991)
	$CaZrO_3$-Ce	90	32	5.6	5.50×10^{-6}	(Kamizono et al., 1991)
Zircon	$ZrSiO_4$	175	14	1.0	5.50×10^{-6}	(Geisler et al., 2005)

sintering, with Gd and Zr leaching rates on the order of 10^{-5}~10^{-6}g·m^{-2}·d^{-1} in 28-day tests (Xu et al., 2020). Similarly, SHS-derived materials also showed good chemical stability, with Gd and Nd leaching rates of 3.28×10^{-5} and 2.27×10^{-5}g·m^{-2}·d^{-1}, respectively, over a 42-day leaching experiment (He et al., 2018).

Zirconolite exhibits chemical stability comparable to that of pyrochlore, with typical Pu leaching rates in the range of 10^{-6}–10^{-4}g·m^{-2}·d^{-1}. In the widely studied Synroc-C synthetic rock waste form, zirconolite is the most chemically stable and leach-resistant phase, followed by rutile, hollandite, and perovskite (Hart et al., 1995). Under neutral conditions (pH = 7), the leaching rate of zirconolite is approximately one order of magnitude lower than that of perovskite (Trocellier, 2001).

During geological disposal, radioactive nuclides in HLW forms continuously decay, causing self-irradiation damage and potential amorphization. Nevertheless, the leaching resistance of natural zirconolite has not been degraded by alpha self-irradiation damage, demonstrating its excellent radiation tolerance and strong potential for the deep geological disposal of HLW.

In contrast to pyrochlore and zirconolite ceramics, perovskite-based ceramics exhibit relatively lower chemical stability. The leaching rate of Ce (as a surrogate for Pu) can be as high as 10^{-4} to 10^{-2} g·m^{-2}·d^{-1}. Furthermore, in studies on the immobilization of ^{90}Sr, the leaching rate of Sr^{2+} reached 1.1×10^{-2} g·m^{-2}·d^{-1} after 28 days (Blackburn et al., 2020), indicating significantly higher release from perovskite-type systems compared to pyrochlore and zirconolite.

Despite this, perovskite ceramics demonstrate notable radiation tolerance. Smith et al. conducted irradiation experiments on zirconolite, pyrochlore, and perovskite using 1.5 MeV Kr^+ ions (Smith et al., 1997). The results showed that the critical amorphization dose for zirconolite and pyrochlore ranged from 3.6 to 6.1×10^{18} ions·m^{-2}, whereas that of perovskite was 1.8×10^{19} ions·m^{-2}—approximately three to five times higher than that of zirconolite.

Zircon ceramics show variability in leaching rates across different studies, though they are generally low. Typical values are around 10^{-7}g·m^{-2}·d^{-1} for Zr. For instance, in a U–Nd–$ZrSiO_4$ ceramic waste form, leaching rates were reported as ~10^{-7}g·m^{-2}·d^{-1} for Zr, ~10^{-5} for U, and 10^{-4}–10^{-5}g·m^{-2}·d^{-1} for Nd (Ding et al., 2015). However, in solutions containing Ca^{2+} ions, the Pu leaching rate increased significantly to 1.79×10^{-1}g·m^{-2}·d^{-1} (Geisler et al., 2005).

Zircon exhibits lower radiation resistance compared to zirconolite, pyrochlore, and perovskite. Additionally, its crystal structure limits the solid solubility of actinides, leading to declining interest in zircon as a host phase for Pu immobilization. The relatively poor amorphization resistance of zircon is attributed to the presence of SiO_2, which promotes the formation of glass-like phases under radiation.

10.8 CHALLENGES AND FUTURE PERSPECTIVES

This chapter reviews a variety of ceramic waste forms with potential for HLW immobilization, focusing on their crystal structures, types of incorporable radionuclides, waste loading capacities, and overall performance. Studies have shown that, compared to glass waste forms, ceramic waste forms exhibit higher incorporation capacity for individual radionuclides, greater densification, superior chemical durability, enhanced thermal stability, and improved radiation resistance—all of which are beneficial for the long-term safe disposal of HLW in deep geological repositories.

Pyrochlore-based ceramics can accommodate uranium and trivalent actinides while also incorporating neutron poisons, demonstrating well-rounded performance. Zirconolite ceramics exhibit excellent comprehensive properties and chemical compatibility, offering inherent advantages for plutonium immobilization and representing a key focus for future research. It is also noted that elemental doping in zirconolite often induces a phase transition from the 2M to the 4M polymorph. The 4M-$CaZrTi_2O_7$ phase can incorporate a higher concentration of actinides, suggesting a promising direction for enhancing waste loading through chemical tailoring. The polytypism of zirconolite further contributes to its high capacity for trivalent and tetravalent actinides.

Perovskite-based waste forms are often used to immobilize ^{90}Sr and display higher radiation tolerance compared to many other ceramics. However, the leaching rates of most actinides in perovskite are generally higher than those in pyrochlore, zirconolite, or zircon, indicating that improving the chemical durability of perovskite ceramics should be a priority in future studies.

Zircon, on the other hand, shows relatively poor amorphization resistance due to the presence of SiO_2, which facilitates the formation of amorphous phases under irradiation. Thus, further research is needed to mitigate the effects of radiation damage in zircon-based ceramics. In addition to the ceramic systems mentioned earlier, other candidate matrices include hollandite, fluorite-type oxides (e.g., ZrO_2), and monazite (e.g., (Ce, La)PO_4).

Ceramic waste forms exhibit selectivity in radionuclide incorporation, requiring matching ionic radii and charges for effective solid solution formation. This selectivity in immobilizing specific HLW components currently limits their broad engineering application.

Nevertheless, their superior chemical stability compared to conventional glass waste forms makes ceramics particularly suitable for immobilizing critical radionuclides, such as Pu, or those poorly accommodated in glasses, such as Cs. Achieving industrial-scale application of ceramic-based HLW immobilization will require extensive fundamental research to ensure reliable long-term safety assessments.

REFERENCES

Bailey, D.J., Lawson, S.M., Sun, S.K., Stennett, M.C., Lee, T.H., Ravel, B., Hyatt, N.C., 2020. A new approach to the immobilisation of technetium and transuranics: Co-disposal in a zirconolite ceramic matrix. *J. Nucl. Mater.* 528, 151885.

Bates, J.K., Ellison, A.J.G., Emery, J.W., Hoh, J.C., 1995. Glass as a waste form for the immobilization of plutonium. *MRS Online Proc. Libr.* 412, 57.

Begg, B.D., Hess, N.J., McCready, D.E., Thevuthasan, S., Weber, W.J., 2001. Heavy-ion irradiation effects in Gd2(Ti2-xZrx)O7 pyrochlores. *J. Nucl. Mater.* 289(1–2), 188–193.

Blackburn, L.R., Sun, S., Gardner, L.J., Maddrell, E.R., Stennett, M.C., Hyatt, N.C., 2020. A systematic investigation of the phase assemblage and microstructure of the zirconolite CaZr1-xCexTi2O7 system. *J. Nucl. Mater.* 535, 152137.

Burakov, B.E., Anderson, E.B., Zamoryanskay, M.V., Yagovkina, M.A., Strykanova, E.E., Nikolaeva, E.V., 2000. Synthesis and study of 239Pu-doped ceramics based on zircon, (Zr, Pu)SiO4, and hafnon, (Hf, Pu)SiO4. *MRS Online Proc. Libr.* 663, 307.

Burakov, B.E., Pazukhin, E.M., 1993. High-uranium zircon from "Chernobyl lavas". *Radiochim. Acta* 60(2–3), 149–152.

Caurant, D., Majerus, O., Loiseau, P., Bardez, I., Baffier, N., Dussossoy, J.L., 2006. Crystallization of neodymium-rich phases in silicate glasses developed for nuclear waste immobilization. *J. Nucl. Mater.* 354(1–3), 143–162.

Clark, B.M., Sundaram, S.K., Misture, S.T., 2017. Polymorphic transitions in cerium-substituted zirconolite (CaZrTi2O7). *Sci. Rep.* 7(1), 5920.

Connelly, A.J., Travis, K.P., Hand, R.J., Hyatt, N.C., Maddrell, E., 2011. Composition–structure relationships in simplified nuclear waste glasses: 1. Mixed alkali borosilicate glasses. *J. Am. Ceram. Soc.* 94(1), 151–159.

Crum, J., Maio, V., McCloy, J., Scott, C., Riley, B., Benefiel, B., Olszta, M., 2014. Cold crucible induction melter studies for making glass ceramic waste forms: A feasibility assessment. *J. Nucl. Mater.* 444(1–3), 481–492.

Deokattey, S., Bhanumurthy, K., Wattal, P.K., 2013. High level waste management in Asia: R&D perspectives. *Prog. Nucl. Energy* 62, 37–45.

Dickson, F.J., Hawkins, K.D., White, T.J., 1989. Calcium uranium titanate—A new pyrochlore. *J. Solid State Chem.* 82(1), 146–150.

Ding, Y., Lu, X., Tu, H., Shu, X., Dan, H., Zhang, S., Duan, T., 2015. Phase evolution and microstructure studies on Nd3+ and Ce4+ co-doped zircon ceramics. *J. Eur. Ceram. Soc.* 35(7), 2153–2161.

Finkeldei, S., Stennett, M.C., Kowalski, P.M., Ji, Y., de Visser-Týnová, E., Hyatt, N.C., Brandt, F., 2020. Insights into the fabrication and structure of plutonium pyrochlores. *J. Mater. Chem. A* 8(5), 2387–2403.

Geisler, T., Burakov, B., Yagovkina, M., Garbuzov, V., Zamoryanskaya, M., Zirlin, V., Nikolaeva, L., 2005. Structural recovery of self-irradiated natural and 238Pu-doped zircon in an acidic solution at 175°C. *J. Nucl. Mater.* 336(1), 22–30.

Hanchar, J.M., Burakov, B.E., Zamoryanskaya, M.V., Garbuzov, V.M., Kitsay, A.A., Zirlin, V.A., 2004. Investigation of Pu incorporation into zircon single crystal. *MRS Online Proc. Libr.* 824, CC4–2.

Hart, K.P., Vance, E.R., Day, R.A., Begg, B.D., Angel, P.J., Jostsons, A., 1995. Immobilization of separated Tc and Cs/Sr in synroc. *MRS Online Proc. Libr.* 412, 281–287.

Hart, K.P., Zhang, Y., Loi, E., Aly, Z., Stewart, M.W.A., Brownscombe, A., Bourcier, W., 1999. Aqueous durability of titanate ceramics designed to immobilize excess plutonium. *MRS Online Proc. Libr.* 608, 353.

He, Z., Zhang, K., Peng, L., Zhao, W., Xue, J., Zhang, H., 2018. Self-propagating plus quick pressing synthesis and characterizations of Gd2-xNdxTi1.3Zr0.7O7 ($0 \leq x \leq 1.4$) pyrochlores. *J. Nucl. Mater.* 504, 61–67.

Helean, K.B., Navrotsky, A., Vance, E.R., Carter, M.L., Ebbinghaus, B., Krikorian, O., Catalano, J.G., 2002. Enthalpies of formation of Ce-pyrochlore, Ca0.93Ce1.00Ti2.035O7.00, U-pyrochlore, Ca1.46U4+0.23U6+0.46Ti1.85O7 and Gd-pyrochlore, Gd2Ti2O7: Three materials relevant to the proposed waste form for excess weapons plutonium. *J. Nucl. Mater.* 303(2–3), 226–239.

Jafar, M., Sengupta, P., Achary, S.N., Tyagi, A.K., 2014. Phase evolution and microstructural studies in CaZrTi2O7–Nd2Ti2O7 system. *J. Am. Ceram. Soc.* 97(2), 609–616.

James, M., Carter, M.L., Zhang, Z., Zhang, Y., Wallwork, K.S., Avdeev, M., Vance, E.R., 2010. Crystal chemistry and structures of (Ca, U) titanate pyrochlores. *J. Am. Ceram. Soc.* 93(10), 3464–3473.

Ji, S., Su, M., Liao, C., Ma, S., Wang, Z., Shih, K., Li, Y., 2020. Synchrotron x-ray spectroscopy investigation of the Ca1-xLnxZrTi2-x(Al, Fe)xO7 zirconolite ceramics (Ln = La, Nd, Gd, Ho, Yb). *J. Am. Ceram. Soc.* 103(2), 1463–1475.

Jung, C.H., Kim, Y.K., Han, Y.M., Lee, S.J., 2016. Effects of fuel to synthesis of CaTiO3 by solution combustion synthesis for high-level nuclear waste ceramics. *J. Nanosci. Nanotechnol.* 16(2), 1676–1679.

Kamizono, H., Hayakawa, I., Muraoka, S., 1991. Durability of zirconium-containing ceramic waste forms in water. *J. Am. Ceram. Soc.* 74(4), 863–864.

Khokhlov, V.F., Yashkin, P.N., Silin, D.I., Djorova, E.S., Lawaczeck, R., 1996. Neutron capture therapy with Gd-DTPA in tumor-bearing rats. In: Hawthorne, M.F., Shelly, K.J., Wiersema, R.J. (Eds.), *Cancer Neutron Capture Therapy*. Springer, Boston, MA, pp. 865–869.

Kholghy, M., Kharatyan, S., Edris, H., 2010. SHS/PHIP of ceramic composites using ilmenite concentrate. *Journal of Alloys and Compounds* 502, 491–494.

Lawson, S.M., Hyatt, N.C., Whittle, K.R., Gandy, A.S., 2019. Resistance to amorphisation in Ca1-xLa2x/3TiO3 perovskites—a bulk ion-irradiation study. *Acta Mater.* 180, 180–188.

Loiseau, P., Caurant, D., Baffier, N., Mazerolles, L., Fillet, C., 2004. Glass–ceramic nuclear waste forms obtained from SiO2–Al2O3–CaO–ZrO2–TiO2 glasses containing lanthanides (Ce, Nd, Eu, Gd, Yb) and actinides (Th): Study of internal crystallization. *J. Nucl. Mater.* 335(1), 14–32.

Lu, X., Fan, L., Shu, X., Su, S., Ding, Y., Yi, F., 2015. Phase evolution and chemical durability of co-doped Gd2Zr2O7 ceramics for nuclear waste forms. *Ceram. Int.* 41(5), 6344–6349.

Maddrell, E., 2013. Hot isostatically pressed wasteforms for future nuclear fuel cycles. *Chem. Eng. Res. Des.* 91(4), 735–741.

Mandal, B.P., Tyagi, A.K., 2007. Preparation and high temperature-XRD studies on a pyrochlore series with the general composition Gd2-xNdxZr2O7. *J. Alloys Compd.* 437(1–2), 260–263.

Martin, C., Ribet, I., Frugier, P., Gin, S., 2007. Alteration kinetics of the glass-ceramic zirconolite and role of the alteration film—Comparison with the SON68 glass. *J. Nucl. Mater.* 366(1–2), 277–287.

Nästren, C., Jardin, R., Somers, J., Walter, M., Brendebach, B., 2009. Actinide incorporation in a zirconia based pyrochlore (Nd1.8An0.2)Zr2O7+x (An = Th, U, Np, Pu, Am). *J. Solid State Chem.* 182(1), 1–7.

Oses, C., Toher, C., Curtarolo, S., 2020. High-entropy ceramics. *Nat. Rev. Mater.* 5(4), 295–309.

Proust, V., Jeannin, R., White, F.D., Albrecht-Schmitt, T.E., 2019. *Tailored Perovskite Waste Forms for Plutonium Trapping*. Inorganic Chemistry 58(5), 3026–3032.

Shu, X., Qing, Q., Li, B., Yang, Y., Li, L., Zhang, H., Lu, X., 2019. Rapid immobilization of complex simulated radionuclides by as-prepared Gd2Zr2O7 ceramics without structural design. *J. Nucl. Mater.* 526, 151782.

Smith, K.L., Zaluzec, N.J., Lumpkin, G.R., 1997. In situ studies of ion irradiated zirconolite, pyrochlore and perovskite. *J. Nucl. Mater.* 250(1), 36–52.
Stefanovsky, S.V., Yudintsev, S.V., Gieré, R., Lumpkin, G.R., 2004. Nuclear waste forms. In: Gieré, R., Stille, P. (Eds.), *Energy, Waste and the Environment: A Geochemical Perspective.* Geological Society, London, Special Publications 236, 37–63.
Strachan, D.M., Scheele, R.D., Buck, E.C., Icenhower, J.P., Kozelisky, A.E., Sell, R.L., Buchmiller, W.C., 2005. Radiation damage effects in candidate titanates for Pu disposition: Pyrochlore. *J. Nucl. Mater.* 345(2–3), 109–135.
Sun, S.K., Stennett, M.C., Corkhill, C.L., Hyatt, N.C., 2018. Reactive spark plasma synthesis of CaZrTi2O7 zirconolite ceramics for plutonium disposition. *J. Nucl. Mater.* 500, 11–14.
Taurines, T., Boizot, B., 2012. Microstructure of powellite-rich glass-ceramics: A model system for high level waste immobilization. *J. Am. Ceram. Soc.* 95(3), 1105–1111.
Trocellier, P., 2001. Chemical durability of high level nuclear waste forms. *Ann. Chim. Sci. Matér.* 26(2), 113–130.
Vance, E.R., Ball, C.J., Day, R.A., Smith, K.L., Blackford, M.G., Begg, B.D., Angel, P.J., 1994. Actinide and rare earth incorporation into zirconolite. *J. Alloys Compd.* 213, 406–409.
Vance, E.R., Lumpkin, G.R., Carter, M.L., Cassidy, D.J., Ball, C.J., Day, R.A., Begg, B.D., 2002. Incorporation of uranium in zirconolite (CaZrTi2O7). *J. Am. Ceram. Soc.* 85(7), 1853–1859.
Weber, W.J., 1991. Self-radiation damage and recovery in Pu-doped zircon. *Radiat. Eff. Defects Solids* 115(4), 341–349.
Williford, R.E., Weber, W.J., Devanathan, R., Gale, J.D., 1999. Effects of cation disorder on oxygen vacancy migration in Gd2Ti2O7. *J. Electroceram.* 3(4), 409–424.
Wu, J., Zhang, M., Li, Z., Huang, M., Wang, H., Xue, L., Chen, H., 2023. High-entropy (Sm0.2Eu0.2Gd0.2Dy0.2Er0.2)2Hf2O7 ceramic with superb resistance to radiation-induced amorphization. *J. Mater. Sci. Technol.* 155, 1–9.
Xu, C., Wang, L., Bai, B., Peng, L., Cai, S., 2020. Rapid synthesis of Gd2Zr2O7 ceramics by flash sintering and its aqueous durability. *J. Eur. Ceram. Soc.* 40(4), 1620–1625.
Yang, K., Bryce, K., Zhu, W., Zhao, D., Lian, J., 2021. Multicomponent pyrochlore solid solutions with uranium incorporation—a new perspective of materials design for nuclear applications. *J. Eur. Ceram. Soc.* 41(4), 2870–2882.
Zhang, Y., Hart, K.P., Blackford, M.G., Thomas, B.S., Aly, Z., Lumpkin, G.R., Brownscombe, A., 2000. Durabilities of pyrochlore-rich titanate ceramics designed for immobilization of surplus plutonium. *MRS Online Proc. Libr.* 663, 325.
Zhang, Y., Stewart, M.W.A., Li, H., Carter, M.L., Vance, E.R., Moricca, S., 2009. Zirconolite-rich titanate ceramics for immobilisation of actinides—Waste form/HIP can interactions and chemical durability. *J. Nucl. Mater.* 395(1–3), 69–74.
Zhao, J., Li, J., Liu, H., Zhang, X., Zheng, K., Yu, H., Huo, J., 2020. Cesium immobilization in perovskite-type Ba1-x(La, Cs)xZrO3 ceramics by sol-gel method. *Ceram. Int.* 46(7), 9968–9971.
Zhou, M., Huang, Z., Ma, N., Qi, J., Zhang, H., Guo, X., Lu, T., 2019. Rapid preparation of dense Gd2Zr2O7 nano-grain ceramics by microwave sintering in air. *Ceram. Int.* 45(8), 10930–10935.

PART IV

Radionuclide Migration in the Disposal and Storage Sites

CHAPTER 11

Aquatic Chemistry and Transport of Key Radionuclides in Repository Environments

Ping Chen

11.1 CONCEPT OF REPOSITORY

Since the discovery of nuclear reactions in the 20th century, nuclear energy has found many applications in both civilian and military sectors, such as the production of nuclear weapons for military purposes (Knorr, 2015), nuclear power for energy generation (Yun et al., 2021), non-destructive testing in industry (IAEA, 2020), nuclear thermal propulsion in aerospace (Gabrielli & Herdrich, 2015), and use in medicine (Ziessman et al., 2013), among others. A major challenge for the peaceful use and sustainable development of nuclear energy is the management of huge amounts of radioactive waste, especially the high-level radioactive waste (HLW) and spent fuel (SF) from nuclear power plants (IAEA, 2022). Many unconventional concepts have been proposed, such as disposal in space (Burns et al., 1978; Coopersmith, 1999) or in the deep sea (Calmet, 1989), beneath ice sheets (Philberth, 1977), fusion within rock formations (Heuze, 1981; Logan, 1974) or conversion (transmutation) into stable nuclides (National Research Council, 1996). However, disposal in a deep geological repository is the only internationally accepted and practical option for the safe disposal of HLW and SF. For instance, the construction of a repository is in progress in Olkiluoto in Finland. The excavation of the first five final disposal tunnels has been completed (POSIVA, 2022). On August 30, 2024, test operations using non-radioactive dummy fuel elements began to demonstrate the functionality of the entire final disposal process (POSIVA, 2024). Additionally, the construction of an SF repository in Forsmark, Osthammar Municipality, began on January 25, 2025, in Sweden (SKB, 2025). In many other countries, field and laboratory investigations are ongoing to characterize potential host rocks, for example, at the Meuse/Haute-Marne underground

DOI: 10.1201/9781003665793-15

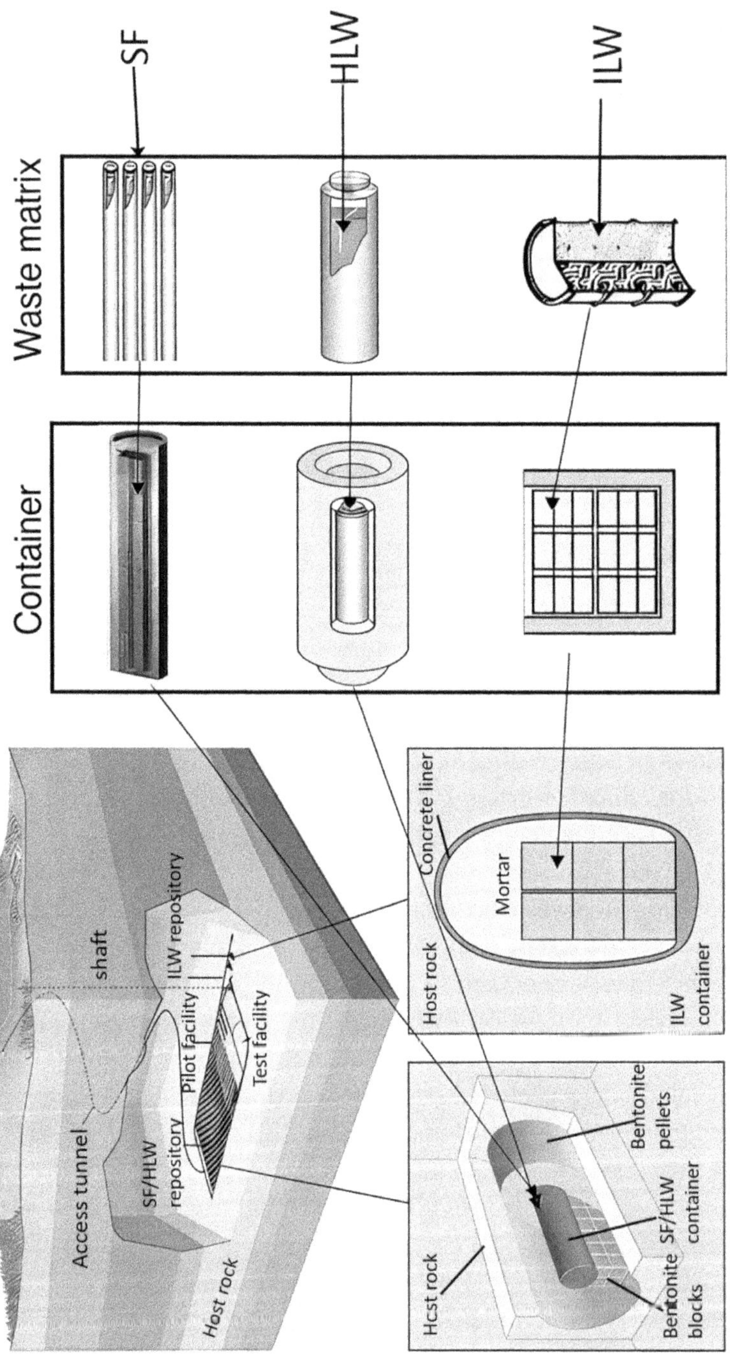

FIGURE 11.1 Conceptual Repository of Switzerland (Chen, 2023).

research laboratory in France (Plúa et al., 2021); the Nizhnekansky Rock Massif in Russia (Neuvazhaev et al., 2020; Rodionova et al., 2019; Rozov et al., 2018); and at many other sites in China (Wang et al., 2018), Switzerland (Nuklearsicherheitsinspektorat, 2018a, 2018b), and other countries (Faybishenko & Swift, 2016).

The sophisticated approach to waste disposal relies on a multi-barrier system, which consists of natural and engineered barriers. The multi-barrier conceptual design of a repository in Switzerland (Chen, 2023) is shown in Figure 11.1. The HLW solidified in glass, or the SF itself, forms the waste matrix, which constitutes the first barrier. The waste matrix is then encapsulated in steel canisters, forming the second barrier. These canisters are further emplaced in tunnels excavated in geological formations at depths of several hundred meters. Bentonite is used as a backfill material to fill the gaps between the canisters and the host rock, forming the third barrier. All the aforementioned barriers are termed "engineered barriers." The host rock (e.g., granite, salt, or claystone) serves as the final barrier, known as the "natural barrier." This multi-barrier system is expected to provide long-term safety for the repository. Radioactive materials are contained within a number of mutually supportive and, as far as possible, independent barriers, so that the failure of one does not jeopardize the overall isolation function. From a scientific perspective, the canisters are assumed to fail due to corrosion after long-term exposure to groundwater. From that point onward, long-lived radioactive nuclides will be released from the waste matrix and will migrate through the backfill material and host rock, and may eventually reach the biosphere.

For some repository concepts, the SF is designed to be directly disposed of without reprocessing, which makes plutonium (Pu) a key radionuclide of concern. Table 11.1 shows the potentially safety-relevant radionuclides in SF 30 years after discharge from a reactor. The major radiological impact of HLW after thousands of years stems from minor actinides (MAs; isotopes of Np, Am, and Cm) and some long-lived fission products (LLFPs; ^{79}Se, ^{93}Zr, ^{99}Tc, ^{107}Pd, ^{129}I, and ^{135}Cs). Therefore, a comprehensive investigation of the aquatic chemistry and migration/retention behavior of these nuclides in the backfill material and host rock is a critical aspect of the safety assessment of a repository. In the context of long-term safety assessment covering at least 10,000 years, radionuclides with short half-lives and/or small inventories may not have a significant effect on the total activity or dose released to the biosphere. This chapter will focus on the properties of four key radionuclides in the repository environment: neptunium (Np), plutonium (Pu), americium (Am), and technetium (Tc).

11.2 CHEMISTRY OF KEY RADIONUCLIDES

The mobility of Np, Pu, Am, and Tc is vital for the long-term safety assessment of a repository. The possible release of these nuclides from the emplaced waste packages is another concern that must be considered. In this scenario, these nuclides may undergo complex processes, including leaching from the waste matrix and interactions with the buffer and backfill materials, the surrounding host rock, and groundwater. The transport and retardation of these radionuclides are determined by geochemical processes, such as sorption, diffusion, precipitation/dissolution (solubility limits), and complexation with various components in the surrounding media, including inorganic and/or organic ligands present in the groundwater.

TABLE 11.1 The Reference Inventory of the Potentially Safety-Relevant Radionuclides in One Tonne of Uranium (tU) at 30 Years after Discharging from the Reactor (Data taken from Posiva Oy (2013)).

Radionuclide	Half-life [a]	Total inventory at 30 years cooling time [GBq/tU]	Partition of activity[%]			
			Fuel matrix	IRF of all components1)	Zirconium alloy	Other mental parts
108mAg	4.38E+02	2.50E+04		100.0%		
241Am	4.32E+02	1.93E+05	100.0%			
243Am	7.37E+03	3.42E+03	100.0%			
10Be	1.51E+06	1.26E−02	95.0%	5.0%		
14C	5.70E+03	1.61E+02	22.4%	5.5%	12.1%	60.0%
36Cl	3.01E+05	2.63E+00	73.6%	8.2%		
245Cm	8.42E+03	1.03E+02	100.0%			
246Cm	4.71E+03	3.57E+01	100.0%			
135Cs	2.30E+06	3.43E+01	95.0%	5.0%		
137Cs	3.01E+01	3.46E+06	95.0%	5.0%		
129I	1.57E+07	1.91E+00	95.0%	5.0%		
93Mo	4.00E+03	2.26E+01	1.4%	0.1%	0.2%	98.3%
91Nb	6.80E+02	2.86E−04	88.8%		11.2%	
92Nb	3.47E+07	2.35E−04	1.2%		98.8%	
93mNb	1.61E+01	5.08E+03	1.7%		98.3%	
94Nb	2.03E+04	7.52E+02			42.8%	57.2%
59Ni	7.60E+04	2.21E+02	0.3%		2.9%	96.8%
63Ni	1.01E+02	2.47E+04	0.3%		3.3%	96.4%
237Np	2.14E+06	2.37E+01	100.0%			
231Pa	3.28E+04	1.39E-03	100.0%			
107Pd	6.50E+06	9.72E+00	99%	1.0%		
238Pu	8.77E+01	2.64E+05	100.0%			
238Pu	2.41E+04	1.42E+04	100.0%			
240Pu	6.56E+03	3.12E+04	100.0%			
241Pu	1.43E+01	1.75E+06	100.0%			
242Pu	3.75E+05	2.17E+02	100.0%			
266Ra	1.60E+03					
79Se	3.27E+05	4.67E+00	99.6%	0.4%		
151Sm	9.00E+01	1.74E+04	100.0%			
126Sn	2.30E+05	3.92E+01	100%			
90Sr	2.88E+01	2.23E+06	99%	1.0%		
99Tc	2.11E+05	8.48E+02	99%	1.0%		
229Th	7.34E+03					
230Th	7.54E+04	1.32E−02	100.0%			
232Th	1.40E+10	2.06E−08	100.0%			
233U	1.59E+05	3.86E−03	100.0%			
234U	2.46E+05	5.53E+01	100.0%			
235U	7.04E+08	8.15E−01	100.0%			
236U	2.34E+07	1.46E+01	100.0%			
238U	4.47E+09	1.17E+01	100.0%			
93Zr	1.61E+06	1.27E+02	87.9%		12.1%	

1 IRF: instant release fraction. The percentages are calculated as a ratio of the IRF activity and the total activity.

These processes depend heavily on the prevailing geochemical conditions within the repository. In this section, the thermodynamic properties of these four key radionuclides are discussed, with a focus on their mobility and speciation in the repository environment.

11.2.1 Np

Neptunium is a key concern in environmental nuclear chemistry due to its long half-life (^{237}Np: 2.14 million years), high radiotoxicity, and complex redox behavior. As shown in Figure 11.2, Np speciation varies depending on the pH and Eh of the aqueous environment, as well as on the chemical composition of the solution. When the Eh is below 500 mV and the pH is higher than 3, Np tends to precipitate as NpO_2(cr). The Beishan groundwater environment is reported to have a pH of around 7 and an Eh of about 200 mV, which is beneficial for the long-term immobilization of Np. In oxidizing environments, Np mainly exists as Np(V), with NpO_2^+ being predominant. In carbonate-rich solutions, carbonate can shift the dominant species from NpO_2^+ to $NpO_2(CO_3)^-$ at pH values higher than 8. At higher pH and Eh, Np(VI) can become dominant, represented in Figure 11.2 as $NpO_2(OH)_4^{2-}$.

Figure 11.3 shows the solubility of Np as a function of pH under an assumed Eh of 500 mV. At pH values between 3 and 8, the solubility is very low, which is beneficial for

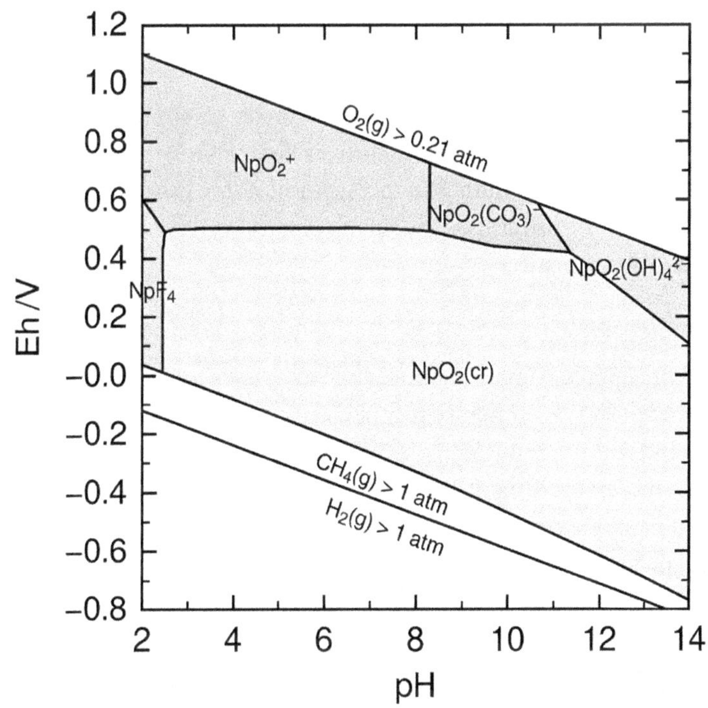

FIGURE 11.2 Np pH–Eh Diagram. This Diagram was Generated by PHREEQC using the ThermoChimie_v12a.dat Database. Np Concentration of 10^{-8} M and Groundwater Composition from Beishan Were Used in the Simulation. NpO_2(cr), NpO_2:$2H_2O$(am), NpO_2CO_3(cr), $NpO_2(OH)_2$:H_2O(cr), and NpO_2OH(am) Were Allowed to Precipitate When Their Saturation Indices Exceeded Zero.

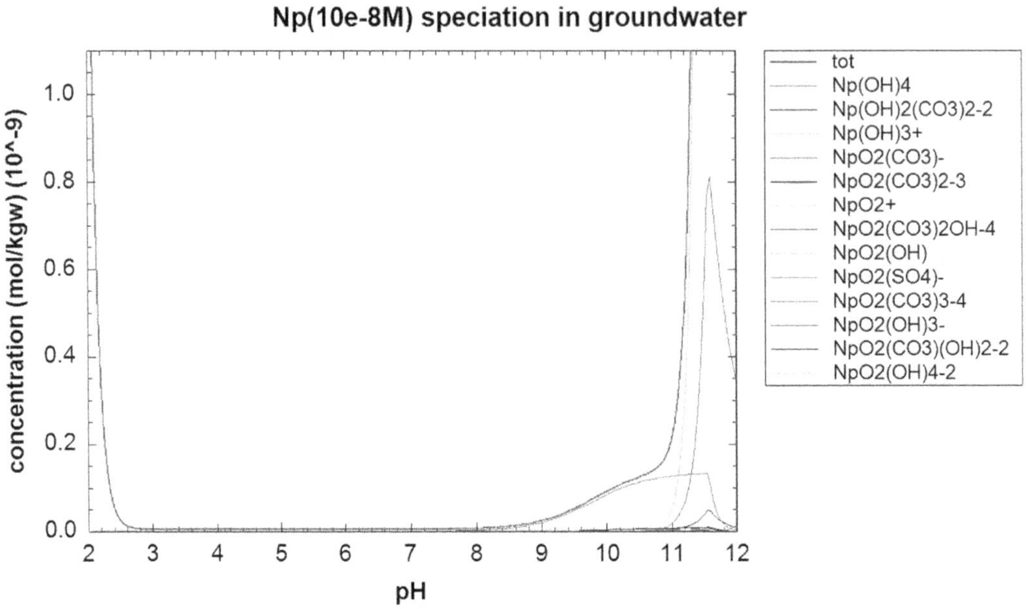

FIGURE 11.3 Np Speciation in Groundwater. Simulation was Performed by PHREEQC using the ThermoChimie_v12a.dat Database. Np Concentration of 10^{-8} M and Groundwater Composition from Beishan Were Used in the Simulation. The Redox Potential (Eh) Was Set at 500 mV.

the repository environment. The presence of carbonate in the solution can significantly increase Np solubility (see the pH range of 8–11 in Figure 11.3). In high-pH environments, such as those with a pH greater than 11—a condition not typically encountered in repositories—Np(VI) becomes dominant in the forms of $NpO_2(OH)_4^{2-}$ and $NpO_2(OH)_{3-}$, and its solubility increases dramatically.

11.2.2 Pu

In the repository concepts of some countries, such as France and China, HLW is reprocessed to recycle plutonium (Pu). In other repository concepts, for instance, those of the USA, Sweden, and Finland, SF from nuclear power plants is disposed of directly without reprocessing. Hence, an understanding of Pu's migration behavior in the subsurface is essential. Plutonium exhibits complex environmental behavior in subsurface systems, governed by redox dynamics, speciation, and interactions with mineral and organic phases. Figure 11.4 shows the Eh–pH Pourbaix diagram. It is clear that at redox potentials below 800 mV and pH values greater than 6, the predominant species is PuO_2(cr), a solid precipitate. This means that Pu in such environments would be immobilized through precipitation, which is beneficial for the long-term safety of a repository. At potentials between 800 and 900 mV, Pu exists in the Pu(V) oxidation state as PuO_2^+. At redox potentials above 900 mV, Pu(VI) becomes the dominant oxidation state. When the redox potential is below 700 mV and the pH is lower than 6, Pu can be reduced to Pu(IV) or Pu(III). The solution composition influences the dominant species under specific conditions, for instance, at high redox potentials, the dominant species may be $PuO_2(SO_4)$ in sulfate-rich solutions or $PuO_2(CO_3)$ in carbonate-rich solutions.

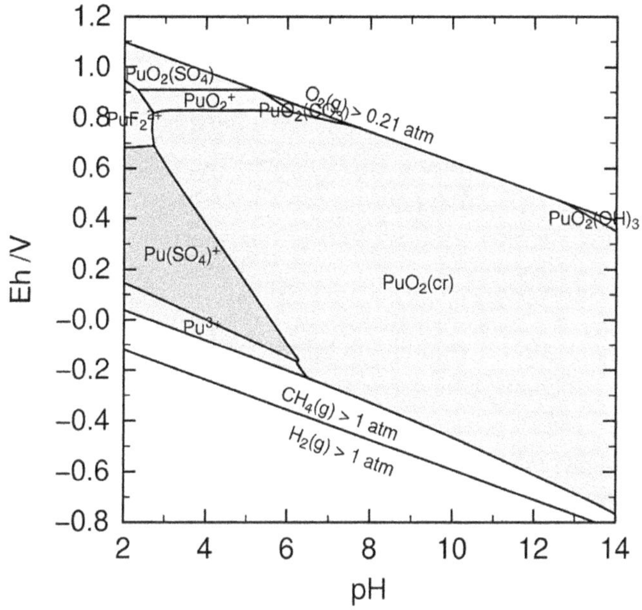

FIGURE 11.4 Pu pH–Eh Diagram. This Diagram Was Generated by PHREEQC using the ThermoChimie_v12a.dat Database. Pu Concentration of 10^{-8} M and Groundwater Composition from Beishan Were Used in the Simulation. PuO_2(cr), PuO_2:$2H_2O$(am), $PuO_2(OH)_2$:H_2O(am), and PuO_2OH(am) Were Allowed to Precipitate When Their Saturation Indices exceeded Zero.

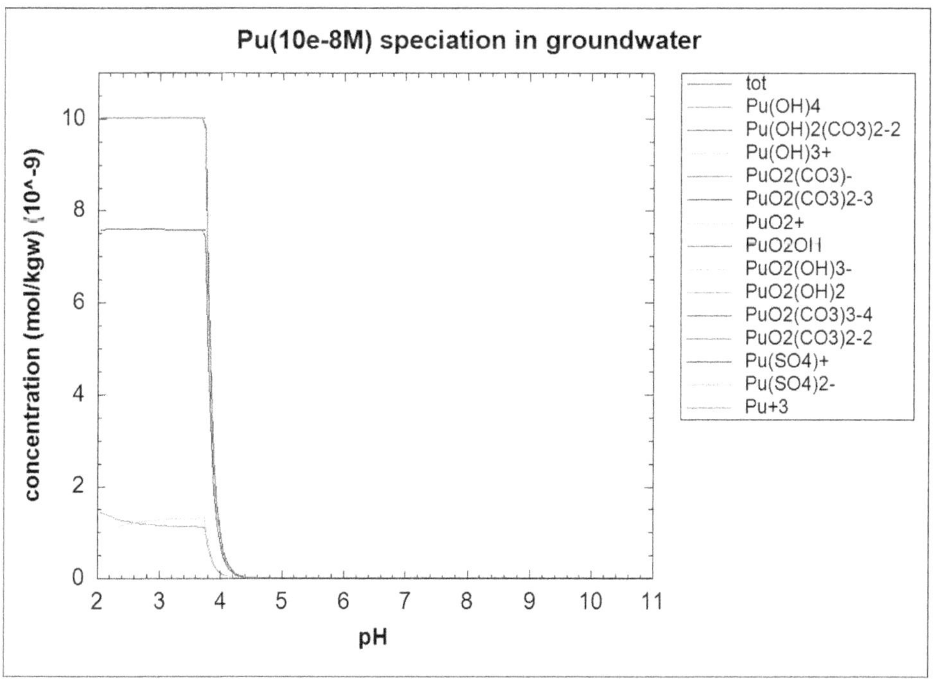

FIGURE 11.5 Pu Speciation in Groundwater. The Simulation Was Performed using PHREEQC with the ThermoChimie_v12a.dat Database. Pu Concentration of 10^{-8} M and Groundwater Composition from Beishan Were Used in the Simulation. The Redox Potential (Eh) Was Set at 250 mV.

The speciation of Pu in solution plays an important role in Pu solubility. As stated earlier, under low redox potentials and high pH conditions, Pu tends to precipitate as PuO_2(cr), leading to very low solubility, as shown in Figure 11.5. In contrast, under low redox potentials and pH values below 3, Pu can be reduced to Pu(III), and species such as $Pu(SO_4)^+$, $Pu(SO_4)_2^-$, and Pu^{3+} become dominant. This increases the solubility and is disadvantageous for the safety of a repository. Studies have reported that Pu can form intrinsic colloids itself or adsorb onto existing colloids, leading to faster transport in the subsurface environment.

11.2.3 Am

Americium (Am), a transuranic element, exhibits complex behavior in subsurface systems governed by redox conditions, geochemical interactions, and environmental parameters. As shown in Figure 11.6, Am(III) is predominant in most cases, only under high redox potentials and high pH conditions can the AmO_2 precipitate form. The behavior of Am(III) in subsurface environments is controlled by pH-driven speciation and carbonate complexation. At pH values below 6, Am^{3+} is the dominant species. At pH values between 6 and 11, carbonate complexation becomes a key factor, and $Am(CO_3)^-$ becomes the main species; furthermore, the $AmCO_3OH$ precipitate begins to form. Under more alkaline conditions,

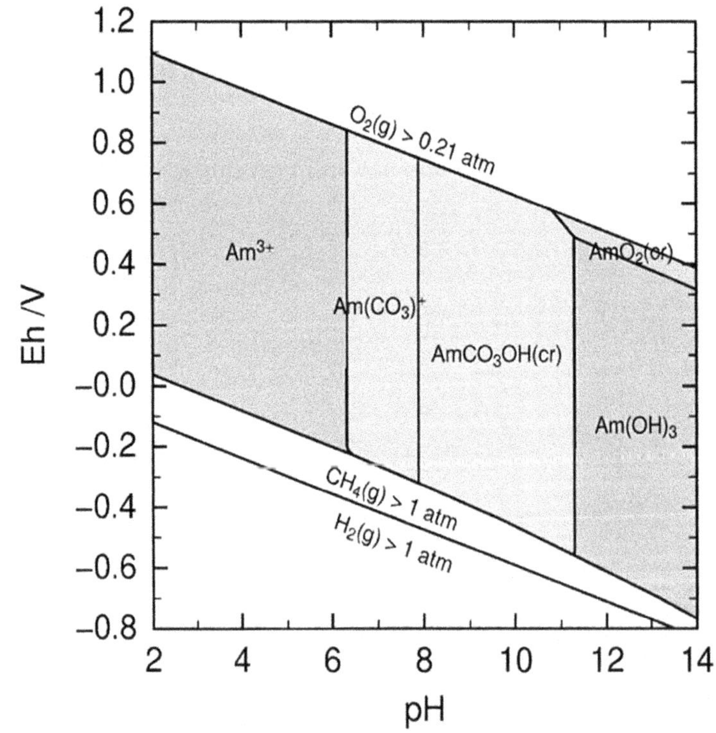

FIGURE 11.6 Am pH–Eh Diagram. This Diagram Was Generated by PHREEQC using the ThermoChimie_v12a.dat database. Am Concentration of 10^{-8} M and Groundwater Composition from Beishan Were Used in the Simulation. $Am(OH)_3$(am), $Am(OH)_3$(cr), AmO_2(cr), $AmCO_3OH$(cr), and $AmCO_3OH{:}0.5H_2O$(s) Were Allowed to Precipitate When Their Saturation Indices Exceeded Zero.

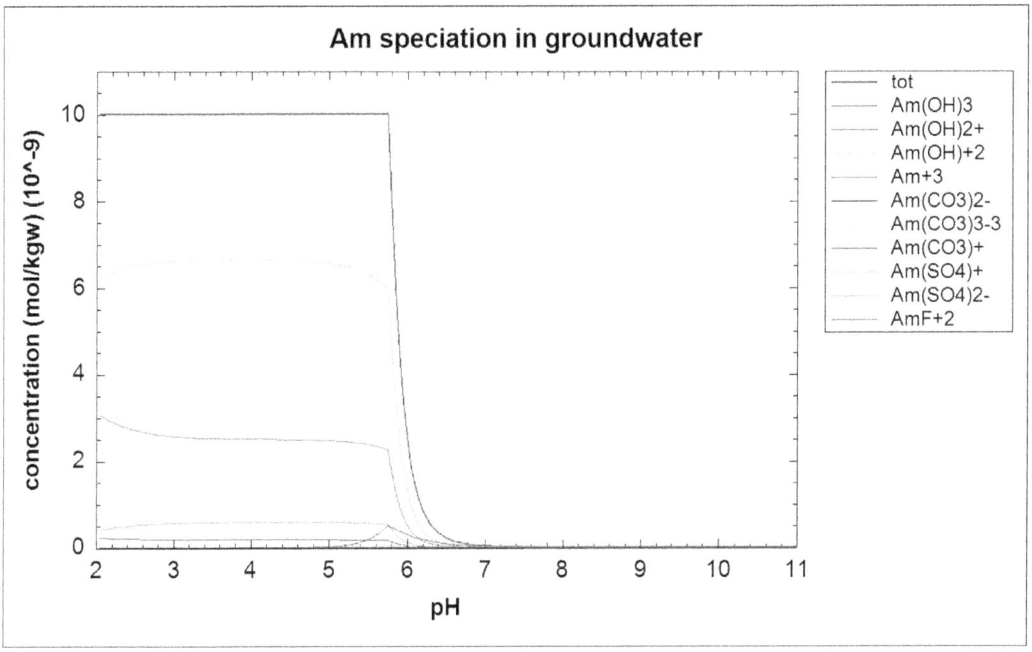

FIGURE 11.7 Am Speciation in Groundwater. The Simulation Was Performed using PHREEQC with the ThermoChimie_v12a.dat Database. Am Concentration of 10^{-8} M and Groundwater Composition from Beishan Were Used in the Simulation. The Redox Potential (Eh) Was Set at 250 mV.

the effect of carbonate complexation diminishes, and the dominant precipitate becomes $Am(OH)_3$. As shown in Figure 11.7, the solubility of Am is high at pH values below 6. The speciation is dependent on solution composition. In the Beishan groundwater environment, $Am(SO_4)^+$ and Am^{3+} are the dominant species. Due to the formation of Am precipitates, the solubility of Am becomes very low at pH values above 6. This is advantageous for Am immobilization in a repository. Although alkaline conditions and high temperatures in repositories like those at Beishan enhance immobilization, carbonate ions and organic ligands pose potential migration risks.

11.2.4 Tc

^{99}Tc is one of the dose-determining radionuclides in HLW disposal, especially because of its long half-life and high mobility potential. Tc is a chemical element with 45 isotopes produced by nuclear reactions. Among these, the long-lived isotopes are ^{97}Tc (2.6×10^6 a), ^{98}Tc (4.2×10^6 a), and ^{99}Tc (2.13×10^5 a). ^{99}Tc is the dominant isotope in the nuclear fuel cycle owing to its high fission product yield (6.03%) and long half-life. It decays via β-emission to stable ruthenium-99 (^{99}Ru), emitting β-particles with an energy of 0.293 MeV. In its metallic form, it is glossy and silver-colored, but it is more commonly produced as a reactive sponge or powder (Greenwood & Earnshaw, 2012). Unlike its neighbors, Nb and Mo, Tc has a [Kr]$4d^5 5s^2$ electron configuration, resulting from the promotion of one electron from the 4*d* to the 5*s* subshell (Siekierski & Burgess, 2002). This electron configuration contributes to its relative stability in the higher oxidation states compared to other transition metals. Its oxidation states range from −1 to +7, with the most common valences being Tc(VII) and Tc(IV), as illustrated in Figure 11.8. Tc is a redox-sensitive element, and its solubility and mobility in the subsurface strongly depend on its oxidation state, as shown in Figure 11.9. Its most stable oxidation state is Tc(VII), with pertechnetate TcO_4^- being the dominant chemical species. TcO_4^- has low affinity for negatively charged mineral surfaces and is therefore highly mobile under oxidizing conditions. In contrast, the poorly soluble Tc(IV) species form under low redox potentials or in the presence of reducing agents in the environment, such as iron (oxyhydr)oxides, iron sulfides, other iron-bearing minerals, sulfide minerals, and organic matter or microorganisms (Huang et al., 2021). This significantly decreases the solubility and mobility of Tc.

In conclusion, these four radionuclides are redox-sensitive elements, and their solubility and mobility in subsurface environments are strongly governed by their oxidation states. The speciation of radionuclides and the prevailing redox conditions exert a significant influence on their migration in the subsurface. Under oxidizing, low-pH conditions, these nuclides are highly mobile as soluble species, while under reducing, high-pH conditions, they tend to form low-oxidation-state precipitates, which are insoluble and immobile, thereby limiting their migration.

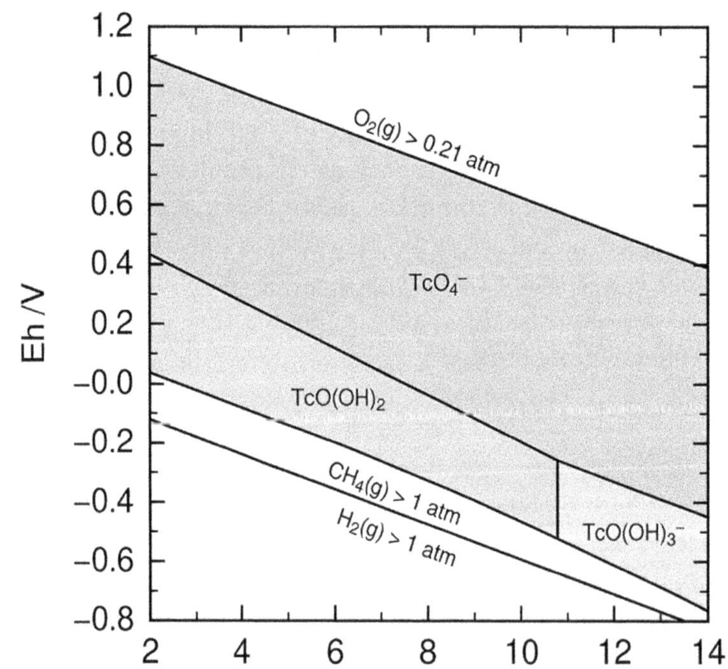

FIGURE 11.8 Tc pH–Eh Diagram Generated Using PHREEQC with the ThermoChimie_v12a.dat Database. The Tc Concentration Was 10^{-8} M, and the Groundwater Composition from Beishan Was Used in the Simulation. TcO_2(cr) Was Allowed to Precipitate When Its Saturation Index Exceeded Zero.

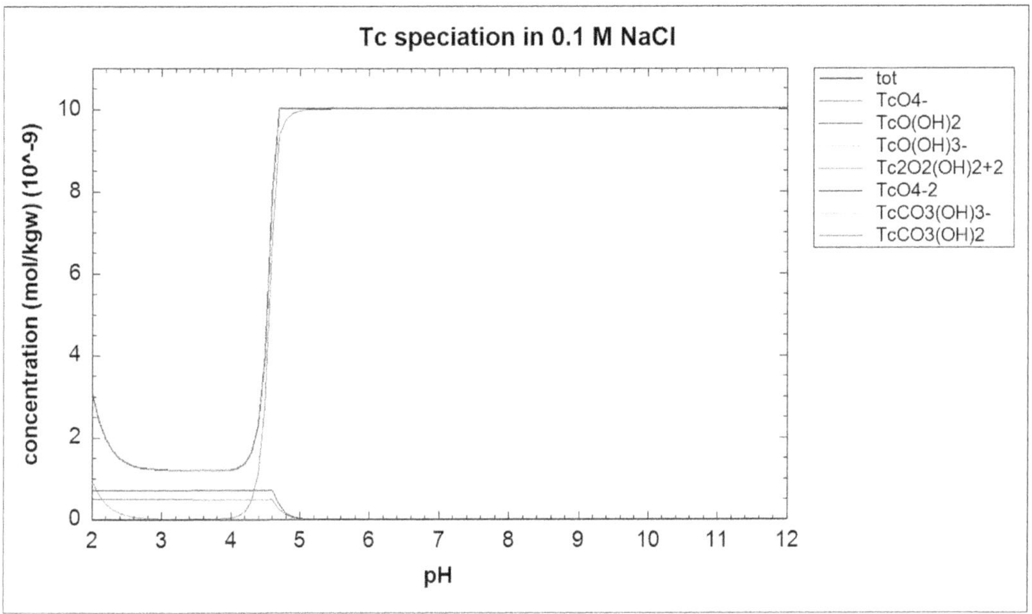

FIGURE 11.9 Tc Speciation in Groundwater Simulated Using PHREEQC with the ThermoChimie_v12a.dat Database. The Tc Concentration Was 10^{-8} M, and the Groundwater Composition from Beishan Was Used in the Simulation. The Redox Potential (Eh) Was Set at 250 mV.

11.3 TRANSPORTATION OF KEY RADIONUCLIDES IN REPOSITORIES

As stated earlier, after canister failure, radionuclides from HLW could be released into the surrounding environment, and they may migrate from the repository near-field to far-field and eventually reach the biosphere. In general, radionuclides first diffuse through the buffer and backfill materials in groundwater, a process driven by concentration gradients. Following this, they are transported by the bulk movement of groundwater, a mechanism known as advection. During this process, factors such as variations in porosity and/or fracture networks, as well as groundwater velocity, can lead to hydrodynamic dispersion (Ma et al., 2021). Furthermore, the retardation of radionuclides within the surrounding geological media is a critical factor controlling their transport (Ram et al., 2021). Various interactions with surrounding media—such as sorption onto buffer/backfill materials or on host rock, chemical reactions with aqueous species, or complexation with organic ligands, radioactive decay, and biotransformation—can either inhibit or facilitate radionuclide migration. Figure 11.10 schematically illustrates the complex interplay of multiple processes contributing to radionuclide retention in a repository system (adapted from the original figure in Zhang X et al., 2022b).

11.3.1 Diffusion

For low-hydraulic-conductivity media, such as buffer materials (typically bentonite), diffusion is considered to be the dominant transport mechanism for radionuclides released from failed canisters in a repository environment. The diffusion process is described by

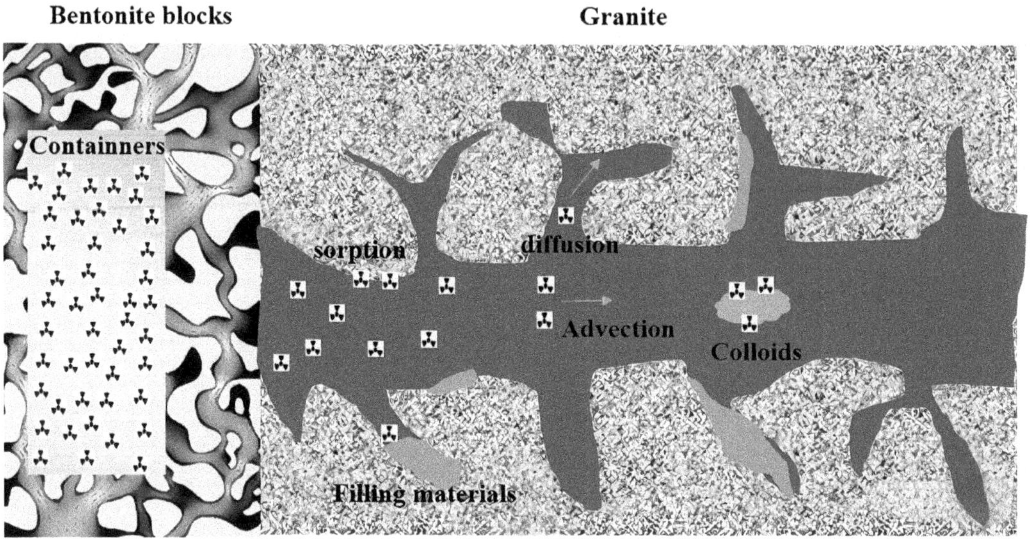

FIGURE 11.10 Schematic Diagram of Various Processes of Solute Transport in a Granitic Repository System.

Fick's second law, and analytical solutions are possible with specific initial and boundary conditions (Van Loon & Soler, 2003). The main parameters describing diffusion are the effective diffusion coefficient (D_e) and the porosity (ε) or rock capacity factor. Diffusion of non-sorbing nuclides, such as HTO, Cl^-, and Na^+ (Van Loon et al., 2003), in clay minerals has been thoroughly studied, and charge effects on diffusion have been observed (Glaus et al., 2010; Van Loon et al., 2003). Extensive research has also been conducted on repository-relevant key radionuclides.

Effective D_e of Tc has been measured in various bentonites (Kuroda et al., 1996; Sawatsky & Oscarson, 1991a; Sawatsky & Oscarson, 1991b; Večerník & Jedináková-Křížová, 2006). It is primarily influenced by dry density and oxidation conditions, with values around 10^{-11} m^2/s under oxidizing conditions and 10^{-12} to 10^{-13} m^2/s under reducing conditions. Apparent diffusivities of Tc in granite are approximately 10^{-14} m^2/s (Ittner et al., 1990). Diffusion studies have also been performed on sediments and clay minerals, yielding D_e values of ~10^{-9} m^2/s in sediment (Mattigod et al., 2012) and ~10^{-11} m^2/s in clay minerals (Martynov et al., 2023). Np diffusion in clay minerals usually involves Np(V); its diffusion coefficients are on the order of 10^{-13} m^2/s for montmorillonite, 10^{-11} m^2/s for illite, and 10^{-11} m^2/s for kaolinite (Staunton et al., 1990). The in-diffusion technique is often used for Np due to its low diffusion coefficients. Dry density is a key factor affecting Np diffusion in compacted bentonite (Sato et al., 1992). Pore size constraints or the size of Np complexes may also contribute to the low diffusion coefficients observed for Np diffusion in compacted bentonite over long timescales (Idemitsu et al., 2023). Np diffusion in Opalinus Clay is similar to that in other clay minerals, with D_e around 6.9×10^{-12} m^2/s and limited sorption (Wu et al., 2009). Limited sorption on granite was also observed during Np diffusion experiments, and bentonite colloids were demonstrated to have a negligible effect on Np diffusion (Elo et al., 2019). However, Np colloids (primarily reduced Np(IV)) have been observed during Np transport in field lysimeters at the Savannah River Site (Di Pietro et al., 2019; Peruski et al., 2018), which could enhance Np mobility. Thus, further research on Np(IV) behavior in clays is necessary. Pu from Pu (V) and Pu (VI) sources migrates faster in sediment than Pu from Pu (III) or Pu (IV) sources (Kaplan et al., 2006). The D_e of Pu(IV) in dense bentonite and sand ranges from 10^{-12} to 10^{-14} m^2/s (Hamada et al., 2016; Idemitsu et al., 1990; Sharma & Oscarson, 1988; TORSTENFELT, 1986). Within this range, a small fraction of Pu migrated faster than the bulk, indicating the presence of a more mobile Pu species (Sharma & Oscarson, 1991). The presence of iron corrosion products in bentonite leads to smaller apparent diffusion coefficients for Pu under oxic conditions (Idemitsu et al., 1999), likely due to corrosion-induced acidification promoting Pu dissolution and migration (Idemitsu et al., 2017). A smaller apparent D_e for Pu in natural clay (Opalinus Clay) has been observed with a value of 3.2×10^{-15} m^2/s (Berg et al., 2025), which may result from the presence of pyrite, siderite, and/or calcite. Its diffusing species is identified as PuO_2^+ by capillary electrophoresis coupled to inductively coupled plasma mass spectrometry (CE-ICP-MS) (Berg et al., 2025) and is progressively reduced to Pu(IV) along its diffusion path (Kaplan, 2013). Synchrotron-based microscopic technique further confirmed Pu(V) is progressively reduced to Pu(IV) and Pu(III) along its diffusion (Kaplan et al., 2024). A first-principle molecular dynamics study reported a D_e value of 1.31×10^{-11} m^2/s for Pu(III) under sedimentary conditions (Barbera, 2023). The effective D_e of Pu in granite is measured as 2.0×10^{-13} m^2/s (Yamaguchi et al., 2002). Colloids can facilitate Pu mobility in fracture granite, both as intrinsic Pu colloids (Abdel-Fattah et al., 2013) and pseudo-colloids (Wolfsberg et al., 2017; Zhang et al., 2022a). Far fewer diffusivity studies exist for Am compared to Tc, Np, and Pu. Apparent D_e of americium is on the order of 10^{-15} m^2/s in both bentonite clay (Christiansen & Torstenfelt, 1988; Torstenfelt, 1986) and granitic rocks (Ittner et al., 1988; Song et al., 2009). Lower penetration depths were observed for Am compared to Pu and U in granite (Yamaguchi & Nakayama, 1998), indicating stronger interactions between Am and granite.

In summary, the diffusive properties of Tc, Np, Pu, and Am have been extensively studied, particularly around the early 2000s. Effective D_e and/or apparent D_e are typically obtained using in-diffusion or through-diffusion techniques. Key factors influencing diffusivity include dry density, media composition, carbonate in solution, and redox conditions. The retarding effect of sorption has been consistently confirmed.

11.3.2 Sorption

Sorption is another key process controlling radionuclide transport in subsurface media. It describes the process of removing radionuclides from solution, encompassing mechanisms such as surface complexation, mineral precipitation, incorporating into or co-precipitation within mineral structures, reduction to sparingly soluble products, and other phenomena. Under oxidizing conditions, Tc shows minimal sorption to subsurface media (Baston et al., 1994). However, its mobility can be significantly decreased when it is reduced by Fe(II)-bearing minerals (such as magnetite, siderite, and pyrite) present in granite, or by Fe(III) oxides, such as ferrihydrite, goethite, and hematite. The availability and steric distribution of these electron donors are critical factors. The speciation of iron has been demonstrated to play a significant role in reactions with redox-sensitive nuclides. For instance, the following affinity series for heterogeneous Tc(VII) reduction by different Fe(II) species has been observed: aqueous Fe(II) ≈ adsorbed Fe(II) in phyllosilicates < structural Fe(II) in phyllosilicates < Fe(II) adsorbed on Fe(III) oxides (Peretyazhko et al., 2008). Tc(VII) is sluggishly reduced by aqueous Fe(II), although the reducing activity of aqueous Fe(II) is high and the Fe(II)-to-Tc electron transfer is thermodynamically feasible (Cui and Eriksen, 1996). The reduced Tc could form hydrous TcO_2_like phases by Fe(II)-bearing minerals (McBeth et al., 2011), or the hydrated Tc dioxide when Fe(II) is incorporated into Fe(III) minerals (hematite and goethite) (Peretyazhko et al., 2008; Skomurski et al., 2010). Because of the similar ionic radii of Tc(IV) and Fe(III) in a six-fold coordination (Skomurski et al., 2010), the incorporated Tc(IV) is more stable against re-oxidation. The effects of biotic or microbial activity have also been studied (Brookshaw et al., 2015; McBeth et al., 2011). In the study of Brookshaw et al, TcO_4^-, UO_2^{2+}, and NpO_2^+ could be removed from solution by microbially reduced biotite and chlorite, with reactivity linked to the redox cycling of a small fraction of Fe in these minerals (Brookshaw et al., 2015). Pyrite, which is a minor component in clay rocks such as Opalinus Clay (Nagra, 2002), is recognized as an effective sorbent for Tc(IV) (Bruggeman et al., 2007).

Similar to Tc, Np exists in multiple oxidation states in the environment, exhibiting high solubility in its higher oxidation states and low solubility in lower oxidation states. However, high-oxidation-state Np exhibits stronger sorption on underground media, for instance, on montmorillonite (Benedicto et al., 2014; Kozai et al., 2014b), on bentonite (Li et al., 2015), on $Al(OH)_3$ (Gückel et al., 2013), on aluminum (hydro)oxide mineral (Baumer & Hixon, 2019), on Fe_2O_3 (Arai et al., 2007; Müller et al., 2015), on TiO_2, and SiO_2 (Müller et al., 2009). Some Np(V) uptake is attributed to cation exchange on the negatively charged planar sites on the montmorillonite surface (Benedicto et al., 2014; Bradbury & Baeyens, 2006; Turner et al., 1998; Zavarin et al., 2012). Np(V) can also form binary and/or ternary inner sphere surface complexes on gibbsite (Gückel et al., 2013) or become incorporated into carbonate and sulfate minerals (Balboni et al., 2015). The formation of neptunium(V)–silicate complexes has also been observed (Pathak & Choppin, 2007; Yusov et al., 2005). Both studies assign the neptunium(V)–silicate complex species to $NpO_2(OSi(OH)_3)$ in the alkaline pH range. The application of different models (Arai et al., 2007; Bradbury & Baeyens, 2006; Gückel et al., 2013; Turner et al., 1998) often correlates with the neptunium(V) surface loading in the uptake studies, where the strong versus weak site models are typically used for low surface loadings (Arai et al., 2007; Bradbury & Baeyens, 2006; Turner et al., 1998), while the diffuse double layer (DDL) model is applied for high surface loadings (Gückel et al., 2013; Virtanen et al., 2016). Sorption of neptunium at Np(IV), Np(V), and Np(VI) oxidation states on TiO_2 was found to depend on the degree of hydrolysis (Tits et al., 2014). Partial reduction of Np(V) to Np(IV) has been indicated during Np sorption on gneiss and dolerite, in which Fe-bearing minerals sorbed more Np than the quartz and feldspar grains, with the sorbed Np present as NpO_2 (Konevnik et al., 2020). Partial reduction of Np(V) has also been observed at illite surface (Banik et al., 2017; Marsac et al., 2015b). Surface-mediated reduction reactions have been demonstrated on montmorillonite for Np, with kinetics potentially affected by structural Fe (Begg et al., 2018). With regard to iron minerals, Np(V) adsorbs to the goethite and ferrihydrite surface likely as an inner-sphere complex, whereas it is reduced to Np(IV) and forms nano-crystalline Np(IV)O_2 when in contact with mackinawite (Townsend et al., 2022). Magnetite has been affirmed to reduce Np(V) to Np(IV) via a surface-mediated reduction mechanism (Bots et al., 2019; Yang et al., 2017), as has green rust, leading to nano-particle precipitation at the edges of crystal platelet (Christiansen et al., 2011). Reaction with siderite can lead to the precipitation of NpO_2^--like nano-particles and the Np(IV) pentacarbonate complex (Scheinost et al., 2016). Microbially mediated metal reduction of Np has been observed in sediments, producing products resistant to oxidative remobilization (Law et al., 2010; Thorpe et al., 2015). Furthermore, Np(IV) may still pose a significant environmental risk through the formation of Np(IV)–silica colloids at near-neutral to slightly alkaline pH (Husar et al., 2015).

Similar to Np and Tc, Pu exists in multiple oxidation states in the environment and exhibits complex physicochemical properties. Changes in Pu oxidation state are frequently reported during interaction with environmental minerals. The rapid reduction of Pu(V) to Pu(IV) on goethite (Keeney-Kennicutt & Morse, 1985; Sanchez et al., 1985) and magnetite (Powell et al., 2004) has been reported early on. Surface-mediated reduction of Pu(V) to Pu(IV) on Mn-containing minerals (Morgenstern & Choppin, 2002) and of Pu(VI) to Pu(IV) correlated with solution pH increasing across its point of zero charge (PZC) of goethite (Wang, 2017) have been confirmed. Pu(V) has been found to sorb predominantly to manganese oxides and smectites (Duff et al., 1999), often following an initial sorption step with subsequent slow reduction to Pu(IV) (Powell et al., 2006; Powell et al., 2005; Romanchuk et al., 2011). The possible product of reduced Pu(IV) is hydrated plutonium oxide precipitates (Kumar et al., 2014; Romanchuk et al., 2013). Modeling predicts that Pu(V) sorbs weakly to kaolinite and never predominates on its surface, whereas Pu(VI) and Pu(IV) predominate under oxidizing and slightly reducing conditions, respectively. In contrast, the Pu(IV)/Pu(III) boundary is hardly affected (Marsac et al., 2015a). Colloid-mediated transport is considered crucial for Pu transport behavior at sites like the Nevada Test Site and downstream from Mayak reprocessing plant. At the Nevada Test Site, Pu is proposed to be transported via adsorption onto clay colloids or as intrinsic colloids (Kersting et al., 1999), whereas at Mayak, transport occurs via adsorption onto amorphous iron hydroxide colloids (Novikov et al., 2006).

Since Am is a redox-stable nuclide, extensive research has focused on Am(III) sorption on clays and clay minerals. Am(III) sorption on kaolinite increases with increasing pH in the absence of CO_2 but decreases in the presence of carbonate due to the formation of soluble Am(III)-carbonate species (Buda et al., 2008). In the presence of humic acid, Am(III) sorption increases at acidic pH, resulting from strong sorption of humic acid on clay, which provides more potential sorption sites, and decreases at near-neutral to basic pH, which is attributed to formation of soluble Am(III)-humate species (Křepelová et al., 2011; Lee et al., 2011; Samadfam et al., 2000). It has been observed that increased ionic strength decreases Am(III) sorption at pH 4 (Ma et al., 2017). Sorption of Am(III) on montmorillonite is dominated by cation exchange at acidic pH, with increasing ionic strength decreasing its sorption (Bradbury & Baeyens, 2006). Outer-sphere complexes form at pH 4 on smectite, transitioning to inner-sphere sorption at higher pH (Stumpf et al., 2004). Am(III) sorption on bentonite resembles that on montmorillonite due to montmorillonite being its dominant component in bentonite (Yu et al., 2012), although a minor component in bentonite may affect the uptake process (Kozai et al., 2014a). Sorption on illite correlates to pH value and involves both cation exchange and surface complexation mechanisms (Bradbury & Baeyens, 2009).

11.4 CONCLUSION AND OUTLOOK

The behavior of radionuclides in a repository is of utmost importance for the long-term safety assessment. Their transport or immobilization closely depends on their speciation in aquatic environment and their interactions with the surrounding media. Both speciation and interactions are significantly influenced by a series of parameters, such as pH, ionic strength, solution composition, redox potential, and the presence of biological matter. Np, Pu, and Tc are redox-sensitive elements. In addition to some degree of sorption onto subsurface media (with the exception of Tc, which is hardly sorbed on media), enhanced immobilization through surface-mediated reduction—forming surface complexes, precipitates, or incorporating into media structure—has been extensively reported. This process can significantly reduce their mobility. Although Am is not redox-sensitive, its sorption onto media is high. Colloids are generally found to facilitate radionuclide transport by forming nuclide–colloid complexes. These processes work together to control radionuclide transport in a repository. Quantitative modeling of radionuclide transport by diffusion or sorption alone is established and performs well. However, modeling studies coupling transport and sorption remain limited, and even more complex simulations coupling thermal, mechanical, chemical, and hydrological (THMC) processes are far less developed. More comprehensive simulation work is needed to support the safety assessment of geological repositories in the future.

REFERENCES

Abdel-Fattah, A.I., et al., 2013. Dispersion stability and electrokinetic properties of intrinsic plutonium colloids: Implications for subsurface transport. *Environ. Sci. Technol.* 47, 5626–5634.

Arai, Y., Moran, P., Honeyman, B., Davis, J., 2007. In situ spectroscopic evidence for neptunium (V)-carbonate inner-sphere and outer-sphere ternary surface complexes on hematite surfaces. *Environ. Sci. Technol.* 41, 3940–3944.

Balboni, E., Morrison, J.M., Wang, Z., Engelhard, M.H., Burns, P.C., 2015. Incorporation of Np (V) and U (VI) in carbonate and sulfate minerals crystallized from aqueous solution. *Geochim. Cosmochim. Acta* 151, 133–149.

Banik, N.L., et al., 2017. Neptunium sorption and redox speciation at the illite surface under highly saline conditions. *Geochim. Cosmochim. Acta* 215, 421–431.

Barbera, B., 2023. Diffusion Mechanism of Pu3+ in Sedimentary Repository Conditions: First Principle Molecular Dynamics Study. *The Canadian Science Fair Journal* 5(1).

Baston, G., et al., 1994. The sorption of uranium and technetium on bentonite, tuff and granodiorite. *MRS Online Proc. Libr.* 353, 989–996.

Baumer, T., Hixon, A.E., 2019. Kinetics of neptunium sorption and desorption in the presence of aluminum (hydr) oxide minerals: Evidence for multi-step desorption at low pH. *J. Environ. Radioact.* 205, 72–78.

Begg, J.D., Edelman, C., Zavarin, M., Kersting, A.B., 2018. Sorption kinetics of plutonium (V)/(VI) to three montmorillonite clays. *Appl. Geochem.* 96, 131–137.

Benedicto, A., et al., 2014. Effect of major cation water composition on the ion exchange of Np (V) on montmorillonite: NpO2+–Na+–K+–Ca2+–Mg2+ selectivity coefficients. *Appl. Geochem.* 47, 177–185.

Berg, F., Sirleaf, C., Lohmann, J., Breckheimer, M., Reich, T., 2025. Investigation of plutonium diffusion profiles in Opalinus Clay rock via TOF-SIMS and rL-SNMS. *Appl. Geochem.* 183, 106332.
Bots, P., et al., 2019. Neptunium (V) and uranium (VI) reactions at the magnetite (111) surface. *Geosciences* 9, 81.
Bradbury, M.H., Baeyens, B., 2006. Modelling sorption data for the actinides Am (III), Np (V) and Pa (V) on montmorillonite. *Radiochim. Acta* 94, 619–625.
Bradbury, M.H., Baeyens, B., 2009. Sorption modelling on illite. Part II: Actinide sorption and linear free energy relationships. *Geochim. Cosmochim. Acta* 73, 1004–1013.
Brookshaw, D.R., et al., 2015. Redox interactions of Tc (VII), U (VI), and Np (V) with microbially reduced biotite and chlorite. *Environ. Sci. Technol.* 49, 13139–13148.
Bruggeman, C., Maes, A., Vancluysen, J., 2007. The identification of FeS2 as a sorption sink for Tc (IV). *Phys. Chem. Earth* 32, 573–580.
Buda, R., Banik, N., Kratz, J.V., Trautmann, N., 2008. Studies of the ternary systems humic substances–kaolinite–Pu (III) and Pu (IV). *Radiochim. Acta* 96, 657–665.
Burns, R.E., Causey, W.E., Galloway, W.E., Nelson, R.W., 1978. *Nuclear Waste Disposal in Space.* NASA Technical Paper 1225. NASA Scientific and Technical Information Office, Washington, DC.
Calmet, D.P., 1989. Ocean disposal of radioactive waste. Status report. *IAEA Bull.* 31, 47–50.
Chen, P., 2023. *Technetium Retention and Migration in Clay Systems in Presence of Iron.* Universität Bern.
Christiansen, B., et al., 2011. Neptunyl (NpO2+) interaction with green rust, GRNa, SO4. *Geochim. Cosmochim. Acta* 75, 1216–1226.
Christiansen, B., Torstenfelt, B., 1988. Diffusion of nickel, strontium, iodine, cesium and americium in loosely compacted bentonite at high pH. *Radiochim. Acta* 44, 219–224.
Coopersmith, J., 1999. Disposal of High-Level Nuclear Waste in Space. *1999 AIAA Annual Technical Symposium.* Space Studies Institute.
Cui, D., Eriksen, T.E., 1996. Reduction of pertechnetate by ferrous iron in solution: Influence of sorbed and precipitated Fe (II). *Environ. Sci. Technol.* 30, 2259–2262.
Di Pietro, S.A., Joseph, C., Zavarin, M., 2019. *Neptunium (IV) Diffusion through Bentonite Clay, Lawrence Livermore National Lab.* LLNL, Livermore, CA.
Duff, M., et al., 1999. Mineral associations and average oxidation states of sorbed Pu on tuff. *Environ. Sci. Technol.* 33, 2163–2169.
Elo, O., Hölttä, P., Kekäläinen, P., Voutilainen, M., Huittinen, N., 2019. Neptunium (V) transport in granitic rock: A laboratory scale study on the influence of bentonite colloids. *Appl. Geochem.* 103, 31–39.
Faybishenko, B., Swift, P., 2016. *International Approaches for Deep Geological Disposal of Nuclear Waste: Geological Challenges in Radioactive Waste Isolation: Fifth Worldwide Review.* Lawrence Berkeley National Laboratory.
Gabrielli, R.A., Herdrich, G., 2015. Review of nuclear thermal propulsion systems. *Prog. Aerosp. Sci.* 79, 92–113.
Glaus, M.A., Frick, S., Rossé, R., Van Loon, L.R., 2010. Comparative study of tracer diffusion of HTO, 22Na+ and 36Cl– in compacted kaolinite, illite and montmorillonite. *Geochim. Cosmochim. Acta* 74, 1999–2010.
Greenwood, N.N., Earnshaw, A., 2012. *Chemistry of the Elements.* Elsevier.
Gückel, K., et al., 2013. Spectroscopic identification of binary and ternary surface complexes of Np (V) on gibbsite. *Environ. Sci. Technol.* 47, 14418–14425.
Hamada, R., et al., 2016. Effect of pH on plutonium migration behavior in compacted bentonite. *MRS Adv.* 1, 4011–4017.
Heuze, F.E., 1981. *Geotechnical Modeling of High-Level Nuclear Waste Disposal by Rock Melting.* Report No. UCRL-53183. Lawrence Livermore National Laboratory, Livermore, CA.
Huang, J., et al., 2021. Fe (II) redox chemistry in the environment. *Chem. Rev.* 121, 8161–8233.

Husar, R., et al., 2015. Formation of neptunium (IV)–silica colloids at near-neutral and slightly alkaline pH. *Environ. Sci. Technol.* 49, 665–671.
IAEA, 2020. *An Introduction to Practical Industrial Tomography Techniques for Non-Destructive Testing (NDT).* International Atomic Energy Agency, Vienna.
IAEA, 2022. *Status and Trends in Spent Fuel and Radioactive Waste Management.* IAEA Nuclear Energy Series International Atomic Energy Agency, Vienna. (No. NW-T-1.14 (Rev. 1)).
Idemitsu, K., et al., 1999. Diffusion of plutonium in compacted bentonites in the reducing condition with corrosion products of iron. *MRS Online Proc. Libr.* 608, 261.
Idemitsu, K., et al., 2017. Plutonium migration in compacted bentonite with iron corrosion for 15 Years. *MRS Adv.* 2, 693–698.
Idemitsu, K., Inagaki, Y., Arima, T., 2023. Long-term diffusion of neptunium and other elements in compacted bentonite reveals filtration. *MRS Adv.* 8, 214–223.
Idemitsu, K., Ishiguro, K., Yusa, Y., Sasaki, N., Tsunoda, N., 1990. Plutonium diffusivity in compacted bentonite. *Eng. Geol.* 28, 455–462.
Ittner, T., Torstenfelt, B., Allard, B., 1988. Diffusion of neptunium, plutonium and americium in granitic rock. *Radiochim. Acta* 44, 171–178.
Ittner, T., Torstenfeit, B., Allard, B., 1990. Diffusion of strontium, technetium, iodine and cesium in granitic rock. *Radiochim. Acta* 49, 101–106.
Kaplan, D.I., et al., 2006. Eleven-year field study of Pu migration from Pu III, IV, and VI sources. *Environ. Sci. Technol.* 40, 443–448.
Kaplan, U., 2013. *Speciation of Plutonium During Sorption and Diffusion in Opalinus Clay.* Mainz University (Germany). Fachbereich 09: Chemie, Pharmazie und Geowissenschaften.
Kaplan, U., Amayri, S., Drebert, J., Grolimund, D., Reich, T., 2024. Plutonium mobility and reactivity in a heterogeneous clay rock barrier accented by synchrotron-based microscopic chemical imaging. Sci. Rep. 14, 3087.
Keeney-Kennicutt, W.L., Morse, J.W., 1985. The redox chemistry of Pu (V) O2+ interaction with common mineral surfaces in dilute solutions and seawater. *Geochim. Cosmochim. Acta* 49, 2577–2588.
Kersting, A., et al., 1999. Migration of plutonium in ground water at the Nevada Test Site. *Nature* 397, 56–59.
Knorr, K.E., 2015. *On the Uses of Military Power in the Nuclear Age.* Princeton University Press.
Konevnik, Y.V., Zakharova, E., Shiryaev, A., Olds, T., Burns, P., 2020. Mineral-specific heterogeneous neptunium sorption onto geological repository rocks in oxic and anoxic conditions and various temperatures. *Chem. Geol.* 545, 119654.
Kozai, N., Yamasaki, S., Ohnuki, T., 2014a. Application of simplified desorption method to study on sorption of americium (III) on bentonite. *J. Radioanal. Nucl. Chem.* 299, 1571–1579.
Kozai, N., Yamasaki, S., Ohnuki, T., 2014b. Application of simplified desorption method to study on sorption of neptunium (V) on montmorillonite-based mixtures. *J. Radioanal. Nucl. Chem.* 299, 1581–1587.
Křepelová, A., Sachs, S., Bernhard, G., 2011. Influence of humic acid on the Am (III) sorption onto kaolinite. *Radiochim. Acta* 99, 253–260.
Kumar, S., et al., 2014. Kinetics of Pu (IV) sorption by smectite-rich natural clay. *J. Radioanal. Nucl. Chem.* 300, 45–49.
Kuroda, Y., Idemitsu, K., Furuya, H., Inagaki, Y., Arima, T., 1996. Diffusion of technetium in compacted bentonites in the reducing condition with corrosion products of iron. *MRS Online Proc. Libr.* 465, 909.
Law, G.T., et al., 2010. Geomicrobiological redox cycling of the transuranic element neptunium. *Environ. Sci. Technol.* 44, 8924–8929.
Lee, M.H., Jung, E.C., Song, K., Han, Y.H., Shin, H.S., 2011. The influence of humic acid on the pH-dependent sorption of americium (III) onto kaolinite. *J. Radioanal. Nucl. Chem.* 287, 639–645.
Li, P., et al., 2015. Effects of pH, ionic strength and humic acid on the sorption of neptunium (V) to Na-bentonite. *J. Mol. Liq.* 206, 285–292.

Logan, S.E., 1974. Deep self-burial of radioactive wastes by rock-melting capsules. *Nucl. Technol.* 21, 111–124.

Ma, F., et al., 2017. Experimental and modelling approaches to Am (III) and Np (V) adsorption on the Maoming kaolinite. *Appl. Geochem.* 84, 325–336.

Ma, Z., et al., 2021. Experimental investigations on scale-dependent dispersivity in three-dimensional heterogeneous porous media. *Environ. Sci. Pollut. Res.* 28, 23336–23348.

Marsac, R., et al., 2015a. Modeling plutonium sorption to kaolinite: Accounting for redox equilibria and the stability of surface species. *Chem. Geol.* 400, 1–10.

Marsac, R., et al., 2015b. Neptunium redox speciation at the illite surface. *Geochim. Cosmochim. Acta* 152, 39–51.

Martynov, K., Konevnik, Y.V., Zakharova, E., 2023. Diffusion of tritium, technetium, cesium and uranium from phosphate glass leachates in a pore solution of clay materials. *Radiochemistry* 65, 457–472.

Mattigod, S.V., Bovaird, C.C., Wellman, D.M., Parker, K.E., Wood, M.I., 2012. An experimental study of diffusivity of technetium-99 in Hanford vadose zone sediments. *Vadose Zone J.* 11, vzj2011.0184.

McBeth, J., et al., 2011. Redox interactions of technetium with iron-bearing minerals. *Mineral. Mag.* 75, 2419–2430.

Morgenstern, A., Choppin, G.R., 2002. Kinetics of the oxidation of Pu (IV) by manganese dioxide. *Radiochim. Acta* 90, 69–74.

Müller, K., et al., 2015. In situ spectroscopic identification of neptunium (V) inner-sphere complexes on the hematite–water interface. *Environ. Sci. Technol.* 49, 2560–2567.

Müller, K., Foerstendorf, H., Brendler, V., Bernhard, G., 2009. Sorption of Np (V) onto TiO2, SiO2, and ZnO: An in situ ATR FT-IR spectroscopic study. *Environ. Sci. Technol.* 43, 7665–7670.

Nagra, 2002. *Project Opalinus Clay: Safety Report: Demonstration of Disposal Feasibility for Spent Fuel, Vitrified High-Level Waste and Long-Lived Intermediate-Level Waste (Entsorgungsnachweis).* Nagra Tech. Rep. NTB 02–05, Nagra, Wettingen, Switzerland.

National Research Council, 1996. *Nuclear Wastes: Technologies for Separations and Transmutation.* National Academy Press, Washington, DC.

Neuvazhaev, G., Rastorguev, A., Morozov, O., Kapyrin, I., Grigorev, F., 2020. *3D Hydrogeological Modeling of Deep Geological Disposal in the Nizhnekansky Rock Massif.* EGU General Assembly Conference Abstracts, p. 21509.

Novikov, A.P., et al., 2006. Colloid transport of plutonium in the far-field of the Mayak Production Association, Russia. *Science* 314, 638–641.

Nuklearsicherheitsinspektorat, E., 2018a. *Recommendations and Notes Resulting from the Assessment of the Nuclear Waste Disposal Program and the 2016 Research and Development Plan.* Eidgenössisches Nuklearsicherheitsinspektorat (ENSI), Brugg (Switzerland).

Nuklearsicherheitsinspektorat, E., 2018b. *Statement on the 2016 Nuclear Waste Disposal Program of Those Responsible for the Waste Disposal.* Eidgenössisches Nuklearsicherheitsinspektorat (ENSI), Brugg (Switzerland).

Pathak, P., Choppin, G., 2007. Silicate complexation of NpO2+ ion in perchlorate media. *J. Radioanal. Nucl. Chem.* 274, 3–7.

Peretyazhko, T., et al., 2008. Heterogeneous reduction of Tc (VII) by Fe (II) at the solid–water interface. *Geochim. Cosmochim. Acta* 72, 1521–1539.

Peruski, K.M., Maloubier, M., Kaplan, D.I., Almond, P.M., Powell, B.A., 2018. Mobility of aqueous and colloidal neptunium species in field lysimeter experiments. *Environ. Sci. Technol.* 52, 1963–1970.

Philberth, K., 1977. The disposal of radioactive waste in ice sheets. *J. Glaciol.* 19, 607–617.

Plúa, C., et al., 2021. A reliable numerical analysis for large-scale modelling of a high-level radioactive waste repository in the Callovo-Oxfordian claystone. *Int. J. Rock Mech. Min. Sci.* 140, 104574.

Posiva Oy, 2013. *Safety Case for the Disposal of Spent Nuclear Fuel at Olkiluoto - Formulation of Radionuclide Release Scenarios 2012*. POSIVA 2012-08, ISBN 978-951-652-189-6. Posiva Oy, Eurajoki, Finland.

POSIVA, 2022. *Excavation of the First Five Disposal Tunnels Completed*. https://www.posiva.fi/en/index/news/pressreleasesstockexchangereleases/2022/excavationofthefirstfivedisposaltunnelscompleted.html.

POSIVA, 2024. *Test Operation—Countdown to Starting Final Disposal*. https://www.posiva.fi/en/index/finaldisposal/finaldisposaltestsfisstytk/testoperation-countdowntostartingfinaldisposal.html.

Powell, B.A., et al., 2006. Plutonium oxidation and subsequent reduction by Mn (IV) minerals in Yucca Mountain tuff. *Environ. Sci. Technol.* 40, 3508–3514.

Powell, B.A., Fjeld, R.A., Kaplan, D.I., Coates, J.T., Serkiz, S.M., 2004. Pu (V) O2+ adsorption and reduction by synthetic magnetite (Fe3O4). *Environ. Sci. Technol.* 38, 6016–6024.

Powell, B.A., Fjeld, R.A., Kaplan, D.I., Coates, J.T., Serkiz, S.M., 2005. Pu (V) O2+ adsorption and reduction by synthetic hematite and goethite. *Environ. Sci. Technol.* 39, 2107–2114.

Ram, R., et al., 2021. Understanding the mobility and retention of uranium and its daughter products. *J. Hazard. Mater.* 410, 124553.

Rodionova, A.A., et al., 2019. The radionuclide distribution onto different mineral phases of the rocks of the exocontact zone of Nizhnekansky granitoid massif. *Perspect. Sci.* 12, 100406.

Romanchuk, A.Y., et al., 2013. Formation of crystalline PuO2+ x· nH2O nanoparticles upon sorption of Pu (V, VI) onto hematite. *Geochim. Cosmochim. Acta* 121, 29–40.

Romanchuk, A.Y., Kalmykov, S.N., Aliev, R.A., 2011. Plutonium sorption onto hematite colloids at femto-and nanomolar concentrations. *Radiochim. Acta* 99, 137–144.

Rozov, K., Rumynin, V., Nikulenkov, A., Leskova, P., 2018. Sorption of 137Cs, 90Sr, Se, 99Tc, 152 (154) Eu, 239 (240) Pu on fractured rocks of the Yeniseysky site (Nizhne-Kansky massif, Krasnoyarsk region, Russia). *J. Environ. Radioact.* 192, 513–523.

Samadfam, M., et al., 2000. Effects of humic acid on the sorption of Am (III) and Cm (III) on kaolinite. *Radiochim. Acta* 88, 717–723.

Sanchez, A.L., Murray, J.W., Sibley, T.H., 1985. The adsorption of plutonium IV and V on goethite. *Geochim. Cosmochim. Acta* 49, 2297–2307.

Sato, H., Ashida, T., Kohara, Y., Yui, M., Sasaki, N., 1992. Effect of dry density on diffusion of some radionuclides in compacted sodium bentonite. *J. Nucl. Sci. Technol.* 29, 873–882.

Sawatsky, N., Oscarson, D., 1991a. Diffusion of technetium in dense bentonite. *Water Air Soil Pollut.* 57, 449–456.

Sawatsky, N., Oscarson, D., 1991b. Diffusion of technetium in dense bentonite under oxidizing and reducing conditions. *Soil Sci. Soc. Am. J.* 55, 1261–1267.

Scheinost, A.C., Steudtner, R., Hübner, R., Weiss, S., Bok, F., 2016. NeptuniumV retention by siderite under anoxic conditions: Precipitation of NpO2–like nanoparticles and of NpIV pentacarbonate. *Environ. Sci. Technol.* 50, 10413–10420.

Sharma, H.D., Oscarson, D., 1988. Diffusion of plutonium (IV) in dense bentonite-based materials. *MRS Online Proc. Libr.* 127, 735.

Sharma, H.D., Oscarson, D., 1991. *Diffusion of Plutonium in Mixtures of Bentonite and Sand at pH 3*. Atomic Energy of Canada Ltd., Pinawa, MB (Canada). Whiteshell Nuclear . . .

Siekierski, S., Burgess, J., 2002. *Concise Chemistry of the Elements*. Elsevier.

SKB, 2025. *Construction Begins on Swedish Spent Fuel Repository*. https://skb.com/nyhet/construction-begins-on-swedish-spent-fuel-repository/.

Skomurski, F.N., Rosso, K.M., Krupka, K.M., McGrail, B.P., 2010. Technetium incorporation into hematite (α-Fe2O3). *Environ. Sci. Technol.* 44, 5855–5861.

Song, L.-W., Feng, X.-g., Liang, J., Zhang, Y.-J., Wang, J., 2009. A simple approach to determine the diffusivity of americium in granite. *Radiochim. Acta* 97, 581–586.

Staunton, S., Clay, P., Rees, L., 1990. Diffusion of neptunium (V) in clays. *Radiochim. Acta* 49, 147–154.

Stumpf, T., Hennig, C., Bauer, A., Denecke, M.A., Fanghänel, T., 2004. An EXAFS and TRLFS study of the sorption of trivalent actinides onto smectite and kaolinite. *Radiochim. Acta* 92, 133–138.

Thorpe, C.L., et al., 2015. Neptunium and manganese biocycling in nuclear legacy sediment systems. *Appl. Geochem.* 63, 303–309.
Tits, J., Gaona, X., Laube, A., Wieland, E., 2014. Influence of the redox state on the neptunium sorption under alkaline conditions: Batch sorption studies on titanium dioxide and calcium silicate hydrates. *Radiochim. Acta* 102, 385–400.
Torstenfelt, B., 1986. Migration of the actinides thorium, protactinium, uranium, neptunium, plutonium and americium in clay. *Radiochim. Acta* 39, 105–112.
Townsend, L.T., et al., 2022. Neptunium and uranium interactions with environmentally and industrially relevant iron minerals. *Minerals* 12, 165.
Turner, D.R., Pabalan, R.T., Bertetti, F.P., 1998. Neptunium (V) sorption on montmorillonite: An experimental and surface complexation modeling study. *Clays Clay Miner.* 46, 256–269.
Van Loon, L.R., Soler, J.M., 2003. *Diffusion of HTO, 36Cl-, 125I-and 22Na+ in Opalinus Clay: Effect of Confining Pressure, Sample Orientation, Sample Depth and Temperature.* Paul Scherrer Institute (PSI), Villigen (Switzerland).
Van Loon, L.R., Soler, J.M., Bradbury, M., 2003. Diffusion of HTO, 36Cl– and 125I– in Opalinus Clay samples from Mont Terri: Effect of confining pressure. *J. Contam. Hydrol.* 61, 73–83.
Večerník, P., Jedináková-Křížová, V., 2006. Diffusion of 99-technetium in compacted bentonite under aerobic and anaerobic conditions. *Czech. J. Phys.* 56, D665–D672.
Virtanen, S., et al., 2016. The specific sorption of Np (V) on the corundum (α-Al2O3) surface in the presence of trivalent lanthanides Eu (III) and Gd (III): A batch sorption and XAS study. *J. Colloid Interface Sci.* 483, 334–342.
Wang, D., 2017. *Interactions of Plutonium and Neptunium with Iron Oxide Minerals Goethite and Hematite.* UC Berkeley.
Wang, J., Chen, L., Su, R., Zhao, X., 2018. The Beishan underground research laboratory for geological disposal of high-level radioactive waste in China: Planning, site selection, site characterization and in situ tests. *J. Rock Mech. Geotech. Eng.* 10, 411–435.
Wolfsberg, A., et al., 2017. Colloid-facilitated plutonium transport in fractured tuffaceous rock. *Environ. Sci. Technol.* 51, 5582–5590.
Wu, T., Amayri, S., Drebert, J., Loon, L.R.V., Reich, T., 2009. Neptunium (V) sorption and diffusion in Opalinus Clay. *Environ. Sci. Technol.* 43, 6567–6571.
Yamaguchi, T., Nakayama, S., 1998. Diffusivity of U, Pu and Am carbonate complexes in a granite from Inada, Ibaraki, Japan studied by through diffusion. *J. Contam. Hydrol.* 35, 55–65.
Yamaguchi, T., Nakayama, S., Okamoto, H., 2002. Effective diffusivity of neptunium and plutonium in granite from Inada, Ibaraki, Japan under anaerobic conditions. *Radiochim. Acta* 90, 863–868.
Yang, H., et al., 2017. Reductive precipitation of neptunium on iron surfaces under anaerobic conditions. *J. Nucl. Mater.* 496, 109–116.
Yu, T., Wu, W.S., Fan, Q.H., 2012. Sorption of Am (III) on Na-bentonite: Effect of pH, ionic strength, temperature and humic acid. *Chin. Chem. Lett.* 23, 1189–1192.
Yun, D., et al., 2021. Current state and prospect on the development of advanced nuclear fuel system materials: A review. *Mater. Rep.: Energy* 1, 100007.
Yusov, A., Fedoseev, A., Isakova, O., Delegard, C.H., 2005. Complexation of Np (V) with silicate ions. *Radiochemistry* 47, 39–44.
Zavarin, M., Powell, B.A., Bourbin, M., Zhao, P., Kersting, A.B., 2012. Np (V) and Pu (V) ion exchange and surface-mediated reduction mechanisms on montmorillonite. *Environ. Sci. Technol.* 46, 2692–2698.
Zhang, X., Wang, Z., Reimus, P., Ma, F., Soltanian, M.R., Xing, B., Dai, Z., 2022a. Plutonium reactive transport in fractured granite: Multi-species experiments and simulations. *Water Res.* 224, 119068.
Zhang, X., Ma, F., Dai, Z., Wang, J., Chen, L., Ling, H., Soltanian, M.R., 2022b. Radionuclide transport in multi-scale fractured rocks: A review. *J. Hazard. Mater.* 424, 127550.
Ziessman, H.A., O'Malley, J.P., Thrall, J.H., 2013. *Nuclear Medicine: The Requisites E-Book: Nuclear Medicine: The Requisites E-Book.* Elsevier Health Sciences.

CHAPTER 12

Radionuclide Diffusion and Other Migration Processes in the Biosphere

Zhengye Feng and Duoqiang Pan

12.1 INTRODUCTION

Utilization of nuclear power inevitably generates radioactive wastes, among which the high-level radioactive waste draws significant concerns due to its toxicity and long-term threat. Its safe disposal is essential for the sustainable development of nuclear energy. The multi-barrier system is widely regarded as the key concept for the geological disposal of high-level radioactive waste. Among these barriers, buffer/backfill materials (e.g., bentonite) are considered the core component due to their low permeability, strong cation exchange capacity, and favorable thermo-hydro-mechanical-chemical coupling properties (Noh et al., 2025, Cao et al., 2025). The International Atomic Energy Agency (IAEA) and several national disposal programs emphasize that understanding radionuclide migration within engineered and geological barriers is the scientific basis for safety assessment and predictive model development (IAEA, 2011, 2012).

Once released into the disposal environment, the migration of radionuclides is governed by the interplay of multiple physicochemical processes, including sorption/desorption at mineral surfaces, diffusion through pore structures, and association with colloidal particles. Batch sorption experiments and molecular-scale mechanistic studies provide important insights into radionuclide interactions at solid–liquid interfaces, whereas diffusion experiments reveal the spatiotemporal evolution of radionuclide transport in porous media under the influence of pore structure, ionic strength, and temperature (Zhao et al., 2024, Feng et al., 2025, Feng et al., 2024b, Tian et al., 2025). In addition, colloids are ubiquitous in natural barriers, and their stability, mobility, and binding capacity with radionuclides may significantly enhance the potential for long-range migration, making them a critical factor in safety assessments (Shi et al., 2025a, Tang et al., 2024, Wei et al., 2023).

DOI: 10.1201/9781003665793-16

In recent years, advances in experimental methods, characterization techniques, and numerical modeling have enabled the introduction of multi-mechanism coupled modeling and machine learning into radionuclide migration studies. These approaches not only elucidate the driving mechanisms of radionuclide transport across molecular, pore, and field scales but also support risk assessment and enhance predictive accuracy, thereby providing robust scientific foundations for the long-term safety of nuclear waste disposal. Therefore, systematically summarizing research progress on sorption/desorption, diffusion, and colloid-facilitated migration, together with emerging modeling and simulation approaches, is of great significance for deepening the understanding of radionuclide migration mechanisms, optimizing the design of multi-barrier systems, and improving the reliability of long-term safety assessments of disposal repositories (Figure 12.1).

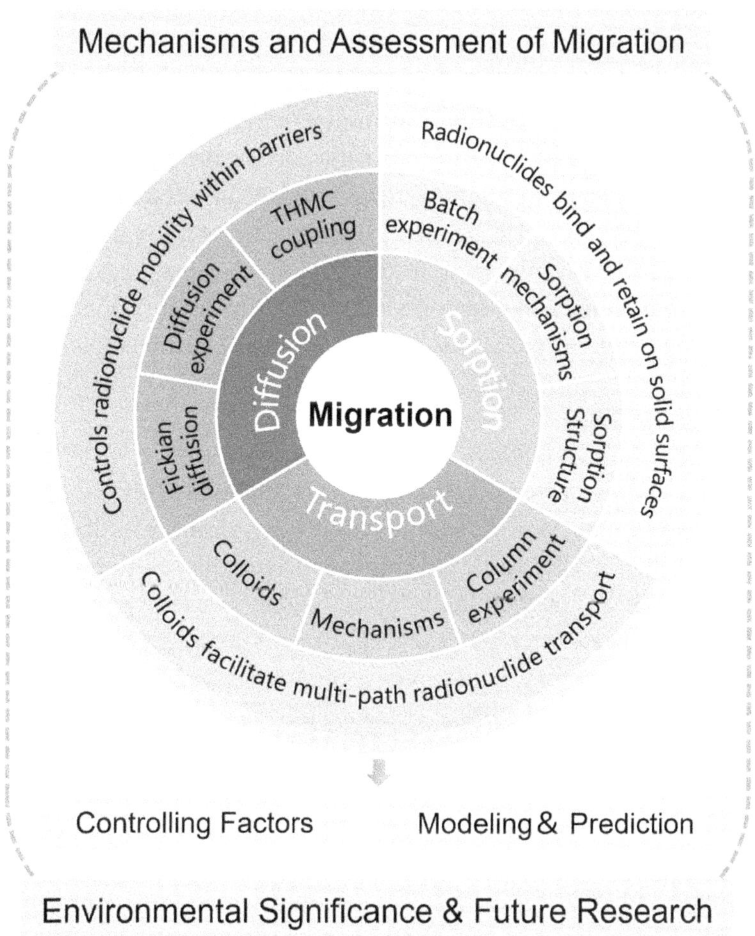

FIGURE 12.1 Schematic Representation of the Mechanisms and Assessment Framework of Radionuclide Migration.

12.2 RADIONUCLIDE MIGRATION

Radionuclide migration refers to the transport of radioactive nuclides within media such as groundwater, waste forms, metals, soils, rocks, and the atmosphere. It involves complex physical, chemical, and biological processes. In essence, radionuclide migration is a coupled process of physical transport and chemical reactions, with its rate and pathways determined by the intrinsic properties of the nuclides, the characteristics of the host medium, and environmental parameters. Understanding these mechanisms is crucial for the safe disposal of radioactive waste, environmental remediation, and risk assessment.

For deep geological disposal of high-level radioactive waste, the predominant concern is radionuclide transport through groundwater, making accurate prediction of migration behavior central to safety evaluation. The migration mechanisms in disposal environments include sorption, diffusion, advection, dispersion, colloid-facilitated transport, and microbially mediated processes. Sorption is the first line of defense for radionuclide retardation. Sorption mechanisms on solid surfaces mainly include ion exchange and surface complexation, with pH and redox potential being the key environmental controls. Diffusion and advection are the primary groundwater-driven transport pathways. In low-permeability host matrices such as bentonite or crystalline rock, diffusion dominates and is governed by diffusion coefficients and tortuosity effects, whereas in highly permeable zones like connected granite fractures, advection prevails and is controlled by flow velocity and thermally driven effects. Colloid-facilitated transport is another important mechanism. Unlike precipitation or complexation, the role of colloids in radionuclide migration is harder to define and strongly depends on colloid type and aqueous chemistry. Colloids may enhance radionuclide mobility.

12.2.1 Sorption/Desorption

12.2.1.1 Batch Sorption Experiments

Since the 1980s, China has started investigations on radionuclide sorption on various environmental media. The interaction of radionuclides with groundwater and barrier materials largely determines their ability to retard migration. The distribution coefficient is commonly used to evaluate the sorption behavior of radionuclides in bentonite, clay minerals, and granite fracture fillings. Batch sorption experiments, extensively applied for determining distribution coefficients, facilitate direct comparisons among different barrier materials and, when integrated with surface characterization (e.g., pore structure, functional groups), can further elucidate sorption mechanisms such as electrostatic attraction and complexation (Philipp et al., 2022, Izosimova et al., 2022, Chen et al., 2022).

Experimental evidence shows that radionuclide sorption on bentonite is strongly influenced by concentration, pH, ionic strength, temperature, dissolved CO_2, and the presence of humic and fulvic acids (Philipp et al., 2022, Izosimova et al., 2022, Parisi, 2020, Wu et al., 2024a). Actinides (U, Np, Pu, Am) mainly sorb through ion exchange and surface coordination, while anions (TcO_4^- , SeO_3^{2-} , IO_3^-) are stabilized via surface modification or reductive processes. Key environmental controls include pH, redox conditions, ionic strength, and mineral type, whereas the heterogeneity of fracture-fillings still requires extensive experimental verification. Future work should emphasize multi-scale experimentation combined with interdisciplinary modeling to better resolve long-term sorption–desorption dynamics, thereby strengthening the scientific basis for the geological disposal of high-level radioactive waste.

12.2.1.2 Sorption Mechanism

Clay minerals and oxides exhibit high specific surface area and surface energies, with hydroxyl groups formed in aqueous environments that enable inner-sphere complexation with radionuclides (Izosimova et al., 2022, Yu et al., 2020). The layered aluminosilicate structure carries permanent negative charges, allowing interlayer cations to participate in ion exchange and confer high cation exchange capacity. Radionuclides may also undergo surface precipitation before reaching thermodynamic equilibrium (Tan et al., 2014). The interactions of radionuclides with mineral surfaces include outer sphere adsorption, inner sphere complexation, incorporation into mineral lattices, and surface-induced reductive precipitation (Figure 12.2) (Chen et al., 2022, Singh & Um, 2023).

12.2.1.2.1 Sorption Kinetics and Influencing Factors U(VI) sorption on montmorillonite is rapid, reaching a maximum at pH 4.0–7.0 and decreasing at higher pH. Competing cations, particularly Mg^{2+}, suppress sorption, while CO_3^{2-} enhances U solubility through uranyl–carbonate complexation, thereby reducing sorption (Yu et al., 2020). Cs(I) and Sr(II) sorption on bentonites is governed by proton competition at low pH (<6) and by background cations (Na^+, K^+, Ca^{2+}, and Mg^{2+}) at higher pH (Izosimova et al., 2022).

12.2.1.2.2 Surface Complexation Mechanisms Ni(II) forms outer- and inner-sphere complexes with diatomite, Ca-montmorillonite, and cryptomelane, depending on pH and reaction conditions (Sheng et al., 2011; Hu et al., 2012). Spectroscopic evidence further

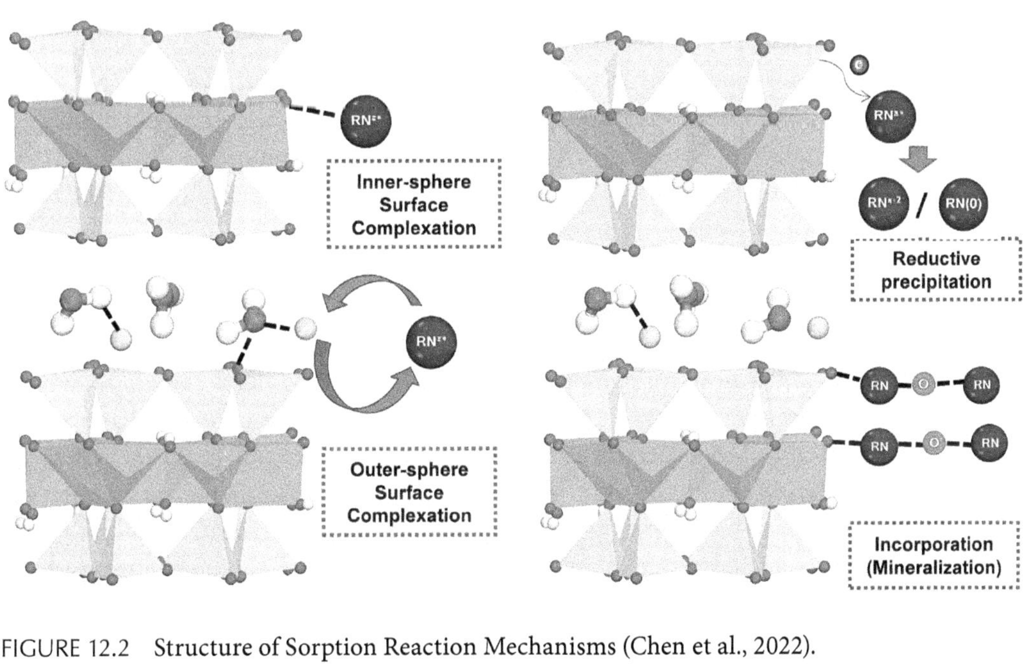

FIGURE 12.2 Structure of Sorption Reaction Mechanisms (Chen et al., 2022).

confirmed outer- and inner-sphere coordination of Ni(II) on cryptomelane (Wu et al., 2018). Wang et al., using time-resolved laser fluorescence spectroscopy (TRLFS), observed that Eu(III) on hydrous alumina forms outer- and inner-sphere complexes, and, in the presence of fulvic acid, additionally generates ternary complexes ($\equiv COO^{-}–Eu–(O–Al\equiv)\cdot 4H_2O$) (Wang et al., 2019b).

12.2.1.2.3 Precipitation and Mineralization Fe(II) behavior in anoxic suspensions of γ-Al_2O_3, montmorillonite, and amorphous silica, was investigated, and it was found that secondary Fe^{2+} precipitates appear above pH 7, with type and rate dependent on adsorbent, pH, and reaction time (Zhu & Elzinga, 2014). Tang et al. demonstrated that U(VI) interaction with alumina at higher pH involves uranyl polymers or hydroxide precipitates, affecting its retention (Tang & Reeder, 2009).

12.2.1.2.4 Surface-Induced Reductive Precipitation Under alkaline repository conditions, Ca^{2+} and Sr^{2+} enhance U(VI) and Np(VI) retention on Ca-bentonite by compensating surface charges (Philipp et al., 2022; Wu et al., 2024a). U(VI) can be reduced to U(IV)/UO_{2+x} by magnetite, which shows higher reductive capacity than goethite or ferrihydrite (Wang et al., 2021b; Ma et al., 2020). Studies have shown that Se(IV) can be reduced by Fe^{2+}- and Fe^{2+}-bearing minerals, such as magnetite and siderite (Kang et al., 2013). Under acidic conditions, Se(0) formation is favored, whereas $FeSe_2$ is more stable under near-neutral to alkaline conditions (Kang et al., 2013; Kang et al., 2016; Ma et al., 2014).

12.2.1.3 Microscopic Structural Analysis of Sorption

Molecular-scale studies of radionuclide adsorption on mineral surfaces are essential for understanding their environmental behavior (Wang et al., 2019b; Philipp et al., 2022). Advanced spectroscopic methods, including X-ray absorption near-edge structure (XANES), extended X-ray absorption fine structure (EXAFS), time-resolved laser fluorescence spectroscopy (TRLFS), and X-ray photoelectron spectroscopy (XPS), are widely used to identify radionuclide adsorption mechanisms. Yet, the complexity of mineral surfaces means that different methods may yield varying interpretations of surface species. Theoretical calculations can provide atomic-level structural information on surface complexes and thus serve as a valuable complement to spectroscopy.

The combination of theoretical calculations and spectroscopic techniques has been widely applied to elucidate adsorption mechanisms (Yan et al., 2013; Tan et al., 2017; Tan et al., 2018; Singh & Um, 2023). Density functional theory has been applied to calculations of uranium and plutonium compounds, and the results indicate that $UO_2^{2+}(H_2O)_5$ and $PuO_2^{2+}(H_2O)_5$ are the most stable hydrated complexes. Further theoretical analysis of UO_2^{2+} adsorption on mineral surfaces examined all possible mononuclear bidentate configurations, identifying the preferential adsorption sites and stable binding geometries. Density functional theory, along with spectroscopic analyses, has revealed distinct U(VI) coordination modes, including edge- and corner-sharing complexes on hematite facets. A study used three-dimensional fluorescence spectroscopy to show that Cs^{+}, Sr^{2+}, and Eu^{3+} affect humic acid polymerization mainly through carboxyl and phenolic hydroxyl groups, with Sr^{2+} and Eu^{3+} causing particularly strong decreases in fluorescence intensity (Yan et al., 2013; Tan et al., 2017; Tan et al., 2018; Mei et al., 2020; Tan et al., 2013).

12.2.2 Diffusion

12.2.2.1 Mineralogical and Physicochemical Basis of Buffer Materials

In geological disposal systems, the diffusion of radionuclides through buffer/backfill materials and natural barriers is one of the most critical processes governing repository performance and long-term safety assessment. The rate and extent of diffusion are strongly influenced by the mineralogical structure and physicochemical properties of the host media, which determine the retardation capacity, sorption behavior, and water transport characteristics of the barriers. Among various engineered materials, buffer and backfill play a dual role: they not only provide mechanical support and sealing around the canister but also act as diffusion-resistant media that significantly reduce radionuclide migration. Bentonite has been widely

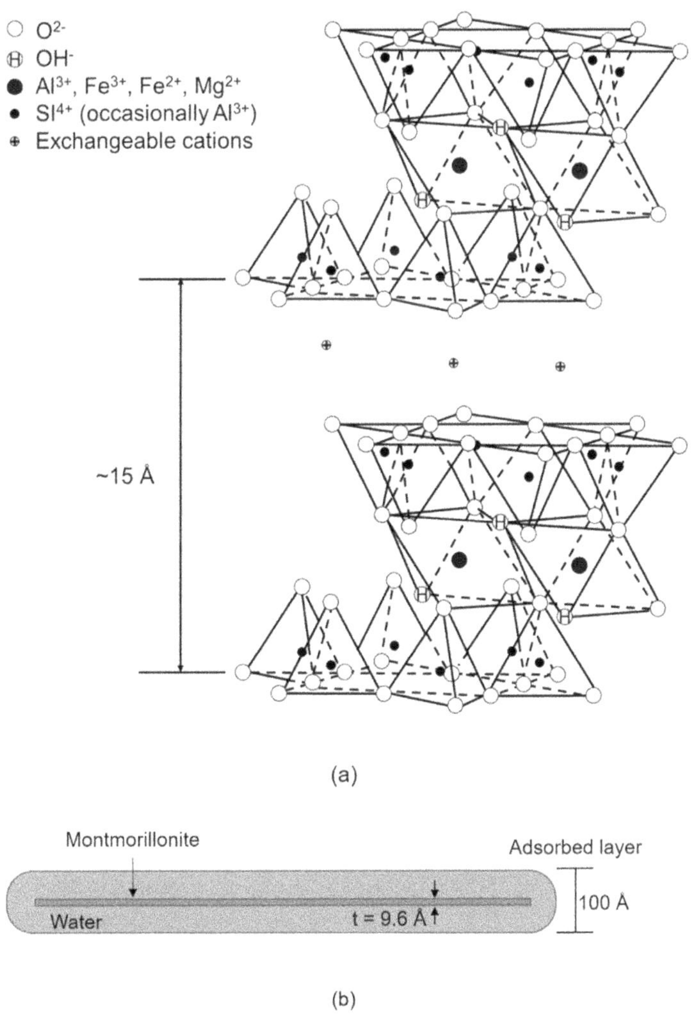

FIGURE 12.3 Molecular Structure of Montmorillonite (Noh et al., 2025).

employed because of its large specific surface area, strong adsorption and cation exchange capacities, low hydraulic conductivity, and remarkable swelling properties (Ye et al., 2010; Ye et al., 2014). Bentonite is a clay material mainly composed of montmorillonite, which belongs to the smectite group of layered silicate minerals. The physicochemical properties of 18 representative clay minerals are summarized in Table 12.1, including external specific surface area (A_{ext}), grain density (ρ_s), cation exchange capacity (CEC), and montmorillonite content (m). Montmorillonite has a layered structure with negatively charged surfaces that attract exchangeable cations, such as Na^+, Ca^{2+}, and Mg^{2+}, into its interlayers to maintain electrostatic neutrality. When water enters bentonite, thermally agitated water molecules hydrate the interlayer cations, neutralize the surface charges of the particles, and promote the formation of an diffuse double layer. The thickness of a montmorillonite layer is about 9.6 Å and its length is approximately 1000 Å. The interlayer accommodates an adsorbed water film approximately 100 Å thick (**Figure 12.3**), and this thin layered morphology confers a high specific surface area conducive to extensive proton and ion adsorption (Noh et al., 2025).

As an engineered barrier material, bentonite is designed to seal the space around the waste canister and to function as a low-permeability barrier. It can immobilize radionuclides through ionic bonding with clay surfaces, thereby limiting their further diffusion into the surrounding environment (Missana et al., 2018; Cao et al., 2019). Nevertheless, potential risks related to radionuclide migration through the bentonite barrier remain.

TABLE 12.1 The Properties of Various Types of Bentonites (Feng et al., 2024a; Wu et al., 2024b)

Clays	A_{ext}* (m^2/g)	m* (−)	ρ_s* (kg/m^3)	CEC (meq/100 g)	Ref.
Wyoming	30–45	0.79–0.85	2,760	79	(Ye et al., 2010)
Gaomiaozi	26	0.75	2,660	72	(Wen, 2008)
Kunipia F	28	0.95–1.00	2,880	108	(Tachi et al., 2010)
FEBEX	56	0.89–0.95	2,700	111	(Lloret & Villar, 2007)
Kunigel V1	30	0.46–0.50	2,700	60	(Sasaki et al., 1994)
Avonlea	55	0.80	2,750	82	(Oscarson et al., 1985)
Anji	60	0.46	2,800	76	(Feng et al., 2023)
Ba-Bent.	27	0.78	2,710	59	(Wu et al., 2023a)
Ca-Bent.	31	0.0	2,800	83	(Sun et al., 2014)
Ca-Mont.	38	1.00	2,800	97	(Li et al., 2012b)
Illite/smectite mixed layer	26	0.33	2,800	38	(Geramian et al., 2016)
Kunipia P	28	0.98	2,800	110	(Sato, 2009)
Mont.	100	1.00	2,800	108	(Wu et al., 2020b; Tachi & Yotsuji, 2014)
Na-Bent.	30	0.81	2,800	76	(Sun et al., 2014)
Na-Mont.	28	0.98	2,900	121	(Li et al., 2012b)
Zhisin	35	0.37	2,670	70	(Chen & Huang, 2013)
Illite	112#	0	2,700#	21	(Cheng & Heidari, 2017)
Kaolinite	46#	0	2,600#	11.3	(Bhattacharyya & Gupta, 2006)

* (Wu et al., 2024b).

\# (González Sánchez et al., 2008).

12.2.2.2 Key Thermo-Hydro-Mechanical Properties of Bentonite Barrier

Thermal conductivity quantifies a material's ability to transmit heat. It is primarily controlled by mineral composition, dry density, water content, and temperature. Experiments show that the thermal conductivity of compacted MX-80 and Gaomiaozi bentonites increases with increasing temperature (Park et al., 2021; Xu et al., 2022). Hydraulic conductivity characterizes the capacity of a porous medium to convey fluid through its interconnected pores when the medium is fully or nearly saturated. It is influenced by factors such as dry density, temperature, and groundwater chemistry. High compaction reduces pore connectivity, while temperature and chemical composition of porewater further alter its hydraulic properties (Ye et al., 2014; Noh et al., 2025). Swelling pressure arises when water molecules enter the interlayer space of montmorillonite and the resulting volumetric expansion is restricted by external stresses or boundary constraints, such as constant volume conditions. This pressure is governed by dry density, temperature, and pore fluid chemistry (Geramian et al., 2016; Sun et al., 2018). Higher compaction produces greater swelling pressure (Kim et al., 2021). The water retention capacity of bentonite is determined by its unsaturated hydraulic properties, typically represented by the soil–water retention curve that relates matric suction to water content or degree of saturation (Noh et al., 2025). At low suction (<15 MPa), higher dry density enhances water retention, whereas under near-saturated conditions, lower-density samples hold more water at the same suction (Yoon et al., 2022).

12.2.2.3 Experimental Determination of Diffusion Coefficients

In groundwater systems, the diffusion behavior of radionuclides within bentonite, clay minerals, granite, and associated fracture-filling materials represents a key parameter for evaluating their long-term migration potential. The principal experimental methods employed to determine these parameters include (i) the capillary method; (ii) the through-diffusion method; and (iii) the in-/out-diffusion method (Van Loon & Müller, 2014; Wu et al., 2020a). In-diffusion capillary methods under varying pH and redox conditions have been used to examine the effects of iron content and its valence states on the diffusion and sorption behavior of $^{75}SeO_3^{2-}$ in Beishan granite (He et al., 2016). Through-diffusion experiments have been conducted to measure effective diffusion coefficient and distribution coefficient values of radionuclides such as HTO, $^{36}Cl^-$, $^{133}Ba^{2+}$, and $^{134}Cs^+$ in crystalline rocks, showing excellent agreement with in situ data. The effective diffusion coefficient followed the trend: HTO > Cl^- > $Ba^{2+} \approx Cs^+$ (Aromaa et al., 2019). In Gaomiaozi bentonite, the effective diffusion coefficient decreases with increasing dry density (Feng et al., 2024c). In contrast, increasing ionic strength compresses the diffuse double layer in compacted bentonite, enlarging pore spaces and facilitating radionuclide diffusion (Wu et al., 2020b). Through-diffusion experiments of Ce(III)–EDTA in Zhisin bentonite were conducted to evaluate the influence of the organic ligand EDTA on the diffusion of trivalent actinides (Wu et al., 2023b).

Extensive experimental studies have been conducted in Beishan granite and Gaomiaozi bentonite. In granite, Am(III) exhibited an apparent diffusion coefficient of $(1.3 \pm 0.16) \times 10^{-15}$ m²/s under groundwater conditions by in-diffusion (Song et al., 2009). For Cs^+, through-diffusion yielded an apparent diffusion coefficient of 1.18×10^{-9} m²/s in fractured granite, while matrix diffusion ranged from 10^{-13} to 10^{-12} m²/s (Wang et al., 2020). Tritiated water (HTO) displayed effective diffusion coefficients ranging from 3.0 × 10–^{12} to 3.4 × 10^{-12} m^2/s (Liu et al., 2003). The diffusion coefficients of I^- varied widely (10^{-13}–10^{-7} m^2/s), depending on the applied method, including capillary, column, and through-diffusion techniques (Wang et al., 2020; Dang et al., 2014; He et al., 2020; Chen et al., 2013). Se(IV) and TcO_4– displayed effective diffusion coefficients of 10^{-13}–10^{-10} m^2/s and approximately 1 × 10^{-12} m^2/s, respectively, as determined by through-diffusion (He et al., 2016; Wang et al., 2019a; Jan et al., 2014; Yang et al., 2017; Li et al., 2012a; Liu et al., 2003). In Gaomiaozi bentonite, through-diffusion experiments indicated that I^- had effective diffusion coefficients in the range of 2.4 × 10^{-12}–2 × 10^{-11} m^2/s (Wu et al., 2012; Zhao et al., 2016). La(III), measured by in-diffusion, showed apparent diffusion coefficients between 2 × 10^{-12} and 1.6×10^{-11} m^2/s (Chen et al., 2016). For Np, the apparent diffusion coefficients obtained from out-diffusion ranged from 3.9 × 10^{-13} to 8.4 × 10^{-13} m^2/s, whereas through-diffusion yielded a higher value of 8.2 × 10^{-12} m^2/s (Yao et al., 2003). Pu demonstrated extremely limited mobility in granite, with a through-diffusion coefficient of (1.33 ± 0.52) × 10^{-13} m^2/s (Dang et al., 2014). Effective diffusion coefficients of Re(VII), obtained by through-diffusion, ranged from 10–^{12} to 10^{-11} m^2/s (Wang et al., 2016; Wu et al., 2014; Wu et al., 2017). For TcO_4^-, through-diffusion gave values of 5.3 × 10^{-12}–1.2 × 10^{-11} m^2/s, while capillary methods produced higher apparent coefficients of 1.3 × 10^{-10}–3.3 × 10^{-10} m^2/s (Zhao et al., 2013; Wu et al., 2013; Tsai et al., 2017). Se(IV) in Gaomiaozi bentonite exhibited diffusion coefficients in the range of 4.5 × 10^{-12}–5.4 × 10^{-11} m^2/s (Wang et al., 2016; Wu et al., 2014; Wu et al., 2017).

12.2.2.4 Diffusion Mechanisms in Porous Media

Diffusion is a fundamental and irreversible process driven by random molecular motion, leading to the spontaneous transfer of matter from regions of high to low chemical potential. Since chemical potential is proportional to concentration, diffusion is typically described as a flux driven by concentration gradients, following Fick's first law. The free-solution diffusion coefficient governs this process and depends on physicochemical factors such as temperature, viscosity, dielectric constant, solute size and charge, electroneutrality, and solute association–dissociation state (Shackelford & Moore, 2013; Feng et al., 2024c; Van Loon & Müller, 2014).

When diffusion occurs in porous media, the classical form of Fick's laws is modified by the presence of the solid matrix. The inclusion of solid particles introduces tortuous diffusion pathways compared with those in free solution, which effectively increases the transport distance over which the concentration gradient is applied (Soler et al., 2019). This results in a reduced diffusive mass flux relative to free solution conditions. Moreover, because diffusing species cannot migrate through crystalline particles, the cross-sectional area available for transport is smaller than the total macroscopic area, further lowering the flux. Consequently, the effective diffusion coefficient in porous media is smaller than the free solution value, and Fick's first and second laws must be reformulated to incorporate effects of tortuosity, porosity, and potential interactions between solutes and the solid matrix. Collectively, these modifications explain why diffusion fluxes in porous media are generally diminished relative to those in aqueous systems without a solid phase (Shackelford & Moore, 2013). The diffusion behavior of Re and Se in compacted Wyoming bentonite was investigated by applying the modified form of Fick's laws in combination with the through-diffusion method, and ReO_4^- exhibited no sorption, whereas $HSeO_3^-$ formed bidentate binuclear inner-sphere surface complexes on montmorillonite, resulting in measurable sorption. A multi-porosity model incorporating discrete water hydration layers was introduced to describe the interlayer hydration behavior of compacted bentonite (Figure 12.4), demonstrating that free porosity constitutes the dominant pathway for anion diffusion (Feng et al., 2024b).

12.2.3 Colloid-Facilitated Radionuclide Transport

12.2.3.1 Types and Sources of Colloids

In hydrogeochemistry, colloids are generally defined as inorganic or organic particles with sizes of 1 to 1000 nm that can remain stably suspended in water (Xu et al., 2018). They possess large specific surface area and usually carry a surface charge, which confers high reactivity. As a result, their environmental behavior, especially their geochemical interactions under near-field conditions of high-level radioactive waste repositories, has received increasing attention (Chen et al., 2022; Tang et al., 2025). Groundwater commonly contains environmental colloids that are broadly classified as inorganic colloids (e.g., aluminosilicate minerals, Fe/Al oxides) and organic colloids derived from natural organic matter. Compacted bentonite in contact with near-field and far-field groundwater near the repository may release bentonite colloids (Xu et al., 2018).

Colloid formation mainly arises from hydrodynamic shear and geochemical disturbance caused by changes in ionic strength, redox potential, or pH. In addition, mineral particles may adsorb charged species or macromolecules, altering their surface properties

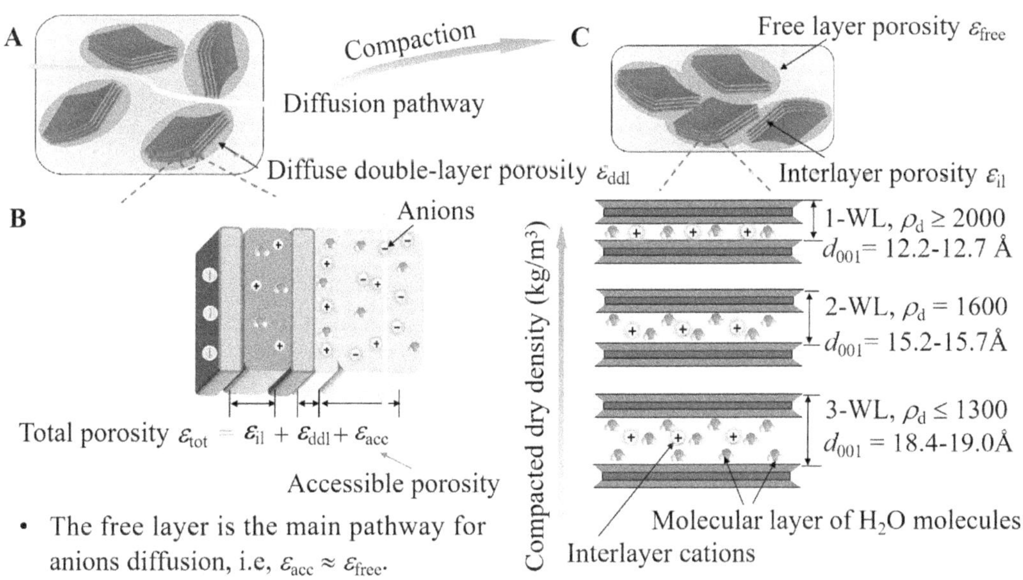

FIGURE 12.4 (A) Schematic Diagram of the Anion Diffusion Pathway and the Microstructure of Compacted Montmorillonite. (B) The Porosity Distribution of Compacted Montmorillonite. (C) The Effect of Compaction on the Interlayer Porosity (Feng et al., 2024b).

and converting to a colloidal state (Zou et al., 2016). Among inorganic colloids, clay mineral colloids are particularly important and complex. They generally carry a high negative charge, have large cation exchange capacity and extensive surface area, enabling strong interactions with radionuclides through sorption, surface complexation, and ion exchange reactions. Under near-field repository conditions, multiple factors can lead to the formation of radioactive colloids.

The properties of composite colloid are not a simple sum of their individual components and synergistic or antagonistic effects may occur among them (Tang et al., 2024). The interactions between composite colloids and radionuclides are therefore complex, and clarifying the binding characteristics and contributions of each component is essential for understanding radionuclide migration and environmental behavior under diverse geochemical conditions. Natural organic matter, represented by humic substances, is ubiquitous in aquifers and plays a major role in the transport of radioactive colloids. Humic substances are negatively charged polyelectrolytes containing numerous functional groups such as carboxyl and phenolic groups. Through various mechanisms of attaching to mineral surfaces, they can alter the surface charge properties of mineral colloids, thereby influencing the environmental behavior of both radionuclides and colloids (Philippe & Schaumann, 2014; Tang et al., 2025; Shi et al., 2025a).

12.2.3.2 Stability and Transformation of Colloids

The stability of colloids is a decisive factor controlling both their potential for environmental migration and their modes of interaction with radionuclides. Aggregation and subsequent sedimentation substantially diminish colloidal mobility; however, under conditions that again favor dispersion, previously aggregated clusters may redisperse, leading to secondary transport and thereby converting colloid-associated radionuclides into latent secondary contamination sources (Xu et al., 2018). Colloidal stability is controlled by a wide range of environmental parameters, including ionic strength, pH, temperature, and colloid concentration (Shi et al., 2025a; Wei et al., 2023; Tang et al., 2025). These parameters alter surface charge density and particle size distribution, which in turn regulate processes of release, aggregation, migration, and transformation, as well as the binding affinity between colloids and radionuclides. In particular, the dissociation state of humic substances can change, which in turn affects their hydrophilicity, charge density, and molecular conformation. At low pH, humic molecules may undergo coiling and aggregation, leading to the formation of humic flocs, while at high pH, they remain more soluble and persist stably in solution (Pan et al., 2019). The mechanisms underlying colloidal aggregation are primarily driven by Brownian motion, hydrodynamic forces, and differential sedimentation. For particles smaller than 300 nm, Brownian motion dominates collision frequency. While colloids are inherently thermodynamically unstable, not all collisions result in aggregation owing to interparticle repulsive forces, predominantly electrostatic repulsion (Wagner et al., 2014). The magnitude of the interparticle potential energy barrier is therefore a critical determinant of stability, and this barrier is strongly influenced by background electrolyte concentration. The electrolyte concentration at which the energy barrier is eliminated is defined as the critical coagulation concentration (Schaefer et al., 2012).

12.2.3.3 Mechanisms of Colloid-Facilitated Transport

The transport of colloids in aqueous environments primarily involves dissolution, filtration, aggregation, adsorption, and diffusion. These processes are mainly controlled by the intrinsic properties of the colloids and the surrounding geochemical conditions (Sun et al., 2021; Xu et al., 2018; Shi et al., 2025a). Colloid-facilitated radionuclide transport can occur in two major ways: (i) radionuclides directly existing as colloidal species (intrinsic colloids), or (ii) radionuclides adsorbed or bound to colloidal surfaces (pseudo-colloids), and its occurrence depends on five conditions (Figure 12.5).

A three-year study was conducted using batch experiments and AF4 to investigate the interaction between montmorillonite colloids and Cs(I), Eu(III), Th(IV), and U(VI) in a natural repository in Sweden (Bouby et al., 2011). Under initial conditions, Eu(III) and

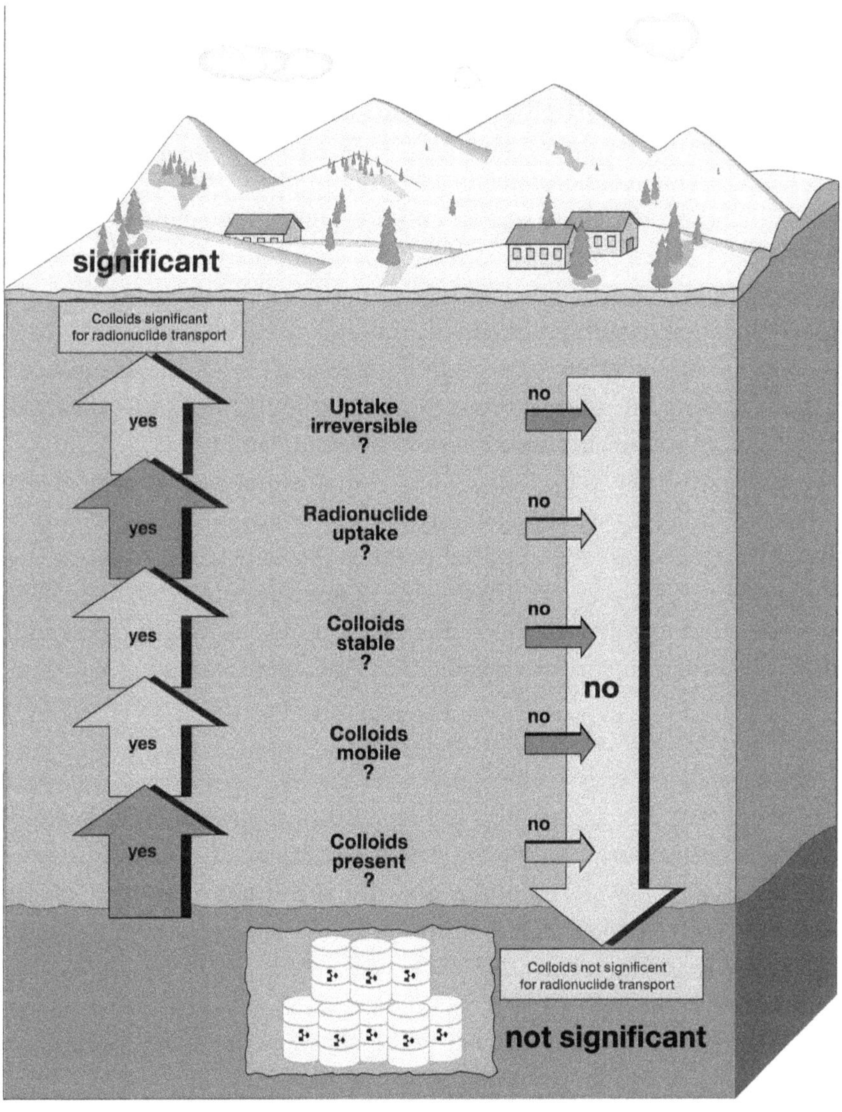

FIGURE 12.5 The Conditions of Colloid-Facilitated Transport (Zaenker & Hennig, 2014).

Th(IV) were significantly adsorbed on montmorillonite colloids, and metal ions preferentially adsorbed on small-sized colloidal particles. The adsorption of Eu(III) and Th(IV) on clay colloids did not increase the aggregation of colloidal particles, while Cs(I) and U(VI) existed in the form of ions and complexed species $UO_2(OH)_3-$, respectively. After one year, the adsorption of Th(IV) on montmorillonite colloids slightly decreased, while the adsorption of Eu(III) on colloids reached 97%. Upon adding humic acid, partial desorption of Eu(III) and Th(IV) occurred due to re-adsorption by humic substances. Eu(III) was almost completely desorbed and bound to humic acid, whereas Th(IV) desorption was much slower, with only about 50% re-adsorbed by humic acid. These results demonstrate that under the presence of humic acid, the adsorption of trivalent Eu(III) onto colloids in groundwater of crystalline rocks is reversible, while the sorption of tetravalent Th(IV) is not. Consequently, stable colloids may facilitate radionuclide transport, whereas unstable colloids under certain geological conditions could effectively retard radionuclide mobility. The adsorption of Th(IV), U(VI), Np(V), Tc(VII), and Pu(IV) onto montmorillonite colloids of various sizes was investigated. Results showed strong sorption of Th(IV) and Pu(IV), but little for U(VI), Np(V), and Tc(VII) (Norrfors et al., 2016). Colloid particle size was shown to influence radionuclide sorption: in open systems with CO_2, radionuclides preferentially adsorbed onto smaller colloids, which offered more sorption sites due to their higher abundance and larger surface area, and thus smaller colloids dominated Eu(III) sorption. In contrast, under closed-system conditions without CO_2, radionuclides such as Am(III) tended to sorb preferentially onto relatively larger colloids (Maher et al., 2016). The reversibility of actinide sorption onto colloids is a critical factor in radionuclide transport, making the speciation of adsorbed radionuclides and its evolution under changing aqueous geochemical conditions a key consideration.

Colloid transport is a key factor in the safety assessment of deep geological repositories. In the presence of colloids, radionuclide migration may be either facilitated or hindered. Unlike dissolved species, colloids migrate mainly through larger pores at higher flow velocities. However, colloids are not always able to traverse porous media freely. They can be filtered, retained, or aggregated, which leads to clogging. Such aggregation and blockage are key physical mechanisms that contribute to radionuclide retention.

12.2.3.4 Experimental Approaches and Case Studies

Column experiments serve as preliminary tests for the migration of radionuclides in porous media, enabling rapid assessment of radionuclide migration and providing insights into general transport patterns and characteristic parameters (e.g., distribution coefficient and retardation factor), thus forming the basis for site-specific studies. Column experiments have been widely applied to investigations of radionuclides, colloids, and radionuclide–colloid complexes (Tang et al., 2025; Wei et al., 2023).

The transport of I^- through crushed granite columns was investigated using a classical advection–dispersion model that considered sorption to interpret the experimental results. An analytical solution was obtained to estimate the dispersion coefficient by the initial and boundary conditions (Yuan et al., 2019). U(VI) transport in saturated sand and crushed granite columns was studied, with the experimental observations fitted by a two-site kinetic non-equilibrium sorption model (Ge et al., 2018). The gradual aggregation of montmorillonite colloids was observed after three years in low-ionic-strength groundwater, with Ca^{2+} in groundwater playing a dominant role in this process. The presence of humic acid accelerated the desorption of Eu(III) from montmorillonite colloids but had little effect on the release of Th(IV) (Bouby et al., 2011). Dittrich et al. found that bentonite colloids facilitated the transport of ^{241}Am but could not cause long-distance migration, and the binding of ^{241}Am to bentonite colloids was weak, and ^{241}Am adsorbed onto the colloid surface desorbed quickly (Dittrich et al., 2015).

12.3 CONTROLLING FACTORS ACROSS PROCESSES

Radionuclide migration in geological disposal systems is governed by geochemical and physical processes influenced by groundwater chemistry, temperature, and host rock mineralogy. These factors affect dissolution, adsorption, redox reactions, and transport pathways, ultimately controlling long-term migration and repository safety (Vu et al., 2025).

12.3.1 Chemical Speciation and Geochemical Controls of Radionuclides

The speciation and physicochemical properties of radionuclides are the key factors determining their migration behavior in geological disposal systems. Geochemically, groundwater pH, redox potential, ionic strength, and ligand composition strongly affect the valence and speciation of radionuclides (Wu et al., 2024a; Perez-Moreno et al., 2020). At $pH < 5$, U and Tc occur as soluble UO_2^{2+} and TcO_4-, which increases their mobility. Alkaline conditions with pH above 9 promote the formation of insoluble carbonate complexes, such as UO_2^{2+} bound to Ca^{2+} and $CO_3^{2}-$, with solubility reduced to about $10-^{11}$ mol/L. Geochemical modeling of Beishan groundwater showed that U(VI) is predominantly present as $Ca_2UO_2(CO_3)_3(aq)$ and $CaUO_2(CO_3)_3^{2-}$, indicating that Ca^{2+} complexation is a key factor controlling U(VI) speciation (Wu et al., 2024a). Changes in Eh govern the redox transformation of radionuclides (Vu et al., 2025). At Eh <−200 mV, mobile Tc(VII) reduces to $Tc(IV)O_2{\cdot}nH_2O$ (solubility <0.1 μg/L). At Eh >200 mV, U(VI) remains as UO_2^{2+} and its mobility increases. Pyrite is a common sulfide mineral widely present in granitic rocks and sedimentary rocks. Previous studies (Kang et al., 2016; Wang et al., 2021a) have demonstrated that pyrite can effectively reduce Se(IV), U(VI), and Tc(VII) in groundwater, making their insoluble forms more likely to precipitate. Ionic strength and ligands also play an important role (Perez-Moreno et al., 2020). In addition to aqueous speciation, colloid transport represents a critical but often overlooked pathway for radionuclide migration. Recent studies have shown that reduced U(IV), traditionally considered immobile, can persist in nano-crystalline or non-crystalline colloidal forms and be transported over long distances when associated with natural colloids such as bentonite or stabilized by humic substances (Figure 12.6). This highlights the need to integrate colloidal processes into safety assessments of geological disposal systems (Shi et al., 2025a).

In deep geological disposal environments, the migration behavior of radionuclides primarily depends on their chemical speciation in groundwater, which is governed by pH, redox potential, ionic strength, and ligand types. These factors collectively determine radionuclide valence states, solubility, and complexation modes, thereby influencing their migration potential.

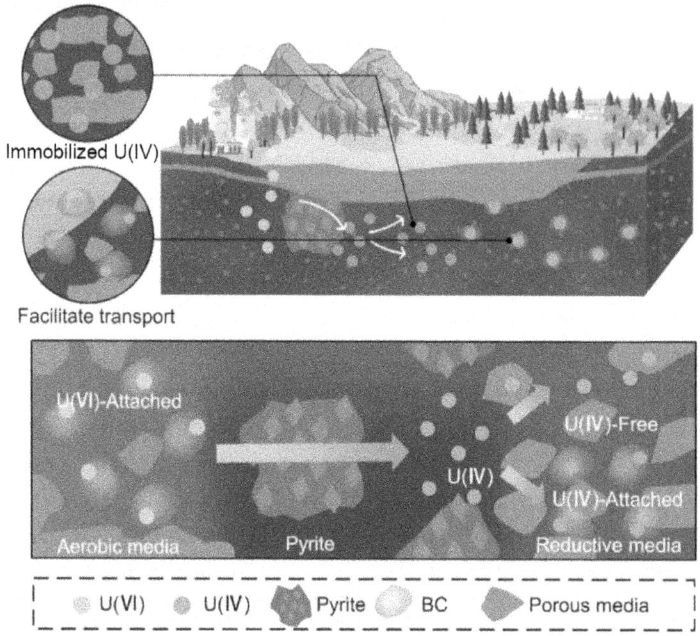

FIGURE 12.6 Colloid-Facilitated U(IV) Transport under Reductive Conditions (Shi et al., 2025a).

12.3.2 Geological Controls on Radionuclide Migration

The migration of radionuclides in rocks and soils is influenced by multiple factors, among which geological conditions play a particularly critical role. In radionuclide transport studies, geological conditions are primarily reflected in the structural and physicochemical properties of soils, rocks, and minerals, including granite, shale, limestone, as well as bentonite, and clay.

Extensive research has been conducted on radionuclide migration under diverse geological settings (Hong et al., 2024; He et al., 2016; Aslannejad et al., 2022). In clay rocks with hydraulic conductivity below 10^{-13} m/s, radionuclide transport is dominated by diffusion, whereas fractured granite with conductivity above 10^{-9} m/s develops preferential flow paths and increases migration by more than two orders of magnitude. A generic repository model with coupled THC and THMC was compared, showing that mechanical effects strongly influenced hydrological and geochemical changes in the early unsaturated stage, but the impact declined as bentonite became saturated (Cao et al., 2025). Radionuclide migration in fractured granite is governed by the coupled effects of advection, diffusion, radioactive decay, and sorption, whereas clay rocks, due to their extremely low permeability and strong sorption capacity, can effectively hinder radionuclide transport (He et al., 2020; Wang et al., 2019a; Wang et al., 2020). In bentonite, montmorillonite plays a crucial role in retarding U(VI) through sorption, while sepiolite clays, owing to their unique molecular structure, can accelerate the migration of Cs. In contrast, radionuclides such as Th, Ni, and Cs exhibit the slowest migration rates in clays because of strong sorption interactions. These observations highlight the significant influence of geological conditions on radionuclide transport. Therefore, effective radioactive waste disposal and environmental protection require comprehensive consideration of factors such as soil and rock types, mineral composition, hydrological conditions, geochemical environment, groundwater pressure, geological structures, and biological processes. Overall, groundwater chemistry, thermal conditions, and host rock properties interact in complex ways to determine radionuclide migration (Perez-Moreno et al., 2020; Cao et al., 2025; Vu et al., 2025).

12.4 MODELING AND SIMULATION ADVANCES

12.4.1 Multi-Mechanism Coupled Modeling

Safety assessment is the core element ensuring the long-term security of high-level radioactive waste repositories over timescales of tens of thousands of years. Its framework follows the technical guidelines of the International Atomic Energy Agency and national regulatory systems (IAEA, 2011, 2012). Radionuclide migration through rocks, groundwater, and soils involves coupled physical, chemical, and biological processes, requiring interdisciplinary modeling that integrates geochemistry, fluid mechanics, numerical simulation, and artificial intelligence (Cao et al., 2025; Wu et al., 2024a; Vu et al., 2025). Since the 1980s, programs have established integrated safety-assessment frameworks combining simplified, process, and system models to enable cross-scale evaluation from mechanistic understanding to repository-scale behavior.

Simplified models provide efficient computational tools for rapid risk evaluation (Hermansson & Rydh, 2013). They serve as benchmarks for complex models, accelerate multi-scenario simulations, and support uncertainty analyses. Representative examples include the near-field compartment model COMP23/NUCTRAN, developed in Sweden for near-field radionuclide transport, and Migration Analysis of Radionuclides in the Far Field (MARFA), developed for far-field radionuclide migration in fractured rock.

Process models simulate coupled multi-physical and geochemical processes across spatial scales (Nole et al., 2022). Typical examples are TOUGHREACT, which couples geochemical reactions with multi-phase flow and heat transfer under non-isothermal conditions, and PFLOTRAN, which models flow and reactive transport in porous and fractured media.

System models integrate multiple processes to evaluate long-term repository performance (Paxton et al., 2025). They simulate repository evolution, consolidate multi-source data, and quantify radiation doses and risk indicators over geological timescales. Examples include GOLDSIM, which performs probabilistic uncertainty analyses, and AMBER, which evaluates radionuclide transport through advection–diffusion–reaction modeling.

12.4.2 Machine Learning

Machine learning has become an increasingly powerful tool for studying radionuclide migration, offering efficient and accurate predictions while revealing complex multi-factor interactions (Feng et al., 2025; Shi et al., 2025b). Machine learning has emerged as a powerful tool for investigating radionuclide sorption on mineral surfaces. An ensemble learning method was applied to predict adsorption rates of multiple metal cations on clay minerals, identifying pH, ionic strength, and surface properties as dominant factors that complement traditional surface complexation models (Zhao et al., 2024). The distribution coefficients of Cs in different solid

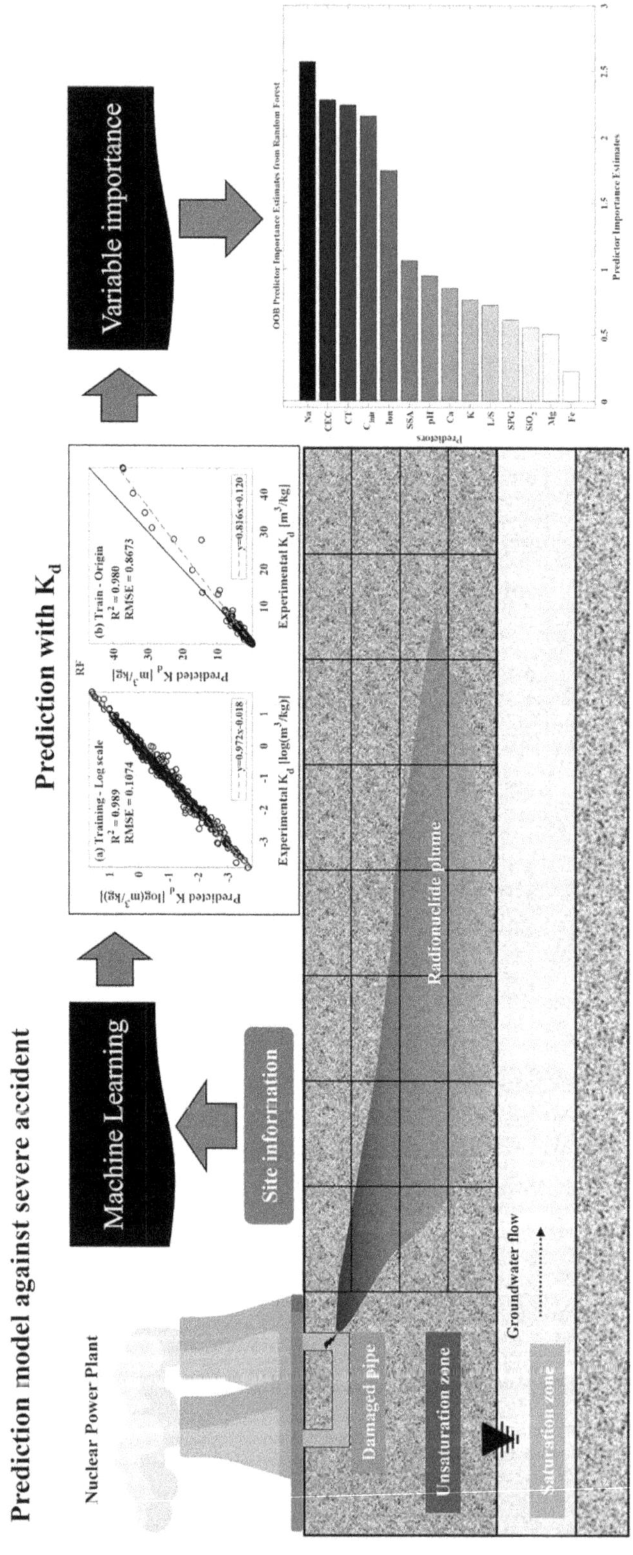

FIGURE 12.7 Predicting the Distribution Coefficient of Cesium in Solid Phase Groups Using Machine Learning (Hong et al., 2024).

phase groups were predicted by three models (Figure 12.7) (Hong et al., 2024). For radionuclide diffusion, various machine learning algorithms have been used to predict the diffusion of ReO_4^-, $HCrO_4^-$, and I^- in compacted bentonite. These studies compared models trained on pseudo-data generated using a porosity model with models trained on experimental datasets (Feng et al., 2023). An integration of generative adversarial networks with machine learning models was developed to predict the hydraulic conductivity of bentonite. Incorporating physically meaningful input features, the hybrid model outperformed both the classical and modified Kozeny–Carman equations (Shi et al., 2025b). In radionuclide transport, integration of Lattice Boltzmann simulations and artificial neural networks accurately predicted colloid migration and clogging in porous media (Aslannejad et al., 2022). A supervised long short-term memory network successfully modeled breakthrough curves of Eu(III) in quartz sand in the presence of *Bacillus subtilis* and bentonite colloids, identifying bentonite colloid concentration and pH as key governing factors (Tang et al., 2025).

12.5 ENVIRONMENTAL SIGNIFICANCE AND FUTURE RESEARCH NEEDS

Radionuclide migration in geological repositories is controlled by sorption, diffusion, and colloid-facilitated transport, all of which directly affect long-term safety assessments. Sorption regulates retardation but varies with geochemical conditions, diffusion in bentonite is influenced by thermo–hydro–mechanical properties, and colloids can act as mobile carriers, accelerating transport. Identifying dominant factors such as pH, ionic strength, temperature, and mineralogy is therefore essential for reliable risk evaluation. Recent advances in multi-mechanism modeling and machine learning have enhanced predictive accuracy, yet challenges remain due to limited datasets and scale effects. Future efforts should prioritize the development of comprehensive migration databases, multi-scale coupled modeling frameworks, and systematic uncertainty analysis, while integrating machine learning with physics-informed models, which holds particular promise for strengthening long-term safety evaluations of high-level radioactive waste repositories.

ACKNOWLEDGMENTS

This work was supported by the National Natural Science Foundation of China (22176077) and the Fundamental Research Funds for the Central Universities (lzujbky-2023-stlt01).

REFERENCES

Aromaa, H., Voutilainen, M., Ikonen, J., Yli-Kaila, M., Poteri, A., Siitari-Kauppi, M., 2019. Through diffusion experiments to study the diffusion and sorption of HTO, 36Cl, 133Ba and 134Cs in crystalline rock. *J. Contam. Hydrol.* 222, 101–111.

Aslannejad, H., Samari-Kermani, M., Nezami, H., Jafari, S., Raoof, A., 2022. Application of machine learning in colloids transport in porous media studies: Lattice Boltzmann simulation results as training data. *Chem. Eng. Sci.* 253, 117548.

Bhattacharyya, K.G., Gupta, S., 2006. Kaolinite, montmorillonite, and their modified derivatives as adsorbents for removal of Cu(II) from aqueous solution. *Sep. Purif. Technol.* 50, 388–397.

Bouby, M., Geckeis, H., Lützenkirchen, J., Mihai, S., Schaefer, T., 2011. Interaction of bentonite colloids with Cs, Eu, Th and U in presence of humic acid: A flow field-flow fractionation study. *Geochim. Cosmochim. Acta* 75, 3866–3880.

Cao, X., Zheng, L., Hou, D., Hu, L., 2019. On the long-term migration of uranyl in bentonite barrier for high-level radioactive waste repositories: The effect of different host rocks. *Chem. Geol.* 525, 46–57.

Cao, X., Zheng, L., Hu, L., 2025. Studying the relevance of mechanical process to the geochemical evolution in bentonite buffer of repositories for high-level radioactive waste. *J. Radioanal. Nucl. Chem.* 334, 1343–1357.

Chen, T., Li, C., Liu, X.Y., Wang, L.H., Zheng, Z., Wang, X.Y., Liu, C.L., 2013. Migration study of iodine in Beishan granite by a column method. *J. Radioanal. Nucl. Chem.* 298, 219–225.

Chen, W.C., Huang, W.H., 2013. Effect of groundwater chemistry on the swelling behavior of a Ca-bentonite for deep geological repository. *Phys. Chem. Earth* 65, 42–49.

Chen, Y.G., Jia, L.Y., Niu, L.H., Ye, W.M., Chen, B., Cui, Y.J., 2016. Effect of dry density and pH on the diffusion behavior of lanthanum in compacted Chinese GMZ bentonite. *J. Radioanal. Nucl. Chem.* 310, 1303–1310.

Chen, Z.Y., Wang, S.Y., Hou, H.J., Chen, K., Gao, P.Y., Zhang, Z., Jin, Q., Pan, D.Q., Guo, Z.J., Wu, W.S., 2022. China's progress in radionuclide migration study over the past decade (2010–2021): Sorption, transport and radioactive colloid. *Chin. Chem. Lett.* 33, 3405–3412.

Cheng, K., Heidari, Z., 2017. A new method for quantifying cation exchange capacity in clay minerals. *Appl. Clay Sci.* 161, 444–455.

Dang, H., Hou, X., Liu, W., Liu, C., Xu, J., 2014. Diffusivity of Sr, I and Pu in granite with through-diffusion experiment. *J. Nucl. Radiochem.* 36, 53–59.

Dittrich, T.M., Boukhalfa, H., Ware, S.D., Reimus, P.W., 2015. Laboratory investigation of the role of desorption kinetics on americium transport associated with bentonite colloids. *J. Environ. Radioact.* 148, 170–182.

Feng, J., Gao, X., Xu, K., Shi, X., Tian, J., Wu, Y., Wu, T., 2025. Predicting distribution coefficient and effective diffusion coefficient of radionuclides in bentonite: Multi-output neural network simulation and diffusion experimental study. *J. Hazard. Mater.* 490, 137787.

Feng, Z., Feng, J., Tian, J., Shi, X., Shao, D., Wu, T., Shen, Q., 2024a. Predicting the diffusion of CeEDTA– and CoEDTA2– in bentonite using decision tree hybridized with particle swarm optimization algorithms. *Appl. Clay Sci.* 262, 107596.

Feng, Z., Gao, Z., Wang, Y., Wu, T., Li, Q., 2023. Application of machine learning to study the effective diffusion coefficient of Re(VII) in compacted bentonite. *Appl. Clay Sci.* 243, 107076.

Feng, Z., Tian, J., Shi, X., Wang, C., Wu, T., 2024b. Analyzing porosity of compacted bentonite via through diffusion method. *J. Radioanal. Nucl. Chem.* 333, 1185–1193.

Feng, Z., Tian, J., Wu, T., Wei, G., Li, Z., Shi, X., Wang, Y., Li, Q., 2024c. Unveiling the Re, Cr, and I diffusion in saturated compacted bentonite using machine-learning methods. *Nucl. Sci. Tech.* 35, 93.

Ge, M., Wang, D., Yang, J., Jin, Q., Chen, Z., Wu, W., Guo, Z., 2018. Co-transport of U(VI) and akaganeite colloids in water-saturated porous media: Role of U(VI) concentration, pH and ionic strength. *Water Res.* 147, 350–361.

Geramian, M., Osacky, M., Ivey, D.G., Liu, Q., Etsell, T.H., 2016. Effect of swelling clay minerals (montmorillonite and illite-smectite) on non-aqueous bitumen extraction from alberta oil sands. *Energ. Fuel* 30, 8083–8090.

González Sánchez, F., Van Loon, L.R., Gimmi, T., Jakob, A., Glaus, M.A., Diamond, L.W., 2008. Self-diffusion of water and its dependence on temperature and ionic strength in highly compacted montmorillonite, illite and kaolinite. *Appl. Geochem.* 23, 3840–3851.

He, J., Ma, B., Kang, M., Wang, C., Nie, Z., Liu, C., 2016. Migration of 75Se(IV) in crushed Beishan granite: Effects of the iron content. *J. Hazard. Mater.* 324, 564–572.

He, J.G., Li, Y., Su, Y., Yang, J.Q., Hou, X.L., Wu, W.S., Shi, K.L., 2020. Influence of γ-irradiation and oxygen conditions on the diffusion of I-125 in crushed Beishan granite. *Appl. Radiat. Isot.* 163, 109224.

Hermansson, H., Rydh, J., 2013. *FARF33: Version 1.1.* Swedish Nuclear Fuel and Waste Management Co., Stockholm.

Hong, S.M., Yoon, I.-H., Cho, K.H., 2024. Predicting the distribution coefficient of cesium in solid phase groups using machine learning. *Chemosphere* 352, 141462.

Hu, J., Tan, X., Ren, X., Wang, X., 2012. Effect of humic acid on nickel(II) sorption to Ca-montmorillonite by batch and EXAFS techniques study. *Dalton Trans.* 41, 10803–10810.

IAEA, 2011. *Geological Disposal Facilities for Radioactive Waste*. IAEA Safety Standards Series No. SSG-14. International Atomic Energy Agency, Vienna.

IAEA, 2012. *The Safety Case and Safety Assessment for the Disposal of Radioactive Waste*. IAEA Safety Standards Series No. SSG-23. International Atomic Energy Agency, Vienna.

Izosimova, Y., Gurova, I., Tolpeshta, I., Karpukhin, M., Zakusin, S., Zakusina, O., Samburskiy, A., Krupskaya, V., 2022. Adsorption of Cs(I) and Sr(II) on bentonites with different compositions at different pH. *Minerals* 12, 862.

Jan, Y., Tsai, S., Li, Y., 2014. Determination of sorption and diffusion parameters of Se(IV) on crushed granite. *J. Radioanal. Nucl. Chem.* 301, 365–371.

Kang, M., Bardelli, F., Ma, B., Charlet, L., Chen, F., Yang, Y., 2016. The influence of pH and reaction time on the formation of FeSe2 upon selenite reduction by nano-sized pyrite-greigite. *Radiochem. Acta* 104, 649–656.

Kang, M., Liu, C., Chen, F., 2013. The reduction of selenite by Fe(II)-containing minerals. *Sci. Sin. Chim.* 43, 536–543.

Kim, M., Lee, S., Cheon, E., Kim, M., Yoon, S., 2021. Thermochemical changes on swelling pressure of compacted bentonite. *Ann. Nucl. Energy* 151, 107882.

Li, C., Liu, X.Y., Chen, T., Tian, W.Y., Zheng, Z., Wang, L.H., Liu, C.L., 2012a. The influence of pH on the sorption and diffusion of 99TcO4− in Beishan granite. *Radiochem. Acta* 100, 449–456.

Li, Y., Wang, X., Wang, J., 2012b. Cation exchange, interlayer spacing, and thermal analysis of Na/Ca-montmorillonite modified with alkaline and alkaline earth metal ions. *J. Therm. Anal. Calorim.* 110, 1199–1206.

Liu, C., Wang, X., Gao, H., Wen, R., 2003. Diffusion behavior of some weakly adsorbed nuclide species in granite. *J. Nucl. Radiochem.* 25, 204–209.

Lloret, A., Villar, M.V., 2007. Advances on the knowledge of the thermo-hydro-mechanical behaviour of heavily compacted "FEBEX" bentonite. *Phys. Chem. Earth* 32, 701–715.

Ma, B., Kang, M., Zheng, Z., Chen, F., Xie, J., Charlet, L., Liu, C., 2014. The reductive immobilization of aqueous Se(IV) by natural pyrrhotite. *J. Hazard. Mater.* 276, 422–432.

Ma, Y., Cheng, X., Kang, M., Yang, G., Yin, M., Wang, J., Gang, S., 2020. Factors influencing the reduction of U(VI) by magnetite. *Chemosphere* 254, 126855.

Maher, Z., Ivanov, P., O'Brien, L., Sims, H., Taylor, R.J., Heath, S.L., Livens, F.R., Goddard, D., Kellet, S., Rand, P., Bryan, N.D., 2016. Americium and plutonium association with magnesium hydroxide colloids in alkaline nuclear industry process environments. *J. Nucl. Mater.* 468, 84–96.

Mei, H., Liu, Y., Tan, X., Feng, J., Ai, Y., Fang, M., 2020. U(VI) adsorption on hematite nanocrystals: Insights into the reactivity of {001} and {012} facets. *J. Hazard. Mater.* 399, 123028.

Missana, T., Alonso, U., Maria Fernandez, A., Garcia-Gutierrez, M., 2018. Colloidal properties of different smectite clays: Significance for the bentonite barrier erosion and radionuclide transport in radioactive waste repositories. *Appl. Geochem.* 97, 157–166.

Noh, D.-H., Kim, S., Calaunan, J.M.F.V., Feng, Y., Eun, J., Kim, Y.-R., 2025. Bentonite as an engineered barrier material for nuclear waste repository—Geotechnical perspectives, key properties, knowledge gaps, and opportunities. *Nucl. Eng. Technol.* 57, 103700.

Nole, M., Beskardes, G.D., Fukuyama, D., et al., 2022. *PFLOTRAN Development FY2022*. Sandia National Lab. (SNL), Albuquerque.

Norrfors, K.K., Marsac, R., Bouby, M., Heck, S., Wold, S., Lützenkirchen, J., Schaefer, T., 2016. Montmorillonite colloids: II. Colloidal size dependency on radionuclide adsorption. *Appl. Clay Sci.* 123, 292–303.

Oscarson, D.W., Taylor, R., Miller, H.G., Cheung, S.C.H., 1985. The effect of bentonite on the interaction of I with PbO. *Appl. Clay Sci.* 1, 115–124.

Pan, D., Xu, Z., Wei, X., Sun, Y., Chen, Z., Jin, Q., Niu, Z., Guo, Z., Wu, W., 2019. Interaction of environmental colloids and radionuclides: A brief review. *Sci. Sin. Chim.* 49, 12–26.
Parisi, F., 2020. Adsorption and separation of crystal violet, cerium(III) and lead(II) by means of a multi-step strategy based on K10-montmorillonite. *Minerals* 10, 466.
Park, S., Yoon, S., Kwon, S., Lee, M.-S., Kim, G.-Y., 2021. Temperature effect on the thermal and hydraulic conductivity of Korean bentonite buffer material. *Prog. Nucl. Energy* 137, 103759.
Paxton, E.A., Wu, J., Hicks, T., Doudou, S., Applegate, D., Mason, R., Price, A., Payne, L., 2025. A computational framework to support probabilistic criticality modelling for the geological disposal of radioactive waste. *Ann. Nucl. Energy* 211, 110965.
Perez-Moreno, S.M., Guerrero, J.L., Mosqueda, F., Gazquez, M.J., Bolivar, J.P., 2020. Hydrochemical behaviour of long-lived natural radionuclides in Spanish groundwaters. *Catena* 191, 104558.
Philipp, T., Huittinen, N., Azzam, S.S.A., Stohr, R., Stietz, J., Reich, T., Schmeide, K., 2022. Effect of Ca(II) on U(VI) and Np(VI) retention on Ca-bentonite and clay minerals at hyperalkaline conditions—New insights from batch sorption experiments and luminescence spectroscopy. *Sci. Total Environ.* 842, 156837.
Philippe, A., Schaumann, G.E., 2014. Interactions of dissolved organic matter with natural and engineered inorganic colloids: A review. *Environ. Sci. Technol.* 48, 8946–8962.
Sasaki, Y., Shibata, M., Yui, M., Ishikawa, H., 1994. Experimental studies on the interaction of groundwater with bentonite. *Mater. Res. Soc. Symp. Proc.* 353, 337.
Sato, H., 2009. Quantification of exchangeable cations in interlayer of tsukinuno sodium-montmorillonite. *Mater. Res. Soc. Symp. Proc.* 1193, 529.
Schaefer, T., Huber, F., Seher, H., Missana, T., Alonso, U., Kumke, M., Eidner, S., Claret, F., Enzmann, F., 2012. Nanoparticles and their influence on radionuclide mobility in deep geological formations. *Appl. Geochem.* 27, 390–403.
Shackelford, C.D., Moore, S.M., 2013. Fickian diffusion of radionuclides for engineered containment barriers: Diffusion coefficients, porosities, and complicating issues. *Eng. Geol.* 152, 133–147.
Sheng, G., Yang, S., Sheng, J., Hu, J., Tan, X., Wang, X., 2011. Macroscopic and microscopic investigation of Ni(II) sequestration on diatomite by batch, XPS, and EXAFS techniques. *Environ. Sci. Technol.* 45, 7718–7726.
Shi, X., Wei, X., Yang, M., Qi, T., Li, J., Zhang, C., Wang, J., Xu, Z., Pan, D., Wu, W., 2025a. Colloid mediated transport of Uranium(IV) in sulfide-rich aquifers: An overlooked pathway for uranium remobilization under the reductive conditions. *J. Hazard. Mater.* 496, 139522.
Shi, X., Zhang, P., Feng, J., Xu, K., Fang, Z., Tian, J., Wu, T., 2025b. Improving hydraulic conductivity prediction of bentonite using machine learning with generative adversarial network-based data augmentation. *Constr. Build. Mater.* 462, 139962.
Singh, B.K., Um, W., 2023. Application of clay materials for sorption of radionuclides from waste solutions. *Minerals* 13, 239.
Soler, J.M., Steefel, C.I., Gimmi, T., Leupin, O.X., Cloet, V., 2019. Modeling the ionic strength effect on diffusion in clay. The DR-A Experiment at Mont Terri. *ACS Earth Space Chem.* 3, 442–451.
Song, L., Feng, X., Liang, J., Zhang, Y., Wang, J., 2009. A simple approach to determine the diffusivity of americium in granite. *Radiochem. Acta* 97, 581–586.
Sun, D., Sun, W., Fang, L., 2014. Swelling characteristics of Gaomiaozi bentonite and its prediction. *J. Rock Mech. Geotech. Eng.* 6, 113–118.
Sun, Y.F., Gao, C., Jin, Q., Chen, Z.Y., Guo, Z.J., 2021. Effects of iron oxide coatings on the mobility and retardation of U(VI) in water saturated media. Colloids Surf. A Physicochem. *Eng. Asp.* 629, 127458.
Sun, Z., Chen, Y., Cui, Y., Xu, H., Ye, W., Wu, D., 2018. Effect of synthetic water and cement solutions on the swelling pressure of compacted Gaomiaozi(GMZ) bentonite: The Beishan site case, Gansu, China. *Eng. Geol.* 244, 66–74.
Tachi, Y., Nakazawa, T., Ochs, M., Yotsuji, K., Suyama, T., Seida, Y., Yamada, N., Yui, M., 2010. Diffusion and sorption of neptunium(V) in compacted montmorillonite: Effects of carbonate and salinity. *Radiochim. Acta* 98, 711–718.

Tachi, Y., Yotsuji, K., 2014. Diffusion and sorption of Cs+, Na+, I– and HTO in compacted sodium montmorillonite as a function of porewater salinity: Integrated sorption and diffusion model. *Geochim. Cosmochim. Acta* 132, 75–93.

Tan, L., Tan, X., Mei, H., Ai, Y., Sun, L., Zhao, G., Hayat, T., Alsaedi, A., Chen, C., Wang, X., 2018. Coagulation behavior of humic acid in aqueous solutions containing Cs+, Sr2+ and Eu3+: DLS, EEM and MD simulations. *Environ. Pollut.* 236, 835–843.

Tan, L., Wang, X., Tan, X., Mei, H., Chen, C., Hayat, T., Alsaedi, A., Wen, T., Lu, S., Wang, X., 2017. Bonding properties of humic acid with attapulgite and its influence on U(VI) sorption. *Chem. Geol.* 464, 91–100.

Tan, X., Ren, X., Chen, C., Wang, X., 2014. Analytical approaches to the speciation of lanthanides at solid-water interfaces. *TrAC Trends Anal. Chem.* 61, 107–132.

Tan, X., Ren, X., Li, J., Wang, X., 2013. Theoretical investigation of uranyl ion adsorption on hydroxylated γ-Al2O3 surfaces. *RSC Adv.* 3, 19551–19559.

Tang, Q.F., Feng, Z., Wang, Y., Wei, X., Pan, D., Wu, W., Xu, Z., 2025. Bacillus subtilis morphology and bentonite colloids govern Eu(III) transport in quartz sand: Mechanisms and machine learning insights. *J. Hazard. Mater.* 496, 139449.

Tang, Q.F., Xu, Z., Tan, Q., Shi, X.Y., Wu, W.S., Pan, D.Q., 2024. Insight into impact of phosphate on the cotransport and corelease of Eu(III) with bentonite colloids in saturated quartz columns. *J. Hazard. Mater.* 461, 132572.

Tang, Y., Reeder, R.J., 2009. Uranyl and arsenate cosorption on aluminum oxide surface. *Geochim. Cosmochim. Acta* 73, 2727–2743.

Tian, J.-L., Feng, J.-X., Shen, J.-C., Yao, L., Wang, J.-Y., Wu, T., Zhao, Y.-L., 2025. Prediction of radionuclide diffusion enabled by missing data imputation and ensemble machine learning. *Nucl. Sci. Tech.* 36, 181.

Tsai, T.-L., Tsai, S.-C., Shih, Y.-H., Chen, L.-C., Lee, C.-P., Su, T.-Y., 2017. Diffusion characteristics of HTO and 99TcO4– in compacted Gaomiaozi (GMZ) bentonite. *Nucl. Sci. Tech.* 28, 67.

Van Loon, L.R., Müller, W., 2014. A modified version of the combined in-diffusion/abrasive peeling technique for measuring diffusion of strongly sorbing radionuclides in argillaceous rocks: A test study on the diffusion of caesium in Opalinus Clay. *Appl. Radiat. Isot.* 90, 197–202.

Vu, M.-N., Armand, G., Zghondi, J., Souley, M., Renaud, V., Alonso, M., Vaunat, J., Gens, A., Yu, Z., Shao, J.F., Corman, G., Collin, F., Camusso, M., Do, D.-P., 2025. A comparative analysis of different approaches to large-scale modeling for a radioactive waste repository in the Callovo-Oxfordian claystone and sensitivity analysis. *Rock Mech. Rock Eng.* 58, 5889–5911.

Wagner, S., Gondikas, A., Neubauer, E., Hofmann, T., von der Kammer, F., 2014. Spot the difference: Engineered and natural nanoparticles in the environment—release, behavior, and fate. *Angew. Chem. Int. Ed.* 53, 12398–12419.

Wang, C., Yang, X., He, J., Wei, F., Zheng, Z., Liu, C., 2019a. The diffusion of 75Se(IV) in Beishan granite—temperature, oxygen condition and ionic strength effects. *Radiochem. Acta* 107, 39–54.

Wang, J., Xie, L., Li, S., Wang, J., Zhang, J., Zeng, H., 2021a. Probing the in situ redox behavior of selenium on a pyrite surface by scanning electrochemical microscopy. *J. Phys. Chem. C* 125, 3018–3026.

Wang, X., Chen, T., Wang, C., Li, C., Liu, C., 2020. Adsorption and diffusion of some important radionuclides in beishan granites and gaomiaozi bentonites. *Sci. Sin. Chim.* 50, 1585–1599.

Wang, X., Yu, S., Chen, Z., Song, W., Chen, Y., Hayat, T., Alsaedi, A., Guo, W., Hu, J., Wang, X., 2019b. Complexation of radionuclide 152+154Eu(III) with alumina-bound fulvic acid studied by batch and time-resolved laser fluorescence spectroscopy. *Sci. China Chem.* 62, 1732–1733.

Wang, Y., Wang, J., Li, P., Qin, H., Liang, J., Fan, Q., 2021b. The adsorption of U(VI) on magnetite, ferrihydrite and goethite. *Environ. Technol. Innov.* 23, 101615.

Wang, Z., Wang, H., Li, Q., Xu, M., Guo, Y., Li, J., Wu, T., 2016. pH effect on Re(VII) and Se(IV) diffusion in compacted GMZ bentonite. *Appl. Geochem.* 73, 1–7.

Wei, X., Pan, D., Tan, Q., Shi, X., Hou, J., Tang, Q., Xu, Z., Wu, W., Ma, B., 2023. Surface charge property governing co-transport of illite colloids and Eu(III) in saturated porous media. *Chin. Chem. Lett.* 34, 108275.

Wen, Z.J., 2008. Selection and basic properties of the buffer material for high-level radioactive waste repository in China. *Acta Geol. Sin. Engl.* 82, 1050–1055.

Wu, C., Chen, L., Yang, S., Cai, Y., Xu, L., Wu, X., Qin, H., Liu, Z., Chen, L., Wang, S., 2018. Macroscopic, theoretical simulation and spectroscopic investigation on the immobilization mechanisms of Ni(II) at cryptomelane/water interfaces. *Chemosphere* 210, 392–400.

Wu, H., Huang, W., Duan, Z., Luo, M., Hua, R., 2020a. Investigation of Se(IV) diffusion in compacted Tamusu clay by capillary method. *J. Radioanal. Nucl. Chem.* 324, 903–911.

Wu, P., Wang, J., Ling, H., Zhou, Z., Santosh, M., Tang, Z., Sun, H., Gui, R., Duan, J., Li, N., Duan, X., 2024a. Safety assessment in the disposal of high-level radioactive wastes (HLWs): A geochemical study of uranium complexes in deep groundwater in granites from Beishan, China. *J. Radioanal. Nucl. Chem.* 333, 1779–1791.

Wu, T., Feng, Z., Geng, Z., Xu, M., Shen, Q., 2023a. Restriction of Re(VII) and Se(IV) diffusion by barite precipitation in compacted bentonite. *Appl. Clay Sci.* 232, 106803.

Wu, T., Hong, Y., Shao, D., Zhao, J., Feng, Z., 2023b. Experimental and modeling study of the diffusion path of Ce(III)–EDTA in compacted bentonite. *Chem. Geol.* 636, 121639.

Wu, T., Li, J., Dai, W., Xiao, G., Shu, F., Yao, J., Su, Y., Shi, L., 2012. Effect of dry density on 125I diffusion in GMZ bentonite. *Sci. China Chem.* 55, 1760–1764.

Wu, T., Tian, J., Shi, X., Li, Z., Feng, J., Feng, Z., Li, Q., 2024b. Predicting anion diffusion in bentonite using hybrid machine learning model and correlation of physical quantities. *Sci. Total Environ.* 946, 174363.

Wu, T., Wang, H., Zheng, Q., Zhao, Y.L., Li, J.Y., 2013. Effect of EDTA on the diffusion behavior of 99TcO4– and ReO4– in GMZ bentonite. *J. Radioanal. Nucl. Chem.* 299, 2037–2041.

Wu, T., Wang, H., Zheng, Q., Zhao, Y.L., Van Loon, L.R., 2014. Diffusion behavior of Se(IV) and Re(VII) in GMZ bentonite. *Appl. Clay Sci.* 101, 136–140.

Wu, T., Wang, Z., Wang, H., Zhang, Z., Van Loon, L.R., 2017. Salt effects on Re(VII) and Se(IV) diffusion in bentonite. *Appl. Clay Sci.* 141, 104–110.

Wu, T., Yang, Y., Wang, Z., Shen, Q., Tong, Y., Wang, M., 2020b. Anion diffusion in compacted clays by pore-scale simulation and experiments. *Water Resour. Res.* 56, e2019WR027037.

Xu, Y., Zhou, X., Sun, D.A., Zeng, Z., 2022. Thermal properties of GMZ bentonite pellet mixtures subjected to different temperatures for high-level radioactive waste repository. *Acta Geotech.* 17, 981–992.

Xu, Z., Pan, D., Sun, Y., Wu, W., 2018. Stability of GMZ bentonite colloids: Aggregation kinetic and reversibility study. *Appl. Clay Sci.* 161, 436–443.

Yan, M., Fu, Q., Li, D., Gao, G., Wang, D., 2013. Study of the pH influence on the optical properties of dissolved organic matter using fluorescence excitation-emission matrix and parallel factor analysis. *J. Lumin.* 142, 103–109.

Yang, X., Ge, X., He, J., Wang, C., Qi, L., Wang, X., Liu, C., 2017. Effects of mineral compositions on matrix diffusion and sorption of 75Se(IV) in granite. *Environ. Sci. Technol.* 52, 1320–1329.

Yao, J., Su, X., Long, H., Fan, X., 2003. Determining of diffusion coefficients of 237Np in bentonite. *J. Nucl. Radiochem.* 25, 199–203.

Ye, W.M., Borrell, N.C., Zhu, J.Y., Chen, B., Chen, Y.G., 2014. Advances on the investigation of the hydraulic behavior of compacted GMZ bentonite. *Eng. Geol.* 169, 41–49.

Ye, W.M., Chen, Y., Chen, B., Wang, Q., Wang, J., 2010. Advances on the knowledge of the buffer/backfill properties of heavily-compacted GMZ bentonite. *Eng. Geol.* 116, 12–20.

Yoon, S., Chang, S., Park, D., 2022. Investigation of soil-water characteristic curves for compacted bentonite considering dry density. *Prog. Nucl. Energy* 151, 104318.

Yu, S., Ma, J., Shi, Y., Du, Z., Zhao, Y., Tuo, X., Leng, Y., 2020. Uranium(VI) adsorption on montmorillonite colloid. *J. Radioanal. Nucl. Chem.* 324, 541–549.

Yuan, G., Chen, T., Liu, P., 2019. Migration characteristics study of iodine in crushed Beishan granite column under dynamic flow condition. *J. Radioanal. Nucl. Chem.* 321, 693–699.
Zaenker, H., Hennig, C., 2014. Colloid-borne forms of tetravalent actinides: A brief review. *J. Contam. Hydrol.* 157, 87–105.
Zhang, J.T., Chen, J.Y., Zhao, Z.H., Chen, S.C., Liu, G.H., Zhao, X.G., Wang, J., Lin, T., Liu, B., 2023. Numerical modeling on nuclide transport around a nuclear waste repository under coupled thermo-hydro-mechanical condition. *Comput. Geotech.* 164, 105776.
Zhang, Z., Geng, W.S., Ye, W.M., He, Y., Su, W., Wang, Q., Chen, Y.G., 2025. Comparative study of hydro-mechanical behaviors of compacted bentonite powder and granular bentonite. *J. Rock Mech. Geotech. Eng.* 17, 1757–1769.
Zhao, S.S., Chen, K., Xiong, B.Y., Guo, C.L., Dang, Z., 2024. Prediction of adsorption of metal cations by clay minerals using machine learning. *Sci. Total Environ.* 924, 171733.
Zhao, Y., Fu, B., Wu, T., Han, X., Wang, H., He, C., Luo, Y., Li, K., Zhang, K., Deng, J., 2016. Transport of 125I in compacted GMZ bentonite containing Fe-oxides, Fe-minerals or Cu2O. *J. Radioanal. Nucl. Chem.* 308, 539–544.
Zhao, Y., Guo, Z., Xu, J., 2013. 99TcO4– diffusion and sorption in compacted GMZ bentonite studied by capillary method. *J. Radioanal. Nucl. Chem.* 298, 147–152.
Zhu, Y., Elzinga, E.J., 2014. Formation of layered Fe(II)-hydroxides during Fe(II) sorption onto clay and metal-oxide substrates. *Environ. Sci. Technol.* 48, 4937–4945.
Zou, Y., Wang, X., Ai, Y., Liu, Y., Li, J., Ji, Y., Wang, X., 2016. Coagulation behavior of graphene oxide on nanocrystallined Mg/Al layered double hydroxides: Batch experimental and theoretical calculation study. *Environ. Sci. Technol.* 50, 3658–3667.

CHAPTER 13

Redox Reactions and Radiolysis of Radionuclide Migration

Hanqin Weng

13.1 INTRODUCTION

In high-level radioactive waste (HLW) repositories, the continuous decay of radionuclides emits α, β, and γ rays that irradiate the near-field groundwater environment, particularly through γ rays, which have strong penetrating ability. Radiolysis of water generates a variety of reactive species, such as e_{aq}^-, H•, •OH, H_2O_2, H^+, H_2, which exhibit distinct redox activities. These species can react with various radionuclides, leading to changes in valence and species. It will undoubtedly alter their migration processes in the geological environment.

Changes in the valence of radionuclides directly affect their speciation, thereby influencing their solubility, complexation affinity, and adsorption behavior at solid–liquid interfaces. For anionic radionuclides like $Tc(VII)O_4^-$ and $Se(VI)O_4^{2-}$, reducing species from water radiolysis (e.g., e_{aq}^-, H•) can reduce them from highly soluble, high-valent forms to low-valent species with lower solubility. As a result, precipitates or colloids are formed, retarding the transportation of radionuclides. Meanwhile, oxidative species like •OH may react with the precipitates, resulting in oxidation and dissolution. Consequently, radioactive anions are re-released to water. For cationic radionuclides like uranium and neptunium, high-valent species $U(VI)O_2^{2+}$ and $Np(V)O_2^+$ readily coordinate with common aqueous anions such as Cl^- and CO_3^{2-}, which form stable soluble complexes with high mobility. Low-valent states like U(IV) and Np (IV) tend to form hydroxide colloids or undergo surface adsorption, leading to immobilization within minerals (Bhalara et al., 2014; Maes et al., 2011).

Alterations in speciation further impact radionuclide migration in geological environments *via* different mechanisms such as adsorption, precipitation, and colloid formation. Soluble species migrate through convection–diffusion processes, while colloids and precipitates may become particulate carriers and durably immobilized, respectively. The transformation of speciation is governed not only by environmental parameters, including pH, Eh,

DOI: 10.1201/9781003665793-17

and ligand concentrations, etc., but also the thermo-hydro-mechanical-chemical (THMC) multi-field coupling effects in HLW repositories. Each physical field regulates radionuclide behavior from unique perspectives. The chemical field influences the radionuclide valence and complexation reaction. The thermal field affects reaction kinetics. The hydraulic field controls the diffusion of redox agents and solutes, indirectly influencing their concentration. The mechanical field affects rock fractures and engineered barriers, thus altering fluid pathways and permeability. Couplings among the these fields collectively determine the migration of radionuclides.

Irradiation will break the existing thermodynamic equilibrium. The radiation-induced redox reactions propel changes in valence and speciation of radionuclides, which reconstruct a new thermodynamic equilibrium. In summary, radiation-induced disruption of thermodynamic equilibrium will change the valence and speciation of radionuclides, which influences their migration behaviors by reconstructing a new equilibrium. The mechanism mentioned earlier renders the migration of radionuclides from HLW in the geological environment far more complex than that of stable nuclides unaffected by irradiation. Thus, a deep comprehension of the interplay of radiation chemical reactions, speciation, and migration is essential for predicting long-term radionuclide transport, which provides a critical scientific foundation for safety assessments and barrier designs in HLW geological disposal systems.

13.2 INFLUENCE OF RADIATION-INDUCED REDOX REACTIONS ON THE MIGRATION OF ACTINIDE ELEMENTS

13.2.1 Uranium (U)

In geological disposal environments, uranium exemplifies actinides whose chemical properties are sensitive to redox conditions. In natural aqueous systems, especially those containing oxygen, uranium predominantly exists in the form of hexavalent uranyl ion (UO_2^{2+}). The linear structure of rigid $[O{=}U{=}O]^{2+}$ arises from the uranium atom's electron configuration ($[Rn]5f^3 6d^1 7s^2$), where two oxygen atoms are bonded to the central uranium through the U=O double bond. The standard redox potential for $E^0(UO_2^{2+}/U^{4+})$ is approximately 0.27 V, enabling the reduction of U(VI) to U(IV) by reducing species from water radiolysis (Dutra et al., 2024). Owing to its Lewis acidity, UO_2^{2+} readily coordinates with various anions in groundwater like Cl^- and CO_3^{2-}, forming stable five- or six-membered chelate complexes that markedly affect uranium solubility (Vodyanitskii, 2011). Although free UO_2^{2+} undergoes hydrolysis and precipitates at neutral pH, resulting in low solubility (approximately 10^{-5} mol/L), complexation with the anions mentioned earlier elevates the solubility of U-species to approximately 10^{-3} mol/L. Furthermore, UO_2^{2+} exhibits rapid water ligand exchange kinetics ($k_{ex} \approx 10^8\ s^{-1}$), facilitating dynamic coordination with environmental ligands and increasing its instability under irradiation.

The redox reactions between uranium and radiolytic products of water can be classified into two types, that is, oxidative dissolution of low-valent U(IV) and reductive precipitation of high-valent U(VI). Strongly oxidizing species generated via water radiolysis can oxidize

U(IV) to highly soluble U(VI). H_2O_2 has been confirmed to be capable of oxidizing UO_2 (Olsson et al., 2024). The •OH radical absorbed on the UO_2 surface plays as a crucial intermediate. In the near-field area of a deep geological disposal repository for HLW, the absorbed dose rate easily exceeds 1 Gy/h. Although the equilibrium concentration of •OH may be only 10^{-9} mol/L, the H_2O_2 will reach a non-negligible concentration around 10^{-6}–10^{-5} mol/L, significantly accelerating oxidative dissolution of UO_2. Reducing species like e_{aq}^- and •H reduce U(VI) to U(V) ($UO_2^{2+} + e_{aq}^- \rightarrow UO_2^+$). The unstable U(V) intermediates rapidly disproportionate into U(IV) and U(VI) ($2UO_2^+ + 4H^+ \rightarrow UO_2^{2+} + U^{4+} + 2H_2O$). Molecular H_2 can decrease the concentration of oxidants by scavenging •OH species (H_2 + •OH → H_2O + •H), thereby inhibiting U(IV) oxidative dissolution and preserving its stable precipitated form. Thus, cycling of uranium species with different valences mainly occurs between U(IV) and U(VI). U(VI) and U(IV) are the dominant species under oxidizing and reducing conditions, respectively.

In real repository environments, the influence of redox reactions in the presence of repository-barrier minerals on uranium speciation becomes more complex. For instance, pyrite (FeS_2) not only immobilizes uranium by reducing soluble U(VI) to U(IV) precipitates, but also efficiently scavenges oxidizing species produced by water radiolysis. It has been reported that γ-irradiation promotes U(VI) reduction on FeS_2 surfaces across pH 4.5–9.0 (Kang et al., 2024). A remarkable decline in the concentration of dissolved uranium was observed. Precipitates abundant in U(IV/V) were also detected on the pyrite surface. In contrast, irradiation in the absence of FeS_2 may oxidize the surface UO_2 into soluble U(VI) species.

Migration of uranium in repository environments is governed by its speciation, including dissolved ions, aggregated colloids, deposits on mineral surfaces, and the corresponding solid solutions, of which the uranium valence is the key factor. During the early to stable stages of a repository, the environment is generally oxidizing or mildly reducing. U(VI) typically exists as UO_2^{2+} and its complexes, forming stable water-soluble species with CO_3^{2-} anions, such as $UO_2(CO_3)_2^{2-}$ and $UO_2(CO_3)_3^{4-}$, which suppress hydrolytic precipitation and sustain uranium concentrations in the solution (10^{-5}–10^{-6} mol/L). The strong complexation with carbonate endows U(VI) with high mobility. $[UO_2(CO_3)_2]^{2-}$ predominates at pH > 6 and $[CO_3^{2-}]$ > 0.5 mmol/L, while highly stable $[UO_2(CO_3)_3]^{4-}$ with higher mobility is dominant at pH > 7 and elevated CO_3^{2-} concentration (Liu et al., 2016). Co-existing cations, such as Ca^{2+}, can also coordinate with UO_2^{2+} to form neutral complexes like $Ca_2UO_2(CO_3)_3$. Moreover, U(VI) may also complex with natural organic compounds or synthetic chelators to enhance dissolution or interact with mineral surfaces (e.g., iron hydroxides, clays) to generate stable U-containing colloids, indirectly amplifying migration risks.

The chemical environment is usually strongly reducing after the repository environment reaches a stable state or within engineered barriers such as FeS_2-containing bentonite backfills. Reduction of U(VI) to U(IV) by e_{aq}^- and •H originated from water radiolysis, and precipitation of U(IV) species became the primary route. Thermodynamically, U(IV) favors the formation of stable UO_2, $UO_2{\cdot}xH_2O$, or $U(OH)_4$ under neutral to alkaline conditions. Crystalline UO_2 displays an exceedingly low solubility around 10^{-8}–10^{-11} mol/L, which should have minimized the uranium migration theoretically. However, retention in repositories is not absolute. Under specific conditions, U(IV) exhibits the potential for complexation with common anions in groundwater. For example, bicarbonate enables the formation of soluble $U(CO_3)_5^{6-}$ ($UO_2 + 5HCO_3^- \rightleftharpoons U(CO_3)_5^{6-} + 3H^+ + H_2O$), elevating the concentration of dissolved uranium by at least one order of magnitude (Tocino et al., 2014). Moreover, U(IV) is readily adsorbed onto ubiquitous natural colloids, including clay minerals, iron hydroxides, and organic-stabilized particles, rendering colloid-bound uranium a primary carrier for U(IV) transport in strongly reducing environments (Grenthe et al., 1992).

It is evident that the radiation field in the near-field areas of the repository reconstructs the speciation and transportation processes of U. First, active species produced by water radiolysis propel valence transitions between U(VI) and U(IV), altering the speciation. Differences in the complexation ability of various uranium species directly impact their existing forms and indirectly regulate dissolution-precipitation equilibrium. Second, radiolytic radicals also react with anions, not only decreasing their concentrations but also generating new species. The uranium complexes become more complicated as a result. This part will be discussed in detail in the following section on anions. Additionally, high-energy ionizing radiation may induce the surface modification of minerals and colloids, influencing colloidal stability. The radiation-induced effects couple with geochemical conditions such as ligand concentrations, pH, and Eh, dynamically reshaping the transformation pathways, speciation, and migration behavior for uranium in groundwater.

13.2.2 Plutonium (Pu)

Compared to the relatively straightforward valence transitions between U(VI) and U(IV), the transuranic element Pu exhibits a more complicated multi-valent chemistry. Pu can stably exist in multiple valences, including III, IV, V, and VI, rendering it exceptionally sensitive to redox perturbations induced by radiation. In natural aqueous media, Pu typically predominates as pentavalent plutonyl ion (PuO_2^+), with the standard redox potential for Pu(VI)/Pu(IV) ($E^0 \approx 0.97$ V) substantially exceeding that of U(VI)/U(IV) (≈ 0.27 V), indicating greater susceptibility of Pu(VI) to reduction to Pu(IV). The PuO_2^+ features a linear O=Pu=O unit, with its unpaired 5f electron configuration ([Rn]$5f^6 7s^2$) conferring weaker Lewis acidity than U(VI), enabling stable complexation with oxalate, sulfate, and similar ligands (e.g., $PuO_2(C_2O_4)^-$, $PuO_2(SO_4)^-$). At neutral pH, PuO_2^+ solubility approximates 10^{-6} mol/L, marginally lower than that of U(VI) (10^{-5} mol/L) (Zhou et al., 2025). However, PuO_2^+ tends to form $Pu(OH)_4$ colloids under alkaline conditions, whereas U(VI) favors soluble complexes, underscoring distinct migratory behaviors between Pu and U. Moreover, although the water exchange rate of PuO_2^+ ($k_{ex} \approx 10^7$ s^{-1}) is slightly slower than that of UO_2^{2+} ($k_{ex} \approx 10^8$ s^{-1}), PuO_2^+ displays heightened sensitivity to α-radiolysis-induced disproportionation due to its intrinsic autoradiolysis. These unique thermodynamic attributes and radiation-redox sensitivities influence Pu's initial speciation within repository near-field barriers, thereby shaping its transport trajectories into irradiated and geochemically complex matrices (Li et al., 2025).

Pu manifests distinctive migratory complexity in repository environments, fundamentally arising from its valence diversity (III, IV, V, VI), which is particularly pronounced in radiation-intense near-field environments. Hydrated electrons (e_{aq}^-) reduce highly mobile

Pu(VI) to the intermediate Pu(V): $PuO_2^{2+} + e_{aq}^- \rightarrow PuO_2^+$ ($k = 1.9 \times 10^{10}$ L/(mol·s)). Pu(V) may subsequently disproportionate ($2PuO_2^+ \rightarrow PuO_2^{2+} + Pu^{4+}$) to yield Pu(VI) and Pu(IV), or undergo further reduction to Pu(IV) or Pu(III). Analogous to U(IV), H_2 effectively scavenges potent oxidants like •OH, safeguarding reductive species such as Pu(III) or Pu(IV) colloids from oxidative dissolution (Dutra et al., 2024). Low-valence Pu is readily oxidized (Pu^{3+} + •OH → Pu^{4+} ($k = 10^9$ L/(mol·s))). Under γ-irradiation at a dose rate of 1 Gy/h, steady-state H_2O_2 concentrations attain 10^{-5} mol/L, facilitating Pu(III) oxidation to Pu(IV) and progression toward soluble high-valence forms (e.g., PuO_2^+ or Pu(VI) complexes), thereby augmenting mobility (Gupta et al., 2016). Consequently, Pu valence shifts in repository environment are critically contingent on the relative abundances of redox-active species, potentially yielding co-existing multi-valent distributions that introduce far greater complexity and uncertainty than observed for uranium in groundwater chemistries (Budantseva et al., 1998).

Stable high-valence Pu(V)/Pu(VI) display hard acid properties, readily forming exceptionally stable soluble anionic carbonate complexes with CO_3^{2-} (e.g., $PuO_2(CO_3)_3^{5-}$, log β > 15). Carbonate complexation markedly enhances PuO_2^+ solubility (>10^{-5} mol/L), with resultant complexes exhibiting minimal surface retention (K_d < 10 L/kg) and accelerated diffusion ($D_e \approx 10^{-9}$ m²/s). In stark contrast, upon transition to stable low-valence states (e.g., Pu(IV)), soft acid characteristics drive vigorous hydrolysis, generating sparingly soluble (<10^{-9} mol/L) amorphous $Pu(OH)_4$ precipitates or nano-scale colloids (10–100 nm). Although Pu(IV) can theoretically form soluble complexes like $[Pu(CO_3)_4]^{4-}$ (log β = 17–18) at exceedingly high carbonate levels, prevailing repository CO_3^{2-} concentrations are insufficient to preclude precipitation/colloid formation (Tanaka et al., 2022); conversely, sulfate ligands promote Pu(IV) aggregation, yielding more stable aggregates that diminish solubility. Under strongly reducing conditions, Pu(III) exhibits solubility (10^{-5}–10^{-7} mol/L) exceeding that of Pu(IV), yet under Fe(II)-rich conditions, it readily co-precipitates or adsorbs. This valence-dependent solubility variation spans over 10^4-fold, surpassing that of uranium (Arab-Chapelet et al., 2008).

The morphology of Pu(IV) precipitates, or colloidal phases, exerts a secondary influence on transport mechanisms. Formed colloids approach charge neutrality near neutral pH (from Pu^{4+} to $Pu(OH)_4$), facilitating evasion of adsorption on negatively charged bentonite or mineral surfaces at elevated pH (R_d declining to 10^{-2}) and bypassing pore filtration, mimicking anionic species (e.g., TcO_4^-). In authentic repository reducing barriers (e.g., bentonite or iron-rich mineral environments), Pu(IV) or Pu(III) primarily undergoes robust surface adsorption (e.g., on iron (hydr)oxides, K_d >10^3 L/kg) or forms inert mineral shielding layers for effective retardation (precipitation/adsorption retardation pathway), confining migration to near-field scales (<10 m) and limiting soluble fractions. Persistent redox cycling under radiation fields imparts ongoing perturbations to immobilization mechanisms, such as localized re-oxidation (e.g., sorbed Pu(IV) oxidized by H_2O_2), inducing intricate dissolution–reprecipitation or re-adsorption cycles. Thus, in contrast to U's comparatively predictable migration dominated by U(VI)–carbonate complexation, Pu's transport is profoundly governed by its steady-state valence distribution under radiolysis, alongside complexation/aggregation modalities, resulting in dynamic competition and uncertainty among the rapid migration of soluble high-valence anionic complexes, the circumvention of barriers by low-valence/neutral colloids, or fixation via precipitation/adsorption in near-field barriers (Budantseva et al., 1998).

13.2.3 Neptunium (Np)

As another pivotal transuranic element, neptunium (Np) warrants significant attention regarding its migratory potential. Distinct from uranium (U) and plutonium (Pu), Np exhibits a unique predominance of intermediate oxidation states in typical groundwater systems, with the pentavalent form (NpO_2^+) representing its most stable dissolved species. This stability stems from Np's distinctive electron configuration ([Rn]$5f^46d^17s^2$) and its moderate standard redox potential ($E^0(NpO_2^+/Np^{4+}) \approx 0.75$ V) (Dutra et al., 2024), allowing soluble NpO_2^+ to prevail across a broad pH spectrum and under low to moderate Eh conditions, thereby conferring enhanced mobility. Given that $E^0(NpO_2^+/Np^{4+})$ is lower than $E^0(PuO_2^{2+}/Pu^{4+})$, the reduction of Np(V) to Np(IV)—yielding sparingly soluble $Np(OH)_4$ or NpO_2—occurs stably only in markedly reducing environments. Although Np(VI) (NpO_2^{2+}) exists, its high $E^0(NpO_2^{2+}/NpO_2^+) \approx 1.2$ V renders it highly unstable under characteristic near-field Eh values, prompting rapid reduction back to the pentavalent state by common reductants. Consequently, the stability and proportional distribution of Np(IV) and Np(VI) are substantially diminished relative to Np(V). The formation of these species hinges on the specific types and concentrations of redox drivers and ligands in localized environments (Pirlet, 2001). This exceptional thermodynamic stability of Np(V) differentiates it from the valence transitions of uranium and the multi-valent co-existence in Pu, establishing a novel focal point in studies of radionuclide transport within geological disposal systems.

Hydrated electrons (e_{aq}^-) reduce NpO_2^+ ($k = 5.0 \times 10^9$ L/(mol·s)), producing insoluble $Np(IV)O_2$. Conversely, $Np(IV)O_2$ can be oxidized by •OH ($k \approx 3.2 \times 10^8$ L/(mol·s)) and H_2O_2 to form soluble $Np(V)O_2^+$, thereby augmenting mobility (Rotermund et al., 2024). Further oxidation of NpO_2^+ by •OH ($k \approx 4.6 \times 10^7$ L/(mol·s)) generates unstable NpO_2^{2+}, which exhibits reactivity akin to uranium analogs yet without significant disproportionation tendencies.

It is evident that the redox cycling of neptunium predominantly occurs between Np(IV) and Np(V). The highly mobile Np(V), characterized by strong ionic character and minimal hydrolytic tendency, forms relatively weak complexes with common anions such as CO_3^{2-} (e.g., $NpO_2CO_3^-$ with $\log \beta \approx 5.5$), enabling soluble NpO_2^+ to dominate across extensive pH ranges and exhibit robust transport. When Np(V) is reduced to Np(IV), the high charge density of Np^{4+} promotes strong complexation with ligands, especially carbonate, as illustrated by $Np(CO_3)_4^{4-}$ with $\log \beta \approx 26.5$. However, the concentration of dissolved Np(IV) is generally limited by its low solubility, often to $\leq 10^{-8}$ mol/L, and Np(IV) may form precipitates or colloids near precipitation thresholds, such as $Np(OH)_4$ (Jegou et al., 2000). These Np(IV) colloids and precipitates can have two contrasting effects on migration. They may be retained by filtration in geological media or by adsorption onto mineral surfaces, with distribution coefficients (Kd) that are 2–4 orders of magnitude higher than those of Np(V), reaching 10^3–10^4 mL/g (Gray et al., 2019). However, under neutral to alkaline conditions with elevated carbonate concentrations ($[CO_3^{2-}] > 10^{-3}$ mol/L), stable Np(IV)–carbonate colloids or slightly more soluble aqueous complexes may still promote limited colloid-assisted transport (Maes et al., 2011).

The coupled effects of radiation fields and geological environment profoundly impact the formation of Np(IV). Under ionizing radiation, interfaces between reducing minerals (e.g., FeS_2, magnetite) and water not only accelerate Np(V) transformation to Np(IV) (substantially more than in non-irradiated scenarios) but may also yield tightly bound mineral surface coatings (e.g., NpO_2 layers) or metastable nano-scale Np(IV) colloidal particles. This process is typically governed by solubility and surface sorption dynamics (experimental evidence indicates Np recovery reductions of 10^2–10^3-fold), establishing migratory barriers. Nonetheless, such radiation-enhanced immobilization is not unequivocal. Under strongly reducing and alkaline conditions (e.g., pH >9 with high CO_3^{2-}), although radiolysis may hasten Np(V) reduction, the resultant soluble Np(IV)–carbonate complexes or stable colloids could still permeate chemical reduction sites for localized transport. Furthermore, localized elevated temperatures induced by HLW decay heat perturb Np species distributions in multi-barrier transport pathways by elevating reaction rate constants, altering complex stability constants, and intensifying adsorption site competition. In summary, the migration behavior of Np in repository near-fields involves a competitive interplay among reduction-driven complexation, precipitation/colloid formation, and interfacial/solid-phase transformation. Within barrier zones, radiochemical reactions predominantly enhance Np(IV) generation to curtail migration, yet under specific regimes, particularly in high-carbonate domains, potential colloid- or complex-mediated transportation may emerge.

13.3 INFLUENCE OF RADIATION-INDUCED REDOX REACTIONS ON THE MIGRATION OF RADIONUCLIDE ANIONS

13.3.1 Technetium (Tc)

In the geological disposal of HLW, ^{99}Tc is one of the primary long-term sources of radiotoxicity due to its long half-life ($T_{1/2} = 2.1 \times 10^5$ y). Tc exists predominantly as the heptavalent pertechnetate ion (TcO_4^-). Its tetrahedral coordination structure originates from the d-electron configuration of the Tc atom ([Kr]$4d^55s^2$), where four oxygen atoms are bonded to the Tc center in a tetrahedral arrangement, forming stable Tc=O bonds. This structure confers high thermal stability and chemical inertness to TcO_4^-. Its standard redox potential $E^0(TcO_4^-/TcO_2)$ is approximately 0.75 V (Eriksen & Cui, 1991), indicating that Tc is readily reduced under reducing conditions, typically to the tetravalent state, but is difficult to reduce further to lower valence states. As a typical radionuclide existing in anionic form, TcO_4^- exhibits high solubility across a wide pH range (0–14), with solubility greater than 1 mol/L under neutral conditions, and does not readily precipitate. Consequently, it possesses very high environmental mobility. TcO_4^- can enter the biosphere through uptake by plants, posing long-term radiological hazards. Although the chemical behavior of Tc is highly regulated by its redox state, unlike radionuclide cations such as U, whose complexation ability changes significantly with valence state, TcO_4^- exhibits weak Lewis basicity. Thus, its ability to coordinate with cations such as alkali metals (Na^+, K^+) and alkaline earth metals (Ca^{2+}, Mg^{2+}) is limited, and it generally does not form stable complex species. Reducing Tc mobility and retarding its diffusion primarily rely on its reduction to insoluble oxides like TcO_2 for immobilization, rather than on complexation or fixation on the surfaces of natural minerals or barrier materials.

The weak complexation ability of TcO_4^- (log $\beta < 1$) means its species distribution is predominantly governed by the system's Eh. Hydrated electrons (e_{aq}^-) and hydrogen atoms (•H) produced by water radiolysis can promote the reduction of TcO_4^- to Tc(IV): $TcO_4^- + 3e_{aq}^- + 4H^+ \rightarrow TcO_2 + 2H_2O$. In anoxic, reducing systems, at an absorbed dose rate of 1 Gy/h, the steady-state concentration of e_{aq}^- can reach 10^{-7} mol/L, favoring the conversion of Tc(VII) to the sparingly soluble $TcO_2{\cdot}nH_2O$. The initial step of Tc reduction in a radiation field is $TcO_4^- + e_{aq}^- \rightarrow Tc(VI)O_4^{2-}$. $Tc(VI)O_4^{2-}$ undergoes a series of disproportionation reactions to yield $Tc(IV)O_2$ (2Tc(VI) → Tc(VII) + Tc(V); 2Tc(V) → Tc(VI) + Tc(IV)) (Sekine et al., 2002). However, intermediate products in this process can be re-oxidized and dissolved by •OH: $Tc(VI)O_4^{2-} + {\bullet}OH \rightarrow TcO_4^- + OH^-$. Therefore, under reducing conditions, TcO_4^- is converted to $TcO_2{\cdot}nH_2O$ precipitate, with solubility dropping below 10^{-8} mol/L, while in oxidizing zones, TcO_4^- maintains high solubility. Studies on the non-radioactive analogue Re indicate that in anoxic environments, the stability of the $Re(VI)O_4^{2-}$ intermediate highly depends on solution pH. Under acidic conditions, $Re(VI)O_4^{2-}$ undergoes rapid disproportionation (2Re(VI) → Re(VII) + Re(V); 2Re(V) → Re(VI) + Re(IV)). Part of the reductive products will be re-oxidized to Re(VII). Meanwhile, OH^- can form stable complexes with $Re(VI)O_4^{2-}$ under alkaline conditions, avoiding the formation of Re(VII) and facilitating further reaction with reductive species like e_{aq}^- to form $Re(IV)O_2$ (Shang et al., 2018). When solutions containing ReO_2 precipitate are exposed to air, rapid oxidation occurs, regenerating soluble $Re(VI)O_4^-$. The reactivity is closely linked to the system's Eh at different pH levels. Tc might likely exhibit similar reactivity.

Although the complexation ability of TcO_4^- with most cations is weak, its reduction process enhances its co-precipitation ability with cations such as Fe^{2+} or Ca^{2+}, promoting the transformation from soluble TcO_4^- to insoluble TcO_2 precipitate. In clay media, over 90% of Tc exists as precipitate. Clays contain significant organic matter, such as humic acids, which can stabilize Tc(IV) colloids, increasing species diversity and complicating migration kinetics (Tang et al., 2025).

Thus, valence changes in Tc also profoundly affect its migration kinetics through alterations in charge and species. Under oxidizing conditions, soluble, negatively charged TcO_4^- has low adsorption distribution coefficients ($K_d < 16$ L/kg) on natural minerals and rocks, and a high diffusion coefficient ($D \approx 10^{-8}$ m²/s) in fractured media, leading to high mobility and risk of far-field transport. Under reducing conditions, Tc(IV) precipitates and colloids are retarded through adsorption onto iron oxides or clays ($K_d > 10^3$ L/kg), limiting Tc migration distances to <1 m. The reduction of TcO_4^- to TcO_2 and the formation of colloids (20–100 nm) result in a charge reversal, reducing electrostatic repulsion with mineral surfaces. In bentonite backfill material, the retardation factor (R_d) for Tc colloids can reach 10^4. The coupled THMC effects collectively shape the long-term chemical environment evolution of the repository, causing critical transitions in Tc behavior between soluble and insoluble states. Its mobility is extremely sensitive to slight changes in Eh, making the assessment of the system's redox capacity and evolution trend a key factor in predicting Tc migration risk.

13.3.2 Selenium (Se)

^{79}Se, with a half-life of $T_{1/2} = 3.8 \times 10^5$ years, is also a key radionuclide in HLW geological disposal. Se exists in five valence states in the environment: –II, –I, 0, IV, and VI. Lower valence states of Se (–II, –I, 0) tend to form insoluble compounds with metal ions or precipitate directly as Se(0), exhibiting weak mobility. In contrast, $Se(VI)O_4^{2-}$ (selenate) and $Se(IV)O_3^{2-}$ (selenite) have high solubility (>0.1 mol/L) in natural waters and limited ability to complex with metal ions, resulting in high environmental mobility. The electron configuration of the Se atom ([Ar]$3d^{10}4s^24p^4$) confers unique structural features. SeO_4^{2-} has a tetrahedral structure with four oxygen atoms bonded to the Se atom, while SeO_3^{2-} has a trigonal pyramidal structure. This structural difference endows them with distinct chemical activities. SeO_4^{2-} is chemically more stable, whereas SeO_3^{2-} more readily participates in interfacial adsorption and complexation reactions. The standard redox potential $E^0(SeO_4^{2-}/SeO_3^{2-})$ is approximately 0.88 V, slightly higher than that for $E^0(TcO_4^-/TcO_2)$ (0.75 V) (Bruggeman et al., 2002). The oxyanions of Se (SeO_4^{2-}, SeO_3^{2-}) display weak Lewis basicity and have limited complexation ability with alkali metals like Na^+, K^+, making stable complex formation difficult. However, SeO_3^{2-} can form stable surface complexes with Fe/Al oxides (e.g., goethite, ferrihydrite, pyrite), clay minerals (e.g., montmorillonite), and organic matter, influencing its retention on solid surfaces. Therefore, the reduction of Se's environmental mobility and the retardation of its diffusion primarily rely on valence state transformation and interfacial adsorption.

Compared to Tc, whose diffusion retardation is mainly achieved through reductive precipitation, Se retardation pathways share similarities but also have differences. Retardation can be achieved through two mechanisms: first, reductive retardation, where soluble Se(IV) or Se(VI) is reduced to less soluble Se(0) or other lower-valent selenides; second, adsorptive retardation, where Se is fixed through interactions with buffer/backfill materials and host rock minerals. For instance, reducing species in montmorillonite and other clay minerals, as well as biotite, can reduce $Se(VI)O_4^{2-}$ to $Se(IV)O_3^{2-}$, forming Se(0) (She et al., 2025). The reduction of Se(VI) to Se(IV) effectively increases the complexation constant ($\log \beta \approx 3$–5). $Se(IV)O_3^{2-}$ complexes with Fe/Al oxides and clay minerals, reducing overall solubility from >0.1 mol/L to 10^{-5} mol/L. Further reduction to Se(0) decreases solubility to 10^{-9} mol/L, influenced by the surface adsorption characteristics of common minerals in granite.

Both $e_{aq}-$ and •H produced by water radiolysis in a radiation field can reduce $Se(IV)O_3^{2-}$ and $Se(VI)O_4^{2-}$. Similar to Tc, Se speciation is dominated by Eh, with SeO_4^{2-} being the dominant species in oxidizing zones and $Se(IV)O_3^{2-}$/Se(0) dominating in reducing zones. In the presence of pyrite (FeS_2), γ-radiation can enhance the reduction rate of Se(VI) from 10^{-4}/s to 10^{-2}/s, generating Se(0) or $FeSe_2$ precipitates on the surface. In anoxic environments, the absorbed dose rate significantly affects the reduction process of Se(VI) and thus its retardation. For example, at medium dose rates (10 Gy/min), pyrite/As(III) can effectively scavenge oxidizing •OH radicals, improving Se(VI) reduction efficiency and preventing Se migration (Wu et al., 2017). At higher dose rates (100 Gy/min), increased •OH concentration raises the proportion of oxidation reactions for lower-valent Se, hindering its fixation.

Since Se(IV) and Se(VI) are highly mobile in aquatic environments, posing significant disposal risks, and lower-valent Se (e.g., Se(0), Se(−II)) has very low solubility and can be effectively retained through precipitation or adsorption, the fundamental approach to controlling Se migration is by manipulating the environmental redox conditions to induce reductive conversion to lower valence states. Under oxidizing conditions, soluble SeO_4^{2-} is negatively charged, has low adsorption distribution coefficients (K_d < 10 L/kg), and a high diffusion coefficient ($D \approx 10^{-8}$ m²/s) in fractured media, leading to high mobility and far-field transport risk. Under reducing conditions, Se(IV/0) precipitates or colloids are retarded through strong adsorption onto iron oxides or clays (K_d > 10^3 L/kg), limiting migration distances to <1 m. Similar to Tc, charge reversal upon colloid formation for Se reduces electrostatic repulsion, increasing the retardation factor R_d to 10^4 in bentonite media. The migration behavior is also comprehensively influenced by coupled THMC effects. The chemical field is core to controlling Se migration; Eh directly determines Se valence state and speciation. In reducing environments, highly mobile Se(VI) and Se(IV) are reduced to insoluble Se(0) or metal selenide precipitates, effectively immobilizing them. In oxidizing environments, immobilized Se can be re-oxidized and dissolved, regaining high mobility.

13.4 REDOX REACTIONS OF ANIONS IN ENVIRONMENTAL MEDIA

Groundwater often contains anions such as carbonate and chloride. These anions undergo more complex valence state and species transformations in radiation fields, an aspect often overlooked when evaluating the long-term migration behavior of radionuclides in disposal sites.

13.4.1 Carbonate (CO_3^{2-})

In the repository environment, CO_3^{2-} is a key anion regulating the environmental behavior of actinide radionuclides, as it can complex with various actinides and profoundly influence their migration. However, the radiolytic reactions of CO_3^{2-} can significantly alter the speciation of carbon–oxygen compounds, thereby affecting actinide migration. In the groundwater system of a repository, radiation-induced reactions of CO_3^{2-} proceed through multiple redox steps, generating stable low-molecular-weight carboxylate anions (e.g., formate, $HCOO^-$; acetate, CH_3COO^-), and finally transforming into oxalate $C_2O_4^{2-}$ (Zhang et al., 2025). The steady-state yield of $C_2O_4^{2-}$ under γ-irradiation can reach 2.0 nmol/J (Belova et al., 2021).

These radiolysis products of CO_3^{2-} influence actinide radionuclides primarily through two mechanisms. The first is reductive regulation. $HCOO^-$, generated during CO_3^{2-} radiolysis, is highly reactive with various water radiolysis products ($HCOO- + {}^\bullet OH \rightarrow CO_2^{\bullet -} + H_2O$, $k = 3.2 \times 10^9$ L/(mol s); $HCOO- + {}^\bullet H \rightarrow CO_2^{\bullet -} + H_2$, $k = 2.1 \times 10^8$ L/(mol s); $HCOO^- + e_{aq}- + H^+ \rightarrow CO_2^{\bullet -} + H_2$, $k = 8.0 \times 10^8$ L/(mol s)). The $CO_2^{\bullet -}$ radical anion can reduce higher-valent actinide radionuclides (e.g., $UO_2^{2+} + CO_2^{\bullet -} \rightarrow$ product, $k = 7.3 \times 10^8$ L/(mol s)), altering their valence states and consequently affecting uranium speciation. The second mechanism is complexation regulation. The radiolysis products $HCOO^-$ and

CH_3COO^- have weak complexation ability with radionuclide cations, primarily because they are monodentate ligands offering limited coordination sites for actinide cations. The formed complexes show very low stability constants (log $\beta < 2$). However, the final product of CO_3^{2-} radiolysis, $C_2O_4^{2-}$, has multiple coordination sites and can form highly stable soluble complexes with actinide ions (e.g., $[UO_2(C_2O_4)_2]^{2-}$). Compared with common uranium-bearing species in natural groundwater, such as the neutral calcium uranyl carbonate complex $Ca_2UO_2(CO_3)_3$, whose dissolved concentration can be lower than 1 μg/L under weakly reducing conditions, uranyl oxalate complexes can significantly increase the solubility and aqueous mobility of actinide radionuclides. The accumulation of $C_2O_4^{2-}$ would undoubtedly affect the complexation behavior and solubility of key actinide radionuclides such as U, Pu, and Np.

13.4.2 Chloride Anion (Cl^-)

Cl^- is the most abundant background anion in groundwater (or pore water). Although the complexation ability of Cl^- with actinide radionuclides (e.g., U, Pu, Np) is generally considered much weaker than that of CO_3^{2-} or organic ligands, its reaction with •OH produced by water radiolysis generates the strongly oxidizing $Cl_2^{\bullet-}$ radical ($Cl^- + {}^\bullet OH \rightarrow ClOH^{\bullet-}$, $k = 4.3 \times 10^9$ L/(mol s); $ClOH^{\bullet-} + Cl^- \rightarrow Cl_2^{\bullet-} + OH^-$, $k = 1 \times 10^4$ L/(mol s)). "$Cl_2^{\bullet-}$ exists in equilibrium with $Cl^\bullet$ and Cl^- ($Cl_2^{\bullet-} \rightleftharpoons Cl^\bullet + Cl^-$), thereby producing highly reactive chlorine atoms ($Cl^\bullet$). $Cl^\bullet$ can recombine to form molecular Cl_2 or react with H_2O/OH^- to regenerate $ClOH^{\bullet-}$, which further sustains $Cl_2^{\bullet-}$ production ($Cl^\bullet + H_2O \rightarrow ClOH^\bullet- + H^+$, $k = 2.5 \times 10^5$ L/(mol s); $Cl^\bullet + OH^- \rightarrow ClOH^\bullet-$, $k = 1.8 \times 10^{10}$ L/(mol s)).

$Cl_2^{\bullet-}$ is a strong oxidant ($E^0 = 2.13$ V) (Zhang et al., 2024) and can oxidize various actinides in different valence states, such as U(IV), U(V), Np(III), Np(V), Pu(III), and Am^{3+} (e.g., $U(V)O_2^+ + Cl_2^{\bullet-} \rightarrow U(VI)O_2^{2+} + 2\,Cl^-$, $k = 4 \times 10^9$ L/(mol s); $Np(V)O_2^+ + Cl_2^{\bullet-} \rightarrow Np(VI)O_2^{2+} + 2\,Cl^-$, $k = 3.1 \times 10^6$ L/(mol s); $Am^{3+} + Cl_2^{\bullet-} \rightarrow Am^{4+} + 2Cl^-$, $k = 3.2 \times 10^5$ L/(mol s)). The strong oxidizing power of $Cl_2^{\bullet-}$ can significantly alter the valence state and species equilibrium of actinide radionuclide cations in the aqueous phase, thereby affecting the stability of actinide redox states, their solubility, and their migration potential.

13.5 SUMMARY

In HLW disposal repositories, radiation from decaying radionuclides induces water radiolysis, generating reactive species (e.g., e_{aq}^-, •OH, H_2O_2) that drive redox reactions. These reactions significantly alter the valence states and speciation of radionuclides, profoundly influencing their migration behavior in geological environments.

For actinides like U, Pu, and Np, redox cycling between higher and lower valence states directly controls solubility, complexation, and adsorption. Uranium predominantly shifts between soluble U(VI) complexes (e.g., carbonate species) and insoluble U(IV) precipitates. Plutonium exhibits complex multi-valent behavior, with Pu(IV) forming colloids or precipitates under reducing conditions, while Pu(V/VI) carbonate complexes enhance mobility. Neptunium is most stable as soluble $Np(V)O_2^+$, but reduction to Np(IV) under strong reducing conditions promotes immobilization via precipitation or adsorption.

Anionic radionuclides such as Tc and Se are highly mobile in oxidized forms (TcO_4^-, SeO_4^{2-}). Reduction to lower valence states (e.g., TcO_2, Se(0)) drastically decreases solubility and enhances retention through precipitation or adsorption on mineral surfaces.

On the other hand, radiolysis of background anions in water also affects the migration of radionuclides, which is easily neglected. Carbonate can be transformed into organic ligands like formate and oxalate. Formate-derived radicals may reduce high-valent actinides. Oxalate will form highly soluble complexes with actinides, potentially increasing mobility. Chloride reacts with radiolytic •OH to form strongly oxidizing $Cl_2•^-$, which can re-oxidize the reduced actinides and counteract immobilization.

Therefore, the overall effect of radiolysis on radionuclide migration strongly depends on the unique condition of disposal repositories including not only pH, Eh, etc., but also reactant concentrations and the dynamic equilibrium among the involving species. Systemic evaluation of the risk for radionuclide migration should also take the aforementioned effects into consideration. Since radiolysis breaks the thermodynamic equilibrium, most of the fast reactions involving active radicals are kinetically controlled. Integrating reaction kinetic simulation into the current THMC multi-field coupling models might provide more precise information of long-term migration, especially in near-field areas.

REFERENCES

Arab-Chapelet, B., Grandjean, S., Nowogrocki, G., Abraham, F., 2008. Synthesis and characterization of mixed An(IV)An(III) oxalates (An(IV)=Th, Np, U or Pu and An(III)=Pu or Am). *J. Nucl. Mater.* 373, 259–268.

Belova, E., Nazin, E., Smirnov, A., Obedkov, A., 2021. Radiation-thermal stability of hydrazine nitrate in nitric acid solutions. *Prog. Nucl. Energy* 141, 103968.

Bhalara, P.D., Punetha, D., Balasubramanian, K.A., 2014. A review of potential remediation techniques for uranium(VI) ion retrieval from contaminated aqueous environment. *J. Environ. Chem. Eng.* 2, 1621–1634.

Bruggeman, C., Vancluysen, J., Maes, A., 2002. New selenium solution speciation method by ion chromatography + gamma counting and its application to FeS2-controlled reducing conditions. *Radiochim. Acta* 90, 629–635.

Budantseva, N., et al., 1998. Capture of Pu(V), Np(V) and Pu(VI) from alkaline solutions by hydroxides of Pu(IV), Th(IV) and La(III). *J. Alloys Compd.* 271–273, 231–235.

Dutra, F.R., Vasiliu, M., Gomez, A.N., Xia, D., Dixon, D.A., 2024. Prediction of redox potentials for U, Np, Pu, and Am in aqueous solution. *J. Phys. Chem. A* 128, 5612–5626.

Eriksen, T.E., Cui, D., 1991. On the Interaction of Granite with Tc(IV) and Tc(VII) in Aqueous Solution. *SKB Technical Report TR 91-47*. Swedish Nuclear Fuel and Waste Management Co., Stockholm.

Gray, A., Chorescu, I., Krüger, S., Rösch, N., 2019. Mononuclear hydroxo carbonato complexes of Np(V), Np(VI), and U(VI): A density functional study. *Eur. J. Inorg. Chem.* 2019, 4516–4526.

Grenthe, I., Stumm, W., Laaksuharju, M., Nilsson, A.C., Wikberg, P., 1992. Redox potentials and redox reactions in deep groundwater systems. *Chem. Geol.* 98, 131–150.

Gupta, R., Gamare, J., Sharma, M.K., Kamat, J.V., 2016. Electrochemical investigations of Pu(IV)/Pu(III) redox reaction using graphene modified glassy carbon electrodes and a comparison to the performance of SWCNTs modified glassy carbon electrodes. *Electrochim. Acta* 191, 530–535.

Jegou, C., Gin, S., Larche, F., 2000. Alteration kinetics of a simplified nuclear glass in an aqueous medium: Effects of solution chemistry and of protective gel properties on diminishing the alteration rate. *J. Nucl. Mater.* 280, 216–229.

Kang, M., et al., 2024. Enhance U(VI) reduction on natural pyrite surfaces by gamma irradiation. *Chem. Eng. J.* 489, 151473.

Kias, F., et al., 2017. Redox properties of monocyclooctatetraenyl uranium(IV) and (V) complexes: Experimental and relativistic DFT studies. *Organometallics* 36, 3841–3853.

Li, N., et al., 2025. Thermodynamic controls on plutonium(Pu) speciation in deep groundwater: Implications for HLW disposal safety at Beishan, China. *J. Environ. Radioact.* 287, 107728.

Liu, W., et al., 2016. Adsorption of U(VI) by multilayer titanate nanotubes: Effects of inorganic cations, carbonate and natural organic matter. *Chem. Eng. J.* 286, 427–435.

Maes, N., et al., 2011. A consistent phenomenological model for natural organic matter linked migration of Tc(IV), Cm(III), Np(IV), Pu(III/IV) and Pa(V) in the Boom Clay. *Phys. Chem. Earth* 36, 1590–1599.

Olsson, D., Aydogan, H., Jonsson, M., 2024. The influence of bicarbonate concentration and ionic strength on peroxide speciation and overall reactivity towards UO2. *RSC Adv.* 14, 16248–16254.

Pirlet, V., 2001. Overview of actinides (Np, Pu, Am) and Tc release from waste glasses: Influence of solution composition. *J. Nucl. Mater.* 298, 47–54.

Rotermund, B.M., et al., 2024. Chemical kinetics for the oxidation of californium(III) ions with select radiation-induced inorganic radicals (Cl2•– and SO4•–). *J. Phys. Chem. A* 128, 590–598.

Sekine, T., et al., 2002. Radiolytic formation of Tc(IV) oxide colloids. *Radiochim. Acta* 90, 611–616.

Shang, Y., et al., 2018. Efficient separation of Re(VII) by radiation-induced reduction from aqueous solution. *Chem. Eng. J.* 341, 317–326.

She, J., et al., 2025. Gamma-irradiation-induced reduction of aqueous Se(VI) by natural pyrite. Geochim. *Cosmochim. Acta* 391, 237–247.

Tanaka, K., Tani, Y., Kozai, N., Ohnuki, T., 2022. Sorption of Pu(IV) on biogenic Mn oxide and complexation of Pu(IV) with organic ligands secreted by fungal cells. *J. Radioanal. Nucl. Chem.* 331, 1109–1114.

Tang, Q.-F., Xu, Z., Sun, Y.-L., Zhang, Q., Tan, Q., Pan, D.-Q., Wu, W.-S., 2025. Progress on effect of clay mineral-biocomposite colloids on radionuclides migration and mechanism. *J. Nucl. Radiochem.* 47, 1–16.

Tocino, F., et al., 2014. Dissolution of uranium mixed oxides: The role of oxygen vacancies vs the redox reactions. *Prog. Nucl. Energy* 72, 101–106.

Verma, P.K., Bhattacharyya, A., Samanta, S., Mohapatra, P.K., 2024. A highly efficient in situ redox stabilization strategy for Am–Cm separation using AgBiO3. *Dalton Trans.* 53, 13583–13590.

Vodyanitskii, Y.N., 2011. Chemical aspects of uranium behavior in soils: A review. *Eurasian Soil Sci.* 44, 862–873.

Wu, X., Kang, M., Cai, Z., Song, Y., Shang, C., Xu, F., Wang, J., Li, Y., Chen, F., 2017. Investigation of redox potential of Beishan site and its impact on mobility of redox-sensitive radionuclides. *J. Nucl. Radiochem.* 39, 227–234.

Zhang, H., et al., 2024. Refinement of kinetic model and understanding the role of dichloride radical (Cl2•–) in radical transformation in the UV/NH2Cl process. *Water Res.* 254, 121440.

Zhang, H., Jiang, Z., Hu, C., Zheng, P., Ma, J., 2025. Tunable hydro/dicarboxylation of ethylene with CO2 via a barrierless radiolytic free-radicals pathway. *J. Am. Chem. Soc.* 147, 10391–10400.

Zhou, D., Wang, B., Shang, C.-M., Chen, P., Zhou, M.-F., Zhang, S.-D., 2025. Current status and challenges in research on chemistry and radionuclide migration behavior of high-level radioactive waste disposal. *J. Nucl. Radiochem.* 47, 419–433.

CHAPTER 14

Radionuclide Adsorption/ Desorption at Mineral–Water Interfaces

Mechanisms, Influencing Factors, and Environmental Implications

Hanyu Wu

14.1 INTRODUCTION

14.1.1 Background

The nuclear age provided a low-carbon energy source and new technologies (Liu et al., 2026; Wang & Li, 2025), but also created an enduring legacy of radioactive waste and environmental contamination (Jiang, 2012). Managing this legacy to ensure long-term safety is a major scientific challenge, requiring prediction of radionuclide behavior over geological timescales (Turner et al., 2023). Central to this is the focus of environmental radiochemistry on the mineral–water interface, a reactive zone that dictates whether a radionuclide is immobilized or becomes mobile (Yu, 2021). The primary anthropogenic sources include the entire nuclear fuel cycle, which generates fission products (e.g., ^{90}Sr, ^{137}Cs, ^{99}Tc, ^{129}I) and actinides (e.g., U, Np, Pu, Am, Cm), alongside releases from accidents and weapons testing. The disposal of high-level waste (HLW) and spent nuclear fuel (SNF) poses the most profound long-term challenge (Maes et al., 2024).

The risks stem from radiological toxicity, which can damage biological tissues (Bryant, 2014; Cases et al., 2023; Fréchard et al., 2025). Long-lived radionuclides such as ^{239}Pu ($T_{1/2}$ = 2.411×10^4 y), ^{237}Np ($T_{1/2} = 2.144 \times 10^6$ y), ^{135}Cs ($T_{1/2} = 2.3 \times 10^6$ y), and ^{99}Tc ($T_{1/2} = 2.111 \times 10^5$ y) are of particular concern for deep geological repositories (DGRs), as their containment must be guaranteed over geological timescales (Brookshaw et al., 2015; Li et al., 2025a). Therefore, a predictive understanding of radionuclide biogeochemistry is essential. Once in the subsurface,

DOI: 10.1201/9781003665793-18

groundwater is the primary transport vector, but the ultimate fate of these contaminants is overwhelmingly controlled by their interactions with geological media (Chen et al., 2023; Zeng et al., 2025; Zhao et al., 2023). The mineral–water interface is the microscopic arena where these critical retention processes occur (Geckeis & Rabung, 2008). This interface is a chemically dynamic zone, where mineral surface functional groups develop a pH-dependent charge, creating an electrical double layer that mediates interactions with dissolved species (Furukawa et al., 2009; Tournassat et al., 2016). Consequently, it acts as a powerful geochemical sink, sequestering radionuclides from the aqueous phase (Wei et al., 2023).

The importance of these interfacial processes is paramount in diverse settings. In contaminated aquifers, minerals act as a natural retardation barrier against contaminant plumes (Hayatifar et al., 2024; Liu et al., 2025; Wang & Putnis, 2020). For DGRs, the multi-barrier safety concept relies on both engineered systems and the natural geological barrier of the host rock. The performance of this natural barrier depends on the quantification of its capacity to retard radionuclide migration over millennia. The mineral–water interface's ability to sequester radionuclides is the single most important factor ensuring the long-term safety of geological disposal, making a predictive understanding of these processes essential (Makarov et al., 2021; Tait, 1984; Unokiwedi et al., 2025).

14.1.2 The Importance of Adsorption and Desorption

The primary mechanism retarding radionuclide migration is adsorption: the accumulation of a dissolved substance (adsorbate) at a solid–liquid interface (adsorbent) (Holgersson & Kumar, 2023). This macroscopic process encompasses microscopic mechanisms like surface precipitation and ion exchange. Desorption, the reverse process, releases radionuclides back into the aqueous phase, often triggered by changing solution chemistry (Wu et al., 2020; Wu et al., 2022; Zhao et al., 2022).

The balance between these processes is quantified by the distribution coefficient (K_d)—the ratio of radionuclide concentration in the solid to the liquid phase at equilibrium (Zhang et al., 2022). A high K_d signifies strong retention and low mobility. As a critical input for reactive transport models (RTMs), K_d determines the retardation factor (R), making adsorption the fundamental brake on radionuclide transport (Fu et al., 2025; Yang et al., 2025).

Early research focused on generating empirical K_d datasets (Baumer & Hixon, 2018; Tinnacher et al., 2016; Um et al., 2004). However, this approach has inherent limitations: it is system-specific, lacks transferability, and provides no mechanistic insight. This has driven a shift from descriptive, macroscopic approaches to predictive, mechanistic ones. This review synthesizes current knowledge, bridging from macroscopic observations to molecular-level understanding.

The objective of modern research is to elucidate precise molecular-level adsorption mechanisms, including radionuclide speciation, surface binding site structures, and chemical bonding types (Yohai et al., 2019). This mechanistic understanding underpins the development of thermodynamically grounded predictive tools, such as surface complexation models (SCMs) and general adsorption models (GAMs) (Figure 14.1), which are applicable across a wide range of geochemical conditions (Paul et al., 2025; Qin et al., 2025; Xiao et al., 2025).

14.2 FUNDAMENTAL PRINCIPLES AND MECHANISMS

A predictive understanding of radionuclide fate and transport in the geosphere is fundamentally based on the principles of aqueous geochemistry and surface chemistry. The macroscopic phenomenon of adsorption is the result of a complex interplay between the properties of the mineral surface, the speciation of the radionuclide in solution, and a suite of distinct microscopic interaction mechanisms (Schabernack et al., 2021). This chapter details the chemical nature of the mineral–water interface, the critical importance of aqueous radionuclide speciation, the spectrum of core adsorption mechanisms from weak physical adsorption to strong chemical bonding and precipitation, and the complexities of desorption and hysteresis.

14.2.1 The Chemistry of the Mineral–Water Interface

The mineral–water interface is a highly dynamic zone where the bulk properties of the solid and liquid phases are disrupted, giving rise to unique chemical reactivity. Virtually all minerals in contact with water develop a surface electrical charge, which is the primary driver for the adsorption of charged radionuclide species (Descostes et al., 2017). This charge originates mainly from two processes: the protonation/deprotonation of surface functional groups and isomorphic substitution within the mineral lattice (Furukawa et al., 2009; Liu et al., 2021).

The surfaces of most common minerals, particularly oxides, hydroxides, and aluminosilicates, are terminated by hydroxyl (-OH) functional groups (Li et al., 2025; Wang et al., 2025). These groups are formed by the hydration of the mineral surface, where surface oxygen atoms or metal cations react with water molecules. These surface hydroxyl groups are amphoteric, meaning they can act as either a Brønsted acid (donating a proton) or a Brønsted base (accepting a proton) depending on the solution pH. This behavior can be represented by the following equilibria, where ≡SOH denotes a generic surface hydroxyl site:

$$\equiv SOH + H^+ \rightleftharpoons \equiv SOH_2^+ \text{ (Protonation at low pH)}$$

$$\equiv SOH \rightleftharpoons \equiv SO^- + H^+ \text{ (Deprotonation at high pH)}$$

Consequently, at low pH, the mineral surface tends to carry a net positive charge, attracting anions. At high pH, it carries a net negative charge, attracting cations. The specific pH value at which the net surface charge is zero is a critical mineral property known as the point of zero charge (PZC). For many radionuclides of concern, which are often cationic (e.g., UO_2^{2+}, Pu^{4+}, Am^{3+}, Sr^{2+}, Cs^+), adsorption is therefore significantly more favorable on negatively charged surfaces at pH values above the mineral PZC (Li et al., 2025; Saeed et al., 2022; Soliman et al., 2017).

In addition to this pH-dependent charge, some minerals, most notably the 2:1 phyllosilicate clays (e.g., montmorillonite, illite), possess a permanent negative charge (de Carvalho Arjona et al., 2025). This charge arises from isomorphic substitution within the crystal lattice, for example, the substitution of Al^{3+} for Si^{4+} in the tetrahedral sheet or Mg^{2+} for Al^{3+} in the octahedral sheet. This permanent charge deficit is structurally fixed and

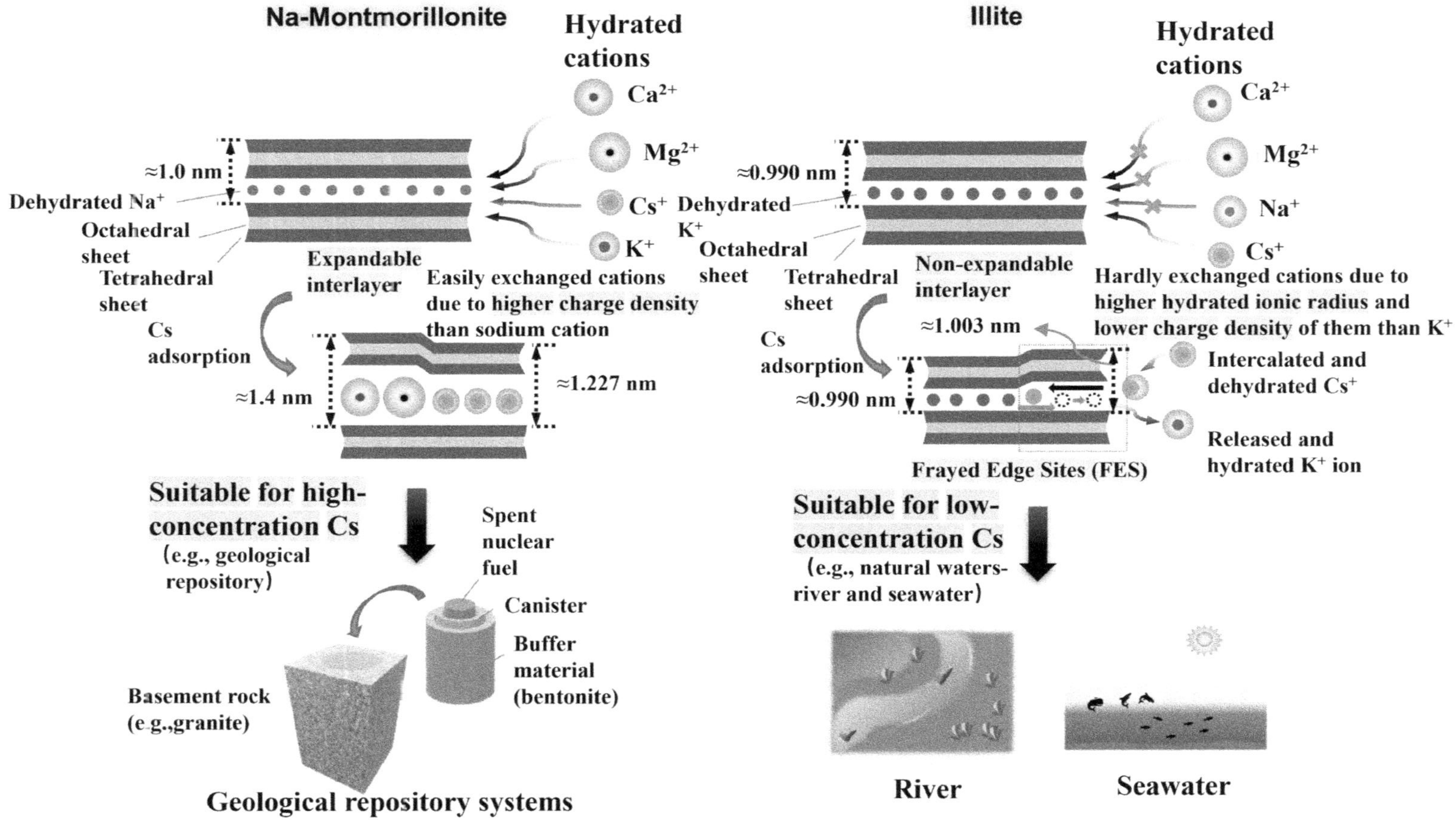

FIGURE 14.1 Schematic Illustration of the Effect of Interlayer Cation and Swelling Behavior of Clay Minerals for ^{137}Cs Adsorption Behavior (Kwon et al., 2023).

independent of solution pH, and it is balanced by the accumulation of hydrated cations (e.g., Na^+, K^+, Ca^{2+}) in the interlayer space and on basal surfaces. This feature makes these clays exceptionally effective adsorbents for cationic radionuclides (Novikau & Lujaniene, 2022; Soliman et al., 2019).

The charged mineral surface and the balancing region of counter-ions in the adjacent solution constitute the electrical double layer (EDL). The structure of the EDL governs the electrostatic interactions (Figure 14.2) that mediate the approach of dissolved ions, including radionuclides, to the surface. The Stern model divides the interfacial region into a compact inner layer (the Stern layer) and an outer diffuse layer (Leroy et al., 2007). The inner Helmholtz plane (IHP) is the locus of centers of specifically adsorbed ions that are dehydrated (at least partially) and form direct chemical bonds with the surface (Figure 14.3). This corresponds to inner-sphere complexation. The outer Helmholtz plane (OHP) is the locus of centers of hydrated ions that are bound electrostatically to the surface. This corresponds to outer-sphere complexation. Beyond the OHP lies the diffuse layer, where ions are influenced by the electrostatic field of the surface but are not regarded as adsorbed. This conceptual knowledge is the foundation for SCMs used to quantify adsorption thermodynamically.

14.2.2 Speciation of Radionuclides in Aqueous Solution

A radionuclide in solution is rarely a simple, free aquo-ion. Its chemical form, or speciation, is highly dependent on the solution properties and dictates its reactivity toward mineral surfaces. The key factors controlling speciation are pH, redox potential (Eh), and the presence of complexing ligands. An accurate prediction of adsorption cannot be made without first accounting for the dominant aqueous species under the relevant conditions.

As pH increases, metal cations undergo hydrolysis, where coordinated water molecules deprotonate to form a series of hydroxo-complexes. For example, the uranyl ion (UO_2^{2+}), the most stable form of uranium in oxic waters, undergoes the following sequence: $UO_2^{2+} \rightarrow UO_2OH^+ \rightarrow (UO_2)_x(OH)y^{2x-y} \rightarrow UO_2(OH)_3^-$. For highly charged actinides like Pu(IV), hydrolysis is so extensive that it can lead to the formation of insoluble hydroxides or

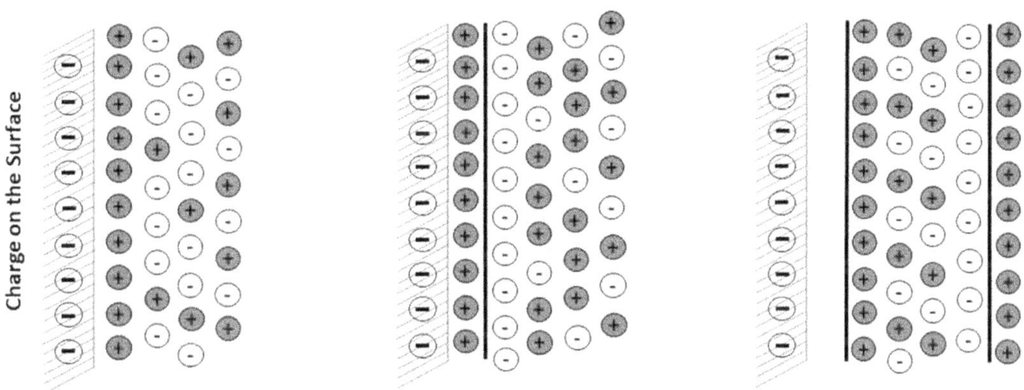

FIGURE 14.2 Illustration of the Three Commonly Used Electrostatic Surface–Water Interface Models (Holgersson & Kumar, 2023).

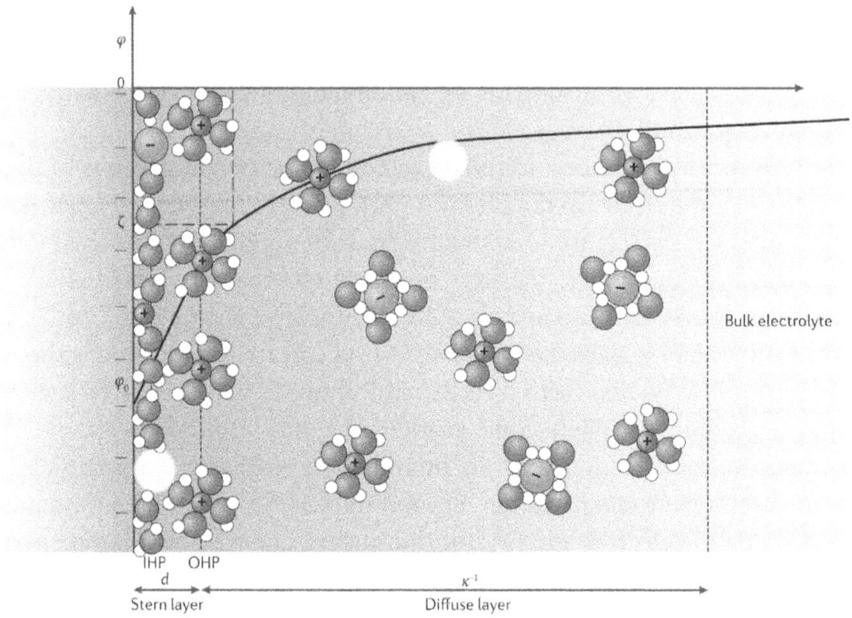

FIGURE 14.3 Classical Mean-Field Description of the Electrical Double Layer (Gonella et al., 2021).

polymeric species even at very low concentrations. Since these different species have different charges and structures, their adsorption behavior varies dramatically.

Many of the critical radionuclides are multi-valent, meaning they can exist in different oxidation states with vastly different chemical properties and mobilities. The prevailing Eh of the groundwater determines which oxidation state is stable (Eun et al., 2025). For instance, uranium exists as the highly soluble and mobile U(VI) (as UO_2^{2+}) under oxidizing conditions, but is reduced to the highly insoluble U(IV) (which precipitates as UO_2(s)) under reducing conditions; technetium exists as the mobile, non-sorbing pertechnetate anion, Tc(VII)O_4^-, in oxic waters, but is readily reduced to the insoluble Tc(IV) oxide, TcO_2(s); and neptunium and plutonium exhibit even more complex redox chemistry, with multiple oxidation states (e.g., Np(IV), Np(V); Pu(III), Pu(IV), Pu(V), Pu(VI)) potentially co-existing under certain conditions. The lower oxidation states (Np(IV), Pu(IV)) are generally much less soluble and more strongly adsorbing than the higher ones (Np(V)O_2^+, Pu(V)O_2^+).

The presence of other dissolved ions (ligands) that can form stable aqueous complexes with radionuclides can profoundly inhibit adsorption by outcompeting the mineral surface for the radionuclide. Carbonate forms strong, soluble, and often anionic complexes with many actinides, such as the uranyl tricarbonato complex $[UO_2(CO_3)_3]^{4-}$, which is highly mobile and adsorbs poorly to negatively charged mineral surfaces. Other important inorganic ligands include phosphate, sulfate, and chloride (Hennig et al., 2020; Shi et al., 2023). Organic ligands, ranging from simple carboxylates to complex macromolecules like humic and fulvic acids, can also form strong aqueous complexes or even ternary surface complexes (mineral–radionuclide–organic), further complicating adsorption behavior.

14.2.3 Microscopic Mechanisms of Adsorption

The macroscopic term adsorption encompasses several distinct microscopic processes, each with unique characteristics regarding bond strength, reversibility, and sensitivity to environmental conditions.

Ion exchange: This process involves the stoichiometric replacement of an ion in the solid phase by an ion from the solution. It is the dominant mechanism for the adsorption of alkali and alkaline earth metals (e.g., $^{137}Cs^{+}$, $^{90}Sr^{2+}$) on minerals with a permanent structural charge, such as illite and montmorillonite. Ion exchange is typically rapid, largely reversible, and highly dependent on the concentration of competing cations (Figure 14.4).

Outer-sphere complexation: This refers to the electrostatic attraction between a hydrated radionuclide ion and a charged mineral surface. The ion remains fully hydrated, and no direct chemical bond is formed with the surface functional groups. The radionuclide is located in the OHP. This type of adsorption is relatively weak, rapid, and highly sensitive to the ionic strength of the solution; an increase in background electrolyte concentration will compress the EDL and enhance competition, leading to decreased adsorption (Zhou & Wang, 2025).

Inner-sphere complexation: This involves the formation of a direct, covalent/ionic chemical bond between the radionuclide and the surface functional groups (e.g., $\equiv SO^{-}$). The ion loses at least some of its coordinated water molecules and is located at the IHP. These bonds are much stronger and more specific than electrostatic interactions. Inner-sphere complexes are distinguished by their geometry, such as monodentate (one bond to the surface), bidentate (two bonds), and binuclear (bridging two surface sites). The

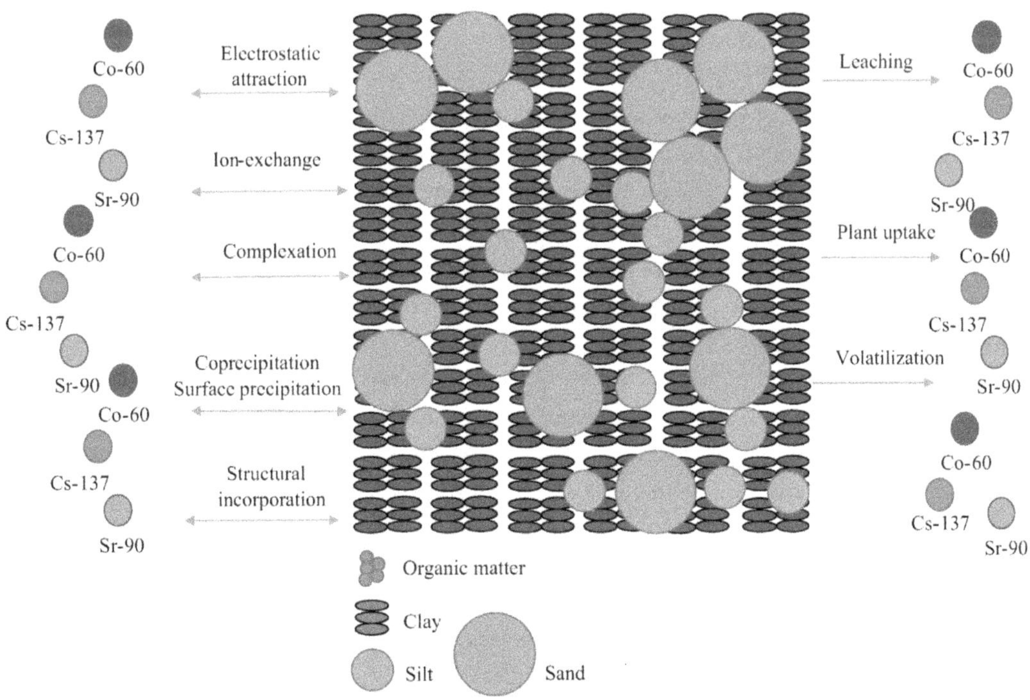

FIGURE 14.4 Adsorption and Desorption Mechanisms of Radionuclides in Soil (Yoon et al., 2021).

formation of inner-sphere complexes is less sensitive to ionic strength than outer-sphere complexation and is considered the primary mechanism for the strong adsorption of many hydrolyzable radionuclides, including U, Np, Pu, and Am, on oxide minerals (Shi et al., 2025). The identification of these specific surface structures, largely through the use of synchrotron-based X-ray absorption spectroscopy (XAS), has been a major focus of modern radiochemistry.

Surface precipitation and co-precipitation: When the concentration of a radionuclide at the mineral–water interface exceeds the solubility limit of one of its solid phases, a new phase can form on the mineral surface. This is termed surface precipitation. For example, at high concentrations, U(VI) can precipitate as schoepite ($UO_3{\cdot}2H_2O$) or other uranyl phases on mineral surfaces. Co-precipitation is a related process where a radionuclide is incorporated into the crystal lattice of a newly forming or recrystallizing mineral. For instance, Ra^{2+} can be incorporated into barite ($BaSO_4$), and U(VI) can be incorporated into calcite ($CaCO_3$) by substituting for Ca^{2+}. These processes can lead to very strong and often irreversible sequestration.

Redox-driven adsorption: This is a coupled process where the adsorption is intrinsically linked to a change in the radionuclide oxidation state. It is particularly important for the immobilization of redox-sensitive radionuclides. The classic example is the reductive immobilization of mobile U(VI) by minerals containing structural Fe(II), such as magnetite (Fe_3O_4), green rust, or certain clays. The Fe(II) at the mineral surface acts as an electron donor, reducing U(VI) to the highly insoluble U(IV), which precipitates as nano-scopic uraninite (UO_2). This mechanism provides a powerful pathway for the natural attenuation of uranium plumes in anoxic environments.

14.2.4 Desorption Processes and Hysteresis

Desorption, the release of adsorbed species back into solution, is as important as adsorption for long-term risk assessment. In an ideal, fully reversible system, desorption would be the exact reverse of adsorption. However, in many real systems, a significant adsorption–desorption hysteresis is observed, meaning that the radionuclide is more strongly retained than predicted by equilibrium adsorption data. Desorption is often found to be much slower or less complete than adsorption when solution conditions are changed (e.g., by lowering the radionuclide concentration or introducing a competing ligand).

This hysteresis can be caused by several factors: (1) slow desorption kinetics (the breaking of strong inner-sphere chemical bonds can be an energetically demanding and thus slow process); (2) surface-induced structural changes (over time (aging), an adsorbed radionuclide may transition from a less stable configuration (e.g., outer-sphere complexes) to a more stable one (e.g., inner-sphere complexes)); (3) diffusion and incorporation (radionuclides may slowly diffuse from surface sites into the mineral lattice or become trapped by surface precipitation/recrystallization processes); and (4) irreversible redox reactions (the re-oxidation of a reduced and precipitated species (e.g., U(IV)) may be kinetically hindered, leading to its long-term stability even if the bulk solution becomes more oxidizing).

The mechanisms and kinetics of desorption are critical because they govern the potential for remobilization of contaminants should geochemical conditions at a disposal or contamination site change in the future. The common observation of hysteresis implies that relying solely on reversible adsorption models can significantly underestimate the long-term sequestration capacity of geological media (Um et al., 2007).

14.3 KEY RADIONUCLIDES AND TYPICAL MINERAL SYSTEMS

The general principles outlined in the previous chapter manifest in a rich diversity of behaviors when applied to specific radionuclide–mineral pairs. The chemical identity of the radionuclide, its charge, size, redox sensitivity, and propensity for hydrolysis, as well as the properties of the mineral, its surface charge characteristics, specific surface area, and structural features, combine to dictate the dominant retention mechanism and its overall effectiveness.

14.3.1 Behavioral Characteristics of Key Radionuclides

The adsorptive properties of geological media are often defined by their most reactive mineral components (Figure 14.5). Key radionuclides of concern are categorized as actinides and fission/activation products, based on their half-lives, toxicity, and mobility. Actinides dominate long-term radiotoxicity and exhibit complex chemistry, including multiple oxidation states, extensive hydrolysis, and strong interactions with environmental ligands like carbonate (Huang C. et al., 2025; Kerisit & Liu, 2014; Zheng et al., 2018).

Actinides:

As the most studied actinide, uranium exists as the hexavalent uranyl cation ($U(VI)O_2^{2+}$) under oxidizing conditions. This form is relatively mobile but adsorbs strongly to minerals via inner-sphere complexes. However, the formation of stable and soluble uranyl-carbonate aqueous complexes can dramatically suppress adsorption and enhance mobility (Kim et al., 2012). Under reducing conditions, U(VI) is reduced to the highly insoluble U(IV) state (e.g., $UO_2(s)$), a critical natural attenuation process.

Neptunium-237 (^{237}Np) is a key contributor to long-term repository dose. Under oxic conditions, it is present mainly as the pentavalent neptunyl cation, $Np(V)O_2^+$. Because of its low charge density, Np(V) interacts weakly with mineral surfaces, mainly through outer-sphere complexation, and is therefore one of the most mobile actinides. Like uranium, however, it can be reduced under anoxic conditions to the less soluble and more strongly adsorbing Np(IV) state (Konevnik et al., 2020).

Plutonium exhibits four oxidation states, from Pu(III) to Pu(VI), which makes its chemistry exceptionally complex. Pu(IV), the tetravalent state, is generally favored and is characterized by extremely low solubility and a strong tendency to adsorb to nearly all mineral surfaces through inner-sphere complexation. However, redox transformations can form the more mobile $Pu(V)O_2^+$ species, making its long-term behavior difficult to predict (Leng et al., 2016).

^{241}Am and ^{243}Cm exist exclusively in the trivalent oxidation state (Am^{3+}, Cm^{3+}). Analogous to trivalent lanthanides, they are hard cations that form strong inner-sphere complexes with mineral surfaces, leading to robust adsorption. However, their adsorption can be reduced at high pH by the formation of soluble carbonate complexes (Yang et al., 2023).

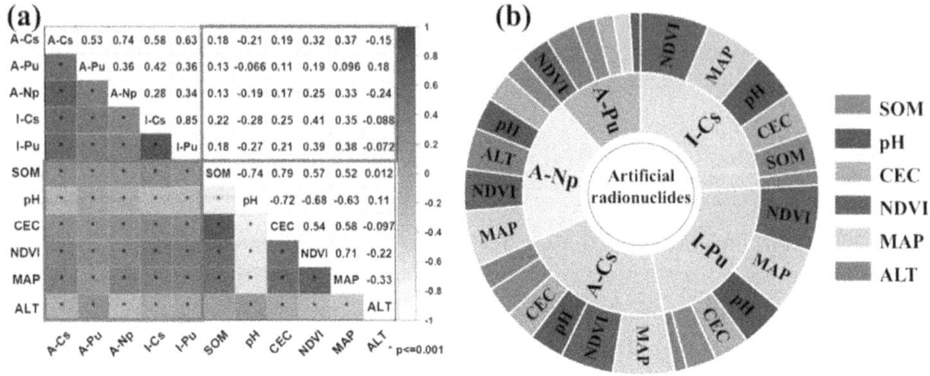

FIGURE 14.5 Correlation Thermograms of ^{137}Cs Activity (A-Cs), $^{239+240}Pu$ Activity (A-Pu), ^{237}Np Activity (A-Np), ^{137}Cs Inventory (I-Cs), and $^{239+240}Pu$ Inventory (I-Pu) with Soil Organic Matter (SOM), pH, Cation Exchange Capacity (CEC), Normalized Difference Vegetation Index (NDVI), Mean Annual Precipitation (MAP), and Active Layer Thickness (ALT) (Huang C. et al., 2025).

Fission and Activation Products:

^{137}Cs is a singly charged cation (Cs^+) whose adsorption is dominated by ion exchange (Figure 14.1). It shows a remarkable and highly selective affinity for the frayed edge sites (FESs) of illite and mica, leading to strong and nearly irreversible fixation. This makes these minerals a critical geological barrier for ^{137}Cs (Goo et al., 2025; Huang Z. et al., 2025).

Strontium (Sr^{2+}), chemically similar to Ca^{2+}, is retained by ion exchange and, importantly, by co-precipitation with calcium minerals. Radium (Ra^{2+}), analogous to Ba^{2+}, is primarily immobilized not by surface adsorption but by highly effective co-precipitation with sulfate minerals, particularly barite (Kim et al., 2021).

Tc and Se are highly redox-sensitive. Under oxic conditions, technetium exists as the highly soluble and mobile pertechnetate anion (TcO_4^-), which interacts weakly with minerals. Under reducing conditions, it is reduced to insoluble Tc(IV) oxides. Similarly, selenium exists as mobile selenate ($Se(VI)O_4^{2-}$) or more strongly adsorbing selenite ($Se(IV)_3^{2-}$) in oxic waters. Under reducing conditions, both can be reduced to highly insoluble elemental selenium (Se(0)) or selenide (Se(-II)), providing effective immobilization pathways.

14.3.2 Adsorption Characteristics of Typical Mineral Systems

The geological medium is a heterogeneous mixture of minerals, but its overall adsorptive properties can often be understood by examining the behavior of its most reactive components toward specific radionuclides, such as clay minerals, oxides and hydroxides, and carbonates and silicates.

Clay minerals (montmorillonite, vermiculite, illite, kaolinite, etc.) are ubiquitous in soils, sediments, and argillaceous rock formations considered for waste disposal. The high specific surface area and charge characteristics make them dominant adsorbents. As a 2:1 swelling clay, montmorillonite (and smectites) possesses a high permanent negative charge from isomorphic substitution, resulting in a high cation exchange capacity (CEC). This makes it an excellent adsorbent for cationic radionuclides like Sr(II) and Am(III) via ion exchange and outer-sphere complexation in its interlayer spaces. It is the primary component of bentonite, the reference buffer and backfill material in many DGR concepts (Savage et al., 2019). Also a 2:1 phyllosilicate, vermiculite is characterized by a very high layer charge, typically even higher than that of montmorillonite, arising from extensive substitution of Al^{3+} for Si^{4+} in the tetrahedral sheets. This results in one of the highest cation exchange capacities among common clay minerals. Vermiculite exhibits limited swelling, placing it structurally between the fully swelling smectites and non-swelling illite. Its primary importance in radiochemistry lies in its exceptional capacity to sequester ^{137}Cs and ^{90}Sr. For ^{90}Sr, adsorption occurs via a highly efficient cation exchange process. For ^{137}Cs, the mechanism is more specific and robust even exceeds that of illite. Upon exchange with interlayer cations like Mg^{2+} or Ca^{2+}, the low hydration energy of the Cs^+ ion allows the high-charge interlayer spaces to collapse, effectively trapping the radionuclide and leading to quasi-irreversible fixation (Figure 14.1). Unlike the FESs of illite, this high-affinity trapping mechanism is available throughout the vermiculite structure, giving it a very large capacity for ^{137}Cs fixation (Shi et al., 2024). Illite has a lower CEC than montmorillonite but is defined by the presence of FESs that are uniquely selective for Cs^+, as discussed earlier. This makes illitic clays a crucial sink for radiocesium in the environment. A non-swelling 2:1 clay, illite has a lower CEC than montmorillonite but is defined by the presence of FESs that are uniquely selective for Cs^+, as discussed earlier. This makes illitic clays a crucial sink for radiocesium in the environment (Ishidera et al., 2016). As a 1:1 non-swelling clay, kaolinite has no permanent structural charge. Its reactivity is dominated by the pH-dependent charge on its edge hydroxyl groups, making its surface chemistry similar to that of aluminum and silicon oxides. Its adsorption capacity is generally lower than that of 2:1 clays, but it is a significant adsorbent for strongly hydrolyzing cations like the actinides (Muslim et al., 2023).

Frequently present as coatings on other minerals (iron, manganese, and aluminum), oxides and hydroxides can exert a disproportionately large influence on bulk rock adsorption due to their high specific surface area and reactive surface hydroxyl groups. With PZCs typically in the range of pH 7–9, they are positively charged in acidic-to-neutral waters, making them effective adsorbents for anions like Se (Jin et al., 2025). At higher pH, their negatively charged surfaces are highly effective at binding cationic actinides (U(VI), Np(V), Pu(IV), Am(III)) through strong inner-sphere complexation. Furthermore, Fe(II)-bearing oxides like magnetite (Fe_3O_4) are strong reductants, capable of immobilizing U, Np, Tc, and Se (Jin et al., 2024). Manganese oxides (e.g., birnessite, cryptomelane) are among the most reactive adsorbents in the natural environment. They typically have very low PZCs (pH 2–3), meaning they carry a high negative charge density across the entire relevant pH range, leading to an exceptionally high capacity for cation adsorption. They are also powerful oxidizing agents, capable of altering the speciation of redox-sensitive elements, which can complicate their role in radionuclide retention (Tan et al., 2008; Tytkowska et al., 2018).

As the most common carbonate mineral, the reactivity of calcite ($CaCO_3$) is distinct. While surface adsorption of radionuclides like Am(III) and U(VI) does occur, its most significant role is often in co-precipitation. The structural incorporation of radionuclides that are chemical analogs for Ca^{2+} (e.g., Sr^{2+}, UO_2^{2+} under certain conditions) during calcite precipitation or recrystallization can lead to extremely stable, long-term sequestration. As the most abundant crustal mineral, quartz is relatively unreactive compared to clays and oxides. It has a low PZC (pH 2–3) and typically a lower specific surface area. While it does adsorb strongly hydrolyzing cations at neutral to high pH, its overall contribution to adsorption in a complex rock is often overshadowed by other minerals.

Ultimately, radionuclides in the geosphere interact with complex soil/sediment, whole rocks (e.g., granite, tuff), or engineered materials like bentonite. Representing the most common medium for near-surface contamination, soils and sediments are the ultimate heterogeneous systems. They are complex mixtures of primary and secondary minerals (quartz, feldspars, clays, and oxides), natural organic matter (NOM), water, air, and microbial communities. The overall adsorption behavior of a soil or sediment is a bulk property reflecting the weighted contributions of all its components. The most critical feature distinguishing soils from deep geological formations is the presence of NOM. NOM, including humic and fulvic acids, possesses a high density of carboxyl and phenolic functional groups, giving it a very high adsorption capacity. It can directly bind radionuclides or coat mineral surfaces, creating complex ternary systems (mineral–NOM–radionuclide) that alter surface reactivity (Wang et al., 2024; Zuo et al., 2019). The role of NOM is dual-natured: it can immobilize radionuclides through adsorption, but it can also form strong, soluble aqueous complexes that enhance their mobility. For these highly complex systems, SCMs become difficult to parameterize. Therefore, contaminant transport modeling for soils often relies on the empirical K_d, defined as the ratio of the contaminant concentration in the solid phase to that in the aqueous phase at equilibrium. While K_d values are inherently site- and condition-specific, they provide a practical means of quantifying retardation for large-scale risk assessments (Koshikawa et al., 2019; Tatsuno et al., 2022).

Granite and tuff rocks are common host rock candidates for DGRs (Chen et al., 2023). Their bulk adsorption properties are a composite of their constituent minerals, quartz, feldspars, micas (like biotite), and accessory minerals. Usually, the adsorption capacity of the intact rock matrix is low and is dominated by the highly reactive secondary minerals (clays, iron oxides) that line fractures, as these are the primary surfaces encountered by flowing groundwater. Used as an engineered barrier, bentonite is a clay-rich material composed primarily of montmorillonite. Its function is to provide a low-permeability, high-swelling, and highly adsorptive barrier around waste canisters. Understanding radionuclide adsorption on bentonite requires considering the effects of high compaction density, elevated temperatures, and high-salinity porewaters, which can alter the clay structure and surface chemistry.

14.3.3 Major Environmental Factors

Radionuclide adsorption is controlled by interconnected environmental variables that alter radionuclide speciation, mineral surface properties, and system chemistry. A predictive understanding requires appreciating master variables like pH, redox potential (Eh), ionic strength, competing ions, complexing ligands, and microbial activity.

Solution pH is the most critical factor that simultaneously controls radionuclide aqueous speciation and mineral surface charge (Nemutudi et al., 2025; Tong et al., 2019). Most mineral surfaces contain amphoteric functional groups (≡SOH) that change charge with pH (Gorski et al., 2013), while polyvalent radionuclides undergo hydrolysis (Figure 14.6). This dual effect creates a characteristic "adsorption edge"—a narrow pH range where cation adsorption sharply increases.

Redox potential (Eh) is a master variable for multi-valent elements such as U, Pu, Np, and Tc. Changes in oxidation state, often driven by dissolved oxygen or redox-active minerals containing Fe(II), can drastically alter a radionuclide's mobility and adsorption behavior. Mineral surfaces themselves can act as direct reductants (Gorski et al., 2013).

Ionic strength and competing ions significantly affect adsorption. High salt concentrations compress the electrical double layer, reducing electrostatically driven outer-sphere complexation. Concurrently, background ions can directly compete for specific adsorption sites, such as K^+ inhibiting ^{137}Cs uptake (Wei et al., 2019) or Ca^{2+} competing with ^{90}Sr. High-salinity environments can severely limit adsorption capacity (Zhao et al., 2014).

Complexing ligands, especially NOM, compete with mineral surfaces for radionuclides (Chervyakov et al., 2023). NOM plays a dual role: it can enhance mobility by forming soluble aqueous complexes, or promote retardation by acting as a powerful adsorbent itself or by forming ternary surface complexes (e.g., mineral-NOM-radionuclide).

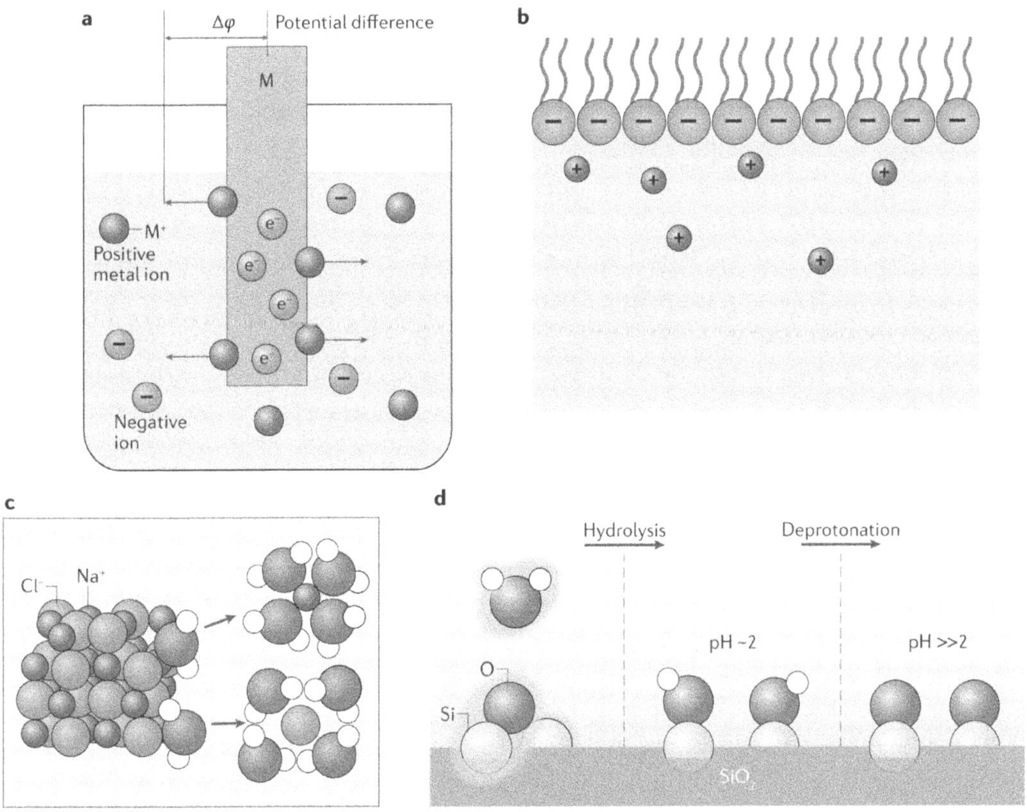

FIGURE 14.6 Possible Sources of Solid Surface Charge (Gonella et al., 2021).

Finally, microorganisms are active geochemical agents. They can directly catalyze radionuclide reduction (Jeon et al., 2004; Li et al., 2016), alter local pH and Eh, excrete complexing agents, and act as effective adsorbents via their cell walls and biofilms. The mineral–microbe–radionuclide ternary system is a key research frontier (Chen et al., 2021; Khan et al., 2019).

14.4 RESEARCH METHODOLOGIES

The evolution of understanding radionuclide adsorption from empirical to molecular level has been driven by developing and applying a suite of analytical techniques and computational methods. While macroscopic batch experiments remain fundamental for quantifying adsorption, they alone cannot reveal the underlying mechanisms. A true, predictive understanding requires a multi-faceted approach integrating thermodynamic data with direct, microscopic-scale evidence and theoretical validation.

14.4.1 Macroscopic and Thermodynamic Approaches

The workhorse of adsorption research is the batch experiment. By measuring the initial and final aqueous concentrations of the radionuclide, the amount adsorbed to the solid can be calculated by mass balance (Munagapati et al., 2026). While batch experiments measure equilibrium partitioning, column experiments are designed to simulate the dynamic process of radionuclide transport in a porous medium. The concentration of the radionuclide in the effluent is monitored over time to obtain a breakthrough curve, which provides crucial information on transport kinetics, retardation factors, and the potential for irreversible adsorption or desorption hysteresis. These experiments provide a critical link between static, lab-scale data and the dynamic conditions of a real geological system (Hamed et al., 2022).

14.4.2 Spectroscopic and Microscopic Techniques

Scanning probe microscopy (SPM) encompasses techniques such as atomic force microscopy (AFM) and scanning tunneling microscopy (STM), which map surface topography with atomic or near-atomic resolution (Figure 14.7). AFM can be operated in situ in aqueous solution, allowing one to watch in real time as mineral surfaces dissolve or as new surface phases nucleate and grow in response to radionuclide adsorption. The use of an in situ fluid cell allows for imaging mineral surfaces while they are fully immersed in an aqueous solution of controlled chemistry (pH, ionic strength, and radionuclide concentration). This approach enables direct imaging of the nucleation and growth of radionuclide-bearing nanoparticles, such as U(VI) or Pu(IV) (hydr)oxides, on mineral surfaces including calcite, mica, and quartz. This technique also allows researchers to visualize how adsorption and precipitation occur preferentially at specific surface features, such as step edges, kinks, defects, and etch pits, thereby linking surface morphology to radionuclide reactivity (Altman & Schwarz, 2010; Medeiros-Costa et al., 2024).

The advent of high-brilliance synchrotron radiation sources has developed environmental science. These techniques are element-specific, non-destructive, and can be used on hydrated samples at environmentally relevant (trace) concentrations. X-ray absorption near edge structure (XANES) is routinely used to determine the redox state of multi-valent radionuclides like uranium (distinguishing U(VI) from U(IV)), neptunium (Np(V) vs. Np(IV)), and technetium (Tc(VII) vs. Tc(IV)) in adsorption samples (Liang et al., 2025). Extended X-ray absorption fine structure (EXAFS) is the oscillatory part of the spectrum well above the edge that contains information about the local coordination environment of the absorbing atom. Analysis of the EXAFS signal can reveal the number, identity, and distance of neighboring atoms. This allows for the direct determination of whether a radionuclide is bound as a hydrated outer-sphere complex or a dehydrated inner-sphere complex, the number of bonds to the surface (monodentate vs. bidentate), and the presence of precipitated nano-particles. Synchrotron X-rays can be focused to micrometer or even nano-meter spot sizes, enabling spatially resolved analysis. Micro-X-ray fluorescence (μ-XRF) is used to create two-dimensional maps of elemental distributions, showing precisely where a radionuclide is concentrated on a heterogeneous surface or within a rock fracture. By combining μ-XRF with μ-XANES, one can create maps of oxidation state distribution, for example, visualizing the reduction of U(VI) at specific hotspots on a mineral surface (Filippova et al., 2025; Wu et al., 2025).

Vibrational spectroscopies probe the energies of molecular vibrations and are highly sensitive to changes in chemical bonding and molecular symmetry. Fourier transform infrared (FTIR) spectroscopy, particularly in the attenuated total reflectance (ATR-FTIR) configuration, is a surface-sensitive technique that can identify the functional groups involved in adsorption. By monitoring shifts in the vibrational frequencies of surface hydroxyl groups (≡S–O–H) or ligands like carbonate and phosphate upon radionuclide adsorption, one can infer the nature of the surface complex. Raman spectroscopy is another vibrational technique that is particularly useful for identifying solid phases. It can be used to distinguish between different surface precipitates (e.g., schoepite vs. uraninite) and is also effective for studying certain aqueous species. Confocal Raman microscopy allows for spatially resolved analysis (Liao et al., 2025).

14.4.3 Computational and Theoretical Modeling

Quantum mechanics (QM) methods can predict the properties of molecules and solids from first principles. Density Functional Theory (DFT) calculations are used to (1) optimize the geometry of proposed surface complexes to determine the most stable bonding configurations (e.g., monodentate vs. bidentate); (2) calculate the reaction energies for adsorption, providing a theoretical basis for stability constants used in SCMs; (3) simulate spectroscopic signatures (e.g., vibrational frequencies, bond lengths) that can be directly compared with FTIR and EXAFS data to validate structural interpretations; and (4) determine electronic properties, such as charge transfer during redox-sensitive adsorption. Classical molecular dynamics (MD) can simulate large systems over nano-seconds, providing insight into the dynamic structure of the mineral–water interface, the hydration shells of ions, and the free energy profiles of adsorption reactions. Ab initio MD (AIMD), which uses DFT to calculate forces on-the-fly, is more computationally expensive but provides a more accurate description of chemical reactions and bond breaking/formation at the interface (Ridzwan et al., 2022).

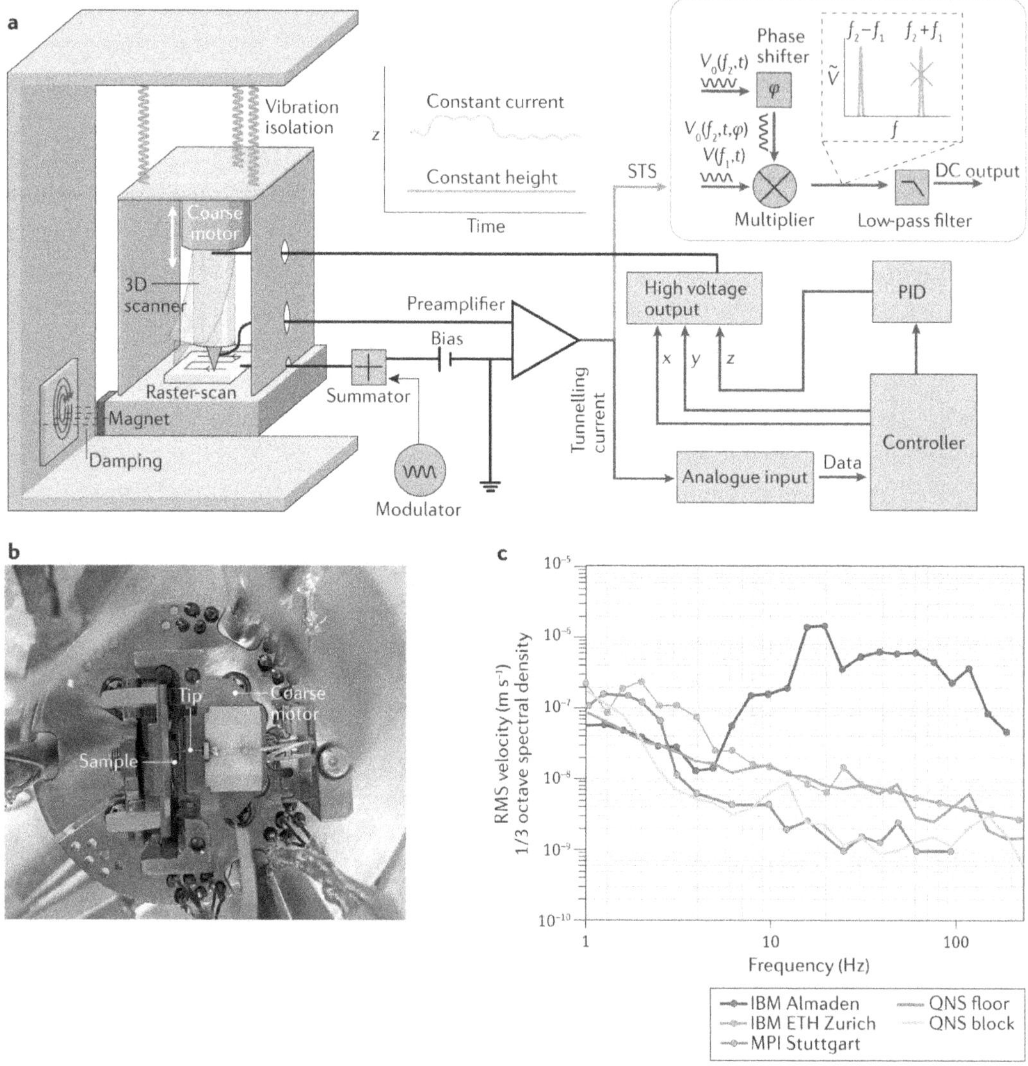

FIGURE 14.7 Scanning Probe Microscopy Application (Bian et al., 2021).

14.4.4 Machine Learning

A new frontier in computational geochemistry is the application of machine learning (ML) to bridge the spatiotemporal scales between molecular processes and field-scale phenomena (Figure 14.8). Traditional physics-based models are often limited because quantum mechanics/molecular dynamics (QM/MD) methods become computationally intractable for heterogeneous systems larger than a few nanometers or simulations extending beyond a few nanoseconds. ML offers a shift by learning complex, non-linear relationships directly from data. ML models, such as random forests or neural networks, can be trained on large databases of experimental adsorption data. The trained model can then rapidly predict adsorption coefficients for new geochemical conditions far faster than experiments (Shelton et al., 2021). This is invaluable for populating the large parameter matrices required for

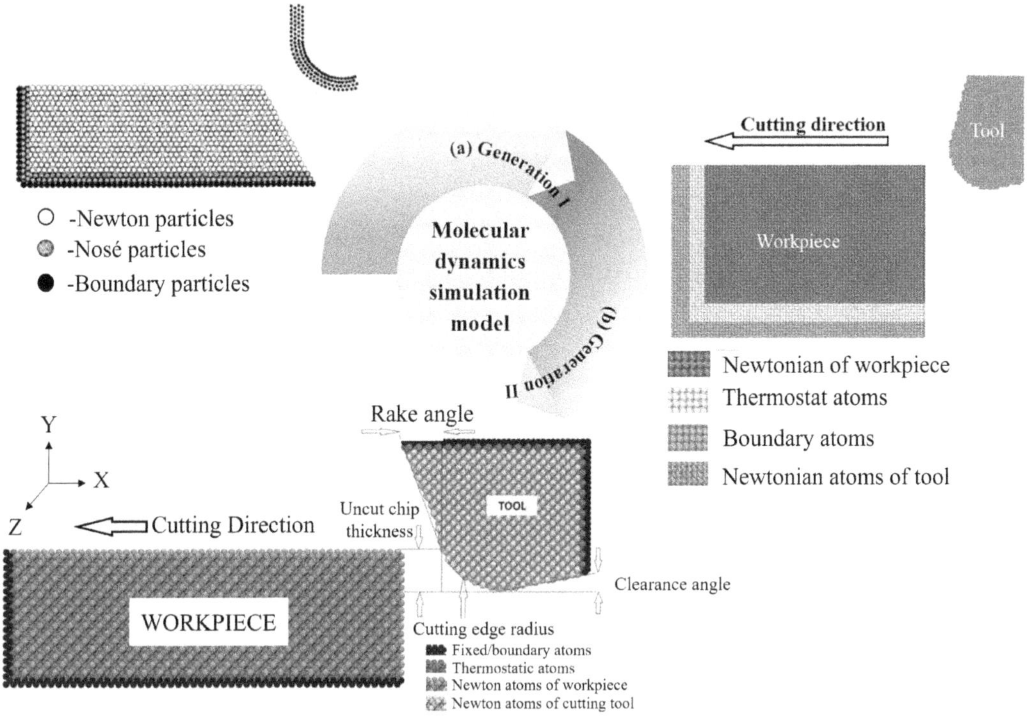

FIGURE 14.8 (a) The First-Generation and (b) the Second-Generation Schematic Model of Micro-Cutting for Molecular Dynamics Simulation (Han et al., 2022).

reactive transport modeling. A major challenge is extending atomistic insights to larger scales. This can be achieved by training ML interatomic potentials (MLPs) on a large set of QM-calculated energies and forces. These MLPs can then be used in classical MD simulations to study systems of millions of atoms over long timescales with near-QM accuracy, bridging the gap between quantum chemistry and mesoscale phenomena (Wang et al., 2016). ML algorithms can analyze complex datasets to identify and rank the environmental factors that most strongly influence radionuclide adsorption. This data-driven approach can reveal unexpected correlations and guide future experimental research toward the most critical parameters (He et al., 2022; Lui et al., 2022; Su et al., 2025).

14.5 ENVIRONMENTAL APPLICATIONS AND IMPLICATIONS

The evolution of radionuclide adsorption studies from empirical observations to predictive, molecular-level mechanisms is fundamental to addressing environmental challenges from nuclear energy. This mechanistic insight is crucial for the environmental impact assessment (EIA) of new facilities, the long-term safety of geological waste disposal, and the remediation of contaminated sites.

For site remediation, this knowledge guides strategies ranging from monitored natural attenuation (MNA) to active interventions. It is directly applied in designing engineered solutions like permeable reactive barriers (PRBs), where specific materials are chosen to reduce, precipitate, or exchange radionuclides (Klammler & Hatfield, 2008; Li et al., 2025).

Despite significant progress, translating molecular understanding to field-scale prediction remains a major challenge. Future research must address the complexity of natural systems. Key frontiers include understanding long-term "aging" processes, where initially reversible adsorption becomes permanent through mechanisms like diffusion or recrystallization. Furthermore, the focus is shifting from simple solid–liquid models to the complex interactions within the quaternary mineral–water–microbe–radionuclide system. Integrating these complexities, potentially with the aid of ML, is essential for accurately predicting the long-term fate of radionuclides in the environment.

REFERENCES

Altman, E.I., Schwarz, U.D., 2010. Mechanisms, kinetics, and dynamics of oxidation and reactions on oxide surfaces investigated by scanning probe microscopy. *Adv. Mater.* 22, 2854–2869.

Baumer, T., Hixon, A.E., 2018. Kinetics of europium sorption to four different aluminum (hydr) oxides: Corundum, γ-alumina, bayerite, and gibbsite. *J. Environ. Radioact.* 195, 20–25.

Bian, K., Gerber, C., Heinrich, A.J., Müller, D.J., Scheuring, S., Jiang, Y., 2021. Scanning probe microscopy. *Nat. Rev. Methods Primers* 1, 36.

Brookshaw, D.R., Pattrick, R.A.D., Bots, P., Law, G.T.W., Lloyd, J.R., Mosselmans, J.F.W., et al., 2015. Redox interactions of Tc(VII), U(VI), and Np(V) with microbially reduced biotite and chlorite. *Environ. Sci. Technol.* 49, 13139–13148.

Bryant, P.A., 2014. Chemical toxicity and radiological health detriment associated with the inhalation of various enrichments of uranium. *J. Radiol. Prot.* 34.

Cases, C., Benegas, M., Sánchez, M., Vollmer, I., Casas, F., Gomà, C., et al., 2023. Biological equivalent dose is associated with radiological toxicity after lung stereotactic ablative radiation therapy. *Radiother. Oncol.* 183, 109552.

Chen, L., Zhao, X., Liu, J., Ma, H., Wang, C., Zhang, H., et al., 2023. Progress on rock mechanics research of beishan granite for geological disposal of high-level radioactive waste in China. *Rock Mech. Bull.* 2, 100046.

Chen, Y., Sun, L., Yang, J., Li, Y., Lin, Q., Wu, H., et al., 2023. Solid-liquid distribution coefficients of Sr(II) at a proposed nuclear power plant site in China and their relations to cation exchange capacity and pH. *J. Radioanal. Nucl. Chem.* 332, 1019–1031.

Chen, Z., An, L., Wei, H., Zhang, J., Zou, Q., Sun, M., et al., 2021. Nitrate alleviate dissimilatory iron reduction and arsenic mobilization by driving microbial community structure change. *Surf. Interfaces* 26, 101421.

Chervyakov, N.M., Boyarintsev, A.V., Kostikova, G.V., Stepanov, S.I., 2023. Dissolution of actinide oxides in carbonate solutions. *Radiochemistry* 65, 276–287.

de Carvalho Arjona, J., Samara, M., Ulsen, C., Valenzuela Diaz, F.R., Demarquette, N.R., 2025. Isoniazid adsorption and release by cloisite and laponite: An effect of surface charge. *Mater. Chem. Phys.* 344, 131097.

Descostes, M., Pointeau, I., Radwan, J., Poonoosamy, J., Lacour, J., Menut, D., et al., 2017. Adsorption and retarded diffusion of EuIII-EDTA− through hard clay rock. *J. Hydrol.* 544, 125–132.

Eun, H., Szabo, P.G., Rehhorn, D., Lee, S., Jo, Y., Altmaier, M., et al., 2025. Uptake of Eu, Th, U, and Pu by granite and biotite gneiss in Korean fresh groundwater under oxidizing and reducing conditions. *Radiochim. Acta* 113, 181–194.

Filippova, I.V., Fekry, A.M., Medany, S.S., Filippov, L.O., 2025. The interaction mechanism of dolomite mineral and phosphoric acid and its impact on the Ti alloy corrosion. *Sci. Rep.* 15, 4802.

Fréchard, T., Bachelot, F., Ménard, V., Brizais, C., Macé, L., Elie, C., et al., 2025. Co-exposure to inhaled tungsten particles and low-dose gamma rays: Neurotoxicological outcome in rats. *Sci. Rep.* 15, 18307.

Fu, H., Yang, L., Dong, L., Ma, M., Deng, Y., Wei, Y., et al., 2025. In-situ generated CoBOx@polyimide aerogel: Autocatalytic, antimicrobial, salt-tolerant and high-resilient adsorbents for U(VI) extraction. *Sep. Purif. Technol.* 361, 131477.

Furukawa, Y., Watkins, J.L., Kim, J., Curry, K.J., Bennett, R.H., 2009a. Aggregation of montmorillonite and organic matter in aqueous media containing artificial seawater. *Geochem. Trans.* 10, 2.

Furukawa, Y., Watkins, J.L., Kim, J., Curry, K.J., Bennett, R.H., 2009b. Aggregation of montmorillonite and organic matter in aqueous media containing artificial seawater. *Geochem. Trans.* 10, 2.

Geckeis, H., Rabung, T., 2008. Actinide geochemistry: From the molecular level to the real system. *J. Contam. Hydrol.* 102, 187–195.

Gonella, G., Backus, E.H., Nagata, Y., Bonthuis, D.J., Loche, P., Schlaich, A., et al., 2021. Water at charged interfaces. *Nat. Rev. Chem.* 5, 466–485.

Goo, J., Kim, J., Lee, S., Kwon, J., Jo, H.Y., 2025. Cesium retention in powdered granite under saline and alkaline conditions: Role of biotite in deep geological repositories. *Nucl. Eng. Technol.* 57, 103417.

Gorski, C.A., Klüpfel, L.E., Voegelin, A., Sander, M., Hofstetter, T.B., 2013. Redox properties of structural Fe in clay minerals: 3. Relationships between smectite redox and structural properties. *Environ. Sci. Technol.* 47, 13477–13485.

Hamed, M.G., Hassan, R.E., Ghazy, O., Borai, E.H., 2022. Novel magnetite-polymer nanocomposite for the removal of some radionuclides: Modeling and performance evaluation of fixed-bed column adsorption. *Ind. Eng. Chem. Res.* 61, 16209–16219.

Han, J., Chen, Y., Wang, J., Zhang, G., Wang, H., 2022. A review of molecular dynamics simulation in studying surface generation mechanism in ultra-precision cutting. *Int. J. Adv. Manuf. Technol.* 122, 1195–1231.

Hayatifar, A., Gravelle, S., Moreno, B.D., Schoepfer, V.A., Lindsay, M.B.J., 2024. Probing atomic-scale processes at the ferrihydrite-water interface with reactive molecular dynamics. *Geochem. Trans.* 25, 10.

He, Y., Zhou, Y., Wen, T., Zhang, S., Huang, F., Zou, X., et al., 2022. A review of machine learning in geochemistry and cosmochemistry: Method improvements and applications. *Appl. Geochem.* 140, 105273.

Hennig, T., Stockmann, M., Kühn, M., 2020. Simulation of diffusive uranium transport and sorption processes in the Opalinus Clay. *Appl. Geochem.* 123, 104777.

Holgersson, S., Kumar, P., 2023. A literature review on thermodynamic sorption models of radionuclides with some selected granitic minerals. *Front. Nucl. Eng.* 2, 1227170.

Huang, C., Guan, Y., Cui, L., Wei, X., Wang, S., Wang, D., et al., 2025. Environmental correlates of the distribution characteristics of artificial radionuclides (137Cs, 239+240Pu, 237Np) in the northeast China. *J. Hazard. Mater.* 490, 137759.

Huang, Z., Tang, D., Dong, S., Wang, H., Zhang, X., Yue, C., Meng, S., Liu, L., 2025. Study on the diffusion behavior of Cs(I) and Eu(III) in Cu-montmorillonite. *J. Radioanal. Nucl. Chem.* 334(8), 5441–5453.

Ishidera, T., Kurosawa, S., Hayashi, M., Uchikoshi, K., Beppu, H., 2016. Diffusion and retention behaviour of Cs in illite-added compacted montmorillonite. *Clay Miner.* 51, 161–172.

Jeon, B., Kelly, S.D., Kemner, K.M., Barnett, M.O., Burgos, W.D., Dempsey, B.A., et al., 2004. Microbial reduction of U(VI) at the solid–water interface. *Environ. Sci. Technol.* 38, 5649–5655.

Jiang, Z.Y., 2012. Radioactive waste minimization for nuclear power sustainable development in China. *Adv. Mater. Res.* 1349–1353.

Jin, W., Kang, M., Kang, Y., She, J., Qin, D., Wu, H., et al., 2024. Kinetics study on the temperature-dependent reduction of aqueous U(VI) by natural pyrite. *Geochim. Cosmochim. Acta* 380, 18–30.

Jin, W., Kang, M., Qin, D., She, J., Huang, Z., Sun, L., et al., 2025. Temperature effects on the reduction of Se(VI) by natural pyrite. *Chem. Geol.* 682, 122747.

Kerisit, S., Liu, C., 2014. Molecular dynamics simulations of uranyl and uranyl carbonate adsorption at aluminosilicate surfaces. *Environ. Sci. Technol.* 48, 3899–3907.

Khan, I., Fahad, S., Wu, L., Zhou, W., Xu, P., Sun, Z., et al., 2019. Labile organic matter intensifies phosphorous mobilization in paddy soils by microbial iron(III) reduction. *Geoderma* 352, 185–196.

Kim, K., Lee, K., Chung, D., Lee, E., Moon, J., Shin, D., 2012. Evaluation of the stability of uranyl peroxo-carbonato complex ions in carbonate media at different temperatures. *J. Hazard. Mater.* 233–234, 213–218.

Kim, Y., Kwon, S., Roh, Y., 2021. Effect of divalent cations (Cu, Zn, Pb, Cd, and Sr) on microbially induced calcium carbonate precipitation and mineralogical properties. *Front. Microbiol.* 12.

Klammler, H., Hatfield, K., 2008. Analytical solutions for flow fields near continuous wall reactive barriers. *J. Contam. Hydrol.* 98, 1–14.

Konevnik, Y.V., Zakharova, E.V., Shiryaev, A.A., Olds, T.A., Burns, P.C., 2020. Mineral-specific heterogeneous neptunium sorption onto geological repository rocks in oxic and anoxic conditions and various temperatures. *Chem. Geol.* 545, 119654.

Koshikawa, M.K., Watanabe, M., Tamaoki, M., Ito, S., Takamatsu, T., Murata, T., et al., 2019. Comparison of 0.1 M stable CsCl and 1 M NH4NO3 as an extraction reagent to evaluate Cs-137 mobility in soils. *Anal. Sci.* 35, 153–158.

Kwon, S., Lim, J., Seoung, D., Cho, Y., Park, B., 2023. Comparative study of the cesium adsorption behavior of montmorillonite and illite based on their mineralogical properties and interlayer cations. *J. Hazard. Mater. Adv.* 10, 100258.

Leng, Y., Henderson, M.J., Courtois, J., Li, H., Xiong, K., Tuo, X., et al., 2016. Sorption of plutonium on geological materials associated with a Chinese radioactive waste repository: Influence of pH. *J. Radioanal. Nucl. Chem.* 308, 895–903.

Leroy, P., Revil, A., Altmann, S., Tournassat, C., 2007. Modeling the composition of the pore water in a clay-rock geological formation (Callovo-Oxfordian, France). *Geochim. Cosmochim. Acta* 71, 1087–1097.

Li, D., Hu, N., Ding, D., Li, S., Li, G., Wang, Y., 2016. An experimental study on the inhibitory effect of high concentration bicarbonate on the reduction of U(VI) in groundwater by functionalized indigenous microbial communities. *J. Radioanal. Nucl. Chem.* 307, 1011–1019.

Li, F., Li, X., Yu, J., Li, S., Guo, L., Xiao, C., et al., 2025. Enhanced removal of phosphate and fluoride from phosphogypsum leachate by cerium/iron-based zeolite: Adsorption behaviors and competition mechanisms. *Sep. Purif. Technol.* 376, 134185.

Li, G., Liu, J., Sun, Z., Nan, J., Kang, M., Zhu, J., 2025. Failure mechanism and risk decision-making process of frozen soil wall for blocking nuclear contaminated water leakage: A laboratory study. *Eng. Fail. Anal.* 169, 109187.

Li, N., Duan, X., Dou, J., Bi, W., Wu, P., Cheemaa, N., et al., 2025a. Thermodynamic controls on plutonium(Pu) speciation in deep groundwater: Implications for HLW disposal safety at Beishan, China. *J. Environ. Radioact.* 287, 107728.

Li, Y., Tuo, K., Liang, C., Shao, C., Hou, W., Sun, Y., et al., 2025b. Construction of hierarchical mesoporous and oxime functionalized metal-organic framework based on Hofmeister ion-mediated template strategy for uranium extraction. *Appl. Surf. Sci.* 712, 164172.

Liang, L., Wu, Y., Tan, W., Xi, F., Zhang, Y., Zhao, J., et al., 2025. Carbon-containing materials coupled with sulfidized nanoscale zero-valent iron effective sequestration of U(VI) and their microbial toxicity to bacteria and fungi. *Colloid Surf. A Physicochem. Eng. Asp.* 726, 138082.

Liao, Y., Zhu, M., Liu, X., Wu, X., Cheng, N., He, C., 2025. Copper-modified carbonized waste amidoxime resin (W-R-Cu) for efficient recovery of tellurium: Adsorption performance and mechanism. *Sep. Purif. Technol.* 379, 135071.

Liu, X., Hu, Z., Gao, J., 2025. Assessment of soil and groundwater pollution risks from coal gangue backfilling: A case study in Inner Mongolia mining areas. *Environ. Res.* 285, 122303.

Liu, X., Pang, H., Liu, X., Li, Q., Zhang, N., Mao, L., et al., 2021. Orderly porous covalent organic frameworks-based materials: Superior adsorbents for pollutants removal from aqueous solutions. *Innovation* 2, 100076.

Liu, X., Sun, M., Ke, Y., Wang, X., Hao, T., Jiang, W., et al., 2026. Solute atom migration and evolution of C-bearing Mn-Si-rich precipitates in RPV steel during the early aging period. *J. Mater. Sci. Technol.* 246, 44–57.

Lui, T.C.C., Gregory, D.D., Anderson, M., Lee, W., Cowling, S.A., 2022. Applying machine learning methods to predict geology using soil sample geochemistry. *Appl. Comput. Geosci.* 16, 100094.

Maes, N., Churakov, S., Glaus, M., Baeyens, B., Dähn, R., Grangeon, S., Charlet, L., Brandt, F., Poonoosamy, J., Hoving, A., Havlova, V., Fischer, C., Noseck, U., Britz, S., Siitari-Kauppi, M., Li, X., Fabritius, O., Missana, T., 2024. EURAD state-of-the-art report on the understanding of radionuclide retention and transport in clay and crystalline rocks. *Front. Nucl. Eng.* 3, 1417827.

Makarov, A.V., Safonov, A.V., Konevnik, Y.V., Teterin, Y.A., Maslakov, K.I., Teterin, A.Y., et al., 2021. Activated carbon additives for technetium immobilization in bentonite-based engineered barriers for radioactive waste repositories. *J. Hazard. Mater.* 401, 123436.

Medeiros-Costa, I.C., Wisser, D., Catita, L., Lesage, A., Lefebvre, V., Gay, A., et al., 2024. Connectivity assessment in hierarchical zeolites by correlating nuclear magnetic resonance, electron microscopy/tomography and adsorption scanning. *Mater. Chem. Phys.* 315, 128969.

Munagapati, V.S., Wen, H., Gutha, Y., Wen, J., Venkateswarlu, S., Yarramuthi, V., et al., 2026. Advances in biocompatible chitosan composites for chromium, arsenic, and radionuclide remediation: Experimental and computational insights. *Coord. Chem. Rev.* 547, 217116.

Muslim, W.A., Al-Nasri, S.K., Albayati, T.M., Hasan Sh Majdi, 2023. Treatment of actual radioactive wastewater containing Cs-137 using kaolinite clay minerals as eco-friendly adsorbents. *Desalin. Water Treat.* 307, 162–170.

Nemutudi, B., Pikinini, S., Mcfadzean, B., Zhang, X., Zhu, Y., Han, L., et al., 2025. pH effect on adsorption and performance of xanthate, dithiocarbamate and s-triazine collectors on sperrylite mineral surface. *Miner. Eng.* 224, 109191.

Novikau, R., Lujaniene, G., 2022. Adsorption behaviour of pollutants: Heavy metals, radionuclides, organic pollutants, on clays and their minerals (raw, modified and treated): A review. *J. Environ. Manage.* 309, 114685.

Paul, D., Pradeep, A., Sujish, D., Ruhela, S.P., 2025. Modeling of fission product adsorption in zeolite packed column for molten salt treatment. *Nucl. Eng. Technol.* 57, 103676.

Qin, W., Lei, Y., Xiao, J., Tang, H., Zhong, W., Chen, M., et al., 2025. MIL-125-NH2(Ti)/etched silicon carbide/carboxymethyl cellulose composite aerogel for efficient cesium adsorption under alkaline conditions. *Sep. Purif. Technol.* 378, 134745.

Ridzwan, M.H., Yaakob, M.K., Zabidi, Z.M., Hamzah, A.S., Shaameri, Z., Rashid, F.N.A.A., et al., 2022. Computational insight into the quantum chemistry, interaction and adsorption energy of aminopolycarboxylic acid chelating agents towards metal cations. *Comput. Theor. Chem.* 1208, 113579.

Saeed, T., Naeem, A., Din, I.U., Farooq, M., Khan, I.W., Hamayun, M., et al., 2022. Synthesis of chitosan composite of metal-organic framework for the adsorption of dyes; Kinetic and thermodynamic approach. *J. Hazard. Mater.* 427, 127902.

Savage, D., Wilson, J., Benbow, S., Sasamoto, H., Oda, C., Walker, C., et al., 2019. Natural systems evidence for the effects of temperature and the activity of aqueous silica upon montmorillonite stability in clay barriers for the disposal of radioactive wastes. *Appl. Clay Sci.* 179, 105146.

Schabernack, J., Kurganskaya, I., Fischer, C., Luttge, A., 2021. Influence of muscovite (001) surface nanotopography on radionuclide adsorption studied by kinetic Monte Carlo simulations. *Minerals* 11(5), 468.

Shelton, J.L., Jubb, A.M., Saxe, S.W., Attanasi, E.D., Milkov, A.V., Engle, M., et al., 2021. Machine learning can assign geologic basin to produced water samples using major ion geochemistry. *Nat. Resour. Res.* 30, 4147–4163.

Shi, L., Qiu, J., Wang, W., Ding, Z., Zhang, W., Liang, J., et al., 2024. Influence of cations and low molecular weight organic acids on Cs(I) adsorption on montmorillonite and vermiculite. *J. Mol. Liq.* 402, 124778.

Shi, L., Wu, Y., Chang, Z., Jiang, P., Hua, Y., Shi, Q., et al., 2025. Lewis acid-base synergy in Co3O4/p-PHI S-scheme heterojunction for photocatalytic aerobic uranium reduction. *Appl. Catal. B Environ.* 378, 125618.

Shi, S., Liu, J., Shu, J., Wu, P., Zhao, C., Liu, N., et al., 2023. Uranium(VI) adsorption from carbonate solutions using cetyltrimethylammonium bromide modified purified-bentonite-MOF composite. *Appl. Clay Sci.* 241, 106986.

Soliman, A.M., Murad, A.A., Sheikh, E.S.E., Massad, A.M., Ali, I.M., 2017. Selective removal of uranium from wastewater using sludge collected from refinery wastewater treatment: Equilibrium, thermodynamic and kinetics studies. *J. Water Process Eng.* 19, 267–276.

Soliman, M.A., Rashad, G.M., Mahmoud, M.R., 2019. Organo-modification of montmorillonite for enhancing the adsorption efficiency of cobalt radionuclides from aqueous solutions. *Environ. Sci. Pollut. Res.* 26, 10398–10413.

Su, K., Xu, Y., Luo, Q., Liu, Y., Li, Y., Yan, G., 2025. Mini-review on petroleum molecular geochemistry: Opportunities with digitalization, machine learning, and artificial intelligence. *Energy Fuels* 39, 5034–5050.

Tait, J.C., 1984. Immobilization and geological disposal of nuclear fuel waste. *Can. J. Physiol. Pharmacol.* 62, 979–985.

Tan, W., Lu, S., Liu, F., Feng, X., He, J., Koopal, L.K., 2008. Determination of the point-of-zero charge of manganese oxides with different methods including an improved salt titration method. *Soil Sci.* 173, 277–286.

Tatsuno, T., Hamamoto, S., Nihei, N., Nishimura, T., 2022. Vertical migration of cesium in weathered granite soil under flowing water condition depending on Cs concentration and states of dissolved organic matter. *J. Environ. Manage.* 306, 114409.

Tinnacher, R.M., Holmboe, M., Tournassat, C., Bourg, I.C., Davis, J.A., 2016. Ion adsorption and diffusion in smectite: Molecular, pore, and continuum scale views. *Geochim. Cosmochim. Acta* 177, 130–149.

Tong, X., Li, Y., Zhang, F., Chen, X., Zhao, Y., Hu, B., et al., 2019. Adsorption of 17β-estradiol onto humic-mineral complexes and effects of temperature, pH, and bisphenol A on the adsorption process. *Environ. Pollut.* 254, 112924.

Tournassat, C., Davis, J.A., Chiaberge, C., Grangeon, S., Bourg, I.C., 2016. Modeling the acid–base properties of montmorillonite edge surfaces. *Environ. Sci. Technol.* 50, 13436–13445.

Turner, J.P., Berry, T.W., Bowman, M.J., Chapman, N.A., 2023. Role of the geosphere in deep nuclear waste disposal—an England and Wales perspective. *Earth-Sci. Rev.* 242, 104445.

Tytkowska, M., Michel, M.M., Reczek, L., Siwiec, T., 2018. Sorption of Ni(II) on surface of bed grains used in iron and manganese removal filters. *Water Supply* 19, 815–822.

Um, W., Serne, R.J., Brown, C.F., Last, G.V., 2007. U(VI) adsorption on aquifer sediments at the Hanford site. *J. Contam. Hydrol.* 93, 255–269.

Um, W., Serne, R.J., Krupka, K.M., 2004. Linearity and reversibility of iodide adsorption on sediments from Hanford, Washington under water saturated conditions. *Water Res.* 38, 2009–2016.

Unokiwedi, O.P., Gao, J., Bethune, T., Awolayo, A.N., 2025. Deep geological repositories—a review of design concepts, near-field evolution, and their implications for nuclear waste containment. *J. Environ. Radioact.* 289, 107750.

Wang, D., Hu, L., Yang, K., Huang, S., Tian, L., 2024. Enhancement of U(VI) sorption in natural soils at prospective high-level radioactive waste geological disposal sites with Fe(OH)3-humic acid composite colloids. *J. Radioanal. Nucl. Chem.* 333, 1203–1212.

Wang, J., Li, J., 2025. Patent mining and diversity assessment of nuclear power technologies. *Nucl. Eng. Technol.* 57, 103272.

Wang, L., Putnis, C.V., 2020. Dissolution and precipitation dynamics at environmental mineral interfaces imaged by in situ atomic force microscopy. *Acc. Chem. Res.* 53, 1196–1205.

Wang, R., Ozhgibesov, M., Hirao, H., 2016. Partial hessian fitting for determining force constant parameters in molecular mechanics. *J. Comput. Chem.* 37, 2349–2359.

Wang, S., Wang, J., Guo, X., 2025. Cesium removal from aqueous solutions using mineral adsorbents: Mechanisms, kinetics, and thermodynamics. *Nucl. Eng. Des.* 445, 114461.
Wei, F., Urashima, S., Nihonyanagi, S., Tahara, T., 2023. Elucidation of the pH-dependent electric double layer structure at the silica/water interface using heterodyne-detected vibrational sum frequency generation spectroscopy. *J. Am. Chem. Soc.* 145, 8833–8846.
Wei, X., Sun, Y., Pan, D., Niu, Z., Xu, Z., Jiang, Y., et al., 2019. Adsorption properties of Na-palygorskite for Cs sequestration: Effect of pH, ionic strength, humic acid and temperature. *Appl. Clay Sci.* 183, 105363.
Wu, H., Lin, S., Cheng, X., Chen, J., Ji, Y., Xu, D., et al., 2020. Comparative study of strontium adsorption on muscovite, biotite and phlogopite. *J. Environ. Radioact.* 225, 106446.
Wu, H., Xu, Z., Zhu, L., Cheng, X., Kang, M., 2022. Adsorption of strontium at K-feldspar-water interface. *Appl. Radiat. Isot.* 181, 110111.
Wu, H., Zhang, J., Chen, L., Zhao, X., Li, L., Dai, Z., 2025. Phytic acid-functionalized polyacrylonitrile nanofibrous membranes via molecular self-assembly for efficient uranium extraction from highly acidic wastewater. *Sep. Purif. Technol.* 377, 134361.
Xiao, J., Yang, F., He, J., Hua, J., Dong, B., Ma, X., et al., 2025. Electrospun dibenzo-18-crown-6 polyimide dendritic nanofiber membranes for efficient adsorption separation of cesium. *J. Environ. Chem. Eng.* 13, 115329.
Yang, L., Fu, H., Dong, L., Ma, M., Deng, Y., Wu, H., et al., 2025. Moldable SnS2/polyimide aerogel as fixed-bed adsorbent: Efficient immobilization of radioiodine vapor with resistance to high-temperature and high-humidity conditions. *Sep. Purif. Technol.* 370, 133159.
Yang, S., Li, Z., Pan, P., 2023. Americium solubility and solubility limiting solid phase (SLSP) analysis in groundwater of Beishan, a proposed site for the disposal of high-level radioactive waste in China. *Appl. Geochem.* 152, 105650.
Yohai, L., Giraldo Mejía, H., Procaccini, R., Pellice, S., Laxman Kunjali, K., Dutta, J., et al., 2019. Nanocomposite functionalized membranes based on silica nanoparticles cross-linked to electrospun nanofibrous support for arsenic(V) adsorption from contaminated underground water. *RSC Adv.* 9, 8280–8289.
Yoon, I.H., Park, C.W., Kim, I., Yang, H.M., Kim, S.M., Kim, J.H., 2021. Characteristic and remediation of radioactive soil in nuclear facility sites: A critical review. *Environ. Sci. Pollut. Res.* 28, 67990–68005.
Yu, S., 2021. *Study on Migration Behavior of Key Nuclides in Rock Cave Repository*. Ph.D. thesis, Southwest University of Science and Technology, Mianyang, China.
Zeng, J., Li, L., Du, Y., Hou, Z., Wu, Y., Li, S., et al., 2025. Public dose assessment for Yangjiang nuclear power plant effluents: A multi-media numerical model of tritium transport in the South China Sea. *J. Environ. Radioact.* 286, 107684.
Zhang, Y., Yu, Y., Qin, H., Peng, D., Chen, X., 2022. Dynamic adsorption characteristics of Cr(VI) in red-mud leachate onto a red clay anti-seepage layer. *Toxics* 10, 606.
Zhao, X., Wu, H., Lu, M., Niu, Z., Pan, D., Wu, W., 2023. Elucidating the binding configuration of uranium at the biotite-water interface. *J. Radioanal. Nucl. Chem.* 332, 3845–3858.
Zhao, X., Wu, W., Pan, D., Wu, H., 2022. Study on the behaviors and mechanism of Ni(II) adsorption at the hydroxyapatite-water interface: Effect of particle size. *Adsorpt. Sci. Technol.* 2022.
Zhao, Y., Shao, Z., Chen, C., Hu, J., Chen, H., 2014. Effect of environmental conditions on the adsorption behavior of Sr(II) by Na-rectorite. *Appl. Clay Sci.* 87, 1–6.
Zheng, M., Chen, F., Tian, J., Pan, Q., 2018. Electron-transfer-enhanced cation–cation interactions in homo- and heterobimetallic actinide complexes: A relativistic density functional theory study. *Inorg. Chem.* 57, 3893–3902.
Zhou, W., Wang, F., 2025. U(VI) sorption onto rutile surface in the presence or absence of EDTA: A combined macroscopic and spectroscopic study. *J. Environ. Radioact.* 283, 107640.
Zuo, R., Chen, M., Lin, Y., Yang, J., Jin, S., Yue, W., et al., 2019. Effect of a humic acid colloid on the sorption behaviour of Sr onto soil in a candidate high-level radioactive waste geological disposal site. *Environ. Sci. Pollut. Res.* 26, 25235–25246.

CHAPTER 15

Effects of Clay–Biocomposite Colloids on Radionuclide Transport Behavior and Mechanisms

Zhen Xu

15.1 INTRODUCTION

Nuclear environmental safety has emerged as a critical issue for human survival and development and constitutes one of the central concerns in building a beautiful and healthy China in the new era. With the advancement and application of nuclear technology, radioactive waste is inevitably generated across various stages of the nuclear fuel cycle. In particular, high-level radioactive waste exhibits high radioactivity, significant radiotoxicity, substantial heat generation, and contains long-half-life radionuclides (Salvatores & Palmiotti, 2011; Tang et al., 2024b). The composition of these radionuclides is highly complex, encompassing species that are both difficult to retain and prone to diffusion in geological settings, as well as long-lived, highly toxic transuranic elements. Once released into the human ecological environment, such radionuclides resist degradation or removal by conventional physical or chemical means, thereby posing a direct threat to human habitats (Duan et al., 2020).

It is widely recognized that radionuclide migration in the environment occurs predominantly in soluble forms, driven mainly by diffusion and adsorption processes. Additionally, radionuclides can undergo complex physicochemical interactions with natural mineral components (e.g., clay minerals) and biological media (e.g., microorganisms), which may retard or even prevent their further migration in natural systems (Salvatores & Palmiotti, 2011; Xiao et al., 2019; Liu et al., 2018a). However, environmental colloids can serve as potential carriers for radionuclides, facilitating their long-distance transport (Tran et al., 2021; Du & Liang, 2019). For instance, studies have shown that colloids enhanced plutonium mobility at the Nevada Nuclear Test Site in the United States, leading to unexpectedly long migration

DOI: 10.1201/9781003665793-19

distances over short timeframes (Kersting et al., 1999). After nuclear tests, a fraction of plutonium was found to migrate in association with colloids, while another fraction traveled as intrinsic PuO_2 true colloids or adsorbed onto environmental colloids (Gerber et al., 2020). Similarly, U(IV) released into the environment following nuclear accidents can migrate not only in the form of uranium oxides (Shiryaev et al., 2018) but also when bound to natural organic matter and other colloidal phases (Romanchuk et al., 2020) (Figure 15.1).

Colloids are commonly defined as suspended particles within a size range of 1 nm to 1 μm, derived from diverse sources such as clay minerals (Tang et al., 2024a; Xu et al., 2022), metal oxides (Yang et al., 2019a), natural organic matter (Tan et al., 2018), and biological entities, including microorganisms and viruses (Lead & Wilkinson, 2006). Additionally, radioactive waste itself can give rise to intrinsic radioactive "true colloids" (Bots et al., 2014). Characterized by their high specific surface area, distinctive surface charge, and abundant functional groups, colloids often exhibit strong affinities for radionuclides (Duan et al., 2020; Bots et al., 2014). In the absence of colloidal consideration, conventional migration models frequently fail to adequately explain certain contaminant transport phenomena, leading to substantial discrepancies in predicting pollutant migration distances. Thus, colloids play an essential role in governing the migration behavior of radionuclides, underscoring their importance in accurately assessing the environmental fate of radioactive species and ensuring reliable nuclear environmental safety evaluation.

15.2 CO-TRANSPORT BEHAVIOR OF COLLOIDS AND RADIONUCLIDES

15.2.1 Co-transport of Clay Mineral Colloids and Radionuclides

Clay mineral colloids such as bentonite, illite, and kaolinite can form pseudocolloids after adsorbing radionuclides (Tang et al., 2024a; Xu et al., 2022), enabling long-distance transport in groundwater, pore water of the unsaturated zone, and flowing groundwater systems. As a typical clay mineral, bentonite is also used as a buffer/backfill material in high-level radioactive waste repositories. Under groundwater erosion, it can generate bentonite colloids (Bots et al., 2014; Bouby et al., 2020; Chen et al., 2021; Cai et al., 2020; Wang et al., 2018a), which may carry radionuclides and facilitate their transport. This process reduces the retention efficiency of engineered barriers and alters the fate of contaminants, posing a potential pollution risk to far-field drinking water sources near nuclear facilities. Studies show that bentonite colloids can carry Eu(III) during transport (Xu et al., 2022), and the extent of this facilitation depends on the stability of the colloidal system (Xu et al., 2022; Xu et al., 2021a). Bentonite colloids remain stable for at least 30 days in deionized water (Xu et al., 2020), and exhibit greater stability—and thus enhanced mobility in porous media—under low ionic strength and high pH conditions (Xu et al., 2022). When salinity or acidity increases, colloids rapidly aggregate; however, under conditions favoring dispersion, deposited aggregates can redisperse and undergo secondary transport (Xu et al., 2021a). It has been demonstrated that bentonite colloids can either enhance or retard the transport of Eu(III) in porous media depending on ionic strength, pH, and the

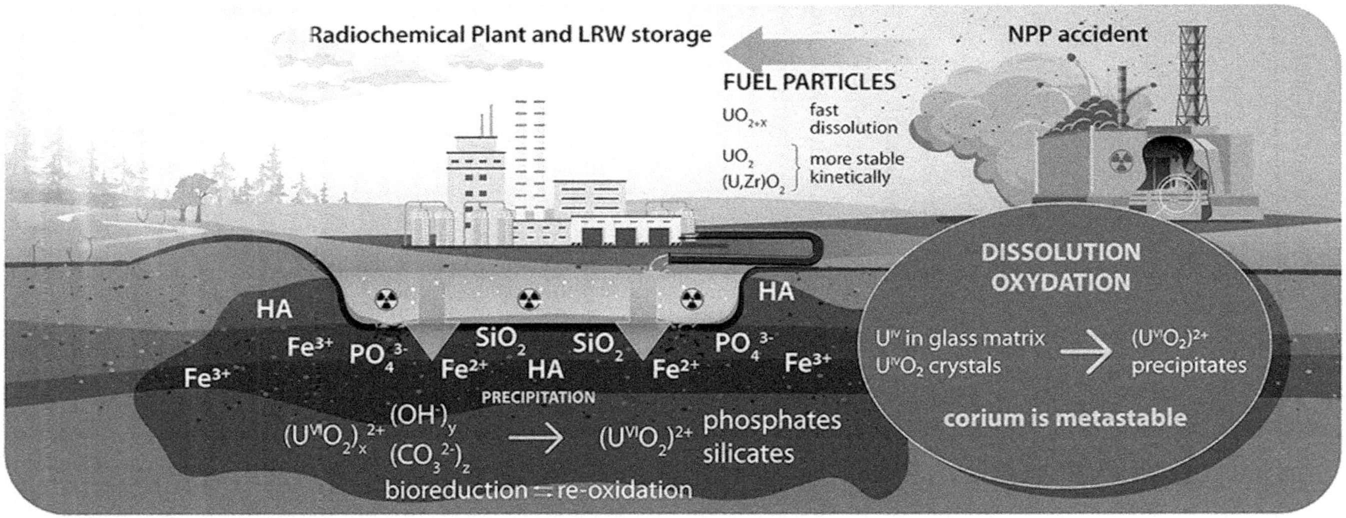

FIGURE 15.1 Uranium Migration into the Environment (Romanchuk et al., 2020).

presence of humic acid (HA) (Xu et al., 2022). Under low ionic strength and high pH, stable dispersed colloids promote Eu(III) migration, whereas aggregated clusters retard it. International researchers have investigated the transport behavior of montmorillonite colloids with respect to ^{238}Pu, ^{233}U, and ^{137}Cs. Montmorillonite colloids do not affect the transport of ^{233}U, but due to their strong affinity for ^{238}Pu, they facilitate its transport in chalk fractures. In contrast, ^{137}Cs is strongly adsorbed by both chalk and montmorillonite colloids; at low colloid concentrations, ^{137}Cs is mainly retained by chalk (Tran et al., 2021). Other researchers (Kolomá et al., 2018) found that a small fraction of cesium (~2%) was irreversibly adsorbed onto bentonite colloids and transported through granite media, while the majority (~98%) was retained in the granitic rock. Nevertheless, bentonite colloids still enhanced cesium mobility.

Illite is a typical layered silicate mineral in natural environments, and its derived colloids are considered effective carriers of radionuclides. Compared with expandable clay minerals such as montmorillonite, non-expandable illite colloids are relatively small and, in principle, more stable (Wei et al., 2021; Fuller et al., 2015). It has been reported (Wei et al., 2021) that illite colloids exhibit good stability under low pH, low temperature, and high ionic strength conditions. The critical coagulation concentration increases exponentially with the r/Z^3 ratio, where r is the colloidal particle radius, and Z is the charge number of electrolyte ions. When conditions favor dispersion, aggregated colloids can redisperse and be released. Illite colloids are effective vectors for radionuclide transport, and their ability to influence radionuclide migration is closely related to their dispersion stability and mobility (Wei et al., 2021; Wei et al., 2023). The adsorption of phosphate onto illite colloids enhances their stability, thereby promoting the transport of both illite colloids and U(VI) (Wei et al., 2024). However, in systems containing iron-coated sand, the promoting effect of illite colloids on radionuclide transport is diminished due to pore throat constriction caused by electrostatic interactions between aggregated illite colloids and iron oxides, which enhance the deposition of illite colloids and radionuclides (Wei et al., 2023).

As another typical 1:1 non-swelling layered silicate, kaolinite, possesses both hydrophilic and hydrophobic basal surfaces but lacks interlayer sites (Lee et al., 2011). Its crystal structure exhibits low isomorphous substitution and minimal permanent structural charge, rendering kaolinite colloids relatively unstable (Sun et al., 2020; Liu et al., 2018). Studies indicate that the dispersion stability of kaolinite colloids is highly dependent on changes in system pH (Sun et al., 2012), with dispersion favored under alkaline and highly acidic conditions. At approximately pH 3.2, the stability of kaolinite colloids is minimal due to alterations in surface electrical properties (Sun et al., 2020). Increased ionic strength and valence reduce colloidal stability, whereas HA enhances it by modifying the surface electrostatic potential and steric effects of the colloids (Yan et al., 2022; Zhao et al., 2018). The stability of kaolinite colloids significantly influences radionuclide transport. Nevertheless, immobilized kaolinite colloids in porous media can effectively retard the transport of contaminants (Won & Burns, 2018).

Additionally, Möri et al. (2003) observed that actinide radionuclides migrate through fractures more rapidly than tracers even in the absence of colloidal loading, a phenomenon attributed to the formation of homogeneous or heterogeneous true colloids. True colloids form through processes such as hydrolysis or complexation of radionuclides and primarily include species like $AnO_n(OH)_{(4-2n)}{\bullet}xH_2O$ (amorphous), $An(OH)_4$ (amorphous), $AnO_2{\bullet}xH_2O$ (amorphous), and AnO_2 (amorphous hydrate) (Altmaier et al., 2013; Zhai et al., 2022), as well as silicate true colloids (Zhang et al., 2022) and phosphate true colloids (Tang et al., 2024). Pu(IV) readily undergoes disproportionation and hydrolysis in aqueous solutions, exhibiting complex chemical behavior, and can form highly stable Pu(IV) true colloids even at elevated nitric acid concentrations (0.5 mol/L) (Micheau et al., 2020). Some radionuclides can also coordinate with ligands such as carbonates to form radioactive true colloids (Tran et al., 2016; Tran et al., 2015). Maher et al. (2016) investigated the colloidal behavior of Pu and Am in feed solutions from the Sellafield reprocessing plant in the UK and found that Am(III) speciation is dominated by carbonate complexes under high carbonate concentrations. Tran et al. (2015) discovered that in bicarbonate-rich environments, Ce(III) can form true colloids through precipitation with carbonates, thereby enhancing Ce(III) transport in fractured bedrock. The introduction of HA or bentonite colloids significantly increases Ce(III) migration (Zhai et al., 2022).

15.2.2 Co-transport of Biocolloids and Radionuclides

Microbial colloids, such as bacteria, are ubiquitous in the environment. The complex activities of microorganisms inevitably alter the chemical and physical properties of the environment, thereby directly or indirectly influencing the biogeochemical processes of radioactive elements—including their speciation, distribution, migration, transformation, and cycling—in various ways. The impact of biocolloids formed by microorganisms on contaminant transport is a complex process involving two phases (mobile and stationary phases) and six systems (Bekhit et al., 2009). Among these, the aqueous phase, dissolved contaminants, and mobile colloids belong to the mobile phase, while the solid matrix, along with contaminants and colloids captured by the matrix, belong to the stationary phase. Contaminants can exist in either the stationary phase or the mobile phase, depending primarily on the chemical characteristics of the system. The process of colloid-facilitated contaminant transport typically involves six systems: (1) mobile colloids; (2) colloids attached to the solid matrix; (3) freely dissolved contaminants in the aqueous phase; (4) contaminants sorbed onto the solid matrix; (5) contaminants adsorbed onto the surfaces of mobile colloids; and (6) contaminants adsorbed onto the surfaces of stationary colloids. Furthermore, for colloids to facilitate contaminant transport in saturated porous media, five conditions must be simultaneously met: (1) the presence of colloids; (2) mobile colloids; (3) stably dispersed colloids; (4) attachment of contaminants to colloid surfaces; and (5) irreversible adsorption of contaminants by the colloids (Zänker & Hennig, 2014).

Biocolloids such as bacteria in the environment are often rich in functional groups like hydroxyl, carboxyl, amine, and phosphate groups. These functional groups can interact intricately with radionuclides (Ruiz-Fresneda et al., 2020), thereby influencing the migration and fate of radionuclides (Fukunaga et al., 2016). The mechanisms of interaction between microorganisms and radionuclides primarily involve four processes: (1) biosorption, which refers to the non-metabolic uptake of radionuclides onto microbial cell surfaces (Hufton et al., 2021); (2) direct or indirect microbial oxidation/reduction of radionuclides, a process closely associated with cellular metabolism (Zhou et al., 2020); (3) bioaccumulation, involving the intracellular uptake and enrichment of radionuclides through metabolic processes (Song et al., 2019); and (4) biomineralization, where radionuclides precipitate and mineralize on microbial cell surfaces with ligands produced by microbial metabolism (e.g., sulfides, phosphates, and carbonates) (Zhang et al., 2020). Studies have found that microbial processes can inhibit the migration of ^{99m}Tc in columns, which can be attributed to the bioreduction of soluble Tc(VII) to less soluble Tc(IV) by bacteria. Additionally, the production of H_2 and CO_2 by bacteria, which may affect the water saturation of the medium, could be a mechanism by which microbial processes alter its mobility (Smith et al., 2017).

Beyond direct interactions between microorganisms and radionuclides affecting radionuclide migration, microorganisms can also alter the surface properties of the medium through their own metabolism. For instance, the formation of biofilms covering medium particles can increase the availability of adsorption sites (Ruiz-Fresneda et al., 2023), or microbial processes can induce mineral dissolution or transformation, which also affects the ability of radionuclides to adsorb onto mineral surfaces (Joo et al., 2023). Furthermore, radionuclides can interact with ligands produced by microbial metabolism and transform from soluble species into low-solubility precipitates, effectively retarding radionuclide migration and bioavailability in the environment, thereby influencing the fate of radionuclides (Zhang et al., 2020). Research has shown that *Bacillus subtilis* can biomineralize U(VI) into uramphite $(NH_4)(UO_2)(PO_4)\cdot 3H_2O$ (Song et al., 2019a). Single-particle inductively coupled plasma mass spectrometry characterization revealed that during the uranium immobilization process by *Shewanella putrefaciens*, U(VI) primarily undergoes non-reductive biomineralization within the extracellular polymeric substances (EPSs) (Huang et al., 2020). *S. putrefaciens* can biomineralize U(VI) into nano-sheet-like minerals distributed around the entire cell. These nano-minerals are identified as hydrogen uranospinite $H_2(UO_2)_2(PO_4)_2\cdot 8H_2O$ (Nie et al., 2022). However, the genuine colloids generated through biomineralization may undergo secondary migration facilitated by biocolloids. Considering the variations in colloidal types and concentrations across different environmental settings, the migration behavior of contaminants such as radionuclides becomes more complex. In particular, research on the role of environmental composite colloids in regulating the adsorption and migration characteristics of radionuclides remains limited. Therefore, further in-depth investigation is required to understand the influence of composite colloids on the retention, distribution, migration, and fate of radionuclides in environmental media.

15.2.3 Co-transport of Environmental Composite Colloids and Radionuclides

In subsurface environments, diverse colloids—such as clay minerals and biocolloids (e.g., bacteria)—often co-exist and collectively influence the transport behavior and mechanisms of contaminants such as radionuclides (Dong et al., 2024). The co-transport of colloids and contaminants involves multi-faceted interactions. For example, in systems containing both clay mineral colloids and bacterial biocolloids, the system can be conceptualized as comprising two phases and 14 distinct components (Bekhit et al., 2009). The two phases include: (1) the mobile phase, which encompasses contaminants, bacteria, and colloids in various forms; and (2) the solid phase, consisting of the solid matrix, deposited colloids, and attached bacteria in different states, as well as the contaminants adsorbed onto them.

Within this framework, colloids primarily occur in two forms: (1) mobile (suspended) colloids, and (2) immobile (deposited) colloids attached to the solid matrix. Bacteria, in turn, exist in four different states: (1) as free-moving cells suspended in the aqueous phase; (2) as immobilized cells directly attached to the solid matrix; (3) as cells attached to mobile colloids; and (4) as cells associated with deposited colloids. Contaminants in such systems display eight distinct forms, half of which are mobile and half immobile, reflecting their distribution between the aqueous and solid phases.

Studies have shown that kaolinite particles significantly inhibit the migration of *Pseudomonas putida* in spherical glass bead columns, where bacterial retention is enhanced by attachment to kaolinite particles deposited in the porous medium (Vasiliadou et al., 2011). Similarly, Vasiliadou et al. (Vasiliadou & Chrysikopoulos, 2011) reported the co-deposition of bacteria and kaolinite from suspension onto glass beads. The adhesion process between bacteria and clay involves an initial reversible attachment phase, followed by firmer adhesion driven by biological activity and biofilm formation (Chandraprabha & Natarajan, 2009). The initial attachment is primarily an abiotic physicochemical process, mediated mainly by electrostatic interactions, hydrogen bonding, and hydrophobic forces (Yong et al., 2022). In a study by Yang et al. (2012a), the co-transport of *B. subtilis* and *Escherichia coli* with bentonite particles was investigated in quartz sand columns; results indicated that bentonite deposits on quartz sand promoted increased bacterial deposition.

Similar to bacterial behavior, montmorillonite colloids exhibit a high affinity for virus adsorption and can act as carriers to enhance viral transport in porous media (Syngouna & Chrysikopoulos, 2015). Zhang et al. (2024a) examined the co-transport of *E. coli* phages with HA or SiO_2 colloids in saturated and unsaturated porous media. They found that HA improved the stability of *E. coli* phage–HA complexes and facilitated phage migration, playing a dominant role in transport. In contrast, SiO_2 reduced the complex stability, leading to the separation of *E. coli* phages from SiO_2 in unsaturated media, which also enhanced phage mobility under those conditions. However, such carrier effects can be modulated by environmental factors. For instance, decreases in pH and dissolved organic matter content enhance virus attachment to both aquifer media and colloids, thereby significantly restraining virus migration in aquifer systems (Syngouna et al., 2017).

15.3 FACTORS AFFECTING COLLOID TRANSPORT

In recent years, column experiments and numerical simulations have been the primary methods for investigating the transport behavior and mechanisms of environmental colloids (e.g., clay minerals and biocolloids) and radionuclides in porous media. Their migration is influenced by a range of physicochemical factors, including mineral surface properties, solution chemistry (pH, ionic strength, and valence), natural organic matter, and microbial activity.

15.3.1 Effect of Physical Factors

Physical factors affecting colloid transport mainly include mineral surface characteristics such as grain size, heterogeneity, and surface roughness, as well as saturation, fluid flow velocity, and temperature. Generally, smaller grain sizes in porous media weaken colloid transport capacity. Smaller grain sizes increase the specific surface area and deposition sites of the medium, enhancing physical retention and thereby hindering colloid transport (Lv et al., 2016). In subsurface fracture environments, preferential flow is common due to variations in geological and stratigraphic conditions (Sheng et al., 2014; Zhang & Yuan, 2019). Preferential flow in heterogeneous porous media accelerates the transport of bentonite colloids carrying Eu(III), thereby increasing the risk of radionuclide migration in subsurface environments (Xu et al., 2023). The flow velocity of pore water is a key factor influencing the advection–dispersion process of colloids in porous media. An increase in flow velocity enhances the shear force between the water and the medium, re-suspending colloids deposited on the medium surface back into the water and promoting colloid transport (Kim et al., 2020). Conversely, flow interruption increases colloidal deposition in porous media, and resumption of flow can lead to sudden colloid release, a mechanism potentially driven by diffusion and fluid momentum (Ning et al., 2021). When the surface roughness of the medium is high, the influence of fluid velocity on colloid transport and release behavior may be weakened (Li et al., 2019).

Additionally, some porous media surfaces contain metallic minerals and biofilms, which can alter surface charge and roughness, thereby affecting colloidal adsorption and transport processes. For instance, biofilm-coated quartz sand can retain more colloids compared to clean quartz sand, primarily because the biofilm enhances the surface roughness of the medium and shortens the flow paths (Jiang et al., 2013). Studies have shown that the promoting effect of illite colloids on Eu(III) migration is reduced in iron-coated quartz sand (Wei et al., 2023), and bacterial migration is inhibited in iron-coated quartz sand (Dong et al., 2014), which may be related to the content of the coated quartz sand. The role of moisture content in the medium in colloid migration cannot be overlooked either. The migration capacity of colloids in saturated media is generally greater than that in unsaturated media, essentially due to more complex interactions among colloids, the gas phase, liquid phase, and solid phase (Knappenberger et al., 2014; Knappenberger et al., 2015). For example, increasing the moisture content of the medium facilitates the migration of *E. coli* phages. Under unsaturated conditions, *E. coli* phages primarily deposit at the solid–liquid interface rather than the gas–liquid interface, and their deposition is strongly influenced by moisture content and hydrochemical conditions (Zhang et al., 2021a).

Temperature can also affect colloid migration in porous media. The actual temperature of groundwater is approximately 10°C (Horne et al., 2004) and varies with seasonal changes (Bense & Kooi, 2004). Therefore, migration patterns observed at room temperature may not be applicable to colloid migration in groundwater environments. Currently, it is widely accepted that high temperatures are unfavorable for colloid migration. This is because elevated temperatures weaken the interaction energy between colloidal particles, and increased random Brownian motion and collision frequency may be the main reasons for rapid aggregation and sedimentation of colloidal particles (Xu et al., 2020). For biological colloids formed by microorganisms, temperature not only affects microbial activity but also influences their surface hydrophobicity/hydrophilicity and growth state. Kim et al. (Kim & Walker, 2009) found that the migration of *E. coli* in $CaCl_2$ solution was insensitive to temperature. However, another study (Li et al., 2019) observed that as the temperature increased from 4°C to 25°C, the negative charge of *Chlorophenolospirillum* FA1 decreased, and bacterial migration weakened. Differences in microbial species and the physicochemical characteristics of reaction systems lead to inconsistent effects of temperature on microbial migration behavior, and further research and exploration are needed to understand the impact of temperature on the migration behavior of environmental colloids.

15.3.2 Effect of Chemical Factors

Generally, hydrochemical factors such as pH, ionic strength, and valence, as well as organic matter, significantly influence the transport behavior of colloids and radionuclides. The zeta potential of colloids is closely related to their point of zero charge (pH_{PZC}). For instance, the pH_{PZC} of kaolinite colloids is 3.2, and their dispersion stability strongly depends on the pH of the colloidal system (Sun et al., 2020). When the pH is below the pH_{PZC}, the zeta potential of colloids is positive; when the pH is above the pH_{PZC}, it becomes negative; and when the pH approaches the pH_{PZC}, the zeta potential tends toward zero. The greater the absolute value of the zeta potential, the stronger the electrostatic repulsion between colloids, leading to higher stability. Conversely, insufficient electrostatic repulsion fails to inhibit colloidal aggregation (Xu et al., 2020). A pH close to the pH_{PZC} induces colloidal aggregation, while a pH farther from the pH_{PZC} enhances colloidal stability, thereby promoting colloid transport in porous media (Sun et al., 2020). In a 10 mmol/L sodium nitrate solution, the pH_{PZC} of nano-titanium dioxide is approximately 6.5, and its point of zero charge plays a critical role in the transport of nano-titanium dioxide (Chen et al., 2018). The composition of groundwater in subsurface environments is rich and highly variable.

The ionic strength and composition of groundwater are also key factors controlling colloid transport in porous media. According to the Derjaguin–Landau–Verwey–Overbeek (DLVO) theory, an increase in ionic strength compresses the thickness of the electrical double layer, reducing the electrostatic repulsion between colloids and porous media and thereby decreasing system stability (Xu et al., 2020; Wei et al., 2021; Lee et al., 2011). This promotes colloidal deposition in porous media. Divalent ions are more effective than monovalent ions in neutralizing the surface charges of colloids and porous media. Divalent ions further diminish electrostatic repulsion between colloids and media, compress the electrical double layer of colloids, and reduce the repulsive potential energy of the system, thereby enhancing colloidal deposition on porous media (Wei et al., 2021; Lee et al., 2011). Higher-valence ions are more likely to form bridges between colloids and media surfaces, a bridging effect that facilitates colloidal adsorption onto media surfaces (Bense & Kooi, 2004; Sun & Lei, 2023). Therefore, variations in ionic strength and ion valence in the system also significantly influence colloid transport and release behavior.

Additionally, substantial amounts of dissolved organic matter are widely present in subsurface environments. These substances are typically negatively charged, possess large specific surface areas, and contain highly reactive functional groups. They influence colloid transport in porous media by altering the surface charges of colloids and media, competing for adsorption sites on media surfaces, and generating steric hindrance (Zhang et al., 2023; Huangfu et al., 2019; Zhong et al., 2017). As a major component of organic matter, HA can interact with other environmental colloids to form composite environmental colloids, significantly affecting their transport behavior. Xu et al. (2021a) investigated the transport behavior and mechanisms of bentonite colloids in saturated porous media in the presence of HA and found that HA enhanced electrostatic repulsion and steric hindrance, substantially improving colloidal dispersion stability and thereby promoting the transport of Eu(III). Wei et al. (2021) demonstrated that HA significantly enhances the stability of illite colloids, with this effect linearly correlated with HA concentration. Zhao Kang et al. (2018) studied the influence of kaolinite colloids on uranium transport in the presence of HA and found that HA increased the adsorption of uranium onto kaolinite colloids, enhancing the co-transport of kaolinite colloids and uranium in porous media. This results in uranium having a greater transport capacity in actual soil and groundwater environments than previously anticipated. Yang et al. (2012b) discovered that HA promotes bacterial transport by competing for adsorption sites on media surfaces, with this promoting effect significantly influenced by HA concentration. The underlying reason may be that as HA concentration increases, the adhesion of bacteria to environmental media is reduced (Park & Kim, 2009).

15.3.3 Effect of Microorganisms

Microbial factors affecting colloid transport primarily include microbial size, shape, growth state, surface charge, surface macromolecular composition, presence of flagella, chemotaxis, and others. Most microorganisms in natural environments typically carry a negative charge. Their surface charge is influenced not only by various physicochemical conditions such as pH, temperature, ionic strength, and microbial type (Zhang et al., 2024), but also by the physiological state of the microorganisms, which plays a significant role in the transport behavior of biocolloids and associated contaminants in porous media. Macromolecules and structures on the bacterial cell surface, including EPSs, lipopolysaccharides (LPS), and flagella, determine the surface properties of the cell (such as surface charge and hydrophobicity), thereby influencing bacterial adsorption and transport in porous media (Kim et al., 2010). For instance, the hydroxyl groups of LPS can form hydrogen bonds with water molecules and bind to mineral surfaces, enhancing bacterial adsorption and subsequently increasing bacterial deposition (Walker et al., 2004). EPSs on the bacterial surface play a crucial role in cell adhesion and aggregate formation (Tsuneda et al., 2003; Tong et al., 2010). The presence of EPS increases collisions between bacterial cells and porous media, as well as cell adhesion to the media, leading to inhibited bacterial transport (Tong et al., 2010). Bacterial flagella are complex helical filamentous organelles responsible for bacterial motility (Nakamura & Minamino, 2019). The number and characteristics of flagella also significantly impact bacterial transport in porous media (Najafi et al., 2018). The presence of flagella enables *E. coli* to adhere more irreversibly to porous media with stronger adhesion, and adhesive flagella can further enhance bacterial deposition on media surfaces (Zhang et al., 2021b). Najafi et al. (2018), based on experimental observations and numerical simulations, found that *B. subtilis* with fewer flagella exhibited longer transport distances and faster propagation rates. Additionally, bacteria with hydrophobic surfaces adhere more effectively to porous media surfaces (Schäfer et al., 1998), while hydrophilic bacteria tend to aggregate more readily (Gargiulo et al., 2008). However, some studies have indicated that bacterial hydrophobicity may not correlate with bacterial transport behavior in porous media (Tsuneda et al., 2003). Currently, research on the mechanisms by which microbial surface structures or macromolecules influence microbial transport remains limited and requires further in-depth investigation. Different types of microorganisms vary in size and shape, with microbial cells exhibiting forms such as rod-shaped, spherical, elliptical, filamentous, and fusiform. Spherical cells or those with smaller hydrodynamic diameters migrate more easily, whereas elongated rod-shaped cells are more prone to retention in porous media (Weiss et al., 1995). Smaller spores migrate faster than mature cells (Li et al., 2018), a phenomenon likely due to the higher collision probability between porous media and larger or elongated bacterial cells (Knappett et al., 2008), as well as the higher strain potential resulting from the disparity between cell size and the average media particle size. Microorganisms are inherently active, and their growth stage, viability, and starvation state can influence their properties, thereby affecting their transport behavior in porous media. Starved cells exhibit higher mobility compared to nutrient-rich cells (Walczak et al., 2012). For example, under nitrogen starvation conditions, bacterial surface hydrophilicity/hydrophobicity significantly decreases, reducing adhesion capacity, whereas under carbon starvation conditions, bacterial surface hydrophobicity remains largely unchanged (Sanin et al., 2003). Due to alterations in cell surface characteristics, non-viable cells demonstrate stronger transport than viable cells (Madumathi et al., 2017). Furthermore, microbial concentration fluctuates with reproduction and death processes, leading to changes in transport behavior. Increasing microbial cell concentration facilitates microbial transport (Haznedaroglu et al., 2009). Microorganisms often actively utilize their inherent characteristics to move toward favorable environments (or avoid harmful ones), a behavior termed chemotaxis. Studies have shown that microbial chemotaxis influences their transport behavior, as microorganisms can sense and respond to chemical gradients in their surroundings, moving toward nutrient-rich areas and away from toxic substances (de Anna et al., 2021; Biria & Balouchi, 2013). Therefore, microbial chemotaxis not only affects their transport within contaminated regions but also influences the transport behavior of biocolloid-associated contaminants.

15.4 COLLOID TRANSPORT MECHANISMS

The transport process of colloids in porous media essentially involves the movement of colloids along with water flow through pore spaces. During transport, colloids may undergo aggregation, deposition, and release, processes regulated by multiple mechanisms. Studies have shown that colloid transport in porous media is primarily governed by physical and physicochemical interactions between moving colloidal particles and collector surfaces, including mechanisms such as advection–diffusion, adsorption–desorption, aggregation, straining, and ripening deposition (Figure 15.2) (Chen et al., 2021; Babakhani et al., 2017; Ling et al., 2021).

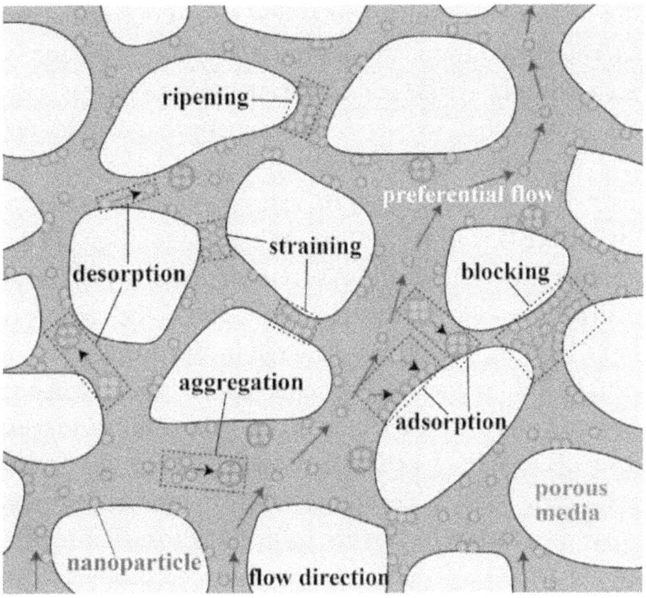

FIGURE 15.2 Primary Mechanisms of Colloid Transport in Porous Media (Ling et al., 2021).

15.4.1 Advection–Diffusion

Under the hydraulic driving force of groundwater flow, the advection–diffusion mechanism of colloids has been recognized as largely similar to that of solutes such as radionuclides (Ling et al., 2021; Auset & Keller, 2004). Colloids suspended in water undergo advection as they move with pore water, influenced by factors such as hydraulic gradient, porosity, and permeability. Advection refers to the movement of colloids along the flow path of the fluid, and under this mode of transport, the velocity profile of colloids within the pore water approximates a parabolic distribution (Sirivithayapakorn & Keller, 2003). Faster flow paths occur at the center of pore throats, while slower flow paths occur at the gas–water or solid–liquid interfaces (Baumann & Werth, 2004). Hydrodynamic dispersion includes both diffusion and mechanical dispersion. Due to the heterogeneity of soil particles and pore structures, flow fields often exhibit complex pore networks, with variations in flow direction and velocity at the pore scale, leading to longitudinal and transverse mixing of particles (mechanical dispersion). Additionally, the non-uniformity of the flow field causes colloids to twist as they pass through porous media, and diffusion also occurs during the advection process. Diffusion is the process by which colloids move from regions of higher concentration to regions of lower concentration via Brownian motion (Keller & Auset, 2007). Smaller colloids are more likely to move along flow paths into retention zones, where deposition occurs, whereas for larger colloids, diffusion blocking typically dominates.

15.4.2 Adsorption–Desorption

Adsorption–desorption is a crucial interaction between colloids and solid–liquid interfaces. Adsorption can be classified as reversible or irreversible. Reversible adsorption involves relatively weak interactions between colloids and the medium, primarily governed by van der Waals forces and hydrogen bonding. Irreversible adsorption is a permanent interaction involving significant energy, mainly controlled by electrostatic interactions and hydrophobic forces. This occurs when local physical and chemical interactions (e.g., van der Waals forces, chemisorption) are strong enough to overcome the energy barrier. Studies have shown that the surface properties of colloids and the medium (e.g., surface roughness of the medium) and solution chemistry (pH and ionic strength) can significantly influence colloidal adsorption on the medium surface (Liang et al., 2020). When the solution pH decreases, the negative charges on both the colloid and medium surfaces are reduced. When the ionic strength increases, the repulsive energy barrier also decreases due to the compression of the diffuse electrical double layer and the reduction of charge in the diffuse layer (Babakhani et al., 2017; Zhang et al., 2012). Both scenarios favor colloid attachment. During colloid transport in the medium, previously adsorbed colloids may significantly hinder further adsorption due to interparticle repulsion. However, due to colloidal adsorption, particularly the deposition of minerals and organic matter, a coating is likely to form on the medium surface, and adsorbed colloids may also influence further colloidal adsorption (Nair & Thampi, 2010). Furthermore, when the DLVO interaction energy between colloids and the medium surface is at the secondary energy minimum, deposition is reversible (Redman et al., 2004). Colloids deposited in porous media are susceptible to chemical or physical disturbances, leading to desorption. Solution chemistry and hydrodynamic conditions are the primary factors influencing colloid desorption. The balance between separation and adsorption forces microscopically determines whether desorption occurs and its extent (Bedrikovetsky et al., 2012). An increase in solution pH and a decrease in ionic strength elevate the energy barrier between deposited particles and the medium. Additionally, increasing fluid velocity enhances the fluid shear force on attached colloids and the torque of the separation force, both of which favor colloid desorption (Zhang et al., 2012).

15.4.3 Aggregation

Aggregation refers to the process by which colloids form clusters of varying sizes, including two types: homogeneous aggregation, which involves clustering between similar colloidal particles (Xu et al., 2018), and heterogeneous aggregation, which involves clustering between different components or colloids through chemical bonds, hydrogen bonding, bridging, and electrostatic interactions—a more common phenomenon in the environment (Wang et al., 2015). Additionally, the presence of HAs and surfactants may increase or decrease the aggregation rate of colloids in the environment (Xu et al., 2021a; Wei et al., 2021; Sun et al., 2020). Colloid aggregation is primarily driven by van der Waals forces and the formation of counterion electrical double layers. Other significant interactions influencing colloid aggregation include magnetic and hydrophobic forces, as well as hydration forces (Sendra et al., 2017). Aggregation typically reduces the chemical reactivity and bioavailable concentration of colloidal particles. Over time, the aggregate size increases, which also enhances the sedimentation rate and the collision rate with solid surfaces. For colloidal particles colliding with surfaces, an increase in aggregate size raises the magnitude of the secondary minimum in DLVO interaction energy, thereby improving the attachment rate. It is also worth noting that observing no change in colloid particle size in the effluent over time or pore volume should not be taken as evidence of no aggregation, as aggregation often causes larger colloids to deposit in porous media, while smaller, non-aggregated colloidal particles are more likely to migrate out of the porous medium.

15.4.4 Pore Blocking

Pore blocking refers to the phenomenon where certain pores in porous media are too small to allow the passage of colloids. Studies (Tang et al., 2024a; Xu et al., 2022) have found that significant pore blocking occurs when the ratio of the diameter of colloidal particles to the diameter of porous media particles is less than 0.0017. In cases of high particle concentration, multiple particles may simultaneously reach the pore openings, leading to "blocking" and "arching," ultimately resulting in the "clogging" of the filtration surface. Straining is typically further subdivided into two mechanisms: wedging and bridging. Wedging occurs when particles are captured between two boundaries without interference from other particles, while bridging involves the simultaneous arrival and aggregation of particles at pore constrictions. Under natural conditions, due to the irregular size and shape of porous media particles, the wide range of pore sizes, and inconsistent surface roughness of media particles, physicochemical factors and hydrodynamic factors collectively influence the straining and pore-blocking effects of colloids in porous media. Auset et al. (Auset & Keller, 2006) demonstrated using optical microscopy that when colloids enter porous media, straining primarily occurs at the first one or two pore throats and is unaffected by flow velocity. Tufenkji et al. (Tufenkji & Elimelech, 2004) also found that straining increases with increasing surface roughness. However, excessive simplification of the complex size and shape distributions of pores and colloids may lead to discrepancies between experimental data and numerical simulations or geometric calculations (Bradford et al., 2013). Therefore, this fact should be considered during experimental analysis.

15.4.5 Colloidal Deposition

Colloidal deposition primarily includes blocking deposition and ripening deposition. Blocking deposition refers to the reduction in the attachment area for other colloidal particles when colloids deposit on the media surface. In contrast, ripening deposition occurs when colloids already deposited on the media surface act as additional collectors, providing deposition sites for mobile colloids in porous media. This typically occurs between colloidal particles and is stronger than the interaction between colloids and the media surface. The rate of blocking deposition gradually decreases over time, whereas ripening deposition exhibits the opposite trend, with the breakthrough curve plateau rising rapidly before sharply declining. In the first scenario, where particle–surface adhesion affinity is higher than particle–particle adhesion affinity, the primary transport mechanisms are physicochemical filtration (described by kinetic or equilibrium adsorption/desorption models), blocking (described by the Langmuir function), and possibly physical retention, that is., straining (described by a depth-dependent function). This mechanism can produce a rising breakthrough curve plateau (Babakhani et al., 2017). The second scenario is dominated by particle–particle attachment tendencies. Here, although physicochemical filtration may still occur, ripening, aggregation, and subsequent retention are the main processes, which may lead to a declining breakthrough curve plateau (Babakhani et al., 2017). This also indicates that if experimental conditions shift transport behavior from blocking to ripening, the substantial retention of colloids due to ripening can partially mitigate the transport of colloids and contaminants. However, colloid aggregation does not necessarily result in a ripening effect. Thus, even if significant colloid aggregation occurs in the flowing phase, ripening may not necessarily occur. It should be noted that blocking and ripening are mutually exclusive phenomena; when physicochemical conditions favor one, the other should not occur. However, due to the spatial heterogeneity of actual porous media, they can still occur simultaneously within a continuous domain (Nascimento et al., 2006).

15.5 RESEARCH METHODS AND APPROACHES FOR COLLOID TRANSPORT

The transport of colloid-bound contaminants primarily involves processes such as advection and diffusion of contaminants, as well as adsorption and desorption of contaminants onto solid medias. Studying the co-transport behavior of colloids and contaminants not only elucidates the transport behavior of colloids but also explains the interaction mechanisms between colloids and contaminants (Petersen et al., 2016). Under laboratory conditions, experimental methods for investigating interactions between colloids and radionuclides mainly include aggregation kinetic experiments, static adsorption batch experiments, and column transport experiments (Figure 15.3). Aggregation kinetics of colloids can be studied using dynamic light scattering. Xu et al. (2018) employed dynamic light scattering to investigate the effects of different solution chemical conditions on the aggregation of bentonite colloids, evaluating colloidal stability through attachment efficiency and critical coagulation concentration during the aggregation process. Their findings revealed that colloid aggregation increased significantly with increasing cation concentration, with Ca^{2+} having a more pronounced effect than Na^{+}, exhibiting critical coagulation concentrations of 0.32 and 3.6 mmol/L, respectively.

Wei et al. (2021) further demonstrated that the critical coagulation concentration of colloids increases exponentially with increasing r/Z^3 values. Sun et al. (2020) also used dynamic light scattering to study the stability of kaolinite colloids, finding that the presence of HA further enhanced colloidal stability. Static adsorption batch experiments are an effective method for studying the adsorption–desorption of radionuclides on colloids, typically employing adsorption isotherms and related models to investigate the adsorbent capacity in solid–liquid systems, and using adsorption kinetics to study the equilibrium time and adsorption rates. For instance, Xu et al. (2021b) explored the adsorption behavior and mechanisms of radionuclides on bentonite colloids, finding that environmental chemical conditions significantly influence adsorption, and that colloids can act as potential carriers and transport vectors for radionuclides. Laboratory column experiments can reveal patterns of colloid transport and deposition, thereby predicting potential risks of radionuclide migration. Studies on the co-transport and release behavior of bentonite colloids and radionuclides in quartz sand columns (Xu et al., 2021a) showed that increasing solution ionic strength reduces bentonite colloidal stability, thereby inhibiting colloid transport, whereas HA enhances colloid electrostatic repulsion and steric hindrance effects, promoting colloid transport. Furthermore, HA and bentonite colloids exhibited a synergistic effect on radionuclide transport, co-migrating through the formation of ternary complexes (Xu et al., 2022). Additionally, in the presence of phosphate, another ternary

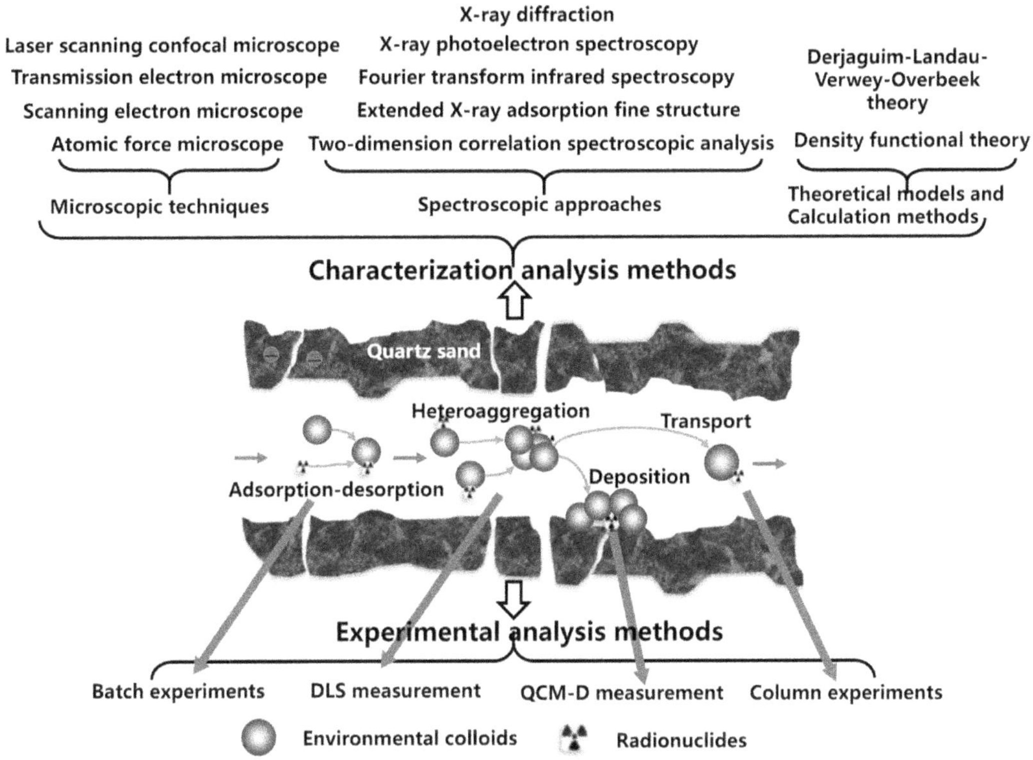

FIGURE 15.3 Experimental Methods, Characterization Methods, and Theoretical Calculation Methods for Studying Interaction between Colloids and Radionuclides.

complex can form in the system, enhancing system stability and thereby promoting radionuclide migration (Tang et al., 2024a).

Analytical techniques for quantifying interactions between colloids and radionuclides typically rely on the unique physicochemical properties of colloids. Different interaction modes may require various advanced characterization methods, theoretical models, and related calculations (Figure 15.3) (Wang et al., 2021; Wang et al., 2022; Pan et al., 2019). Characterization methods for studying interactions between colloids and contaminants mainly include microscopy and spectroscopy. Commonly used characterization techniques include microscopy such as scanning electron microscopy (SEM) and transmission electron microscopy (TEM), as well as spectroscopic methods like Fourier-transform infrared spectroscopy (FTIR) and time-resolved laser fluorescence spectroscopy (TRLFS). For example, Xu et al. (2021a) used FTIR to determine that HA likely adhered to bentonite colloids, and this speculation was corroborated by SEM characterization. Tang et al. (2024a) further used SEM to confirm that radionuclides were primarily retained at the inlet of the quartz sand column in the presence of phosphate, and demonstrated through TEM and TRLFS that the presence of $EuPO_4$ true colloids is a key factor influencing Eu(III) migration. Wei et al. (2024) also employed TEM and TRLFS characterization to confirm that enhanced uranium migration is attributed to phosphate-induced stabilization of illite colloids, the carrier effect of illite colloids for uranyl phosphate true colloids, and the shielding effect that inhibits collision–coagulation of uranyl phosphate true colloids. Unlike experimental analysis and spectroscopic characterization methods, DLVO theory can be used to qualitatively describe the stability characteristics of colloids. It provides the total interaction potential energy in colloidal systems, including van der Waals forces and electrostatic repulsion. However, since most colloids in real environments have rough surfaces and irregular, heterogeneous geometric structures, DLVO-based qualitative descriptions can yield significant deviations (Wei et al., 2021). Furthermore, density functional theory (DFT) can be used to calculate changes in electron density and atomic charges, thereby analyzing interactions between co-existing cations and colloidal particles (Xu et al., 2021a; Wang et al., 2017; Zou et al., 2016). Xu et al. (2021a) performed DFT calculations and found that as ionic charge increases, the trend of electron transfer between montmorillonite and cations becomes more pronounced, indicating that increased ionic charge promotes aggregation of colloidal particles.

15.6 CONCLUSION AND PROSPECTS

Biocolloidal composite particles in the environment possess complex and distinctive surface structures. Their inherent biological attributes—such as metabolic activity—lead to markedly different migration patterns compared with inorganic colloids. The migration behavior and underlying mechanisms of these composites are governed by the interplay of physical, geochemical, and biological processes. While it is acknowledged that biocolloids influence the adsorption and transport of radionuclides, their role in natural systems is further complicated by microbial life cycles. In complex and dynamic environmental conditions, microorganisms undergo natural metabolic aging, death, and decomposition, yielding microbial residues and decay products that also play a non-negligible role in the migration and transformation of radionuclides. Therefore, elucidating the interactions and microscopic mechanisms among microbial activity, their decay products, and radionuclides in environmental media is of critical importance.

Currently, research on the influence of clay mineral–biocolloidal composites on radionuclide transport remains fragmented, lacking a systematic and consistent theoretical framework. This review summarizes recent advances in understanding the factors and mechanisms by which clay–biocolloids affect radionuclide migration and offers some preliminary insights. It should be noted that the studies discussed here are largely confined to the laboratory scale and do not fully account for the complexity of real-world field conditions. In actual natural environments, diverse types of clay minerals and bacterial colloids co-exist, along with complex water chemistry involving carbonates, phosphates, sulfates, and silicates. These components may exhibit synergistic or antagonistic effects, collectively altering the environmental behavior of radioactive contaminants and posing potential risks to nuclear environmental safety. Hence, future research should prioritize the study of the interaction mechanisms, migration patterns, and environmental fate of multi-component environmental colloids within integrated field systems—spanning from the surface and vadose zone through various media down to groundwater. Such efforts will be essential for developing predictive models and effective safety interventions.

REFERENCES

Altmaier, M., Gaona, X., Fanghänel, T., 2013. Recent advances in aqueous actinide chemistry and thermodynamics. *Chem. Rev.* 113, 901–943.

Auset, M., Keller, A., 2004. Pore-scale processes that control dispersion of colloids in saturated porous media. *Water Resour. Res.* 40, 114–125.

Auset, M., Keller, A., 2006. Pore-scale visualization of colloid straining and filtration in saturated porous media using micromodels. *Water Resour. Res.* 42, W12S02.

Babakhani, P., Bridge, J., Doong, R., et al., 2017. Continuum-based models and concepts for the transport of nanoparticles in saturated porous media: A state-of-the-science review. *Adv. Colloid Interface Sci.* 246, 75–104.

Baumann, T., Werth, C., 2004. Visualization and modeling of polystyrol colloid transport in a silicon micromodel. *Vadose Zone J.* 3, 434–443.

Bedrikovetsky, P., Zeinijahromi, A., Siqueira, F., et al., 2012. Particle detachment under velocity alternation during suspension transport in porous media. *Transp. Porous Media* 91, 173–197.

Bekhit, H., El-Kordy, M., Hassan, A., 2009. Contaminant transport in groundwater in the presence of colloids and bacteria: Model development and verification. *J. Contam. Hydrol.* 108, 152–167.

Bense, V., Kooi, H., 2004. Temporal and spatial variations of shallow subsurface temperature as a record of lateral variations in groundwater flow. *J. Geophys. Res. Solid Earth* 109, B04103.

Biria, D., Balouchi, A., 2013. Investigation of the role of chemotaxis in bacterial transport through saturated porous media using Taguchi approach. *Colloids Surf. A Physicochem. Eng. Asp.* 436, 542–548.

Bots, P., Morris, K., Hibberd, R., et al., 2014. Formation of stable uranium(VI) colloidal nanoparticles in conditions relevant to radioactive waste disposal. *Langmuir* 30, 14396–14405.

Bouby, M., Kraft, S., Kuschel, S., et al., 2020. Erosion dynamics of compacted raw or homo ionic MX80 bentonite in a low ionic strength synthetic water under quasi-stagnant flow conditions. *Appl. Clay Sci.* 198, 105797.

Bradford, S., Verónica, L., Zhang, W., et al., 2013. Transport and fate of microbial pathogens in agricultural settings. *Crit. Rev. Environ. Sci. Technol.* 43, 775–893.

Cai, Y., Chen, Y., Ye, W., et al., 2020. Research progress on the generation and stability of bentonite colloids in the near-field of a repository. *Chin. J. Geotech. Eng.* 42, 1996–2005.

Chandraprabha, M., Natarajan, K., 2009. Microbially induced mineral beneficiation. *Miner. Process. Extr. Metall. Rev.* 31, 1–29.

Chen, C., Yang, S., Zhu, J., et al., 2024. Comparison of migration behavior of 90Sr and 241Am in unsaturated silty clay. *Nucl. Chem. Radiochem.* 46, 264–270.

Chen, M., Xu, N., Christodoulatos, C., et al., 2018. Synergistic effects of phosphorus and humic acid on the transport of anatase titanium dioxide nanoparticles in water-saturated porous media. *Environ. Pollut.* 243, 1368–1375.

Chen, Y., Cai, Y., Ye, W., et al., 2021. Research progress on the adsorption migration characteristics of bentonite colloids and co-migration properties of radionuclides in disposal repositories. *Chin. J. Geotech. Eng.* 43, 2149–2158.

de Anna, P., Pahlavan, A.A., Yawata, Y., Stocker, R., Juanes, R., 2021. Chemotaxis under flow disorder shapes microbial dispersion in porous media. *Nat. Phys.* 17, 68–73.

Dong, H., Zeng, Q., Liu, D., et al., 2024. Mechanisms of clay mineral–microorganism interactions and their applications in environmental fields. *Earth Sci. Front.* 31, 467–485.

Dong, Z., Yang, H., Wu, D., et al., 2014. Influence of silicate on the transport of bacteria in quartz sand and iron mineral-coated sand. *Colloids Surf. B Biointerfaces* 123, 995–1002.

Du, Y., Liang, Z., 2019. Research progress on the transport of nanoparticles and their carried heavy metal contaminants in porous media. *Henan Sci. Technol.* 38, 133–136.

Duan, X., Wang, C., Tang, Z., Wang, J., Ling, H., Ma, A., Niu, S., 2020. Research on safety assessment of geological disposal of high-level radioactive waste. *J. Univ. South China (Sci. Technol.)* 34(2), 42–49.

Fukunaga, S., Miyasaka, I., Ishii, K., et al., 2016. Column experiment for assessing microbial behavior around radioactive waste repositories, including migration of potentially radionuclide-accumulating bacteria. *Appl. Geochem.* 71, 99–109.

Fuller, A., Shaw, S., Ward, M., et al., 2015. Caesium incorporation and retention in illite interlayers. *Appl. Clay Sci.* 108, 128–134.

Gargiulo, G., Bradford, S., Simunek, J., et al., 2008. Bacteria transport and deposition under unsaturated flow conditions: The role of water content and bacteria surface hydrophobicity. *Vadose Zone J.* 7, 406–419.

Gerber, E., Romanchuk, A., Pidchenko, I., et al., 2020. The missing pieces of the PuO2 nanoparticle puzzle. *Nanoscale* 12, 18039–18048.

Haznedaroglu, B., Kim, H., Bradford, S., et al., 2009. Relative transport behavior of Escherichia coli O157:H7 and Salmonella enterica serovar pullorum in packed bed column systems: Influence of solution chemistry and cell concentration. *Environ. Sci. Technol.* 43, 1838–1844.

Horne, B., Rutherford, E., Wehrly, K., 2004. Simulating effects of hydro-dam alteration on thermal regime and wild steelhead recruitment in a stable-flow Lake Michigan tributary. *River Res. Appl.* 20, 185–203.

Huang, F., Zhang, H., Wang, Y., et al., 2020. Uranium speciation and distribution in Shewanella putrefaciens and anaerobic granular sludge in the uranium immobilization process. *J. Radioanal. Nucl. Chem.* 326, 393–405.

Huangfu, X., Ma, C., Huang, R., et al., 2019. Deposition kinetics of colloidal manganese dioxide onto representative surfaces in aquatic environments: The role of humic acid and biomacromolecules. *Environ. Sci. Technol.* 53, 146–156.

Hufton, J., Harding, J., Smith, T., et al., 2021. The importance of the bacterial cell wall in uranium(VI) biosorption. *Phys. Chem. Chem. Phys.* 23, 1566–1576.

Jiang, X., Wang, X., Tong, M., et al., 2013. Initial transport and retention behaviors of ZnO nanoparticles in quartz sand porous media coated with Escherichia coli biofilm. *Environ. Pollut.* 174, 38–49.

Joo, H., Kwon, T., Dai, S., 2023. Clay-bacteria interaction: Effect of bacterial cell density on sedimentation behavior and fabric map of kaolinite clay. *Appl. Clay Sci.* 241, 106973.

Keller, A., Auset, M., 2007. A review of visualization techniques of biocolloid transport processes at the pore scale under saturated and unsaturated conditions. *Adv. Water Resour.* 30, 1392–1407.

Kersting, A., Efurd, D., Finnegan, D., et al., 1999. Migration of plutonium in ground water at the Nevada Test Site. *Nature* 397, 56–59.

Kim, C., Pennell, K., Fortner, J., 2020. Delineating the relationship between nanoparticle attachment efficiency and fluid flow velocity. *Environ. Sci. Technol.* 54, 13992–13999.

Kim, H., Walker, S., 2009. Escherichia coli transport in porous media: Influence of cell strain, solution chemistry, and temperature. *Colloids Surf. B Biointerfaces* 71, 160–167.

Kim, H., Walker, S., Bradford, S., 2010. Macromolecule mediated transport and retention of Escherichia coli O157:H7 in saturated porous media. *Water Res.* 44, 1082–1093.

Knappenberger, T., Aramrak, S., Flury, M., 2015. Transport of barrel and spherical shaped colloids in unsaturated porous media. *J. Contam. Hydrol.* 180, 69–79.

Knappenberger, T., Flury, M., Mattson, E., et al., 2014. Does water content or flow rate control colloid transport in unsaturated porous media? *Environ. Sci. Technol.* 48, 3791–3799.

Knappett, P., Emelko, M., Zhuang, J., et al., 2008. Transport and retention of a bacteriophage and microspheres in saturated, angular porous media: Effects of ionic strength and grain size. *Water Res.* 42, 4368–4378.

Kolomá, K., Červinka, R., Hanusová, I., 2018. 137Cs transport in crushed granitic rock: The effect of bentonite colloids. *Appl. Geochem.* 96, 55–61.

Lead, J., Wilkinson, K., 2006. Aquatic colloids and nanoparticles: Current knowledge and future trends. *Environ. Chem.* 3, 159.

Lee, M., Jung, E., Song, K., et al., 2011. The influence of humic acid on the pH-dependent sorption of americium(III) onto kaolinite. *J. Radioanal. Nucl. Chem.* 289, 639–645.

Liu, Y., Alessi, D.S., Flynn, S.L., Alam, M.S., Hao, W., Gingras, M., Zhao, H., Konhauser, K.O., 2018a. Acid-base properties of purified and Na-exchanged montmorillonites: A combined potentiometric titration and surface complexation modeling study. *Chem. Geol.* 483, 191–200.

Li, Q., Yang, J., Fan, W., et al., 2018. Different transport behaviors of Bacillus subtilis cells and spores in saturated porous media: Implications for contamination risks associated with bacterial sporulation in aquifer. *Colloids Surf. B Biointerfaces* 162, 35–42.

Li, X., Xu, H., Gao, B., et al., 2019. Transport of a PAH-degrading bacterium in saturated limestone media under various physicochemical conditions: Common and unexpected retention and remobilization behaviors. *J. Hazard. Mater.* 380, 120858.

Liang, Y., Zhou, J., Dong, Y., et al., 2020. Evidence for the critical role of nanoscale surface roughness on the retention and release of silver nanoparticles in porous media. *Environ. Pollut.* 258, 113803.

Ling, X., Yan, Z., Liu, Y., et al., 2021. Transport of nanoparticles in porous media and its effects on the co-existing pollutants. *Environ. Pollut.* 283, 117098.

Liu, Y., Alessi, D., Flynn, S., et al., 2018. Acid-base properties of kaolinite, montmorillonite and illite at marine ionic strength. *Chem. Geol.* 483, 191–200.

Liu, Y., Wei, Q., Gao, B., et al., 2018. Current status and progress in the study of radionuclide migration in different media. *Nonferrous Met. (Extract. Metall.)* 6, 76–82.

Lv, X., Gao, B., Sun, Y., et al., 2016. Effects of grain size and structural heterogeneity on the transport and retention of nano-TiO2 in saturated porous media. *Sci. Total Environ.* 563, 987–995.

Madumathi, G., Philip, L., Bhallamudi, S., 2017. Transport of E. coli in saturated and unsaturated porous media: Effect of physiological state and substrate availability. *Sādhanā* 42, 1007–1024.

Maher, Z., Ivanov, P., O'Brien, L., et al., 2016. Americium and plutonium association with magnesium hydroxide colloids in alkaline nuclear industry process environments. *J. Nucl. Mater.* 468, 84–96.

Micheau, C., Virot, M., Dourdain, S., et al., 2020. Relevance of formation conditions to the size, morphology and local structure of intrinsic plutonium colloids. *Environ. Sci. Nano* 7, 2252–2266.

Möri, A., Alexander, W., Geckeis, H., et al., 2003. The colloid and radionuclide retardation experiment at the Grimsel Test Site: Influence of bentonite colloids on radionuclide migration in a fractured rock. *Colloids Surf. A Physicochem. Eng. Asp.* 217, 33–47.

Nair, V., Thampi, S., 2010. Numerical modelling of colloid transport in sets of parallel fractures with fracture skin. *Colloids Surf. A Physicochem. Eng. Asp.* 364, 109–115.

Najafi, J., Shaebani, M., John, T., et al., 2018. Flagellar number governs bacterial spreading and transport efficiency. *Sci. Adv.* 4, eaar6425.

Nakamura, S., Minamino, T., 2019. Flagella-driven motility of bacteria. *Biomolecules* 9, 279.

Nascimento, A., Tótola, M., Souza, C., et al., 2006. Temporal and spatial dynamics of blocking and ripening effects on bacterial transport through a porous system: A possible explanation for CFT deviation. *Colloids Surf. B Biointerfaces* 53, 241–244.

Ning, Z., Li, R., Lian, H., Zhou, S., Liao, P., Liu, C., 2021. Effects of flow-interruption on the bacteria transport behavior in porous media. *J. Hydrol.* 595, 125677.

Nie, X., Lin, Q., Dong, F., et al., 2022. Surface biomineralization of uranium onto Shewanella putrefaciens with or without extracellular polymeric substances. *Ecotoxicol. Environ. Saf.* 241, 113719.

Pan, D., Xu, Z., Wei, X., et al., 2019. Research progress on interactions between environmental colloids and radionuclides. *Sci. China Chem.* 49, 12–26.

Park, S., Kim, S., 2009. Adhesion of Escherichia coli to iron-coated sand in the presence of humic acid: A column experiment. *Water Environ. Res.* 81, 125–130.

Petersen, E., Flores-Cervantes, D., Bucheli, T., et al., 2016. Quantification of carbon nanotubes in environmental matrices: Current capabilities, case studies, and future prospects. *Environ. Sci. Technol.* 50, 4587–4605.

Redman, J., Walker, S., Elimelech, M., 2004. Bacterial adhesion and transport in porous media: Role of the secondary energy minimum. *Environ. Sci. Technol.* 38, 1777–1785.

Romanchuk, A., Vlasova, I., Kalmykov, S., 2020. Speciation of uranium and plutonium from nuclear legacy sites to the environment: A mini review. *Front. Chem.* 8, 630.

Ruiz-Fresneda, M., Lopez-Fernandez, M., Martinez-Moreno, M., et al., 2020. Molecular binding of Eu(III)/Cm(III) by Stenotrophomonas bentonitica and its impact on the safety of future geodisposal of radioactive waste. *Environ. Sci. Technol.* 54, 15180–15190.

Ruiz-Fresneda, M., Martinez-Moreno, M., Povedano-Priego, C., et al., 2023. Impact of microbial processes on the safety of deep geological repositories for radioactive waste. *Front. Microbiol.* 14, 1134078.

Salvatores, M., Palmiotti, G., 2011. Radioactive waste partitioning and transmutation within advanced fuel cycles: Achievements and challenges. *Prog. Part. Nucl. Phys.* 66, 144–166.

Sanin, S., Sanin, F., Bryers, J., 2003. Effect of starvation on the adhesive properties of xenobiotic degrading bacteria. *Process Biochem.* 38, 909–914.

Schäfer, A., Ustohal, P., Harms, H., et al., 1998. Transport of bacteria in unsaturated porous media. *J. Contam. Hydrol.* 33, 149–169.

Sendra, M., Yeste, M., Gatica, J., et al., 2017. Homoagglomeration and heteroagglomeration of TiO2, in nanoparticle and bulk form, onto freshwater and marine microalgae. *Sci. Total Environ.* 592, 403–411.

Sheng, F., Liu, H., Wang, K., et al., 2014. Investigation into preferential flow in natural unsaturated soils with field multiple-tracer infiltration experiments and the active region model. *J. Hydrol.* 508, 137–146.

Shiryaev, A., Vlasova, I., Yapaskurt, V., et al., 2018. Forensic study of early stages of the Chernobyl accident: Story of three hot particles. *J. Nucl. Mater.* 511, 83–90.

Sirivithayapakorn, S., Keller, A., 2003. Transport of colloids in unsaturated porous media: A pore-scale observation of processes during the dissolution of air-water interface. *Water Resour. Res.* 39, 1346.

Smith, S., Boothman, C., Williams, H., et al., 2017. Microbial impacts on 99mTc migration through sandstone under highly alkaline conditions relevant to radioactive waste disposal. *Sci. Total Environ.* 575, 485–495.

Song, J., Han, B., Song, H., et al., 2019a. Nonreductive biomineralization of uranium by Bacillus subtilis ATCC-6633 under aerobic conditions. *J. Environ. Radioact.* 208, 106027.

Song, W., Wang, X., Sun, Y., et al., 2019b. Bioaccumulation and transformation of U(VI) by sporangiospores of Mucor circinelloides. *Chem. Eng. J.* 362, 81–88.

Sun, H., Yin, X., Wang, Y., 2012. Effect of pH on the transport of clay mineral colloids in saturated porous media. *Acta Sci. Circumstantiae* 32, 419–424.

Sun, Y., Lei, A., 2023. Ca2+-facilitated adhesion of bacteria on the Na-montmorillonite surface. *ACS Omega* 8, 3385–3395.

Sun, Y., Pan, D., Wei, X., et al., 2020. Insight into the stability and correlated transport of kaolinite colloid: Effect of pH, electrolytes and humic substances. *Environ. Pollut.* 266, 115189.

Syngouna, V., Chrysikopoulos, C., 2015. Experimental investigation of virus and clay particles cotransport in partially saturated columns packed with glass beads. *J. Colloid Interface Sci.* 440, 140–150.

Syngouna, V., Chrysikopoulos, C., Kokkinos, P., et al., 2017. Cotransport of human adenoviruses with clay colloids and TiO2 nanoparticles in saturated porous media: Effect of flow velocity. *Sci. Total Environ.* 598, 160–167.

Tan, L., Tan, X., Mei, H., et al., 2018. Coagulation behavior of humic acid in aqueous solutions containing Cs+, Sr2+ and Eu3+: DLS, EEM and MD simulations. *Environ. Pollut.* 236, 835–843.

Tang, Q.F., Xu, Z., Tan, Q., Shi, X.Y., Wu, W.S., Pan, D.Q., 2024a. Insight into impact of phosphate on the cotransport and corelease of Eu(III) with bentonite colloids in saturated quartz columns. *J. Hazard. Mater.* 461, 132572.

Tang, Z.P., Liu, X.Y., Zhang, P., Duan, W.Q., Chen, Y.H., Wang, X.X., Fan, Q.H., 2024b. Research progress on radionuclide migration in geological disposal of high-level radioactive waste. *At. Energy Sci. Technol.* 58, 1704–1715.

Tong, M., Long, G., Jiang, X., Kim, H., 2010. Contribution of extracellular polymeric substances on representative gram negative and gram positive bacterial deposition in porous media. *Environ. Sci. Technol.* 44(7), 2393–2399.

Tran, E., Klein-Ben David, O., Teutsch, N., et al., 2015. Influence of intrinsic colloid formation on migration of cerium through fractured carbonate rock. *Environ. Sci. Technol.* 49, 13275–13282.

Tran, E., Klein Ben-David, O., Teutsch, N., et al., 2016. Influence of heteroaggregation processes between intrinsic colloids and carrier colloids on cerium(III) mobility through fractured carbonate rocks. *Water Res.* 100, 88–97.

Tran, E., Zavrin, M., Kersting, A., et al., 2021. Colloid-facilitated transport of 238Pu, 233U and 137Cs through fractured chalk: Laboratory experiments, modelling, and implications for nuclear waste disposal. *Sci. Total Environ.* 757, 143818.

Tsuneda, S., Aikawa, H., Hayashi, H., Yuasa, A., Hirata, A., 2003. Extracellular polymeric substances responsible for bacterial adhesion onto solid surface. *FEMS Microbiol. Lett.* 223(2), 287–292.

Tufenkji, N., Elimelech, M., 2004. Correlation equation for predicting single-collector efficiency in physicochemical filtration in saturated porous media. *Environ. Sci. Technol.* 38, 529–536.

Vasiliadou, I., Chrysikopoulos, C., 2011. Cotransport of Pseudomonas putida and kaolinite particles through water-saturated columns packed with glass beads. *Water Resour. Res.* 47, W02543.

Vasiliadou, I., Papoulis, D., Chrysikopoulos, C., et al., 2011. Attachment of Pseudomonas putida onto differently structured kaolinite minerals: A combined ATR-FTIR and 1H NMR study. *Colloids Surf. B Biointerfaces* 84, 354–359.

Walczak, J., Wang, L., Bardy, S., et al., 2012. The effects of starvation on the transport of Escherichia coli in saturated porous media are dependent on pH and ionic strength. *Colloids Surf. B Biointerfaces* 90, 129–136.

Wang, D., Zhang, J., Cao, R., et al., 2022. The detection and characterization techniques for the interaction between graphene oxide and natural colloids: A review. *Sci. Total Environ.* 808, 151906.

Wang, H., Dong, Y., Zhu, M., et al., 2015. Heteroaggregation of engineered nanoparticles and kaolin clays in aqueous environments. *Water Res.* 80, 130–138.

Wang, J., Yu, S., Zhao, Y., et al., 2017. Experimental and theoretical studies of ZnO and MgO for the rapid coagulation of graphene oxide from aqueous solutions. *Sep. Purif. Technol.* 184, 88–96.

Wang, K., Lin, Z., Yang, Z., et al., 2018. Basic properties of Gaomiaozi bentonite colloid. *J. Nucl. Radiochem.* 40, 189–195.

Wang, P., Bao, T., Hu, B., et al., 2021. Aquatic environmental behaviors of natural colloids. *J. Lake Sci.* 33, 28–48.

Wei, X., Pan, D., Tan, Q., et al., 2023. Surface charge property governing co-transport of illite colloids and Eu(III) in saturated porous media. *Chin. Chem. Lett.* 34, 108275.

Wei, X., Pan, D., Xu, Z., et al., 2021. Colloidal stability and correlated migration of illite in the aquatic environment: The roles of pH, temperature, multiple cations and humic acid. *Sci. Total Environ.* 768, 144174.

Wei, X., Shi, X., Yang, M., et al., 2024. Phosphate and illite colloid pose a synergistic risk of enhanced uranium transport in groundwater: A challenge for phosphate immobilization remediation of uranium contaminated environmental water. *Water Res.* 255, 121514.

Weiss, T., Mills, A., Hornberger, G., et al., 1995. Effect of bacterial cell shape on transport of bacteria in porous media. *Environ. Sci. Technol.* 29, 1737–1740.

Won, J., Burns, S.E., 2018. Colloid-facilitated transport of metals in kaolinite-coated systems. *Environ. Sci. Technol.* 52(5), 2735–2741.

Xiao, L., Wang, Z., Hua, R., et al., 2019. Research progress on radionuclide migration in geological materials. *Environ. Sci. Technol.* 42, 53–60.

Xu, Z., Niu, Z., Pan, D., et al., 2021a. Mechanisms of bentonite colloids and humic acid affecting the transport of Eu(III) in saturated porous media. *Sci. Total Environ.* 797, 149208.

Xu, Z., Niu, Z., Tang, Q., et al., 2021b. Adsorption characteristics of Eu(III) on bentonite colloids under different physicochemical conditions. *J. Radioanal. Nucl. Chem.* 330, 765–773.

Xu, Z., Pan, D., Sun, Y., et al., 2018. Stability of GMZ bentonite colloids: Aggregation kinetic and reversibility study. *Appl. Clay Sci.* 161, 436–443.

Xu, Z., Pan, D., Tang, Q., et al., 2022. Co-transport and co-release of Eu(III) with bentonite colloids in saturated porous sand columns: Controlling factors and governing mechanisms. *Environ. Pollut.* 298, 118842.

Xu, Z., Sun, Y., Niu, Z., et al., 2020. Kinetic determination of sedimentation for GMZ bentonite colloids in aqueous solution: Effect of pH, temperature and electrolyte concentration. *Appl. Clay Sci.* 184, 105393.

Xu, Z., Tang, Q., Pan, D., et al., 2023. Co-transport of bentonite colloids and Eu(III) transport in saturated heterogeneous porous media. *J. Radioanal. Nucl. Chem.* 332, 1125–1133.

Yan, C., Li, Y., Sharma, P., et al., 2022. Influence of dissolved organic matter, kaolinite, and iron oxides on aggregation and transport of biochar colloids in aqueous and soil environments. *Chemosphere* 306, 135555.

Yang, H., Kim, H., Tong, M., 2012a. Influence of bentonite particles on representative gram negative and gram positive bacterial deposition in porous media. *Environ. Sci. Technol.* 46, 11627–11634.

Yang, H., Tong, M., Kim, H., 2012b. Influence of humic acid on the transport behavior of bacteria in quartz sand. *Colloids Surf. B Biointerfaces* 91, 122–129.

Yang, J., Wang, X., Zhu, M., Liu, H., Ma, J., 2019a. Investigation of U(VI) adsorption on the gibbsite/water interface by EXAFS and DFT calculation. *ACS Earth Space Chem.* 3, 1019–1029.

Yang, S., Yang, L., Zhang, L., et al., 2019b. Impact of natural soil colloids on the transport and transformation of U(VI) in saturated porous media: Co-transport or heavy metal contaminant retardation? *Sci. Total Environ.* 684, 190–199.

Yong, S., Lim, S., Ho, C., et al., 2022. Mechanisms of microbial-based iron reduction of clay minerals: Current understanding and latest developments. *Appl. Clay Sci.* 228, 106653.

Zhai, B., Tian, Q., Li, N., et al., 2022. SAXS study of the formation and structure of polynuclear thorium(IV) colloids and thorium dioxide nanoparticles. *J. Synchrotron Radiat.* 29, 281–287.

Zhang, D., Chen, X., Larson, S., et al., 2020. Uranium biomineralization with phosphate—biogeochemical process and its application. *ACS Earth Space Chem.* 4, 2205–2214.

Zhang, D., Wang, Y., Heng, J., et al., 2022. Stability of Eu(III)-silicate colloids: Effect of Eu content, pH, electrolyte and fulvic acid. *J. Hazard. Mater.* 438, 129363.

Zhang, M., He, L., Jin, X., et al., 2021a. Flagella and their properties affect the transport and deposition behaviors of Escherichia coli in quartz sand. *Environ. Sci. Technol.* 55, 4964–4973.

Zhang, W., Guo, X., Jiang, M., 2023. Influence of humic acid and bovine serum albumin on colloid-associated heavy metal transport in saturated porous media. *Environ. Technol.* 44, 3965–3974.

Zhang, W., Li, S., Zhao, K., et al., 2024. E. coli phage transport in porous media: Response to colloid types and water saturation. *Sci. Total Environ.* 906, 167635.

Zhang, W., Tang, X., Weisbrod, N., et al., 2012. A review of colloid transport in fractured rocks. *J. Mt. Sci.* 9, 770–787.

Zhang, W., Wu, S., Qin, Y., et al., 2021b. Deposition and mobilization of viruses in unsaturated porous media: Roles of different interfaces and straining. *Environ. Pollut.* 270, 116072.

Zhang, W., Yuan, S., 2019. Characterizing preferential flow in landfilled municipal solid waste. *Waste Manag.* 84, 20–28.

Zhao, K., Chen, C., Shang, J., 2018. Effect of humic acid-kaolinite colloids on the transport of uranium(VI) in saturated porous media. *J. Soil Water Conserv.* 32, 286–293.

Zhong, H., Liu, G., Jiang, Y., et al., 2017. Transport of bacteria in porous media and its enhancement by surfactants for bioaugmentation: A review. *Biotechnol. Adv.* 35, 490–504.

Zhou, X., Kang, F., Qu, X., et al., 2020. Role of extracellular polymeric substances in microbial reduction of arsenate to arsenite by Escherichia coli and Bacillus subtilis. *Environ. Sci. Technol.* 54, 6185–6193.

Zou, Y., Wang, X., Ai, Y., et al., 2016. Coagulation behavior of graphene oxide on nanocrystallined Mg/Al layered double hydroxides: Batch experimental and theoretical calculation study. *Environ. Sci. Technol.* 50, 3658–3667.

Zänker, H., Hennig, C., 2014. Colloid-borne forms of tetravalent actinides: A brief review. *J. Contam. Hydrol.* 157, 87–105.

PART V

Future Prospects and Conclusions

CHAPTER 16

Research Gaps, Challenges, and Prospects for Nuclear Waste Management

Hong Li and Bin Ma

16.1 INTRODUCTION

Nuclear energy, as a low-carbon alternative to fossil fuels, plays a critical role in the global energy transition (Denysov et al., 2024; Mathew, 2022). It provides a stable, reliable source of electricity that is increasingly recognized for its potential to mitigate climate change by reducing greenhouse gas emissions (Alwaeli & Mannheim, 2022). With nuclear power contributing around 10% of the world's electricity and ongoing investments in new reactor technologies, nuclear energy is expected to grow, especially as countries seek to diversify their energy portfolios and meet net-zero emission targets (Rajagukguk et al., 2024; Wagner, 2021). However, the widespread use of nuclear energy comes with a significant challenge: the management and disposal of the radioactive waste generated by nuclear power plants (Ojovan & Steinmetz, 2022; Voss et al., 2025). Effective nuclear waste management (NWM) is a cornerstone of nuclear energy sustainability, influencing environmental protection, public health, and technological development (McCloy et al., 2024).

The generation of nuclear waste is an inherent consequence of the nuclear fission process (Guo, 2025). During operation, nuclear reactors produce large quantities of spent nuclear fuel that remains highly radioactive for thousands to millions of years (Ojovan & Steinmetz, 2022). According to radioactivity and retention time, the generated waste is classified into five categories by the French National Radioactive Waste Management Agency (ANDRA): very low-level waste (VLLW), very short-lived waste (VSLW), low-level waste (LLW), and high-level waste (HLW). VLLW and VSLW are lower categories of waste that are typically managed using simpler, less costly disposal techniques such as shallow land burial (Darda et al., 2021). LLW includes contaminated discarded clothing, tools, and filters, etc., but the radioactive isotopes contained in it have a relatively short half-life. LLW represents the largest volume of nuclear waste but accounts for a very small portion of

DOI: 10.1201/9781003665793-21

the total radioactivity. ILW includes materials that are more radioactive than LLW but do not generate significant heat. Examples include reactor components, resins, and chemical sludges. HLW includes spent nuclear fuel and materials from reprocessing spent fuel. This category is the most hazardous and generates significant heat, requiring special handling and storage. The long-lived isotopes in HLW, such as ^{137}Cs, ^{90}Sr, and ^{239}Pu, pose substantial risks to human health and the environment if not properly contained (Cao et al., 2021).

As the use of nuclear energy expands, managing its byproducts becomes ever more crucial. The technical challenges associated with nuclear waste extend across the entire back end of the fuel cycle, from the conditioning and characterization of HLWs to their long-term isolation in engineered facilities (Darda et al., 2021). Without reliable and scientifically validated technological solutions, the safety, efficiency, and sustainability of nuclear power generation could be compromised (Alwaeli & Mannheim, 2022). Advancements in waste form development, partitioning and transmutation, deep geological disposal systems, and long-term monitoring and containment methods are therefore critical to ensuring that radioactive materials are effectively immobilized, their radiological hazards minimized, and their migration into the biosphere prevented over extended timescales. In this context, continual innovation and refinement of NWM technologies constitute a pivotal requirement for sustaining and expanding the role of nuclear energy in future energy systems.

16.2 GLOBAL TRENDS IN NUCLEAR WASTE MANAGEMENT

Globally, the need for effective NWM systems has led to the development of several strategies, with varying degrees of success. One of the most common solutions is deep geological disposal, where nuclear waste is stored in stable geological formations deep underground (Bajestani et al., 2024; Kim et al., 2025). This method has been extensively researched and implemented in countries like Finland, Sweden, Canada, and China. In Finland, the newly put into operation Onkalo repository has become the world's first deep geological repository for spent nuclear fuel (Clayton et al., 2025). The Swedish Nuclear Fuel and Waste Management Company (SKB) has constructed the Äspö repository in the Forsmark area in eastern Sweden, for the long-term storage of used nuclear waste (Lisjak et al., 2023). Meanwhile, the Kärnbränslesäkerhet-3 (KBS-3; Nuclear Fuel Safety-3) has been adopted for geological disposal of spent nuclear fuel both in Sweden and Finland, as illustrated in Figure 16.1 (Zou & Cvetkovic, 2023). The feasibility and safety of the KBS-3 approach have been extensively studied and tested in the underground laboratories both at Äspö and ONKALO over the past several decades (SKB, 2006; Selroos & Gylling, 2023). However, while geological disposal is widely regarded as the safest option, it is not without challenges. The safety of repositories must be assured for thousands of years, posing a significant engineering and scientific challenge in terms of monitoring, containment, and material degradation (Harithra et al., 2024).

In addition to geological disposal, spent fuel reprocessing and waste recycling are gaining attention as potential solutions to reduce the volume of HLW (Holdsworth et al., 2023; Terranova & Tavares, 2024). The partially closed, thermal recycle, or twice-through cycle approach, based on reprocessing spent nuclear fuel to recover fissile material (i.e., uranium and plutonium), which is then recycled as mixed oxide (MOX) fuel, is sometimes referred to as plutonium mono-recycling (Kiegiel et al., 2022; Pucciarelli et al., 2024). The fully closed cycle, in which spent nuclear fuel is reprocessed, and fissile materials are usually recycled in a fast reactor multiple times to maximize the energy value of the fuel components, is also referred to as plutonium multi-recycling, as shown in Figure 16.2 (Kiegiel

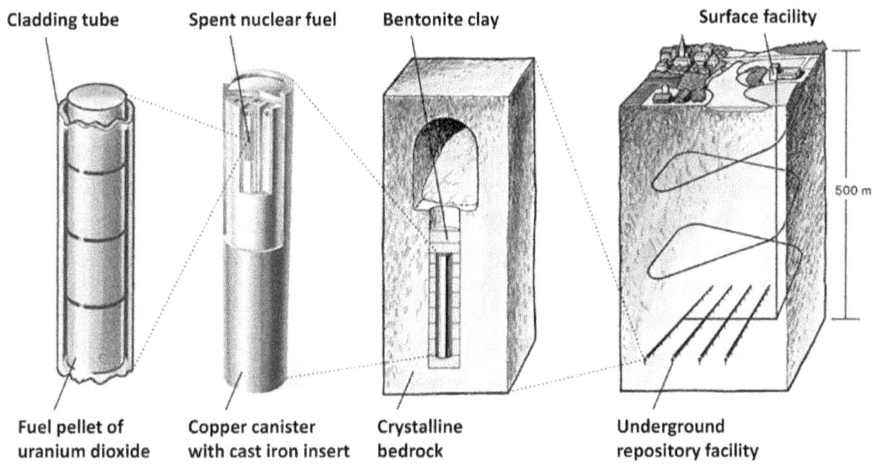

FIGURE 16.1 KBS-3 Approach Developed and Adopted for the Safe Disposal of HLW in Sweden and Finland (Zou & Cvetkovic, 2023).

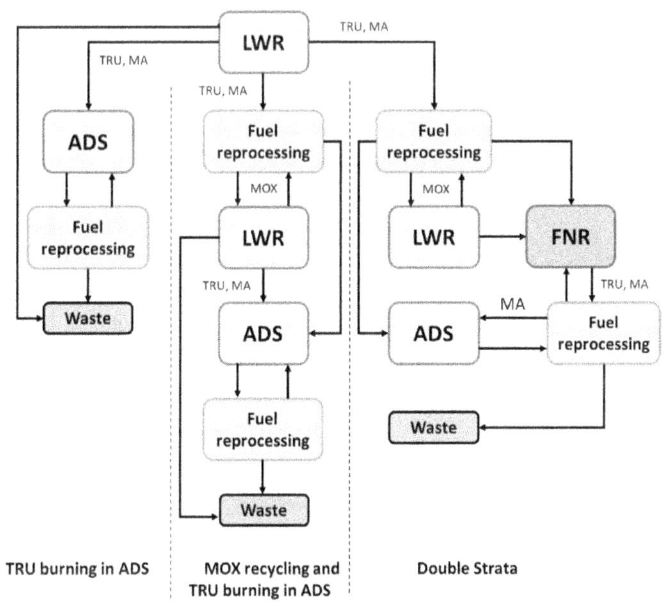

FIGURE 16.2 A Strategy for Closing the Nuclear Fuel Cycle using Accelerator-Driven Systems (ADSs) (Kiegiel et al., 2025).

et al., 2025). In the partitioning and transmutation (P&T) scenario, minor actinides are also recycled for burning (usually) in fast reactors or in accelerator-driven systems (Merk et al., 2025). The primary aim of the P&T scenario is to reduce the burden of minor actinides destined for deep geological repositories, rather than to recover their energy value (Cammi et al., 2025).

More recently, innovative reactor designs, such as molten salt reactors (MSRs) and fast breeder reactors (FBRs), have been proposed as part of a solution to reduce nuclear waste (Mausolff et al., 2021). MSRs utilize a liquid fuel form—typically a eutectic mixture of fluoride salts containing dissolved actinides, which enables continuous reprocessing and dynamic fuel management (Figure 16.3) (Kiegiel et al., 2025). By operating at atmospheric pressure with liquid fuel, MSRs can continuously fission transuranic elements and LLFPs, thereby shrinking both the radiotoxicity and the half-life of residual waste streams and easing future repository burdens (Alang & Rashid, 2024b). Fast breeder reactors (FBRs) are a class of advanced nuclear reactors that offer several advantages over water-cooled reactors (WCRs) in terms of safety, sustainability, and nuclear waste disposal (Mathew, 2022). A typical flow diagram of an FBR is shown in Figure 16.4. FBRs breed more fuel than they consume, which is a basic feature of renewable energy sources. These reactors not only generate power but can also reduce the quantity of HLW by utilizing spent nuclear fuel, thus providing a potential means of waste mitigation.

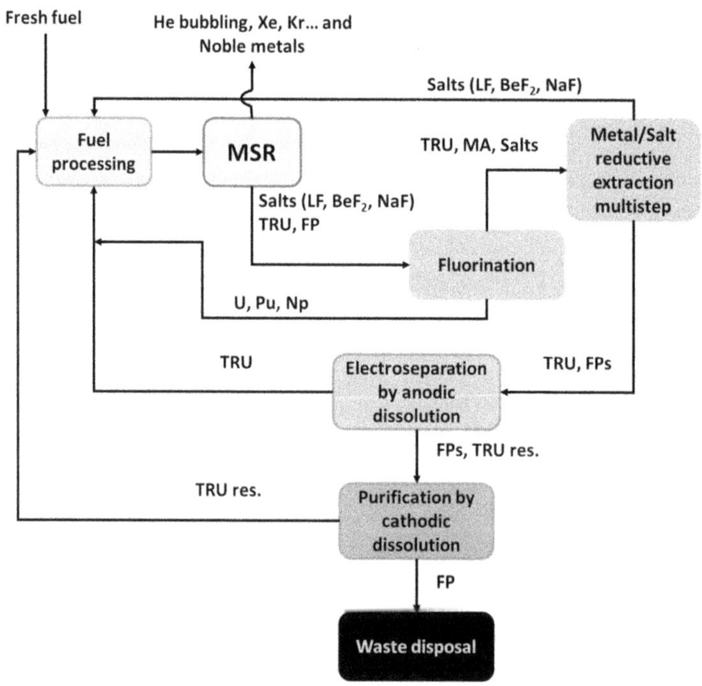

FIGURE 16.3 Simplified Scheme of Fuel Cycle of Molten Salt Transmutation Reactor (Kiegiel et al., 2025).

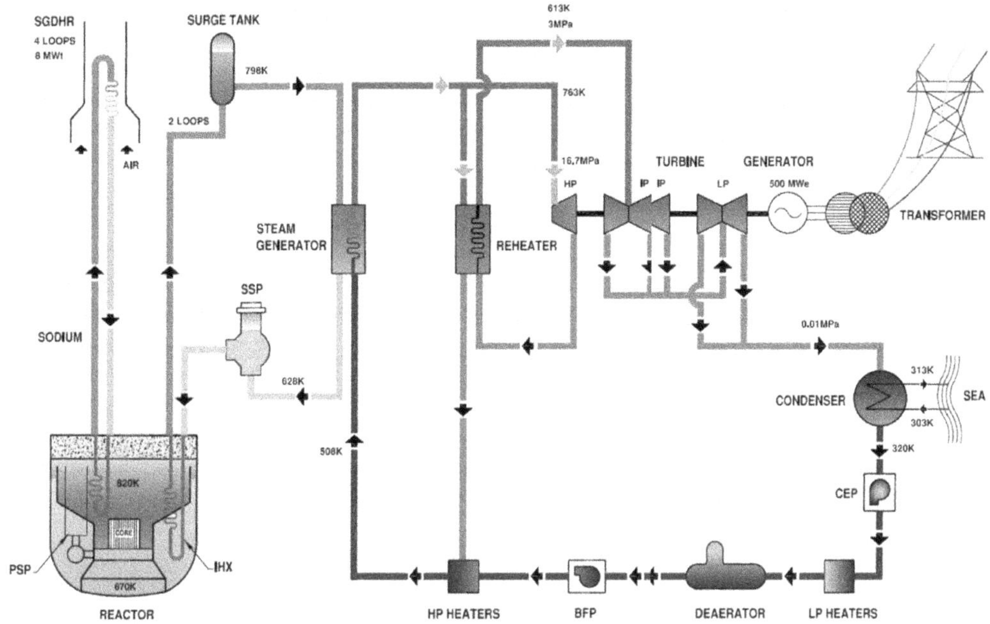

FIGURE 16.4 Flow Diagram of India's 500 MWe Prototype Fast Breeder Reactor Undergoing Commissioning (Mathew, 2022).

16.3 KEY RESEARCH GAPS AND CHALLENGES

The effective management of nuclear waste represents one of the greatest technological challenges for the sustainable expansion of nuclear energy. Although nuclear fission produces comparatively small volumes of waste relative to fossil fuel combustion, the longevity, toxicity, and radiological hazards of spent nuclear fuel and HLW demand solutions that can guarantee safety across geological timescales (Deng et al., 2020). Addressing this challenge requires advances in multiple domains of science and engineering, from geological isolation systems to chemical processing and advanced reactor development. While progress has been achieved in each of these areas, persistent gaps in research, coupled with technological and material uncertainties, highlight the need for sustained innovation.

16.3.1 Geological Storage Technology

Deep geological disposal has emerged as the reference solution for the long-term management of high-level radioactive waste and spent nuclear fuel. The rationale for this approach is grounded in the recognition that no engineered system can remain intact indefinitely and that geological formations, if carefully selected, provide a natural barrier capable of isolating radionuclides over timescales extending to hundreds of thousands of years. Geological disposal projects, such as the Onkalo repository in Finland, the Cigéo project in France, and the Swedish Forsmark site, illustrate the technical feasibility of such an approach. Yet, despite decades of research, significant scientific and technological uncertainties remain that complicate the design, validation, and long-term monitoring of repositories.

One of the most significant challenges in geological storage is the accurate prediction of repository behavior over geological timescales. The interactions between the engineered barriers and the host rock are governed by coupled thermal–hydrological–mechanical–chemical (THMC) processes. The interaction between these fields is not a simple additive process but rather a dynamic interplay where each field influences the others (Figure 16.5) (Wang et al., 2025). Chang et al. investigated various levels of thermal impacts on the permeability of intact core samples, explicitly accounting for the modulating roles of anisotropic fabric and in situ stress state (Chang et al., 2025). The experimental results show that at 90°C, radial compaction of intact rock induced 75% irreversible permeability loss via mineral expansion, whereas axial compaction caused only 15%; cracked specimens

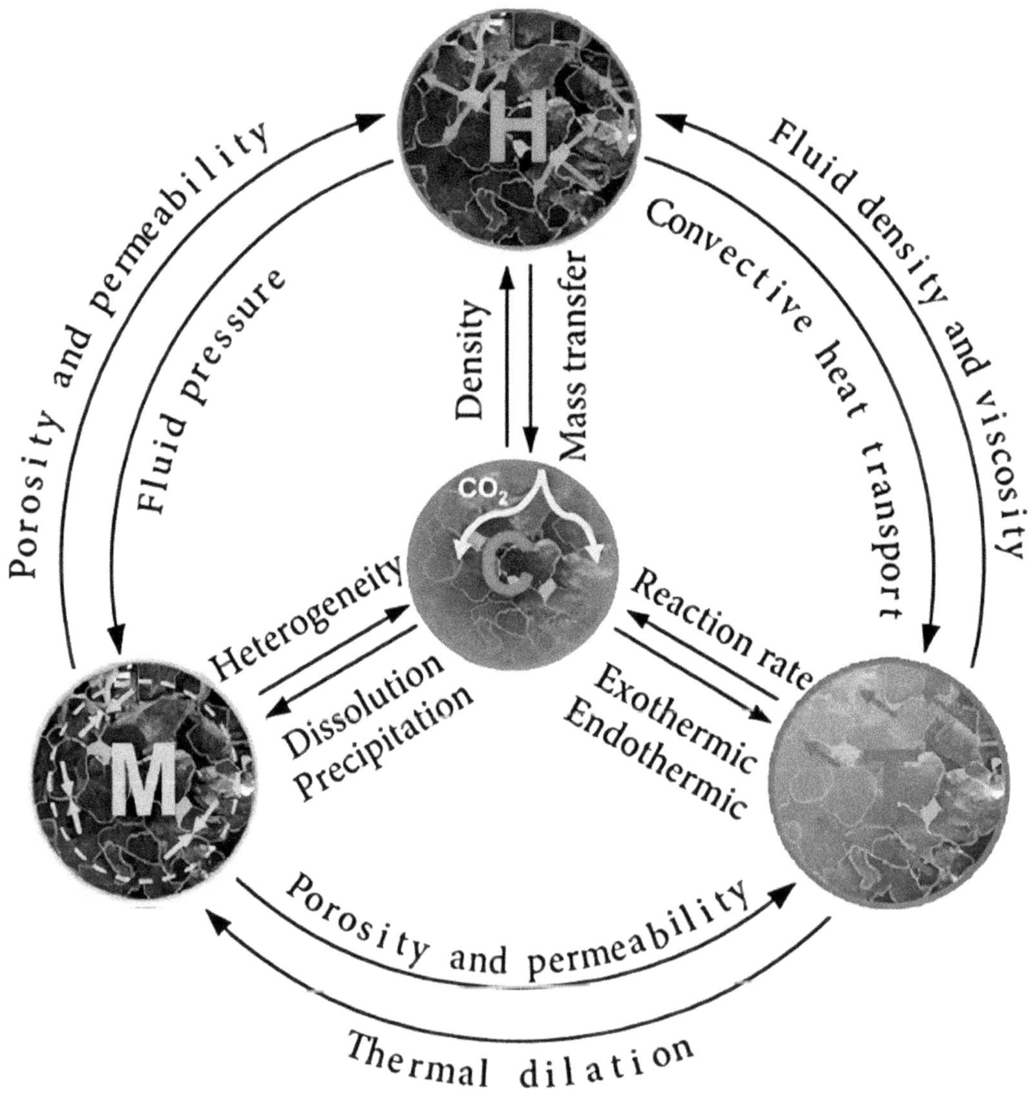

FIGURE 16.5 Schematic Illustration of the Attribution of THMC Coupling (Wang et al., 2025).

exhibited fracture-driven permeability jump and rapid collapse. Effluent chemistry revealed enhanced dissolution during shear-sliding tests compared with compaction-dominated tests. To understand these coupling mechanisms, numerous material models and numerical codes have been developed. Yoon et al. present a thermo–hydro–mechanical (THM) model to observe the effect of the soil–water retention curve (SWRC) and the thermal conductivity function (TCF) under high temperatures (Yoon et al., 2024). Model results provide a decent match of measured temporal and spatial changes, but the density distribution at the early stage of the hydration front (<10 days) cannot be explained. The lack of validated models that fully capture THMC coupling remains one of the central research gaps in geological disposal.

Another critical issue concerns the durability of engineered barriers. Copper, steel, and titanium alloys have been proposed as container materials, yet their performance under repository conditions is still not fully understood (Morales-Hidalgo et al., 2024). Microbial activity, sulfide-rich groundwater, and radiolysis of water can accelerate corrosion, undermining long-term containment (Haynes et al., 2021). Abramova et al. examined microbial processes in engineered safety barrier (ESB) materials, including steel, cement, and clay, under DGR conditions, demonstrating that elevated temperatures or contact with hydrogen and iron-containing products can accelerate microbial processes (Abramova et al., 2023). Similarly, bentonite clay, widely regarded as a reliable buffer material due to its swelling and sorption properties, may lose effectiveness under prolonged thermal stress. Alzamel experimentally demonstrated that the coupled effect of salinity and temperature decreases the swelling potential of the bentonite–sand mixture (Alzamel, 2022). Ensuring the robustness of these engineered barriers requires long-term experimental programs, accelerated testing methodologies, and advanced material innovations that can better withstand repository conditions. Research into nano-engineered coatings and multi-layer composite canisters has begun, but their scalability and performance remain unresolved.

The question of site characterization further complicates repository development. Geological variability introduces significant uncertainty into the assessment of long-term stability. Even in geologically stable regions, faulting, groundwater circulation, and seismic risks pose challenges (Na et al., 2025; Yu et al., 2023). Moreover, future climate change, including potential glaciation cycles, could alter hydrological pathways and stress regimes in ways that are difficult to predict. Pizarro et al.'s comprehensive review identified critical knowledge gaps in climate-adaptation strategies for nuclear-waste repositories (Pizarro & Sainsbury, 2023). Consequently, research must focus not only on improving site characterization methods but also on developing adaptive repository designs that can accommodate a broader range of geological uncertainties.

Finally, long-term monitoring remains an unresolved challenge. Once a repository is closed, the ability to observe and verify its continued safety becomes limited. Conventional instrumentation cannot reliably operate for the required durations, leaving uncertainty about whether early warning signs of failure can be detected. Advances in durable sensor technologies, fiber optics, and remote data transmission systems are urgently needed to enable real-time monitoring of repository conditions over extended timescales. Perrot et al. tested radiation tolerance of seven optical fiber cable types, and among them, a poly-ether-ether-ketone (PEEK) structured cable showed notable resilience to γ and neutron/γ exposures (Perrot et al., 2025). This enhancement not only aids in nuclear waste monitoring but also offers a comprehensive qualification process for evaluating optical fiber cable performance for nuclear waste monitoring. Complementing these technologies with advanced machine learning models capable of interpreting monitoring data to forecast repository behavior represents another promising but underdeveloped research avenue (Abulifa et al., 2025).

16.3.2 Spent Fuel Reprocessing Technology

Spent nuclear fuel reprocessing is central to the strategy of reducing waste volumes and recovering valuable fissile materials. Conventional reprocessing, exemplified by the plutonium uranium redox extraction (PUREX) process, separates uranium and plutonium from fission products through aqueous chemical methods (Huang et al., 2025; Wang et al., 2021). While PUREX has been implemented at industrial scales in countries such as France and the UK, it presents several technological and sustainability challenges, particularly in relation to the generation of secondary waste, process efficiency, and material robustness.

The principal technological gap lies in the inefficiency of current separation techniques for minor actinides and certain LLFPs. PUREX was designed to recover plutonium and uranium but does not address americium, curium, or neptunium, which contribute significantly to the long-term radiotoxicity of waste. Research has therefore turned toward advanced aqueous processes such as TRUEX, DIAMEX, TALSPEAK, and SANEX, which aim to selectively separate these elements (Xu et al., 2023; Yang et al., 2024). Cao et al. engineered the hydrophilic lanthanide/actinide separation ligand Phen-2DIC4NH2 by introducing amine as the terminal solubilizing group onto the phenanthroline diimide backbone and delicately adjusting the alkyl chain spacing, yielding exceptional extraction performance (Cao et al., 2024). Phen-2DIC4NH2 achieved high separation factors of ≈300 for Eu(III)/Am(III) in 1.25 M HNO_3. While laboratory-scale demonstrations have shown promise, challenges remain in solvent stability, radiolytic degradation, and process scalability. The chemical complexity of spent fuel, combined with the need for extreme selectivity, makes these processes inherently difficult to optimize for industrial application (Wang Y. et al., 2024b).

An alternative method, pyroprocessing, has attracted increasing interest due to its ability to handle high-burn-up fuels and its potential for integration with advanced reactors. Conducted in molten salt or metal electrolytes, pyroprocessing allows the electrochemical separation of actinides from fission products (Galashev, 2022; Mirza et al., 2023). Its advantages include compact facility design, reduced criticality risk, and compatibility with metallic fuels. Yuan et al. used boron (B) as a solid cathode to remove active fission products from waste molten salt (Yuan et al., 2023). Under 3.4 V and 720°C, the percentage removal of Sr^{2+} can be as high as nearly 90%, while the percentage removal of La^{3+}, Nd^{3+}, Sm^{3+}, and Ce^{3+} can reach more than 99% from the waste NaCl-KCl. However, technological gaps hinder its deployment, particularly in the development of corrosion-resistant materials that can withstand the aggressive environments of molten salts and the management of volatile fission products such as cesium and iodine (Roper et al., 2022). Furthermore, the scalability of pyroprocessing has yet to be demonstrated, and its integration into a closed fuel cycle remains largely theoretical.

The generation of secondary waste during reprocessing remains a fundamental problem. Both aqueous and pyrochemical methods produce liquid and solid byproducts that themselves require long-term management. Research into advanced waste forms, such as glass–ceramics, crystalline ceramics, and geopolymers, has shown potential for immobilizing these byproducts in more stable matrices (Tameni & Bernardo, 2025). Figure 16.6 illustrates the immobilization process of radioactive nuclides in the geopolymer matrix (Singh et al., 2021). Yet the performance of these materials under repository conditions, particularly under high radiation and thermal flux, requires further validation. A holistic approach to reprocessing must therefore integrate waste minimization with advances in immobilization science to ensure that the benefits of recycling do not come at the cost of additional disposal burdens.

Economic and scalability issues compound the technological challenges. Reprocessing facilities require significant capital investment and complex infrastructure. Current methods are energy-intensive and costly, raising doubts about economic sustainability in the

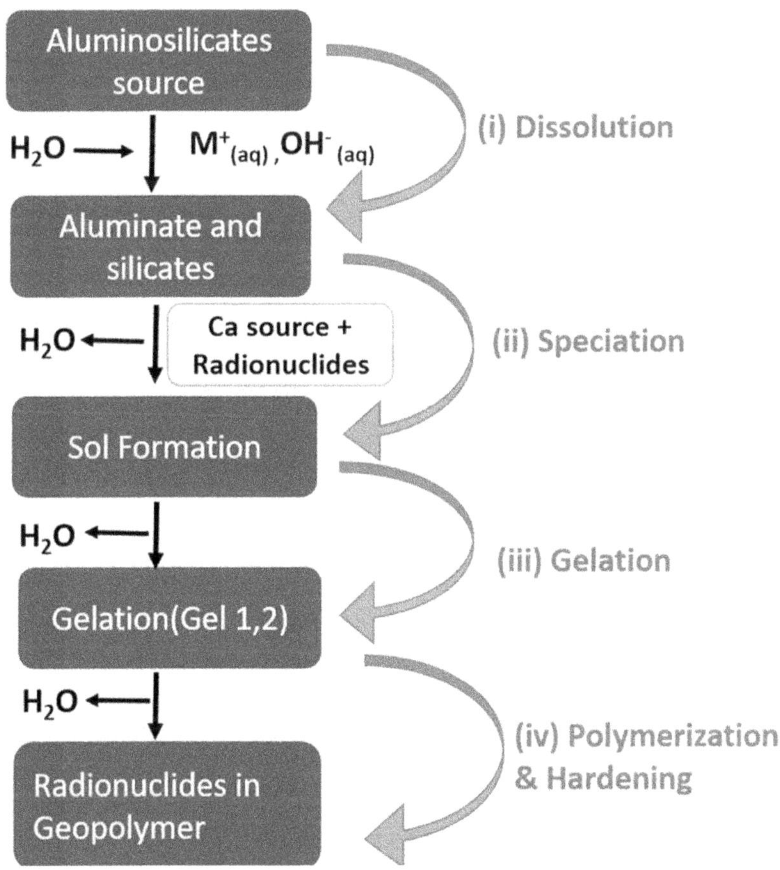

FIGURE 16.6 Flowchart for the Immobilization of Radionuclides in Geopolymer Matrix (Singh et al., 2021).

absence of clear market incentives for recovered plutonium and uranium. Research into modular, small-scale reprocessing facilities could offer pathways to reducing costs and enhancing flexibility, but the technological feasibility of such approaches remains unproven (Gaster, 2025).

16.3.3 Spent Fuel Recycling Technology

While reprocessing refers primarily to chemical separation, recycling encompasses the broader reuse of recovered materials within the nuclear fuel cycle. Recycling technologies aim to extend the utilization of nuclear resources, reduce demand for fresh uranium, and decrease the volume of HLW requiring disposal. At the same time, recycling introduces technical challenges associated with fuel fabrication, reactor performance, and material behavior.

The recycling of plutonium into MOX fuel has been demonstrated in several countries, most notably in France and Japan (Minari et al., 2023; Torvund et al., 2025). MOX fuel allows plutonium recovered from reprocessing to be blended with depleted uranium, creating a viable fuel for light water reactors. However, the performance of MOX fuel is not without limitations. Burn-up rates are generally lower than those of enriched uranium fuel, and issues such as reduced thermal conductivity, increased fission gas release, and altered neutronic characteristics complicate its use (Magni et al., 2023). Additionally, the repeated recycling of plutonium through MOX fabrication results in the accumulation of higher isotopes such as ^{237}Pu and ^{242}Pu, which degrade fuel quality and increase radiological hazards during handling (Li et al., 2024). This "plutonium degradation" problem underscores the need for advanced reactor systems capable of more effectively utilizing recycled fuels.

Recycling minor actinides, such as americium and curium, poses even greater challenges. These isotopes exhibit high heat generation and strong neutron emission, complicating their incorporation into fuel matrices (Kooyman, 2021). Research into transmutation, where minor actinides are irradiated to convert them into shorter-lived isotopes, offers a pathway for managing these materials (Figure 16.7) (Kiegiel et al., 2025). However, successful implementation requires reactors with fast neutron spectra, such as FBRs or accelerator-driven systems (Taiwo & Wigeland, 2021). The development of suitable fuel forms for minor actinide recycling, such as inert-matrix fuels or ceramic composites, remains an area of active investigation, but their performance under irradiation is not yet well characterized (Yudintsev et al., 2025).

Recycling technologies also raise challenges in fuel fabrication. Handling highly radioactive recycled materials requires remote fabrication techniques and advanced shielding. The development of robotic systems, hot-cell facilities, and advanced powder metallurgy techniques is essential for enabling safe and efficient fabrication (Poozhiyil et al., 2025). Moreover, the economic viability of recycling depends on achieving high throughput and maintaining quality assurance standards in environments where radiological constraints are severe.

Ultimately, while recycling offers clear benefits for resource utilization and waste reduction, its successful implementation depends on overcoming the technological hurdles of

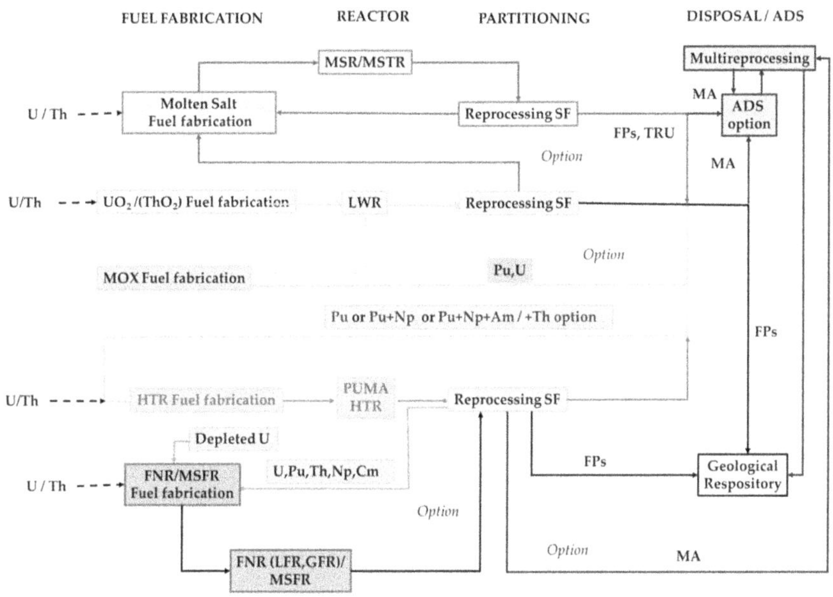

FIGURE 16.7 Transmutation Strategy Using Various Fuel-Cycle Options, Including Light Water Reactors (LWRs), High-Temperature Reactors (HTRs), Molten Salt Reactors (MSRs), Fast Neutron Reactors (FNRs), and Accelerator-Driven Systems (ADS) (Kiegiel et al., 2025).

fuel performance, actinide transmutation, and remote fabrication. Without significant progress in these areas, the promise of recycling as a cornerstone of sustainable nuclear energy will remain unrealized.

16.3.4 Advanced Reactor Design

The design of advanced nuclear reactors represents perhaps the most transformative approach to addressing the challenges of NWM. By altering the fundamental parameters of the nuclear fuel cycle, advanced reactors offer the potential to reduce waste volumes, transmute long-lived isotopes, and enhance fuel utilization efficiency. Several reactor concepts are under development, including FBRs, MSRs, and thorium-based systems, each with its own technological challenges and research gaps.

FBRs operate with a fast neutron spectrum that allows the conversion of fertile isotopes such as ^{238}U into fissile ^{239}Pu. This not only extends the energy extracted from uranium resources but also enables the transmutation of minor actinides, thereby reducing the long-term radiotoxicity of waste. Despite decades of experimental operation, including the French Phénix and Superphénix reactors and the Russian BN series, the commercial deployment of FBRs has been limited by technical and economic challenges (Cheng et al., 2024; Tashlykov et al., 2025). The development of fuels and cladding materials capable of withstanding high neutron flux and temperature remains a major obstacle. Conventional oxide fuels face limitations in thermal conductivity and swelling, while metallic and nitride fuels offer improved properties but raise fabrication and safety concerns. Netto produced Cr and CrN coatings on Zr-alloy nuclear fuel rod cladding material with enhanced mechanical resistance using magnetron sputtering (Netto, 2025). However, the CrN and Cr + CrN coatings demonstrated poor adhesion, toughness, and no wettability advantage over Cr coatings, making them less practical for accident-tolerant fuel (ATF) applications. Cladding materials must resist radiation-induced swelling, creep, and corrosion, yet no universally suitable material has yet been identified.

MSRs offer another promising pathway. Using molten salt as both coolant and fuel medium, MSRs enable operation at low pressure and high temperature, enhancing thermal efficiency (Wang Y. et al., 2024c). Furthermore, the liquid-fuel nature of MSRs permits continuous online reprocessing, which could facilitate the removal of fission products and the recycling of actinides (Figure 16.8) (Baron et al., 2019). This feature offers significant advantages for waste reduction, as MSRs could, in principle, consume spent fuel from conventional reactors. However, challenges remain in developing corrosion-resistant structural materials capable of withstanding prolonged exposure to aggressive molten salts (Okonkwo et al., 2024). Lee et al. evaluated the corrosion resistance of three Fe- and Ni-based alloys as structural materials in NaCl–KCl molten salt for 48 h at 800°C and 40% RH conditions in an air environment (Lee et al., 2025). The corrosion rate was high in SS316L (16Cr-Fe), N10003 (7Cr-Ni), and ^{276}C (16Cr-Ni), in decreasing order. This indicated that Ni-based alloys with high Cr content exhibited high corrosion resistance in unpurified molten salt environments. Additionally, the chemistry of molten salts under irradiation

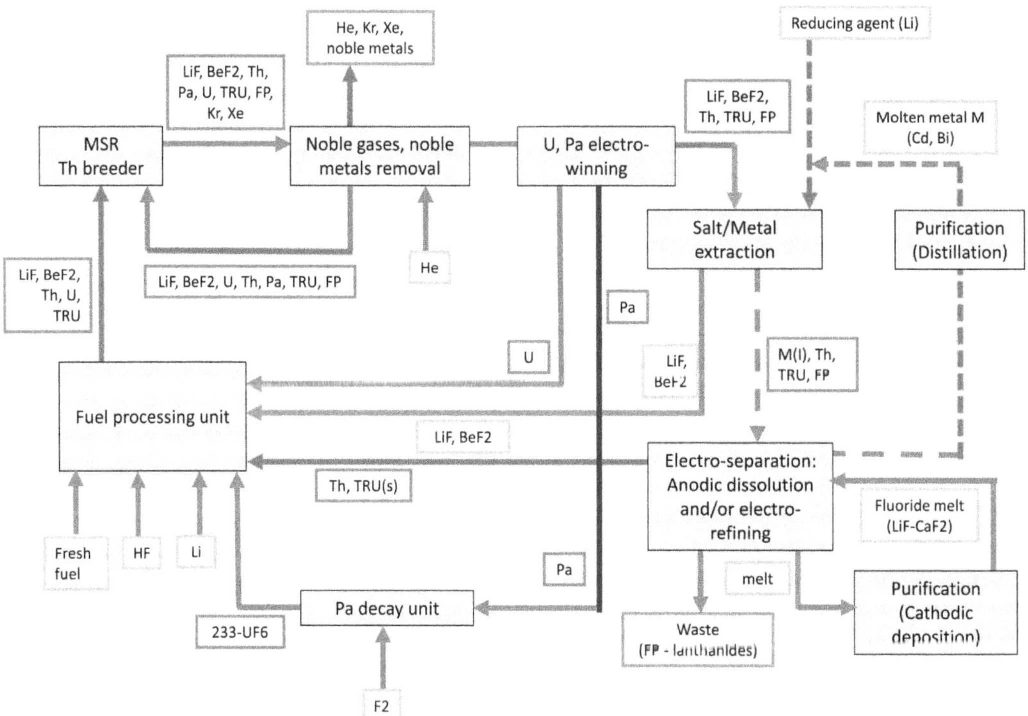

FIGURE 16.8 The MSR Fuel Cycle Concept for Thorium Breeder Reactor with Single-Fluid Online Reprocessing Technology (Baron et al., 2019).

is complex and not fully understood, creating uncertainties in fuel behavior and fission product management. Research into salt chemistry control, redox potential management, and advanced structural alloys is urgently needed.

Thorium-based reactors represent a different paradigm, leveraging the abundance of ^{232}Th as a fertile material. When irradiated, thorium breeds ^{233}U, a fissile isotope that can sustain nuclear reactions. Thorium cycles produce less plutonium and fewer long-lived actinides than conventional uranium cycles, making them attractive from a waste management perspective (Alang & Rashid, 2024a). Nonetheless, thorium fuel cycles face substantial technological gaps, particularly in fuel fabrication, reprocessing, and the handling of ^{233}U, which is highly radiotoxic and proliferation-sensitive (Silva et al., 2024). The lack of operational experience with thorium reactors further underscores the need for comprehensive experimental programs to validate their performance.

Across all advanced reactor concepts, a recurring challenge is the need for materials capable of operating under extreme conditions of temperature, radiation, and chemical reactivity. Advances in materials science, such as the development of oxide dispersion-strengthened alloys, silicon carbide composites, and high-entropy alloys, hold promise but require extensive validation. Furthermore, the integration of advanced monitoring and diagnostic systems, leveraging artificial intelligence and robotics, will be essential to ensure safe and reliable operation under these demanding conditions (Christopher Selvam et al., 2025).

16.4 PROSPECTS AND FUTURE DIRECTIONS

As nuclear power continues to assert itself as a vital pillar of global decarbonization strategies, the management of radioactive waste remains one of the most decisive factors shaping public trust, policy frameworks, and the future trajectory of the technology itself. While substantial progress has been achieved in geological repository design, reprocessing methodologies, and advanced fuel cycles, the persistence of scientific uncertainties and technological constraints underlines the necessity of sustained innovation. Looking forward, a series of prospective developments is emerging that may redefine the contours of NWM. These range from the design of modular, standardized storage architectures and the deployment of artificial intelligence for predictive monitoring to the integration of next-generation reactor technologies that actively minimize the long-lived radiotoxic inventory of waste. Complementing these engineering frontiers is the growing promise of advanced materials science in developing ultra-durable containment solutions, alongside a broader cultural and intellectual shift toward cross-disciplinary research that transcends traditional silos. Together, these future-oriented directions represent not merely incremental refinements but potential paradigm shifts in how the nuclear community envisions and executes waste stewardship across centuries.

16.4.1 Modular and Standardized Waste Storage Systems

The pursuit of modular, standardized storage systems arises from the recognition that NWM must balance safety, adaptability, and economic efficiency. Traditional storage strategies have often been site-specific, designed around the unique geological, regulatory, or infrastructural characteristics of particular facilities. While this bespoke approach ensures tailored safety performance, it has also led to fragmented architectures, logistical inefficiencies, and challenges in scaling up to meet increasing volumes of spent nuclear fuel. In contrast, modular storage concepts envision the development of highly engineered, replicable units—whether in dry cask storage or interim above-ground facilities—that can be manufactured under controlled conditions, transported, and deployed as needed in diverse contexts (Yessenbayev et al., 2024).

This modularity carries multiple advantages. First, it enhances flexibility: standardized units can be adapted for different reactor types, waste forms, and geographical conditions, reducing the need for bespoke engineering solutions in each case. Second, it creates economies of scale by enabling mass production, thus reducing per-unit costs and promoting wider deployment, particularly in regions with nascent nuclear programs. Third, it promotes retrievability and system adaptability. A modular design can be reconfigured, expanded, or upgraded as new technologies emerge, thereby preventing technological lock-in. Furthermore, modular systems can facilitate international collaboration by establishing interoperable standards that allow cross-border sharing of storage infrastructure—an attractive prospect for states with limited geological or technical resources for building permanent repositories (Ahmad & Usman, 2025).

Yet, implementing modularity also introduces challenges. Standardization must reconcile divergent national regulatory frameworks, public perceptions, and varying waste compositions. More critically, the materials used in modular casks and containers must demonstrate resilience not only to radiation and thermal loads, but also to environmental variability across deployment sites (Bagheri et al., 2025). Hence, while modular systems promise to streamline logistics and cost, their success ultimately hinges on advances in materials engineering and monitoring, as well as harmonization of global governance frameworks. This vision is not merely a technical proposition but a redefinition of NWM as a globally networked, interoperable infrastructure.

16.4.2 AI-Driven Risk Assessment and Real-Time Monitoring

If modular storage defines the physical architecture of future waste management, artificial intelligence (AI) represents the epistemic infrastructure—reshaping how risks are assessed, monitored, and anticipated across the entire lifecycle of waste isolation. Conventional monitoring strategies have relied heavily on periodic sampling, deterministic modeling, and human expertise. While effective in many contexts, these methods face limitations in detecting emergent risks, predicting non-linear system behaviors, and integrating vast, multi-modal datasets. AI and machine learning, by contrast, offer the capacity to synthesize real-time sensor streams, historical data, and predictive modeling outputs into adaptive, continuously updating risk assessments.

The application of AI extends across multiple layers of NWM. In near-surface storage, AI algorithms can integrate sensor networks tracking parameters such as temperature, humidity, radiation flux, and structural strain, thereby enabling predictive maintenance of casks and storage modules (Abulifa et al., 2025). At geological repository depths, AI may assist in simulating the coupled THMC processes that define repository evolution over millennia, identifying early-warning signatures of potential barrier degradation (Churakov et al., 2024). Moreover, AI-driven image recognition can enhance remote inspection of canisters, while reinforcement learning techniques may optimize repository design layouts to minimize coupled stress regimes (Mohamed, 2024).

Importantly, AI does not merely automate existing processes—it reshapes epistemology. By uncovering hidden correlations within massive datasets, AI has the potential to reveal system dynamics that elude conventional deterministic modeling. For instance, predictive analytics may identify complex feedback loops between groundwater chemistry and canister corrosion that traditional modeling overlooks (Abulifa et al., 2025). Similarly, probabilistic AI models can quantify uncertainty in long-term predictions, offering decision-makers a more nuanced appreciation of risk.

Nevertheless, embedding AI into NWM raises profound challenges. Data scarcity—especially for long-term geological processes—poses a significant barrier to training robust algorithms (Chenniappan & Devarajan, 2024). Sensor durability and cybersecurity also present vulnerabilities, as real-time monitoring systems must withstand hostile underground environments while ensuring protection from malicious interference. Furthermore, AI outputs must remain transparent and interpretable to regulatory bodies, ensuring that automated decisions do not undermine accountability. The integration of AI into nuclear waste stewardship, therefore, requires not only technological refinement but also ethical frameworks, governance protocols, and human oversight mechanisms (Cha, 2024).

16.4.3 Next-Generation Nuclear Reactors and Waste Minimization

While storage and monitoring strategies focus on managing existing waste inventories, a parallel frontier lies in reducing the production of long-lived waste at its source. Next-generation nuclear reactors—encompassing fast neutron systems, MSRs, and accelerator-driven subcritical systems—embody this aspiration by fundamentally altering the isotopic composition of spent fuel. Advanced reactors are designed to either recycle these isotopes as fuel or transmute them into shorter-lived fission products. The implications for waste management are profound. If widely deployed, next-generation reactors could reshape the inventory of waste destined for geological disposal, reducing repository heat loads, radiotoxicity, and long-term hazard timescales. However, technical hurdles persist: fast reactor technologies must overcome challenges of coolant chemistry, fuel fabrication, and materials resilience under intense neutron fluxes. MSRs face equally formidable obstacles in corrosion resistance, salt chemistry control, and the development of robust reprocessing infrastructures. Moreover, the proliferation risks associated with separating actinides, even in advanced fuel cycles, necessitate rigorous safeguards and governance mechanisms.

Despite these challenges, the broader trajectory is clear: the future of waste management is inseparable from the evolution of nuclear reactor technology. The degree to which next-generation systems succeed in minimizing long-lived waste will fundamentally determine the scale, design, and urgency of geological repositories, linking upstream innovation with downstream disposal strategies in a tightly coupled technological ecosystem.

16.4.4 Advanced Materials Science for Ultra-Durable Waste Forms

At the heart of all waste management strategies lies a central question: how can radioactive isotopes be immobilized for timescales spanning tens of thousands of years? This question is, ultimately, a challenge in materials science. While vitrification into borosilicate glass remains the dominant approach, concerns over long-term stability under groundwater leaching, radiation damage, and thermal cycling have motivated the exploration of alternative materials (Al-Buriahi et al., 2025). The future of waste immobilization may be defined by breakthroughs in ceramics, glass–ceramics, crystalline matrices, and nano-engineered composites that offer unprecedented durability (Elsalamawy et al., 2024; Nevolina et al., 2024; Wang P. et al., 2024a). Glass–ceramic composites, which combine the homogeneity of glass with the structural robustness of crystalline phases, represent another promising avenue (Wu et al., 2024). Additionally, research into high-entropy oxides and novel engineered minerals suggests that entirely new classes of materials may be synthesized to mimic or exceed the stability of natural analogs (Sun et al., 2025). Advances in computational materials science—particularly atomistic simulations and AI-driven materials discovery—are accelerating this exploration, allowing researchers to screen vast chemical spaces for candidate materials with optimized thermodynamic and kinetic stability (Arhouni et al., 2025).

Equally important is the development of ultra-durable canister and container materials. Copper, titanium alloys, and corrosion-resistant steels remain standard candidates, but their performance under repository conditions must be validated over geological timescales (Qiao et al., 2024; Reda et al., 2025). Novel coatings, self-healing alloys, and nanostructured composites may significantly extend container lifetimes. For instance, the use of graphene-based protective layers has been proposed as a barrier to both water infiltration and hydrogen embrittlement (Rabbani et al., 2024). The integration of such advanced materials into modular storage casks and geological repository designs represents a critical convergence of materials science and systems engineering.

Yet, challenges remain in bridging laboratory results with repository-scale conditions (Shao et al., 2024). Accelerated aging tests cannot fully replicate the complex, coupled environments of deep geological storage, and natural analogs, while informative, provide incomplete guidance. The future of materials research in NWM must therefore embrace not only innovation in synthesis but also rigor in validation, scaling, and cross-disciplinary integration with geoscience and repository engineering.

16.4.5 Synergy of Multidisciplinary Research

Finally, the future of NWM is not defined solely by individual technological breakthroughs, but by the integration of knowledge across disciplines. The complexity of waste systems—spanning nuclear physics, geochemistry, hydrology, structural engineering, materials science, and computational modeling—demands a mode of research that is both interdisciplinary and transdisciplinary. Indeed, the most promising innovations often emerge at the intersections: where geochemical insights inform materials design, where computational modeling enhances geological understanding, or where engineering pragmatism intersects with scientific discovery.

For instance, the study of coupled THMC processes in repositories requires simultaneous expertise in rock mechanics, groundwater hydrology, and corrosion chemistry. Similarly, the development of next-generation reactor fuels necessitates contributions from nuclear engineers, metallurgists, and radiation chemists. Integrative frameworks such as the "reactor-to-repository" concept illustrate this necessity, explicitly linking upstream reactor design decisions with downstream repository performance assessments. Such holistic perspectives not only optimize individual technologies but also ensure coherence across the nuclear energy system as a whole.

The rise of data-driven methods further reinforces the value of cross-disciplinary synergy. AI applications in waste monitoring, for example, require collaboration between nuclear engineers, data scientists, and cybersecurity experts. Likewise, the development of ultra-durable materials is increasingly reliant on computational techniques drawn from condensed matter physics and informatics. These intersections are not incidental but foundational: the ability to integrate insights across domains will determine the pace and scope of innovation in nuclear waste stewardship.

In practice, achieving such integration requires institutional and cultural shifts. Funding structures must incentivize collaborative projects that bridge traditional disciplinary boundaries. International consortia, such as those organized under the European EURAD program, exemplify such collaborative frameworks, bringing together geologists, chemists, and engineers in shared research platforms. The future trajectory of NWM will thus depend as much on the cultivation of collaborative ecosystems as on the advancement of any single technology.

16.5 CONCLUSION

The safe management of nuclear waste remains one of the most formidable challenges facing the global nuclear energy community. As the world accelerates its transition to low-carbon energy sources, nuclear power is expected to play an increasingly significant role. However, the issue of nuclear waste—its long-lived radiotoxicity, storage, and disposal—continues to be a point of contention that demands innovative and sustainable solutions. The present chapter has highlighted the most prominent technological strategies currently under investigation, along with the significant research gaps that persist.

From geological storage to spent fuel reprocessing, recycling technologies, and the evolution of advanced reactor designs, each approach brings its own set of benefits and limitations. While geological repositories offer a seemingly permanent solution, they are burdened by uncertainties in long-term performance predictions. Spent fuel reprocessing technologies promise waste volume reduction but often at high economic and environmental costs. Recycling technologies and advanced reactors represent promising paradigms for reducing long-lived waste, yet they remain in the early stages of development and require substantial investment in research, materials, and regulatory frameworks.

Looking forward, the integration of modular, standardized storage systems, coupled with the potential for real-time, AI-driven monitoring, offers an exciting avenue for advancing NWM. Similarly, next-generation reactors that minimize waste generation, along with the continuous refinement of ultra-durable materials for waste containment, offer transformative possibilities. However, achieving these goals will require unprecedented levels of multidisciplinary collaboration across materials science, nuclear engineering, geoscience, and computational modeling.

In conclusion, while the technological landscape for NWM is far from complete, significant strides are being made in advancing the necessary infrastructure and scientific foundations. The continued integration of innovative technologies, combined with a commitment to research and collaboration, offers a path toward a safer, more sustainable future for nuclear energy. The ability to manage nuclear waste effectively will not only determine the long-term viability of nuclear power but will also be a defining factor in shaping the future of global energy systems.

REFERENCES

Abramova, E., Popova, N., Artemiev, G., Boldyrev, K., Kazakov, K., Kryuchkov, D., Safonov, A., 2023. Biological factors affecting the evolution of safety barrier materials in the Yeniseisky deep geological repository. *Eng. Geol.* 312, 106931.

Abulifa, A.A., Enbais, F.A., Shwehdy, D.M., 2025. Artificial intelligence for nuclear waste management: Opportunities, challenges, and future prospects. *Int. J. Electr. Eng. Sustain.* 19–30.

Ahmad, F., Usman, S., 2025. Advanced and small modular reactors' supply chain: Current status and potential for global cooperation. *Prog. Nucl. Energy* 184, 105661.

Alang, I., Rashid, M., 2024a. Comparative analysis of uranium and thorium fuel for nuclear reactor. *Future* 74, 83.

Alang, I., Rashid, M., 2024b. *Sustainable Electricity Generation Using Molten Salt Nuclear Reactors (MSR)*. Universiti Teknologi Malaysia, Pagoh, Malaysia.

Al-Buriahi, M., Olarinoye, I., Yılmaz, E., Çalıskan, F., Sriwunkum, C., 2025. Evaluation of the structural and radiation transmission parameters of recycled borosilicate waste glass system: An effective material for nuclear shielding. *Ann. Nucl. Energy* 213, 111136.

Alwaeli, M., Mannheim, V., 2022. Investigation into the current state of nuclear energy and nuclear waste management—A state-of-the-art review. *Energies* 15(12), 4275.

Alzamel, M., 2022. *Swelling, Thermal, and Hydraulic Properties of a Bentonite-Sand Barrier in a Deep Geological Repository for Radioactive Wastes: Effect of Groundwater Chemistry, Temperature and Physical Factors*. University of Ottawa.

Arhouni, F.E., Abdo, M.A.S., Ouakkas, S., Bouhssa, M.L., Boukhair, A.J., 2025. Artificial intelligence-driven advances in nuclear technology: Exploring innovations, applications, and future prospects. *Ann. Nucl. Energy* 213, 111151.

Bagheri, S., Khalafi, H., Bahrami, A., 2025. An optimal radiation shield design of a small modular nuclear power reactor by genetic algorithm. *Nucl. Technol.* 211(5), 940–952.

Bajestani, M.S., Nasir, O., Coulson, C., 2024. Analysis of long-term thermo–hydro–mechanical behavior in the near-field of a deep geological repository system. *Minerals* 14(12), 1262.

Baron, P., Cornet, S., Collins, E., DeAngelis, G., Del Cul, G., Fedorov, Y., Glatz, J., Ignatiev, V., Inoue, T., Khaperskaya, A., 2019. A review of separation processes proposed for advanced fuel cycles based on technology readiness level assessments. *Prog. Nucl. Energy* 117, 103091.

Cammi, A., Loi, L., Missaglia, A., Tumminelli, L., Giacobbo, F., Padovani, E., 2025. Photonuclear treatment for spent fuel radiotoxicity reduction: A case study investigation on minor actinides. *Nucl. Eng.* 442, 114204.

Cao, H., Kang, Y., Li, B., Liu, Y., Bao, M., Li, H., Zheng, Y., Wang, L., Weng, C., Tang, X., 2024. Amine-terminated phenanthroline diimides as aqueous masking agents for Am (III)/Eu (III) separation: An alternative ligand design strategy for water-soluble lanthanide/actinide chelating ligands. *Inorg. Chem.* 63(23), 10511–10518.

Cao, Y., Zhao, Z., Wang, P., Yu, S., Lai, Z., Zhang, M., Gao, X., Zhao, Y., Xuan, Z., Ren, H., Zhang, D., Lou, X., 2021. Long-term variation of 90Sr and 137Cs in environmental and food samples around Qinshan nuclear power plant, China. *Sci. Rep.* 11(1), 20903.

Cha, S., 2024. Towards an international regulatory framework for AI safety: Lessons from the IAEA's nuclear safety regulations. *Humanit. Soc. Sci. Commun.* 11(1), 1–13.

Chang, C., Nakagawa, S., Dong, W., Zheng, L., 2025. Coupled thermo-hydrological-mechanical-chemical behavior of anisotropic granite for geologic disposal of high-level radioactive waste: A core-scale laboratory investigation. *Rock Mech. Rock Eng.* 58, 12969–12983.

Cheng, H., Cheng, S., Chen, S., 2024. Sodium fast reactors. In: *Nuclear Power Reactor Designs*. Elsevier, pp. 185–208.

Chenniappan, T., Devarajan, Y., 2024. A critical analysis of the role of artificial intelligence and machine learning in enhancing nuclear waste management. *Kerntechnik* 89(5), 586–594.

Christopher Selvam, D., Devarajan, Y., Raja, T., 2025. Exploring the potential of artificial intelligence in nuclear waste management: Applications, challenges, and future directions. *Nucl. Eng. Des.* 431, 113719.

Churakov, S.V., Claret, F., Idiart, A., Jacques, D., Govaerts, J., Kolditz, O., Prasianakis, N., Samper, J., 2024. Position paper on high fidelity simulations for coupled processes, multi-physics and chemistry in geological disposal of nuclear waste. *Environ. Earth Sci.* 83(17), 521.

Clayton, R., Kirk, J., Banford, A., Stamford, L., 2025. A review of radioactive waste processing and disposal from a life cycle environmental perspective. *Clean Technol. Environ. Policy* 27(2), 665–682.

Darda, S.A., Gabbar, H.A., Damideh, V., Aboughaly, M., Hassen, I., 2021. A comprehensive review on radioactive waste cycle from generation to disposal. *J. Radioanal. Nucl. Chem.* 329(1), 15–31.

Deng, D., Zhang, L., Dong, M., Samuel, R.E., Ofori-Boadu, A., Lamssali, M., 2020. Radioactive waste: A review. *Water Environ. Res.* 92(10), 1818–1825.

Denysov, V., Kulyk, M., Babak, V., Zaporozhets, A., Kostenko, G., 2024. Modeling nuclear-centric scenarios for Ukraine's low-carbon energy transition using diffusion and regression techniques. *Energies* 17(20), 5229.

Elsalamawy, M., Gouda, M.M., Abdalmawla, I.G., Abbas, M.I., El-Khatib, A.M., 2024. Radiation and mechanical performance of cementitious materials containing ecofriendly nano laboratory waste glass. *Sci. Rep.* 14(1), 21887.

Galashev, A.Y., 2022. Recovery of actinides and fission products from spent nuclear fuel via electrolytic reduction: Thematic overview. *Int. J. Energy Res.* 46(4), 3891–3905.

Gaster, R., 2025. *Small Modular Reactors: A Realist Approach to the Future of Nuclear Power*. Information Technology and Innovation Foundation.

Guo, X., 2025. Perspective chapter: Safe disposal and storage of nuclear waste. In: Tsvetkov, P., Sabharwall, P. (Eds.), *Nuclear Fission Energy—Deployment, Applications, and Beyond*. IntechOpen, Rijeka.

Harithra, V., Lakshmi Priya, S., Vivek, P., Ivo Romauld, S., Meenambiga, S., Rajakumari, K., 2024. Long-term storage and safety of radioactive waste storage facilities. In: *Radioactive Pollutant: Sources, Issues and Remediation*. Springer, pp. 219–238.

Haynes, H.M., Bailey, M.T., Lloyd, J.R., 2021. Bentonite barrier materials and the control of microbial processes: Safety case implications for the geological disposal of radioactive waste. *Chem. Geol.* 581, 120353.

Holdsworth, A.F., Eccles, H., Sharrad, C.A., George, K., 2023. Spent nuclear fuel—waste or resource? The potential of strategic materials recovery during recycle for sustainability and advanced waste management. *Waste* 1(1), 249–263.

Huang, X., Li, X.-B., Wu, Q.-Y., Shi, W.-Q., 2025. Experimental and theoretical progress on the reduction of Np (vi) with salt-free reagents in the PUREX process. *Ind. Chem. Mater.* 3, 412–430.

Kiegiel, K., Herdzik-Koniecko, I., Fuks, L., Zakrzewska-Kołtuniewicz, G., 2022. Management of radioactive waste from HTGR reactors including spent TRISO fuel—state of the art. *Energies* 15(3), 1099.

Kiegiel, K., Smoliński, T., Herdzik-Koniecko, I., 2025. Advanced nuclear reactors—challenges related to the reprocessing of spent nuclear fuel. *Energies* 18(15), 4080.

Kim, J.-S., Lee, C., Yoon, S., Lee, M., Lee, Y.H., Kim, J.-W., Lee, M., Cha, Y., Kim, J.-T., Hong, C.-H., Park, T., Kim, M., Kim, T., Ha, S.-J., Kim, K.-I., Kwon, S., Choi, S., Lee, Y., Kwon, J.-S., 2025. Advancements in long-term safety evaluation technology for engineered barrier system (EBS): A comprehensive review of Korea's high-level waste disposal R&D program. *Korean J. Chem. Eng.* 42(6), 1301–1316.

Kooyman, T., 2021. Current state of partitioning and transmutation studies for advanced nuclear fuel cycles. *Ann. Nucl. Energy* 157, 108239.

Lee, U., Kim, M.W., Na, J., Lee, M., Kim, S.J., Kim, D.-J., Yoon, Y.S., 2025. A study on the corrosion behavior of Fe/Ni-based structural materials in unpurified molten chloride salt. *Materials* 18(7), 1653.

Li, T., Liu, F., Jia, Z., Luo, F., Yan, T., Zheng, W., 2024. Dissolution of mixed oxide (MOX) fuel in nitric acid: A review. *Heliyon* 10(6), e27502.

Lisjak, A., Mahabadi, O., Ha, J., Mas Ivars, D., 2023. Analysis of thermo-mechanical damage around tunnel and deposition boreholes of an underground nuclear waste disposal facility at the Forsmark site (Sweden) by 3D coupled FDEM simulations. *Int. J. Rock Mech. Min. Sci.* 171, 105586.

Magni, A., Di Gennaro, M., Guizzardi, E., Pizzocri, D., Zullo, G., Luzzi, L., 2023. Analysis of the performance of driver MOX fuel in the MYRRHA reactor under Beam Power Jump transient irradiation conditions. *Nucl. Eng.* 414, 112589.

Mathew, M.D., 2022. Nuclear energy: A pathway towards mitigation of global warming. *Prog. Nucl. Energy* 143, 104080.

Mausolff, Z., DeHart, M., Goluoglu, S., 2021. Design and assessment of a molten chloride fast reactor. *Nucl. Eng. Des.* 379, 111181.

McCloy, J., Riley, B., Dixon Wilkins, M., Evarts, J., Bussey, J., Vienna, J., Bingham, P., Gregg, D., Ojovan, M., Schuller, S., Uruga, K., Perret, D., Regnier, E., Giboire, I., Um, W., Xu, K., Goel, A., Kruger, A., 2024. International perspectives on glass waste form development for low-level and intermediate-level radioactive waste. *Mater. Today* 80, 594–618.

Merk, B., Jain, L., Noori-kalkhoran, O., Derrer-Merk, E., Cartland-Glover, G., Powell, L., Jones, A., Detkina, A., Litskevich, D., Patel, M.J.E.E., 2025. Does partitioning & transmutation have a future as a nuclear waste management strategy? In: *Everlasting Energy—In Memoriam Prof. Dr.-Ing. Robert Stieglitz*. KIT Scientific Publishing.

Minari, E., Kabasawa, S., Mihara, M., Makino, H., Asano, H., Nakase, M., Takeshita, K., 2023. Impact of MOX fuel use in light-water reactors: Long-term radiological consequences of disposal of high-level waste in a geological repository. *J. Nucl. Sci. Technol.* 60(7), 793–803.

Mirza, M., Abdulaziz, R., Maskell, W.C., Wilcock, S., Jones, A.H., Woodall, S., Jackson, A., Shearing, P.R., Brett, D.J., 2023. Electrochemical processing in molten salts–a nuclear perspective. *Energy Environ. Sci.* 16(3), 952–982.

Mohamed, A.M.O., 2024. Enhancing radioactive waste management with cutting-edge digital technologies: A review. *Academia Eng.* 1(4).

Morales-Hidalgo, M., Povedano-Priego, C., Martinez-Moreno, M.F., Ruiz-Fresneda, M.A., Lopez-Fernandez, M., Jroundi, F., Merroun, M.L., 2024. Insights into the impact of physicochemical and microbiological parameters on the safety performance of deep geological repositories. *Microorganisms* 12(5), 1025.

Na, T., Chae, B.-G., Kim, M.-J., Park, E. S., Nahm, W.-H., Hyun, S.P., Choi, J., 2025. A review of site selection criteria and investigation factors for HLW disposal: Insights from leading countries and proposals for Korea. *Geosci. J.* 29, 446–467.

Netto, T.R., 2025. *Development of Accident Tolerant Fuel Cladding via Magnetron Sputtering Chromium Depositions for Light Water Reactors*. Manchester Metropolitan University.

Nevolina, L.A., Shtenberg, M.V., Gladkochub, E.A., Koroleva, O.N., 2024. Structural features and crystallization of Na2O-Cs2O-B2O3-SiO2 glasses for radioactive waste immobilization. *Materialia* 36, 102134.

Ojovan, M.I., Steinmetz, H.J., 2022. Approaches to disposal of nuclear waste. *Energies* 15(20), 7804.

Okonkwo, B.O., Kim, C., An, T., Jang, C., Kang, D., Yang, W., Choi, S., 2024. Corrosion behaviour of Al-containing alloys in Cl-based molten salt environment. *J. Nucl. Mater.* 599, 155207.

Perrot, J., Morana, A., Marin, E., Ouerdane, Y., Boukenter, A., Bertrand, J., Poirier, S., Houjeij, H., Novak, E., Peltan, T., Girard, S., 2025. Radiation vulnerability of optical fiber cables for underground nuclear waste monitoring. *Opt. Fiber Technol.* 92, 104181.

Pizarro, J., Sainsbury, B., 2023. Climate adaptation in the spent fuel storage and disposal facilities: A literature review. *Ann. Nucl. Energy* 192, 109954.

Poozhiyil, M., Argin, O.F., Rai, M., Esfahani, A.G., Hanheide, M., King, R., Saunderson, P., Moulin-Ramsden, M., Yang, Wen García, L.P., 2025. A framework for real-time autonomous robotic sorting and segregation of nuclear waste: Modelling, identification and control of DexterTM robot. *Machines* 13(3), 214.

Pucciarelli, M., Palethorpe, S.J., Spencer, J., Banford, A., Lettieri, P., Paulillo, A., 2024. Using life cycle assessment to quantify the environmental benefits of circular economy strategies in the nuclear industry. *Prog. Nucl. Energy* 168, 105026.

Qiao, Y., Wang, T., Chen, Z., Wang, J., Li, C., Chen, J., 2024. Corrosion techniques and strategies for used fuel containers with copper corrosion barriers under deep geological disposal conditions: A review. *Int. J. Miner. Metall. Mater.* 31(12), 2582–2606.

Rabbani, S., Qureshi, Z.A., Alfantazi, A., Alkaabi, A.K., Alameri, S.A., Addad, Y., Samad, Y.A., Afgan, I., 2024. Graphene-nuclear nexus: A critical review. *2D Mater.* 11(4), 042001.

Rajagukguk, M., Siagian, R., Gunawan, B., 2024. The role of nuclear energy in achieving sustainable energy solutions. *J. Lemhannas RI* 12, 131–148.

Reda, A., Shahin, M.A., Montague, P., 2025. Review of material selection for corrosion-resistant alloy pipelines. *Eng. Sci.* 33, 1373.

Roper, R., Harkema, M., Sabharwall, P., Riddle, C., Chisholm, B., Day, B., Marotta, P., 2022. Molten salt for advanced energy applications: A review. *Ann. Nucl. Energy* 169, 108924.

Selroos, J.-O., Gylling, B., 2023. How findings from a multi-annual international modeling initiative are implemented in a nuclear waste management organization. *Energies* 16(2), 684.

Shao, H., Radeisen, E., Hesser, J., Wang, W., Kolditz, O., 2024. Coupled processes at micro-and macroscopic levels for long-term performance assessment studies of nuclear waste repositories. *Minerals* 14(5), 453.

Silva, C.A., Magalhães, I.R., Lorduy-Alós, M., Gallardo, S., Pereira, C., Costa, A.L., Verdú, G., 2024. A neutronic evaluation of a thorium-based molten salt breeder reactor. *Nucl. Eng. Des.* 421, 113049.

Singh, B.K., Hafeez, M.A., Kim, H., Hong, S., Kang, J., Um, W., 2021. Inorganic waste forms for efficient immobilization of radionuclides. *ACS EST Eng.* 1(8), 1149–1170.

SKB, 2006. *Long-Term Safety for KBS-3 Repositories at Forsmark and Laxemar—A First Evaluation: Main Report of the SR-Can Project* (Vol. 6). SKB.

Sun, J., Liu, W., Liang, F., Liu, R., Yao, Y., Zou, Q., Dong, P., Wang, S., 2025. Insight into high entropy compounds: Advances, challenges and energy applications. *Adv. Funct. Mater.* e10855.

Taiwo, T., Wigeland, R., 2021. Fuel cycle considerations for uranium, plutonium and minor actinide partitioning and transmutation. *Ann. Nucl. Energy* 156, 108182.

Tameni, G., Bernardo, E., 2025. Alkali activation of glass for sustainable upcycling: An overview. *Ceramics* 8(3), 108.

Tashlykov, O., Sesekin, A., Klimova, V., Mahmoud, K., 2025. Optimization of the route of the refueling machine to reduce the refueling time: Case study of the BN-800 reactor. *Nucl. Eng. Des.* 431, 113725.

Terranova, M.L., Tavares, O.A.P., 2024. Trends and perspectives on nuclear waste management: Recovering, recycling, and reusing. *J. Nucl. Eng.* 5(3), 299–317.
Torvund, G., Ernoult, M., Courtin, F., Doligez, X., Gjestvang, D., Siem, S., Thiollière, N.E., 2025. Multi-MOX: Facilitating plutonium multi-recycling in the French PWR fleet. *Ann. Nucl. Energy* 223, 111641.
Voss, M., Maurer-Rurack, U., Poller, A., 2025. Processes and interactions of potential relevance for deep geological disposal of HLW and L/ILW at the same site—international approaches and transferability to the German site selection. *Nucl. Technol.* 211(5), 889–904.
Wagner, F., 2021. CO2 emissions of nuclear power and renewable energies: A statistical analysis of European and global data. *Eur. Phys. J. Plus* 136, 562.
Wang, J., Ahmed, B., Huang, J., Nong, X., Xiao, R., Abbasi, N.S., Alidekyi, S.N., Li, H., 2025. Physical and mechanical properties and constitutive model of rock mass under THMC coupling: A comprehensive review. *Appl. Sci.* 15(4), 2230.
Wang, P., Teng, Z., Zeng, S., Chen, C., Feng, W., Zhou, X., 2024. The mechanical properties and chemical durability of granite wastes based glass-ceramics for immobilization of high-level nuclear wastes. *Ceram. Int.* 50(13), 24907–24912.
Wang, Y., Shield, K.M., Abergel, R.J., 2024. Hydrophilic chelators for aqueous reprocessing of spent nuclear fuel. *Sep. Purif. Rev.* 53(2), 119–137.
Wang, Y., Zhang, Z., Abergel, R.J., 2021. Hydroxypyridinone-based stabilization of Np(IV) enabling efficient U/Np/Pu separations in the adapted PUREX process. *Sep. Purif. Technol.* 259, 118178.
Wang, Y., Zhu, C., Zhang, M., Zhou, W., 2024. Molten salt reactors. In: *Nuclear Power Reactor Designs*. Elsevier, pp. 163–183.
Wu, B., Ning, H., Zhu, H., Chen, J., Wang, K., Zhang, D., Wang, F., Liao, Q., 2024. Synthesis and characterization of sodium-aluminophosphate based glass-ceramics containing NZP phase for HLW immobilization. *J. Nucl. Mater.* 588, 154832.
Xu, L., Yang, X., Zhang, A., Xu, C., Xiao, C., 2023. Separation and complexation of f-block elements using hard-soft donors combined phenanthroline extractants. *Coord. Chem. Rev.* 496, 215404.
Yang, X., Xu, L., Fang, D., Zhang, A., Xiao, C., 2024. Progress in phenanthroline-derived extractants for trivalent actinides and lanthanides separation: Where to next? *Chem. Commun.* 60(81), 11415–11433.
Yessenbayev, O., Nguyen, D.C.D., Jeong, T., Kang, K.J., Kim, H.R., Ko, J., Park, J.Y., Roh, M.S., Comuzzi, M., 2024. Combining blockchain and IoT for safe and transparent nuclear waste management: A prototype implementation. *J. Ind. Inf. Integr.* 39, 100596.
Yoon, S., Zheng, L., Chang, C., Borglin, S., Chou, C., 2024. Coupled thermo-hydro-mechanical modeling of bentonite under high temperature heating and hydration for a bench-scale laboratory experiment. *Rock Mech. Rock Eng.* 57(8), 5483–5504.
Yu, Y.-C., Chen, C.-J., Chung, C.-C., Ni, C.-F., Lee, I.-H., Wu, Y.-C., Lin, T.-Y., 2023. A multimodel framework for quantifying flow and advective transport controlled by earthquake-induced canister failures in a reference case for radioactive waste geological disposal. *Energies* 16(13), 5081.
Yuan, Y., Zhang, Y., Chen, X., Zhao, Z., Xie, H., Wang, D., Yin, H., 2023. Electrochemical purification of waste salt from pyro-processing of spent nuclear fuels. *Sep. Purif. Technol.* 326, 124805.
Yudintsev, S.V., Ojovan, M.I., Stefanovsky, O.I., 2025. Composite forms in the REE2O3–ZrO2–TiO2 system for minor actinides (Am, Cm) and REE immobilization. *J. Compos. Sci.* 9(2), 58.
Zou, L., Cvetkovic, V., 2023. Disposal of high-level radioactive waste in crystalline rock: On coupled processes and site development. *Rock Mech. Bull.* 2(3), 100061.

Index

Note: Page numbers in *italics* indicate a figure and page numbers in **bold** indicate a table on the corresponding page.

J

K

L

M

S

V

W

X

Y

Z

For Product Safety Concerns and Information please contact our EU
representative GPSR@taylorandfrancis.com
Taylor & Francis Verlag GmbH, Kaufingerstraße 24, 80331 München, Germany

www.ingramcontent.com/pod-product-compliance
Ingram Content Group UK Ltd.
Pitfield, Milton Keynes, MK11 3LW, UK
UKHW052218190926
475702UK00006B/1088
9781041126256